D0722872

A HISTORY OF

MATHEMATICS

BY

FLORIAN CAJORI, PH. D.

PROFESSOR OF HISTORY OF MATHEMATICS IN THE
UNIVERSITY OF CALIFORNIA

"I am sure that no subject loses more than mathematics by any attempt to dissociate it from its history."—J. W. L. GLAISHER

CHELSEA PUBLISHING COMPANY
NEW YORK, N. Y.

FOURTH EDITION, 1985

PRINTED ON 'LONG-LIFE' ACID-FREE PAPER

FIRST EDITION, NEW YORK, 1893
SECOND, REVISED AND ENLARGED EDITION, 1919
REPRINTED 1931, 1938, 1942, 1949, 1960
THIRD EDITION, NEW YORK, 1980

LIBRARY OF CONGRESS CATALOGING IN PUBLICATION DATA

Cajori, Florian, 1859–1930.

A history of mathematics.

Includes Bibliographical references and index.
1. Mathematics—History. I. Title.
QA21.C15 510'.9 70-113120
ISBN 0-8284-1303-X

PRINTED IN THE UNITED STATES OF AMERICA

PREFACE TO THE THIRD EDITION

The Second Edition of *A History of Mathematics* covers the period from Antiquity to the close of World War I. It is a work which, with its emphasis on advanced mathematics, and in particular the advanced mathematics of the nineteenth and early twentieth centuries, has found a very special place in the mathematical literature. Like Cajori's two-volume *History of Mathematical Notations,* it is a unique work — but unique for a very different reason.

The clue to its uniqueness lies in the very fact that a history of mathematical notations — which is merely a single theme in the history of mathematics — requires two substantial volumes; that a standard history of just one period, Heath's *History of Greek Mathematics,* is in two large volumes; that a history of just a single branch of mathematics, Dickson's *History of the Theory of Numbers,* is in three large volumes; and that, indeed, one quite minor topic, that of Determinants, is the subject of a multi-volume history.

With so vast and complex a subject to set forth within so small a compass, the author of a one-volume history of Mathematics must decide what facts to select, what interrelations to point out, whom to name, how much to explain, and the like. There is no optimum, no one way of doing this. Every one-volume history of mathematics is therefore necessarily very different. Indeed, each of the few that exist are, as the reader can easily verify, strikingly different.

Florian Cajori, the author of this *History*, was Professor of the History of Mathematics at the University of California at Berkeley. His books include *A History of Elementary Mathematics,* Second Ed. (a different work from the present one), *A History of the Logarithmic Slide Rude, A History of Physics, William Oughtred, Early Mathematical Science in North and South America,* the already mentioned *History of Mathematical Notations,* in two volumes, *The Teáching and History of Mathematics in the United States,* and many others. He also wrote Chapter XX of Cantor's four-volume *Geschichte Der Mathematik.*

The revised and enlarged Second Edition of this *History* was written more than a quarter century after the First Edition and after almost a lifetime of devotion to the History of Mathematics.

It remains to indicate how this Third Edition differs from the Second. Professor Cajori left instructions for a number of small revisions and corrections and these have been made in the text. Some supplementary dates have been added. In the Index, a star has been placed next to a scientist's name to indicate that his collected papers have been published; many such collected papers have been published, some only quite recently (Eisenstein, 1975; Kummer, 1975). A few Editor's Notes, mostly of a bibliographical nature, have been added (pages 487ff.); they are indicated in the text by the symbol ✡.

A.G.

PREFACE TO THE FOURTH EDITION

Our knowledge of Babylonian mathematics has been virtually revolutionized in recent years by the work of a group of eminent scholars—Otto Neugebauer, A. Sachs, and others. It seemed desirable, therefore, to rewrite the chapter on Babylonian mathematics to take account of this recent research. Apart from this, a few minor improvements have been made in the text and in the Editor's Notes.

A. G.

PREFACE TO THE SECOND EDITION

IN preparing this second edition the earlier portions of the book have been partly re-written, while the chapters on recent mathematics are greatly enlarged and almost wholly new. The desirability of having a reliable one-volume history for the use of readers who cannot devote themselves to an intensive study of the history of mathematics is generally recognized. On the other hand, it is a difficult task to give an adequate bird's-eye-view of the development of mathematics from its earliest beginnings to the present time. In compiling this history the endeavor has been to use only the most reliable sources. Nevertheless, in covering such a wide territory, mistakes are sure to have crept in. References to the sources used in the revision are given as fully as the limitations of space would permit. These references will assist the reader in following into greater detail the history of any special subject. Frequent use without acknowledgment has been made of the following publications: *Annuario Biografico del Circolo Matematico di Palermo,* 1914; *Jahrbuch über die Fortschritte der Mathematik,* Berlin; *J. C. Poggendorff's Biographisch-Literarisches Handwörterbuch,* Leipzig; *Gedenktagebuch für Mathematiker,* von Felix Müller, 3. Aufl., Leipzig und Berlin, 1912; *Revue Semestrielle des Publications Mathématiques,* Amsterdam.

The author is indebted to Miss Falka M. Gibson of Oakland, Cal., for assistance in the reading of the proofs.

FLORIAN CAJORI

University of California,
March, 1919.

TABLE OF CONTENTS

A HISTORY OF MATHEMATICS

INTRODUCTION

THE contemplation of the various steps by which mankind has come into possession of the vast stock of mathematical knowledge can hardly fail to interest the mathematician. He takes pride in the fact that his science, more than any other, is an *exact* science, and that hardly anything ever done in mathematics has proved to be useless. The chemist smiles at the childish efforts of alchemists, but the mathematician finds the geometry of the Greeks and the arithmetic of the Hindus as useful and admirable as any research of to-day. He is pleased to notice that though, in course of its development, mathematics has had periods of slow growth, yet in the main it has been pre-eminently a *progressive* science.

The history of mathematics may be instructive as well as agreeable; it may not only remind us of what we have, but may also teach us how to increase our store. Says A. De Morgan, "The early history of the mind of men with regard to mathematics leads us to point out our own errors; and in this respect it is well to pay attention to the history of mathematics." It warns us against hasty conclusions; it points out the importance of a good notation upon the progress of the science; it discourages excessive specialisation on the part of investigators, by showing how apparently distinct branches have been found to possess unexpected connecting links; it saves the student from wasting time and energy upon problems which were, perhaps, solved long since; it discourages him from attacking an unsolved problem by the same method which has led other mathematicians to failure; it teaches that fortifications can be taken in other ways than by direct attack, that when repulsed from a direct assault it is well to reconnoitre and occupy the surrounding ground and to discover the secret paths by which the apparently unconquerable position can be taken.[1] The importance of this strategic rule may be emphasised by citing a case in which it has been violated. An untold amount of intellectual energy has been expended on the quadrature of the circle, yet no conquest has been made by direct assault. The circle-squarers have existed in crowds ever since the period of Archimedes. After innumerable failures to solve the problem at a time, even, when in-

[1] S. Günther, *Ziele und Resultate der neueren Mathematisch-historischen Forschung*. Erlangen, 1876.

vestigators possessed that most powerful tool, the differential calculus, persons versed in mathematics dropped the subject, while those who still persisted were completely ignorant of its history and generally misunderstood the conditions of the problem. "Our problem," says A. De Morgan, "is to square the circle with the *old allowance of means:* Euclid's postulates and nothing more. We cannot remember an instance in which a question to be solved by a *definite method* was tried by the best heads, and answered at last, *by that method*, after thousands of complete failures." But progress was made on this problem by approaching it from a different direction and by newly discovered paths. J. H. Lambert proved in 1761 that the ratio of the circumference of a circle to its diameter is irrational. Some years ago, F. Lindemann demonstrated that this ratio is also transcendental and that the quadrature of the circle, by means of the ruler and compasses only, is *impossible.* He thus showed by actual proof that which keen-minded mathematicians had long suspected; namely, that the great army of circle-squarers have, for two thousand years, been assaulting a fortification which is as indestructible as the firmament of heaven.

Another reason for the desirability of historical study is the value of historical knowledge to the teacher of mathematics. The interest which pupils take in their studies may be greatly increased if the solution of problems and the cold logic of geometrical demonstrations are interspersed with historical remarks and anecdotes. A class in arithmetic will be pleased to hear about the Babylonians and Hindus and their invention of the "Arabic notation"; they will marvel at the thousands of years which elapsed before people had even thought of introducing into the numeral notation that Columbus-egg—the zero; they will find it astounding that it should have taken so long to *invent* a notation which they themselves can now *learn* in a month. After the pupils have learned how to bisect a given angle, surprise them by telling of the many futile attempts which have been made to solve, by elementary geometry, the apparently very simple problem of the trisection of an angle. When they know how to construct a square whose area is double the area of a given square, tell them about the duplication of the cube, of its mythical origin—how the wrath of Apollo could be appeased only by the construction of a cubical altar double the given altar, and how mathematicians long wrestled with this problem. After the class have exhausted their energies on the theorem of the right triangle, tell them the legend about its discoverer—how Pythagoras, jubilant over his great accomplishment, sacrificed a hecatomb to the Muses who inspired him. When the value of mathematical training is called in question, quote the inscription over the entrance into the academy of Plato, the philosopher: "Let no one who is unacquainted with geometry enter here." Students in analytical geometry should know something of Descartes, and, after taking up the differential and integral calculus, they should become familiar with the parts that Newton, Leibniz, and Lagrange played in creating that science. In his historical talk it is possible for the teacher to make it plain to the student that mathematics is not a dead science, but a living one in which steady progress is made.[1]

A similar point of view is taken by Henry S. White: [2] "The accepted truths of to-day, even the commonplace truths of any science, were the doubtful or the novel theories of yesterday. Some indeed of prime importance were long esteemed of slight importance and almost forgotten. The first effect of reading in the history of science is a naïve astonishment at the darkness of past centuries, but the ultimate effect is a fervent admiration for the progress achieved by former generations, for the triumphs of persistence and of genius. The easy credulity with which a young student supposes that of course every algebraic equation must have a root gives place finally to a delight in the slow conquest of the realm of imaginary numbers, and in the youthful genius of a Gauss who could demonstrate this once obscure fundamental proposition."

The history of mathematics is important also as a valuable contribution to the history of civilisation. Human progress is closely identified with scientific thought. Mathematical and physical researches are a reliable record of intellectual progress. The history of mathematics is one of the large windows through which the philosophic eye looks into past ages and traces the line of intellectual development.

[1] Cajori, F., *The Teaching and History of Mathematics in the United States.* Washington, 1890, p. 236.

[2] *Bull. Am. Math. Soc.*, Vol. 15, 1909, p. 325.

[3] M. Cantor, *Vorlesungen über Geschichte der Mathematik*, 3. Aufl., Leipzig, 1907. This work has appeared in four large volumes and carries the history down to 1799. The fourth volume (1908) was written with the coöperation of nine scholars from Germany, Italy, Russia and the United States. *Moritz Cantor* (1829–1920) ranks as the foremost general writer of the nineteenth century on the history of mathematics. Born in Mannheim, a student at Heidelberg, at Göttingen under Gauss and Weber, at Berlin under Dirichlet, he lectured at Heidelberg where in 1877 he became ordinary honorary professor His first historical article was brought out in 1856, but not until 1880 did the first volume of his well-known history appear.

THE BABYLONIANS ☼

THE fertile valley of the Tigris and Euphrates was one of the primeval seats of human society. It was there that the art of writing was invented—wedge-shaped ('cuneiform') marks impressed into soft clay with a reed stylus and then baked, in the sun or in kilns, for permanence. Some half-million clay tablets have been unearthed, of which fewer than five hundred are devoted to mathematics. The latter are from two periods, 1800-1600 B.C. (the 'Old-Babylonian' period, roughly that of the Hammurapi dynasty) and 300-0 B.C. (the Seleucid period).

The exigencies of commerce required the development of a special class of *scribes,* who received formal training in writing (in Akkadian and in the earlier Sumerian) and in mathematics (arithmetic and—for the abler among them—higher arithmetic, including quadratic and sometimes certain higher order equations, linear equations in several variables, and the like, and of course, practical applications). The highly rational number system used by the scribes enabled them to handle with ease numerical computations which other cultures (even the later Greeks) found difficult, and it led in the Seleucid period to a highly developed (mathematical) astronomy.

In this number system, the symbols

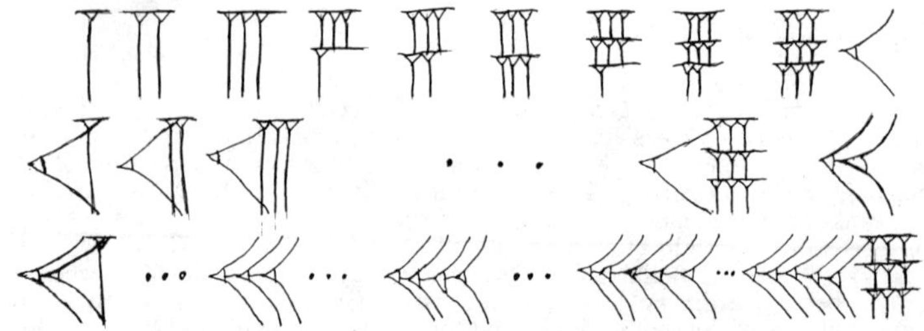

which we shall denote by (1), (2), . . . , (59), represent the numbers 1, 2, . . . , 59. Clearly, these numerals are derived from two symbols

the first representing units and the other, tens.

It might seem that if one is to be able to represent ever larger and larger numbers, one must invent symbols for larger and larger groupings (as in the Roman system, where we have I (= 1), V (= 5), X

(= 10), L (= 50), C (= 100), . . .). The Babylonian system escaped this apparent necessity by a highly important invention which, in modified form, we use today: the place-value system, in which a *limited* number of symbols suffices to represent numbers of *unlimited* size. Thus, just as in our decimal system 222 represents $2 \cdot 10^2 + 2 \cdot 10 + 2$, so in the sexagesimal system (2)(2)(2) represents $2 \cdot 60^2 + 2 \cdot 60 + 2 =$ 7,322. More generally, (2)(2)(2) represents $2 \cdot 60^{n+2} + 2 \cdot 60^{n+1} + 2 \cdot 60^n$, where n is any integer, positive, negative, or zero. For example, (1)(30) represents 1-1/2, as well as 90, 5,400, and so forth—the context indicating which of these is meant. No real ambiguity existed in actual practice. This 'sophisticated' concept of numeration is obviously of great value in multiplication and division and in the handling of fractions. However, this system did not 'take' with the ordinary Babylonian citizen and it did not replace the myriad other systems, sometimes of a decimal or partly decimal nature, that were in use in different contexts and in different localities. Nor did it 'take' with any successors of the Babylonians (certain individual mathematicians excepted): the Greeks used the sexagesimal system only for the fractional part but (for numbers over 60) not the integral part of a number. We ourselves use the decimal system in representing whole degrees of arc (or hours); we follow the Babylonians in dividing these into minutes (1/60-th) and seconds (1/60-th of 1/60-th), but we then continue with decimal subdivisions.

No symbol for zero was used in the Old-Babylonian period, so that no formal distinction existed between, say, (5)(7) = 307 and (5) . (7) = 18,007. Sometimes, it is true, the two numerals were separated clearly to indicate that a sexagesimal place was missing, but often a separation of the numerals or the absence of such a separation was without significance. Let us examine some implications of a sexagesimal system as compared with a decimal system. First, numbers with a missing sexagesimal place are rare; fewer than 1.7% of the numbers from 1 to 216,000 are of this kind, as compared with roughly 40% that require a zero in the decimal system. (And of these 40% quite a few require not one, but several zeros.) Second, it is not likely—to take the above example—that 18,007 and 307 should have been mistaken for each other in a computation, or even in a numerical table. In the Seleucid period, however, when astronomical tables were extensively computed, a symbol for zero came into use; from about 300 B.C. on, a symbol that had appeared earlier as a separation mark between sentences was in regular use as a zero. This symbol, which we shall denote by a period, was used between non-zero symbols, as in (5) . (10) = 18,010 and in (5) . . (20) = 1,080,020, and it sometimes appeared as a leftmost symbol, as in . (20)(6) = 1,206. But, interestingly, it appears never to have been used as a rightmost symbol.

Multiplication and division were carried out by means of *multiplication tables* and (since division by a is equivalent to multiplication by $1/a$) a *table of reciprocals*. The multiplication tables were a set of clay tablets each giving the multiples of a different fixed number and the whole intended to give the product of any number from 1 to 59 by any number from 1 to 59. Each tablet listed the numbers 1, 2, . . . , 20, 30, 40, 50 (rather than 1, . . . , 59) in one column and the products of each of these by the fixed number in an adjacent column. The product of any number from 1 to 59 by the fixed number could be obtained from the table by at most one addition, so that the abridgement of the set 1, . . . 59 was merely a space-saving device, making the tablets smaller, lighter in weight, and easier to handle.

A discussion of the table of reciprocals calls for some preliminary remarks. In our own decimal system, the reciprocals of some numbers can be written as terminating decimal fractions and the reciprocals of other numbers cannot. The numbers 8 and 50 are of the first kind (1/8 = 0.125, 1/50 = 0.02); the numbers 3 and 7 are of the second kind. Similarly, in a sexagesimal system, some numbers—let us call them 'regular' numbers—can be written as terminating sexagesimal fractions and others cannot. The criterion is simple enough: the regular numbers are those of the form $2^r3^s5^t$, since those are the numbers that divide evenly into some power of 60. The regular numbers smaller than 59 are the following: 2, 3, 4, 5, 6, 8, 9, 10, 12, 15, 16, 18, 20, 24, 27, 30, 32, 36, 40, 45, 48, 50, 54. The reciprocals of the first three and the last three of these numbers, written as sexagesimal fractions, are (30), (20), (15); (1)(15), (1)(12), (1)(6)(40). We recall that (30) here means 30/60 (= 1/2) and that (1)(15) means $1/60 + 15/60^2 = 1/60 + 15/3600 = 75/3600 = 1/48$.

The table of reciprocals was a tablet listing the *regular* numbers 2, 3, 4, 5, 6, 8, 9, 10, . . . , 75, 80, 81 and their corresponding reciprocals. In the standard table, there are no entries for the irregular numbers 7, 11, . . . , 78, 79. There do exist, however, tables of reciprocals for complete sequences of numbers, both regular and irregular, the reciprocals of the irregulars being given to three or four sexagesimal places. In the Seleucid period came truly extensive tables of reciprocals. These gave the reciprocals of the *regular* numbers up to 7 sexagesimal places, that is, up to 60^7 = 2,799,360,000,000. The reciprocals themselves were given to 17 sexagesimal places, which is the equivalent of thirty decimal places. That the entries in the table extended into the quadrillions is due to the sparseness of the regular numbers: up to 81, there are 30 regular numbers; up to 60^2 = 3,600, there are only 129; . . . ; up to 60^6, there are 2,460; and up to 60^7, only 3,802. Such a table made it possible to obtain, by interpolation, the reciprocals of irregular numbers.

Instead of multiplication tablets for each of the numbers 2, 3, . . . , 59, there was a smaller, but more useful set. There were tablets only for each of the regular numbers and sometimes also a tablet for the irregular number 7 (thus completing the sequence 1, . . . , 10). In addition, there were tablets for each of the *reciprocals* in the standard table of reciprocals. With the inclusion of these latter, division as well as multiplication became possible. Multiplication could be effected by a simple addition, where necessary, of the results from two tablets.

Although the Babylonians never developed an algebraic symbolism and, in this sense, never created the subject that we call algebra, they did deal with and, in some cases, mastered a number of topics that today fall under the heading of elementary algebra. Such topics were dealt with in two major classes of texts, which we shall call problem texts and solution texts. Let us consider quadratic equations, a subject that they cultivated extensively. Let us follow the solution of one problem: the area of a rectangle is unity, and the sum of the length and the width is the number (2) (1)(12)(12). In modern notation,

$$xy = 1, \qquad x + y = b \qquad (b = (2)\ (1)(12)(12) = 2.02\ldots).$$

The text instructs the reader somewhat as follows. Take one-half the sum of the length and the width and square it, obtaining

$$(b/2)^2 = (1)\ (1)(12)(33)(43)(12)(36).$$

[What we now have is $x^2/4 + xy/2 + y^2/4$.] Subtract 1, the product of the length and the width, and take the square root, obtaining

$$(0)\ (8)(31)(6).$$

[What we have computed is the square root of $x^2/4 - xy/2 + y^2/4$, or $x/2 - y/2$.]

We check our last result by squaring. We then add it to, and also subtract it from one-half the sum of the length and the width [i.e., $x/2 + y/2 = b/2$], obtaining

$$x = b/2 + \sqrt{\ } = (1)\ (0)(36)(6) + (0)\ (8)(31)(6) = (1)\ (9)(7)(12),$$

$$y = b/2 - \sqrt{\ } = (1)\ (0)(36)(6) - (0)\ (8)(31)(6) = (0)\ (52)(5).$$

This result is exactly equivalent to the standard algebraic formula for the solution of a quadratic equation.

We note several typical features. First, the generality of the method is made clear throughout, even though it is a particular numerical problem that is under discussion. In our example, although 1 is to be subtracted at one stage, it is described as the number 1, 'the product of the length and the width.' Similarly, if a multiplication is called for in the general case and the multiplier in the example under discussion happened to be 1, the multiplication by 1 was actually carried out. And the solution text often ends with the words 'such is the procedure.' Second, although the process concerns the particular numbers that arise in the course of the solution, the verbal description of the numbers—'the sum of the length and the width,' for example—is the exact equivalent of our own $x + y$; indeed, the verbal description is often directly translatable into algebraic symbols. Third, there is no hesitancy in making use of numbers having many sexagesimal digits. Fourth, the square root comes out 'even'—the number whose square root is to be taken proves to be a perfect square. The Babylonians knew how to solve the problem of producing Pythagorean numbers—integers (or, more generally, rational sexagesimal numbers) such that the difference of two squares is itself a perfect square. Countless problems in problem texts teach how to reduce quadratic problems to a normal form: two numbers are to be found if their product and their sum (or their difference) are given. In our illustrative example (already in normal form) the numbers represent lengths. However, in general they may represent disparate quantities—number of days and number of workmen, area and volume, and so forth. The artificiality of a problem in which the number of days and the number of workmen are required to have a certain sum was of no concern. Nor need the number of workmen prove to be a whole number. The technique of solving such problems was all that mattered.

Beyond this, certain equations of the fourth, sixth, and eighth degree were solved; these were, however, essentially quadratics in powers of the unknown. Problems leading to equations of the third and fifth degrees were solved; the existence of tables for the sum $x^2 + x^3$ indicates that they were solved purely numerically. Texts posing problems on compound interest and the existence of tables for consecutive powers of certain numbers indicate that the Babylonians knew how to deal with many problems of this kind. Texts exist which deal with the determination of exponents to a given base—in other words, a special case of the use of logarithms. The solving of several equations in several unknowns was, like quadratics, a topic extensively cultivated. Elementary Geometry existed but it was not a separate discipline, and the concept of proof was entirely absent. The approximation $\pi = 3\text{-}1/8 = 3.125$ was known, although for many purposes the approximation 3 was used. The Pythagorean theorem was known as far back as the Old-Babylonian period. The concept of prime number was, curiously, not recognized. (Was this because of a possible avoidance of the *terra incognita* of the irregular numbers?)

How did the sexagesimal place-value system originate? In the very earliest writing, in which the 'words' were pictographs made with the sharp end of the stylus, numbers were represented by impressing the clay with the *round* end of the stylus. Held in a slanted position, the stylus created a roughly elliptical-shaped impression; held vertically, a circle. The former represented units, the latter, tens. The number 24, for example, would be represented by two circles followed by four ellipses. As the need arose to represent larger numbers, various number systems built upon this basic system came into being; these systems might differ in various trades for various classes of objects and might vary from place to place. We describe two such systems: [1] A decimal system which, besides the 1 and the 10, had a symbol for 100, a circle resembling a 10-symbol, but larger: a 'big 10.' [2] The 1 and the 10, as before. In addition, a 'big 1,' representing 60. The number 120 was represented by two big-1 signs written in opposite directions; and 1,200 was represented by the same pair of big-1 signs with a 10-symbol between them. A very big 10-sign represented 3,600. In the course of time, the distinction between the 'big' symbol and its normal-size counterpart became unnecessary, as an example will make clear. A big-1 (representing 60) followed by a 10-symbol represented 60 + 10 = 70, whereas a 10-symbol followed by an ordinary 1-symbol represented 10 + 1 = 11. Thus the position alone of the 1 indicated whether it stood for 60 or 1; the larger size of the former was redundant, and it was ultimately written as an ordinary 1. As to the choice of 60, the *mina,* a unit of weight (and of money: a *mina* of silver) was equal to 60 *shekels.* This ratio of 60 to 1 is the most likely origin of the 60:1 ratio of the sexagesimal system.

THE EGYPTIANS

Though there is difference of opinion regarding the antiquity of Egyptian civilisation, yet all authorities agree in the statement that, however far back they go, they find no uncivilised state of society. "Menes, the first king, changes the course of the Nile, makes a great reservoir, and builds the temple of Phthah at Memphis." The Egyptians built the pyramids at a very early period. Surely a people engaging in enterprises of such magnitude must have known something of mathematics—at least of practical mathematics.

All Greek writers are unanimous in ascribing, without envy, to Egypt the priority of invention in the mathematical sciences. Plato in *Phœdrus* says: "At the Egyptian city of Naucratis there was a famous old god whose name was Theuth; the bird which is called the Ibis was sacred to him, and he was the inventor of many arts, such as arithmetic and calculation and geometry and astronomy and draughts and dice, but his great discovery was the use of letters."

Aristotle says that mathematics had its birth in Egypt, because there the priestly class had the leisure needful for the study of it. Geometry, in particular, is said by Herodotus, Diodorus, Diogenes Laertius, Iamblichus, and other ancient writers to have originated in Egypt.[1] In Herodotus we find this (II. c. 109): "They said also that this king [Sesostris] divided the land among all Egyptians so as to give each one a quadrangle of equal size and to draw from each his revenues, by imposing a tax to be levied yearly. But every one from whose part the river tore away anything, had to go to him and notify what had happened; he then sent the overseers, who had to measure out by how much the land had become smaller, in order that the owner might pay on what was left, in proportion to the entire tax imposed. In this way, it appears to me, geometry originated, which passed thence to Hellas."

We abstain from introducing additional Greek opinion regarding Egyptian mathematics, or from indulging in wild conjectures. We rest our account on documentary evidence. A hieratic papyrus, included in the Rhind collection of the British Museum, was deciphered by Eisenlohr in 1877, and found to be a mathematical manual containing problems in arithmetic and geometry. It was written by **Ahmes** some time before 1700 B. C., and was founded on an older work believed by Birch to date back as far as 3400 B. C.! This curious

[1] C. A. Bretschneider *Die Geometrie und die Geometer vor Euklides*. Leipzig, 1870, pp. 6–8. Carl Anton Bretschneider (1808–1878) was professor at the Realgymnasium at Gotha in Thuringia.

papyrus—the most ancient mathematical handbook known to us—puts us at once in contact with the mathematical thought in Egypt of three or five thousand years ago. It is entitled "Directions for obtaining the Knowledge of all Dark Things." We see from it that the Egyptians cared but little for theoretical results. Theorems are not found in it at all. It contains "hardly any general rules of procedure, but chiefly mere statements of results intended possibly to be explained by a teacher to his pupils." [1] In geometry the forte of the Egyptians lay in making constructions and determining areas. The area of an isosceles triangle, of which the sides measure 10 *khets* (a unit of length equal to 16.6 *m.* by one guess and about thrice that amount by another guess [2]) and the base 4 *khets*, was erroneously given as 20 square *khets*, or half the product of the base by one side. The area of an isosceles trapezoid is found, similarly, by multiplying half the sum of the parallel legs by one of the non-parallel sides. The area of a circle is found by deducting from the diameter $\frac{1}{9}$ of its length and squaring the remainder. Here π is taken $=(\frac{16}{9})^2=3.1604...$, a very fair approximation. The papyrus explains also such problems as these,—To mark out in the field a right triangle whose sides are 10 and 4 units; or a trapezoid whose parallel sides are 6 and 4, and the non-parallel sides each 20 units.

Some problems in this papyrus seem to imply a rudimentary knowledge of proportion.

The base-lines of the pyramids run north and south, and east and west, but probably only the lines running north and south were determined by astronomical observations. This, coupled with the fact that the word *harpedonaptæ*, applied to Egyptian geometers, means "rope-stretchers," would point to the conclusion that the Egyptian, like the Indian and Chinese geometers, constructed a right triangle upon a given line, by stretching around three pegs a rope consisting of three parts in the ratios 3 : 4 : 5, and thus forming a right triangle.[3] If this explanation is correct, then the Egyptians were familiar, 2000 years B. C., with the well-known property of the right triangle, for the special case at least when the sides are in the ratio 3: 4: 5.☆

On the walls of the celebrated temple of Horus at Edfu have been found hieroglyphics, written about 100 B. C., which enumerate the pieces of land owned by the priesthood, and give their areas. The area of any quadrilateral, however irregular, is there found by the formula $\frac{a+b}{2}\cdot\frac{c+d}{2}$. Thus, for a quadrangle whose opposite sides are 5 and 8, 20 and 15, is given the area $113\frac{1}{2}\ \frac{1}{4}$.[4] The incorrect for-

[1] James Gow, *A Short History of Greek Mathematics*. Cambridge, 1884, p. 16.

[2] A. Eisenlohr, *Ein mathematisches Handbuch der alten Aegypter*, 2. Ausgabe, Leipzig, 1897, p. 103; F. L. Griffith in *Proceedings of the Society of Biblical Archæology*, 1891, 1894.

[3] M. Cantor, *op. cit.* Vol. I, 3. Aufl., 1907, p. 105.

[4] H. Hankel, *Zur Geschichte der Mathematik in Alterthum und Mittelalter*, Leipzig, 1874, p. 86.

mulæ of Ahmes of perhaps 3000 years B. C. yield generally closer approximations than those of the Edfu inscriptions, written 200 years after Euclid!

The fact that the geometry of the Egyptians consists chiefly of constructions, goes far to explain certain of its great defects. The Egyptians failed in two essential points without which a *science* of geometry, in the true sense of the word, cannot exist. In the first place, they failed to construct a rigorously logical system of geometry, resting upon a few axioms and postulates. A great many of their rules, especially those in solid geometry, had probably not been proved at all, but were known to be true merely from observation or as matters of fact. The second great defect was their inability to bring the numerous special cases under a more general view, and thereby to arrive at broader and more fundamental theorems. Some of the simplest geometrical truths were divided into numberless special cases of which each was supposed to require separate treatment.

Some particulars about Egyptian geometry can be mentioned more advantageously in connection with the early Greek mathematicians who came to the Egyptian priests for instruction.

An insight into Egyptian methods of numeration was obtained through the ingenious deciphering of the hieroglyphics by Champollion, Young, and their successors. The symbols used were the following: for 1, for 10, for 100, for 1000, for 10,000, for 100,000, for 1,000,000, for 10,000,000.[1] The symbol for 1 represents a vertical staff; that for 10,000 a pointing finger; that for 100,000 a burbot; that for 1,000,000, a man in astonishment. The significance of the remaining symbols is very doubtful. The writing of numbers with these hieroglyphics was very cumbrous. The unit symbol of each order was repeated as many times as there were units in that order. The principle employed was the *additive*. Thus, 23 was written

Besides the hieroglyphics, Egypt possesses the *hieratic* and *demotic* writings, but for want of space we pass them by.

Herodotus makes an important statement concerning the mode of computing among the Egyptians. He says that they "*calculate with pebbles* by moving the hand from right to left, while the Hellenes move it from left to right." Herein we recognise again that *instrumental* method of figuring so extensively used by peoples of antiquity. The Egyptians used the decimal scale. Since, in figuring, they moved their hands horizontally, it seems probable that they used ciphering-boards with vertical columns. In each column there must have been not more than nine pebbles, for ten pebbles would be equal to one pebble in the column next to the left.

[1] M. Cantor, *op. cit.* Vol. I, 3. Aufl., 1907, p. 82.

The *Ahmes papyrus* contains interesting information on the way in which the Egyptians employed fractions. Their methods of operation were, of course, radically different from ours. Fractions were a subject of very great difficulty with the ancients. Simultaneous changes in both numerator and denominator were usually avoided. In manipulating fractions the Babylonians kept the denominators (60) constant. The Romans likewise kept them constant, but equal to 12. The Egyptians and Greeks, on the other hand, kept the numerators constant, and dealt with variable denominators. Ahmes used the term "fraction" in a restricted sense, for he applied it only to *unit-fractions*, or fractions having unity for the numerator. It was designated by writing the denominator and then placing over it a dot. Fractional values which could not be expressed by any one unit-fraction were expressed as the *sum* of two or more of them. Thus, he wrote $\frac{1}{3}\,\frac{1}{15}$ in place of $\frac{2}{5}$. While Ahmes knows $\frac{2}{3}$ to be equal to $\frac{1}{2}\,\frac{1}{6}$, he curiously allows $\frac{2}{3}$ to appear often among the unit-fractions and adopts a special symbol for it. The first important problem naturally arising was, how to represent any fractional value as the sum of unit-fractions. This was solved by aid of a table, given in the papyrus, in which all fractions of the form $\frac{2}{2n+1}$ (where n designates successively all the numbers up to 49) are reduced to the sum of unit-fractions. Thus, $\frac{2}{7}=\frac{1}{4}\,\frac{1}{28}$; $\frac{2}{99}=\frac{1}{66}\,\frac{1}{198}$. When, by whom, and how this table was calculated, we do not know. Probably it was compiled empirically at different times, by different persons. It will be seen that by repeated application of this table, a fraction whose numerator exceeds two can be expressed in the desired form, provided that there is a fraction in the table having the same denominator that *it* has. Take, for example, the problem, to divide 5 by 21. In the first place, $5=1+2+2$. From the table we get $\frac{2}{21}=\frac{1}{14}\,\frac{1}{42}$. Then $\frac{5}{21}=\frac{1}{21}+(\frac{1}{14}\,\frac{1}{42})+(\frac{1}{14}\,\frac{1}{42})=\frac{1}{21}+(\frac{2}{14}\,\frac{2}{42})=\frac{1}{21}\,\frac{1}{7}\,\frac{1}{21}=\frac{1}{7}\,\frac{2}{21}=\frac{1}{7}\,\frac{1}{14}\,\frac{1}{42}$. The papyrus contains problems in which it is required that fractions be raised by addition or multiplication to given whole numbers or to other fractions. For example, it is required to increase $\frac{1}{4}\,\frac{1}{8}\,\frac{1}{10}\,\frac{1}{30}\,\frac{1}{45}$ to 1. The common denominator taken appears to be 45, for the numbers are stated as $11\frac{1}{4}$, $5\frac{1}{2}\,\frac{1}{8}$, $4\frac{1}{2}$, $1\frac{1}{2}$, 1. The sum of these is $23\frac{1}{2}\,\frac{1}{4}\,\frac{1}{8}$ forty-fifths. Add to this $\frac{1}{9}\,\frac{1}{40}$, and the sum is $\frac{2}{3}$. Add $\frac{1}{3}$, and we have 1. Hence the quantity to be added to the given fraction is $\frac{1}{3}\,\frac{1}{9}\,\frac{1}{40}$.

Ahmes gives the following example involving an *arithmetical progression:* "Divide 100 loaves among 5 persons; $\frac{1}{7}$ of what the first three get is what the last two get. What is the difference?" Ahmes gives the solution: "Make the difference $5\frac{1}{2}$; 23, $17\frac{1}{2}$, 12, $6\frac{1}{2}$, 1. Multiply by $1\frac{2}{3}$; $38\frac{1}{3}$, $29\frac{1}{6}$, 20, $10\frac{2}{3}\,\frac{1}{6}$, $1\frac{2}{3}$." How did Ahmes come upon

$5\frac{1}{2}$? Perhaps thus:[1] Let a and $-d$ be the first term and the difference in the required arithmetical progression, then $\frac{1}{7}[a+(a-d)+(a-2d)]=(a-3d)+(a-4d)$, whence $d=5\frac{1}{2}(a-4d)$, *i. e.* the difference d is $5\frac{1}{2}$ times the last term. *Assuming* the last term 1, he gets his first progression. The sum is 60, but should be 100; hence multiply by $1\frac{2}{3}$, for $60\times1\frac{2}{3}=100$. We have here a method of solution which appears again later among the Hindus, Arabs and modern Europeans—the famous method of *false position.*

Ahmes speaks of a ladder consisting of the numbers 7, 49, 343, 2401, 16807. Adjacent to these powers of 7 are the words *picture, cat, mouse, barley, measure.* What is the meaning of these mysterious data? Upon the consideration of the problem given by Leonardo of Pisa in his *Liber abaci*, 3000 years later: "7 old women go to Rome, each woman has 7 mules, each mule carries 7 sacks, etc.", Moritz Cantor offers the following solution to the Ahmes riddle: 7 persons have each 7 cats, each cat eats 7 mice, each mouse eats 7 ears of barley, from each ear 7 measures of corn may grow. How many persons, cats, mice, ears of barley, and measures of corn, altogether? Ahmes gives 19607 as the sum of the geometric progression. Thus, the Ahmes papyrus discloses a knowledge of both arithmetical and geometrical progression.

Ahmes proceeds to the solution of equations of one unknown quantity. The unknown quantity is called 'hau' or heap. Thus the problem, "heap, its $\frac{1}{7}$, its whole, it makes 19," *i. e.* $\frac{x}{7}+x=19$. In this case, the solution is as follows: $\frac{8x}{7}=19$; $\frac{x}{7}=2\frac{1}{4}\,\frac{1}{8}$; $x=16\frac{1}{2}\,\frac{1}{8}$. But in other problems, the solutions are effected by various other methods. It thus appears that the beginnings of algebra are as ancient as those of geometry.

That the period of Ahmes was a flowering time for Egyptian mathematics appears from the fact that there exist other papyri (more recently discovered) of the same period. They were found at Kahun, south of the pyramid of Illahun. These documents bear close resemblance to Ahmes. They contain, moreover, examples of quadratic equations, the earliest of which we have a record. One of them is:[2] A given surface of, say, 100 units of area, shall be represented as the sum of two squares, whose sides are to each other as $1:\frac{3}{4}$. In modern symbols, the problem is, to find x and y, such that $x^2+y^2=100$ and $x:y=1:\frac{3}{4}$. The solution rests upon the method of false position. Try $x=1$ and $y=\frac{3}{4}$, then $x^2+y^2=\frac{25}{16}$ and $\sqrt{\frac{25}{16}}=\frac{5}{4}$. But $\sqrt{100}=10$ and $10\div\frac{5}{4}=8$. The rest of the solution cannot be made

[1] M. Cantor, *op. cit.*, Vol. I, 3. Aufl., 1907, p. 78.
[2] Cantor, *op. cit.* Vol. I, 1907, pp. 95, 96.

out, but probably was $x=8\times 1$, $y=8\times\frac{3}{4}=6$. This solution leads to the relation $6^2+8^2=10^2$. The symbol ▛ was used to designate square root.

In some ways similar to the Ahmes papyrus is also the Akhmim papyrus,[1] written over 2000 years later at Akhmim, a city on the Nile in Upper Egypt. It is in Greek and is supposed to have been written at some time between 500 and 800, A. D. It contains, besides arithmetical examples, a table for finding "unit-fractions," like that of Ahmes. Unlike Ahmes, it tells how the table was constructed. The rule, expressed in modern symbols, is as follows: $\frac{z}{pq}=\frac{1}{q\frac{p+q}{z}}+\frac{1}{p\frac{p+q}{z}}$. For $z=2$, this formula reproduces part of the table in Ahmes.

The principal defect of Egyptian arithmetic was the lack of a simple, comprehensive symbolism—a defect which not even the Greeks were able to remove.

The Ahmes papyrus and the other papyri of the same period represent the most advanced attainments of the Egyptians in arithmetic and geometry. It is remarkable that they should have reached so great proficiency in mathematics at so remote a period of antiquity. But strange, indeed, is the fact that, during the next two thousand years, they should have made no progress whatsoever in it. The conclusion forces itself upon us, that they resemble the Chinese in the *stationary character*, not only of their government, but also of their learning. All the knowledge of geometry which they possessed when Greek scholars visited them, six centuries B. C., was doubtless known to them two thousand years earlier, when they built those stupendous and gigantic structures—the pyramids.

[1] J. Baillet, "Le papyrus mathématique d'Akhmim," *Mémoires publiés par les membres de la mission archéologique française au Caire*, T. IX, 1r fascicule, Paris, 1892, pp. 1–88. See also Cantor, *op. cit.* Vol. I, 1907, pp. 67, 504.

THE GREEKS

Greek Geometry

About the seventh century B. C. an active commercial intercourse sprang up between Greece and Egypt. Naturally there arose an interchange of ideas as well as of merchandise. Greeks, thirsting for knowledge, sought the Egyptian priests for instruction. Thales, Pythagoras, Œnopides, Plato, Democritus, Eudoxus, all visited the land of the pyramids. Egyptian ideas were thus transplanted across the sea and there stimulated Greek thought, directed it into new lines, and gave to it a basis to work upon. Greek culture, therefore, is not primitive. Not only in mathematics, but also in mythology and art, Hellas owes a debt to older countries. To Egypt Greece is indebted, among other things, for its elementary geometry. But this does not lessen our admiration for the Greek mind. From the moment that Hellenic philosophers applied themselves to the study of Egyptian geometry, this science assumed a radically different aspect. "Whatever we Greeks receive, we improve and perfect," says Plato. The Egyptians carried geometry no further than was absolutely necessary for their practical wants. The Greeks, on the other hand, had within them a strong speculative tendency. They felt a craving to discover the reasons for things. They found pleasure in the contemplation of *ideal* relations, and loved science *as* science.

Our sources of information on the history of Greek geometry before Euclid consist merely of scattered notices by ancient writers. The early mathematicians, Thales and Pythagoras, left behind no written records of their discoveries. A full history of Greek geometry and astronomy during this period, written by Eudemus, a pupil of Aristotle, has been lost. It was well known to Proclus, who, in his commentaries on Euclid, gives a brief account of it. This abstract constitutes our most reliable information. We shall quote it frequently under the name of *Eudemian Summary*.

The Ionic School

To **Thales** (640–546 B. C.), of Miletus, one of the "seven wise men," and the founder of the Ionic school, falls the honor of having introduced the study of geometry into Greece. During middle life he engaged in commercial pursuits, which took him to Egypt. He is said to have resided there, and to have studied the physical sciences and mathematics with the Egyptian priests. Plutarch declares that Thales soon excelled his masters, and amazed King Amasis by measuring the heights of the pyramids from their shadows. According to Plutarch, this was done by considering that the shadow cast by a vertical staff of known length bears the same ratio to the shadow of the pyramid as the height of the staff bears to the height of the pyramid. This solution presupposes a knowledge of proportion, and the Ahmes papyrus actually shows that the rudiments of proportion were known to the Egyptians. According to Diogenes Laertius, the pyramids were measured by Thales in a different way; viz. by finding the length of the shadow of the pyramid at the moment when the shadow of a staff was equal to its own length. Probably both methods were used.

The *Eudemian Summary* ascribes to Thales the invention of the theorems on the equality of vertical angles, the equality of the angles at the base of an isosceles triangle, the bisection of a circle by any diameter, and the congruence of two triangles having a side and the two adjacent angles equal respectively. The last theorem, combined (we have reason to suspect) with the theorem on similar triangles, he applied to the measurement of the distances of ships from the shore. Thus Thales was the first to apply theoretical geometry to practical uses. The theorem that all angles inscribed in a semicircle are right angles is attributed by some ancient writers to Thales, by others to Pythagoras. Thales was doubtless familiar with other theorems, not recorded by the ancients. It has been inferred that he knew the sum of the three angles of a triangle to be equal to two right angles, and the sides of equiangular triangles to be proportional.[1] The Egyptians must have made use of the above theorems on the straight line, in some of their constructions found in the Ahmes papyrus, but it was left for the Greek philosopher to give these truths, which others saw, but did not formulate into words, an explicit, abstract expression, and to put into scientific language and subject to proof that which others merely felt to be true. Thales may be said to have created the geometry of lines, essentially abstract in its character, while the Egyptians studied only the geometry of surfaces and the rudiments of solid geometry, empirical in their character.[2]

With Thales begins also the study of scientific astronomy. He acquired great celebrity by the prediction of a solar eclipse in 585 B.C. Whether he predicted the day of the occurrence, or simply the year, is not known. It is told of him that while contemplating the stars during an evening walk, he fell into a ditch. The good old woman attending him exclaimed, "How canst thou know what is doing in the heavens, when thou seest not what is at thy feet?"

The two most prominent pupils of Thales were **Anaximander** (b. 611

[1] G. J. Allman, *Greek Geometry from Thales to Euclid*. Dublin, 1889, p. 10. George Johnston Allman (1824–1904) was professor of mathematics at Queen's College, Galway, Ireland.

[2] G. J. Allman, *op. cit.*, p. 15.

B. C.) and **Anaximenes** (b. 570 B. C.). They studied chiefly astronomy and physical philosophy. Of **Anaxagoras** (500–428 B. C.), a pupil of Anaximenes, and the last philosopher of the Ionic school, we know little, except that, while in prison, he passed his time attempting to square the circle. This is the first time, in the history of mathematics, that we find mention of the famous problem of the quadrature of the circle, that rock upon which so many reputations have been destroyed. It turns upon the determination of the exact value of π. Approximations to π had been made by the Chinese, Babylonians, Hebrews, and Egyptians. But the invention of a method to find its *exact* value, is the knotty problem which has engaged the attention of many minds from the time of Anaxagoras down to our own. Anaxagoras did not offer any solution of it, and seems to have luckily escaped paralogisms. The problem soon attracted popular attention, as appears from the reference to it made in 414 B. C. by the comic poet Aristophanes in his play, the "Birds." [1]

About the time of Anaxagoras, but isolated from the Ionic school, flourished **Œnopides** of Chios. Proclus ascribes to him the solution of the following problems: From a point without, to draw a perpendicular to a given line, and to draw an angle on a line equal to a given angle. That a man could gain a reputation by solving problems so elementary as these, indicates that geometry was still in its infancy, and that the Greeks had not yet gotten far beyond the Egyptian constructions.

The Ionic school lasted over one hundred years. The progress of mathematics during that period was slow, as compared with its growth in a later epoch of Greek history. A new impetus to its progress was given by Pythagoras.

The School of Pythagoras

Pythagoras (580?–500? B. C.) was one of those figures which impressed the imagination of succeeding times to such an extent that their real histories have become difficult to be discerned through the mythical haze that envelops them. The following account of Pythagoras excludes the most doubtful statements. He was a native of Samos, and was drawn by the fame of Pherecydes to the island of Syros. He then visited the ancient Thales, who incited him to study in Egypt. He sojourned in Egypt many years, and may have visited Babylon. On his return to Samos, he found it under the tyranny of Polycrates. Failing in an attempt to found a school there, he quitted home again and, following the current of civilisation, removed to Magna Græcia in South Italy. He settled at Croton, and founded the famous Pythagorean school. This was not merely an academy for

[1] F. Rudio in *Bibliotheca mathematica*, 3 S., Vol. 8, 1907–8, pp. 13–22.

the teaching of philosophy, mathematics, and natural science, but it was a brotherhood, the members of which were united for life. This brotherhood had observances approaching masonic peculiarity. They were forbidden to divulge the discoveries and doctrines of their school. Hence we are obliged to speak of the Pythagoreans as a body, and find it difficult to determine to whom each particular discovery is to be ascribed. The Pythagoreans themselves were in the habit of referring every discovery back to the great founder of the sect.

This school grew rapidly and gained considerable political ascendency. But the mystic and secret observances, introduced in imitation of Egyptian usages, and the aristocratic tendencies of the school, caused it to become an object of suspicion. The democratic party in Lower Italy revolted and destroyed the buildings of the Pythagorean school. Pythagoras fled to Tarentum and thence to Metapontum, where he was murdered.

Pythagoras has left behind no mathematical treatises, and our sources of information are rather scanty. Certain it is that, in the Pythagorean school, mathematics was the principal study. Pythagoras raised mathematics to the rank of a science. Arithmetic was courted by him as fervently as geometry. In fact, arithmetic is the foundation of his philosophic system.

The *Eudemian Summary* says that "Pythagoras changed the study of geometry into the form of a liberal education, for he examined its principles to the bottom, and investigated its theorems in an immaterial and intellectual manner." His geometry was connected closely with his arithmetic. He was especially fond of those geometrical relations which admitted of arithmetical expression.

Like Egyptian geometry, the geometry of the Pythagoreans is much concerned with areas. To Pythagoras is ascribed the important theorem that the square on the hypotenuse of a right triangle is equal to the sum of the squares on the other two sides. He had probably learned from the Egyptians the truth of the theorem in the special case when the sides are 3, 4, 5, respectively. The story goes, that Pythagoras was so jubilant over this discovery that he sacrificed a hecatomb. Its authenticity is doubted, because the Pythagoreans believed in the transmigration of the soul and opposed the shedding of blood. In the later traditions of the Neo-Pythagoreans this objection is removed by replacing this bloody sacrifice by that of "an ox made of flour!" The proof of the law of three squares, given in Euclid's *Elements*, I. 47, is due to Euclid himself, and not to the Pythagoreans. What the Pythagorean method of proof was has been a favorite topic for conjecture.

The theorem on the sum of the three angles of a triangle, presumably known to Thales, was proved by the Pythagoreans after the manner of Euclid. They demonstrated also that the plane about a point is completely filled by six equilateral triangles, four squares, or three regular hexagons, so that it is possible to divide up a plane into figures of either kind.

From the equilateral triangle and the square arise the regular solids: the tetraedron, octaedron, icosaedron, and the cube. These solids were, in all probability, known to the Egyptians, excepting, perhaps, the icosaedron. In Pythagorean philosophy, they represent respectively the four elements of the physical world; namely, fire, air, water, and earth. Later another regular solid was discovered, namely, the dodecaedron, which, in absence of a fifth element, was made to represent the universe itself. Iamblichus states that Hippasus, a Pythagorean, perished in the sea, because he boasted that he first divulged "the sphere with the twelve pentagons." The same story of death at sea is told of a Pythagorean who disclosed the theory of irrationals. The star-shaped pentagram was used as a symbol of recognition by the Pythagoreans, and was called by them Health.

Pythagoras called the sphere the most beautiful of all solids, and the circle the most beautiful of all plane figures. The treatment of the subjects of proportion and of irrational quantities by him and his school will be taken up under the head of arithmetic.

According to Eudemus, the Pythagoreans invented the problems concerning the application of areas, including the cases of defect and excess, as in Euclid, VI. 28, 29.

They were also familiar with the construction of a polygon equal in area to a given polygon and similar to another given polygon. This problem depends upon several important and somewhat advanced theorems, and testifies to the fact that the Pythagoreans made no mean progress in geometry.

Of the theorems generally ascribed to the Italian school, some cannot be attributed to Pythagoras himself, nor to his earliest successors. The progress from empirical to reasoned solutions must, of necessity, have been slow. It is worth noticing that on the circle no theorem of any importance was discovered by this school.

Though politics broke up the Pythagorean fraternity, yet the school continued to exist at least two centuries longer. Among the later Pythagoreans, Philolaus and Archytas are the most prominent. **Philolaus** wrote a book on the Pythagorean doctrines. By him were first given to the world the teachings of the Italian school, which had been kept secret for a whole century. The brilliant **Archytas** (428–347 B. C.) of Tarentum, known as a great statesman and general, and universally admired for his virtues, was the only great geometer among the Greeks when Plato opened his school. Archytas was the first to apply geometry to mechanics and to treat the latter subject methodically. He also found a very ingenious mechanical solution to the problem of the duplication of the cube. His solution involves clear notions on the generation of cones and cylinders. This problem reduces itself to finding two mean proportionals between two given lines. These mean proportionals were obtained by Archytas from the section of a half-cylinder. The doctrine of proportion was advanced through him.

There is every reason to believe that the later Pythagoreans exercised a strong influence on the study and development of mathematics at Athens. The Sophists acquired geometry from Pythagorean sources. Plato bought the works of Philolaus, and had a warm friend in Archytas.

The Sophist School

After the defeat of the Persians under Xerxes at the battle of Salamis, 480 B. C., a league was formed among the Greeks to preserve the freedom of the now liberated Greek cities on the islands and coast of the Ægæan Sea. Of this league Athens soon became leader and dictator. She caused the separate treasury of the league to be merged into that of Athens, and then spent the money of her allies for her own aggrandisement. Athens was also a great commercial centre. Thus she became the richest and most beautiful city of antiquity. All menial work was performed by slaves. The citizen of Athens was well-to-do and enjoyed a large amount of leisure. The government being purely democratic, every citizen was a politician. To make his influence felt among his fellow-men he must, first of all, be educated. Thus there arose a demand for teachers. The supply came principally from Sicily, where Pythagorean doctrines had spread. These teachers were called *Sophists*, or "wise men." Unlike the Pythagoreans, they accepted pay for their teaching. Although rhetoric was the principal feature of their instruction, they also taught geometry, astronomy, and philosophy. Athens soon became the headquarters of Grecian men of letters, and of mathematicians in particular. The home of mathematics among the Greeks was first in the Ionian Islands, then in Lower Italy, and during the time now under consideration, at Athens.

The geometry of the circle, which had been entirely neglected by the Pythagoreans, was taken up by the Sophists. Nearly all their discoveries were made in connection with their innumerable attempts to solve the following three famous problems:—

(1) To trisect an arc or an angle.

(2) To "double the cube," *i. e.*, to find a cube whose *volume* is double that of a given cube.

(3) To "square the circle," *i. e.* to find a square or some other rectilinear figure exactly equal in area to a given circle.

These problems have probably been the subject of more discussion and research than any other problems in mathematics. The bisection of an angle was one of the easiest problems in geometry. The trisection of an angle, on the other hand, presented unexpected difficulties. A right angle had been divided into three equal parts by the Pythagoreans. But the general construction, though easy in appearance, cannot be effected by the aid only of ruler and compasses. Among the first to wrestle with it was **Hippias of Elis,** a contemporary of Socrates, and born about 460 B. C. Unable to reach a solution by ruler and compasses only, he and other Greek geometers resorted to the use of other means. Proclus mentions a man, Hippias, presumably Hippias of Elis, as the inventor of a transcendental curve which served to divide an angle not only into three, but into any number of equal parts. This same curve was used later by Dinostratus and others for the quadrature of the circle. On this account it is called the *quadratrix*. The curve may be described thus: The side AB of the square shown in the figure turns uniformly about A, the point B moving along the circular arc BED. In the same time, the side BC moves parallel to itself and uniformly from the position of BC to that of AD. The locus of intersection of AB and BC, when thus moving, is the quadratrix BFG. Its equation we now write $y=x\cot\frac{\pi x}{2r}$. The ancients considered only the part of the curve that lies inside the quadrant of the circle; they did not know that $x=\pm 2r$ are asymptotes, nor that there is an infinite number of branches. According to Pappus, Dinostratus effected the quadrature by establishing the theorem that BED:AD =AD:AG.

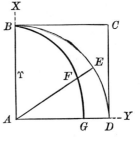

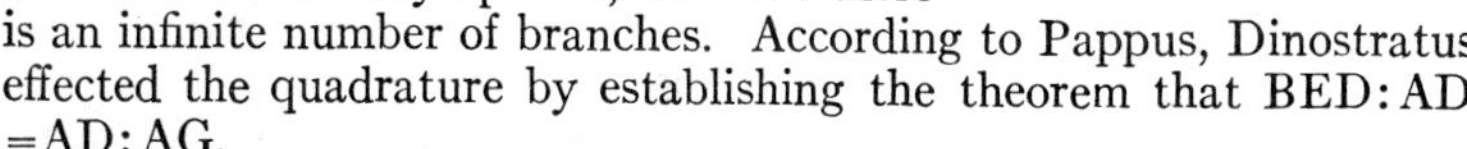

The Pythagoreans had shown that the diagonal of a square is the side of another square having double the area of the original one. This probably suggested the problem of the duplication of the cube, *i. e.*, to find the edge of a cube having double the volume of a given cube. Eratosthenes ascribes to this problem a different origin. The Delians were once suffering from a pestilence and were ordered by the oracle to double a certain cubical altar. Thoughtless workmen simply constructed a cube with edges twice as long, but brainless work like that did not pacify the gods. The error being discovered, Plato was consulted on the matter. He and his disciples searched eagerly for a solution to this "Delian Problem." An important contribution to this problem was made by **Hippocrates of Chios** (about 430 B. C.). He was a talented mathematician but, having been defrauded of his property, he was pronounced slow and stupid. It is also said of him that he was the first to accept pay for the teaching of mathematics. He showed that the Delian Problem could be reduced to finding two mean proportionals between a given line and another twice as long. For, in the proportion $a:x=x:y=y:2a$, since $x^2=ay$ and $y^2=2ax$ and $x^4=a^2y^2$, we have $x^4=2a^3x$ and $x^3=2a^3$. But, of course, he failed to find the two mean proportionals by geometric construction with ruler and compasses. He made himself celebrated by squaring certain lunes. According to Simplicius, Hippocrates believed he actually succeeded in applying one of his lune-quadratures to the quadrature of the circle. That Hippocrates really committed this fallacy is not generally accepted.

In the first lune which he squared, he took an isosceles triangle ABC, right-angled at C, and drew a semi-circle on AB as a diameter, and passing through C. He drew also a semi-circle on AC as a diameter and lying outside the triangle ABC. The lunar area thus formed is half the area of the triangle ABC. This is the first example of a curvilinear area which admits of exact quadrature. Hippocrates squared other lunes, hoping, no doubt, that he might be led to the quadrature of the circle.[1] In 1840 Th. Clausen found other quadrable lunes, but in 1902 E. Landau of Göttingen pointed out that two of the four lunes which Clausen supposed to be new, were known to Hippocrates.[2]

In his study of the quadrature and duplication-problems, Hippocrates contributed much to the geometry of the circle. He showed that circles are to each other as the squares of their diameters, that similar segments in a circle are as the squares of their chords and contain equal angles, that in a segment less than a semi-circle the angle is obtuse. Hippocrates contributed vastly to the logic of geometry. His investigations are the oldest "reasoned geometrical proofs in existence" (Gow). For the purpose of describing geometrical figures he used letters, a practice probably introduced by the Pythagoreans.

The subject of similar figures, as developed by Hippocrates, involved the theory of proportion. Proportion had, thus far, been used by the Greeks only in numbers. They never succeeded in uniting the notions of numbers and magnitudes. The term "number" was used by them in a restricted sense. What we call irrational numbers was not included under this notion. Not even rational fractions were called numbers. They used the word in the same sense as we use "positive integers." Hence numbers were conceived as *discontinuous*, while magnitudes were *continuous*. The two notions appeared, therefore, entirely distinct. The chasm between them is exposed to full view in the statement of Euclid that "incommensurable magnitudes do not have the same ratio as numbers." In Euclid's *Elements* we find the theory of proportion of magnitudes developed and treated independent of that of numbers. The transfer of the theory of proportion from numbers to magnitudes (and to lengths in particular) was a difficult and important step.

[1] A full account is given by G. Loria in his *Le scienze esatte nell'antica Grecia*, Milano, 2 edition, 1914, pp. 74–94. Loria gives also full bibliographical references to the extensive literature on Hippocrates.

[2] E. W. Hobson, *Squaring the Circle*, Cambridge, 1913, p. 16.

Hippocrates added to his fame by writing a geometrical text-book, called the *Elements*. This publication shows that the Pythagorean habit of secrecy was being abandoned; secrecy was contrary to the spirit of Athenian life.

The sophist **Antiphon,** a contemporary of Hippocrates, introduced the *process* of exhaustion for the purpose of solving the problem of the quadrature. He did himself credit by remarking that by inscribing in a circle a square or an equilateral triangle, and on its sides erecting isosceles triangles with their vertices in the circumference, and on the sides of these triangles erecting new triangles, etc., one could obtain a succession of regular polygons, of which each approaches nearer to the area of the circle than the previous one, until the circle is finally *exhausted*. Thus is obtained an inscribed polygon whose sides coincide with the circumference. Since there can be found squares equal in area to any polygon, there also can be found a square equal to the last polygon inscribed, and therefore equal to the circle itself. **Bryson of Heraclea,** a contemporary of Antiphon, advanced the problem of the quadrature considerably by circumscribing polygons at the same time that he inscribed polygons. He erred, however, in assuming that the area of a circle was the arithmetical mean between circumscribed and inscribed polygons. Unlike Bryson and the rest of Greek geometers, Antiphon seems to have believed it possible, by continually doubling the sides of an inscribed polygon, to obtain a polygon coinciding with the circle. This question gave rise to lively disputes in Athens. If a polygon can coincide with the circle, then, says Simplicius, we must put aside the notion that magnitudes are divisible *ad infinitum*. This difficult philosophical question led to paradoxes that are difficult to explain and that deterred Greek mathematicians from introducing ideas of infinity into their geometry; rigor in geometric proofs demanded the exclusion of obscure conceptions. Famous are the arguments against the possibility of motion that were advanced by **Zeno** of Elea, the great dialectician (early in the 5th century B. C.). None of Zeno's writings have come down to us. We know of his tenets only through his critics, Plato, Aristotle, Simplicius. Aristotle, in his *Physics*, VI, 9, ascribes to Zeno four arguments, called "Zeno's paradoxes": (1) The "Dichotomy": You cannot traverse an infinite number of points in a finite time; you must traverse the half of a given distance before you traverse the whole, and the half of that again before you can traverse the whole. This goes on *ad infinitum*, so that (if space is made up of points) there is an infinite number in any given space, and it cannot be traversed in a finite time. (2) The "Achilles": Achilles cannot overtake a tortoise. For, Achilles must first reach the place from which the tortoise started. By that time the tortoise will have moved on a little way. Achilles must then traverse that, and still the tortoise will be ahead. He is always nearer, yet never makes up to it. (3) The "Arrow":

An arrow in any given moment of its flight must be at rest in some particular point. (4) The "Stade": Suppose three parallel rows of points in juxtaposition, as in Fig. 1. One of these (B) is immovable,

```
A . . . .            ←A . . . .
B . . . .             B     . . . .
C . . . .             C         . . . .→
   Fig. 1                  Fig. 2
```

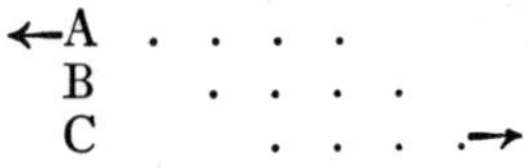

while A and C move in opposite directions with equal velocity, so as to come into the position in Fig. 2. The movement of C relatively to A will be double its movement relatively to B, or, in other words, any given point in C has passed twice as many points in A as it has in B. It cannot, therefore, be the case that an instant of time corresponds to the passage from one point to another.

Plato says that Zeno's purpose was "to protect the arguments of Parmenides against those who make fun of him"; Zeno argues that "there is no *many*," he "denies plurality." That Zeno's reasoning was wrong has been the view universally held since the time of Aristotle down to the middle of the nineteenth century. More recently the opinion has been advanced that Zeno was incompletely and incorrectly reported, that his arguments are serious efforts, conducted with logical rigor. This view has been advanced by Cousin, Grote and P. Tannery.[1] Tannery claims that Zeno did not deny motion, but wanted to show that motion was impossible under the Pythagorean conception of space as the sum of points, that the four arguments must be taken together as constituting a dialogue between Zeno and an adversary and that the arguments are in the form of a double dilemma into which Zeno forces his adversary. Zeno's arguments involve concepts of continuity, of the infinite and infinitesimal; they are as much the subjects of debate now as they were in the time of Aristotle. Aristotle did not successfully explain Zeno's paradoxes. He gave no reply to the query arising in the mind of the student, how is it possible for a variable to reach its limit? Aristotle's continuum was a sensuous, physical one; he held that, since a line cannot be built up of points, a line cannot actually be subdivided into points. "The continued bisection of a quantity is unlimited, so that the unlimited exists potentially, but is actually never reached." No satisfactory explanation of Zeno's arguments was given before the creation of Georg Cantor's continuum and theory of aggregates.

The process of exhaustion due to Antiphon and Bryson gave rise to the cumbrous but perfectly rigorous "method of exhaustion." In determining the ratio of the areas between two curvilinear plane figures, say two circles, geometers first inscribed or circumscribed similar polygons, and then by increasing indefinitely the number of

[1] See F. Cajori, "The History of Zeno's Arguments on Motion" in the *Americ. Math. Monthly*, Vol. 22, 1915, p. 3.

sides, nearly exhausted the spaces between the polygons and circumferences. From the theorem that similar polygons inscribed in circles are to each other as the squares on their diameters, geometers may have divined the theorem attributed to Hippocrates of Chios that the circles, which differ but little from the last drawn polygons, must be to each other as the squares on their diameters. But in order to exclude all vagueness and possibility of doubt, later Greek geometers applied reasoning like that in Euclid, XII, 2, as follows: Let C and c, D and d be respectively the circles and diameters in question. Then if the proportion $D^2 : d^2 = C : c$ is not true, suppose that $D^2 : d^2 = C : c^1$. If $c^1 < c$, then a polygon p can be inscribed in the circle c which comes nearer to it in area than does c^1. If P be the corresponding polygon in C, then $P : p = D^2 : d^2 = C : c^1$, and $P : C = p : c^1$. Since $p > c^1$, we have $P > C$, which is absurd. Next they proved by this same method of *reductio ad absurdum* the falsity of the supposition that $c^1 > c$. Since c^1 can be neither larger nor smaller than c, it must be equal to it, Q.E.D. Hankel refers this Method of Exhaustion back to Hippocrates of Chios, but the reasons for assigning it to this early writer, rather than to Eudoxus, seem insufficient.

Though progress in geometry at this period is traceable only at Athens, yet Ionia, Sicily, Abdera in Thrace, and Cyrene produced mathematicians who made creditable contributions to the science. We can mention here only **Democritus of Abdera** (about 460–370 B. C.), a pupil of Anaxagoras, a friend of Philolaus, and an admirer of the Pythagoreans. He visited Egypt and perhaps even Persia. He was a successful geometer and wrote on incommensurable lines, on geometry, on numbers, and on perspective. None of these works are extant. He used to boast that in the construction of plane figures with proof no one had yet surpassed him, not even the so-called harpedonaptæ ("rope-stretchers") of Egypt. By this assertion he pays a flattering compliment to the skill and ability of the Egyptians.

The Platonic School

During the Peloponnesian War (431–404 B. C.) the progress of geometry was checked. After the war, Athens sank into the background as a minor political power, but advanced more and more to the front as the leader in philosophy, literature, and science. **Plato** was born at Athens in 429 B. C., the year of the great plague, and died in 348. He was a pupil and near friend of Socrates, but it was not from him that he acquired his taste for mathematics. After the death of Socrates, Plato travelled extensively. In Cyrene he studied mathematics under Theodorus. He went to Egypt, then to Lower Italy and Sicily, where he came in contact with the Pythagoreans. Archytas of Tarentum and Timæus of Locri became his intimate friends. On his return to Athens, about 389 B. C., he founded his school in the groves of the *Academia*, and devoted the remainder of his life to teaching and writing.

Plato's physical philosophy is partly based on that of the Pythagoreans. Like them, he sought in arithmetic and geometry the key to the universe. When questioned about the occupation of the Deity, Plato answered that "He geometrises continually." Accordingly, a knowledge of geometry is a necessary preparation for the study of philosophy. To show how great a value he put on mathematics and how necessary it is for higher speculation, Plato placed the inscription over his porch, "Let no one who is unacquainted with geometry enter here." Xenocrates, a successor of Plato as teacher in the Academy, followed in his master's footsteps, by declining to admit a pupil who had no mathematical training, with the remark, "Depart, for thou hast not the grip of philosophy." Plato observed that geometry trained the mind for correct and vigorous thinking. Hence it was that the *Eudemian Summary* says, "He filled his writings with mathematical discoveries, and exhibited on every occasion the remarkable connection between mathematics and philosophy."

With Plato as the head-master, we need not wonder that the Platonic school produced so large a number of mathematicians. Plato did little real original work, but he made valuable improvements in the logic and methods employed in geometry. It is true that the Sophist geometers of the previous century were fairly rigorous in their proofs, but as a rule they did not reflect on the inward nature of their methods. They used the axioms without giving them explicit expression, and the geometrical concepts, such as the point, line, surface, etc., without assigning to them formal definitions.[1] The Pythagoreans called a point "unity in position," but this is a statement of a philosophical theory rather than a definition. Plato objected to calling a point a "geometrical fiction." He defined a point as the "beginning of a line" or as "an indivisible line," and a line as "length without breadth." He called the point, line, surface, the "boundaries" of the line, surface, solid, respectively. Many of the definitions in Euclid are to be ascribed to the Platonic school. The same is probably true of Euclid's axioms. Aristotle refers to Plato the axiom that "equals subtracted from equals leave equals."

One of the greatest achievements of Plato and his school is the invention of *analysis* as a method of proof. To be sure, this method had been used unconsciously by Hippocrates and others; but Plato, like a true philosopher, turned the instinctive logic into a conscious, legitimate method.

[1] "If any one scientific invention can claim pre-eminence over all others, I should be inclined myself to erect a monument to the unknown inventor of the mathematical point, as the supreme type of that process of abstraction which has been a necessary condition of scientific work from the very beginning." *Horace Lamb's* Address, Section A, Brit. Ass'n, 1904.

The terms *synthesis* and *analysis* are used in mathematics in a more special sense than in logic. In ancient mathematics they had a different meaning from what they now have. The oldest definition of mathematical analysis as opposed to synthesis is that given in Euclid, XIII, 5, which in all probability was framed by Eudoxus: "Analysis is the obtaining of the thing sought by assuming it and so reasoning up to an admitted truth; synthesis is the obtaining of the thing sought by reasoning up to the inference and proof of it." The analytic method is not conclusive, unless all operations involved in it are known to be reversible. To remove all doubt, the Greeks, as a rule, added to the analytic process a synthetic one, consisting of a reversion of all operations occurring in the analysis. Thus the aim of analysis was to aid in the discovery of synthetic proofs or solutions.

Plato is said to have solved the problem of the duplication of the cube. But the solution is open to the very same objection which he made to the solutions by Archytas, Eudoxus, and Menæchmus. He called their solutions not geometrical, but mechanical, for they required the use of other instruments than the ruler and compasses. He said that thereby "the good of geometry is set aside and destroyed, for we again reduce it to the world of sense, instead of elevating and imbuing it with the eternal and incorporeal images of thought, even as it is employed by God, for which reason He always is God." These objections indicate either that the solution is wrongly attributed to Plato or that he wished to show how easily non-geometric solutions of that character can be found. It is now rigorously established that the duplication problem, as well as the trisection and quadrature problems, cannot be solved by means of the ruler and compasses only.

Plato gave a healthful stimulus to the study of stereometry, which until his time had been entirely neglected by the Greeks. The sphere and the regular solids have been studied to some extent, but the prism, pyramid, cylinder, and cone were hardly known to exist. All these solids became the subjects of investigation by the Platonic school. One result of these inquiries was epoch-making. **Menæchmus,** an associate of Plato and pupil of Eudoxus, invented the conic sections, which, in course of only a century, raised geometry to the loftiest height which it was destined to reach during antiquity. Menæchmus cut three kinds of cones, the "right-angled," "acute-angled," and "obtuse-angled," by planes at right angles to a side of the cones, and thus obtained the three sections which we now call the parabola, ellipse, and hyperbola. Judging from the two very elegant solutions of the "Delian Problem" by means of intersections of these curves, Menæchmus must have succeeded well in investigating their properties. In what manner he carried out the graphic construction of these curves is not known.

Another great geometer was **Dinostratus,** the brother of Menæchmus and pupil of Plato. Celebrated is his mechanical solution of the quadrature of the circle, by means of the *quadratrix* of Hippias.

Perhaps the most brilliant mathematician of this period was **Eudoxus.** He was born at Cnidus about 408 B. C., studied under Archytas, and later, for two months, under Plato. He was imbued with a true spirit of scientific inquiry, and has been called the father of scientific astronomical observation. From the fragmentary notices of his astronomical researches, found in later writers, Ideler and Schiaparelli succeeded in reconstructing the system of Eudoxus with its celebrated representation of planetary motions by "concentric spheres." Eudoxus had a school at Cyzicus, went with his pupils to Athens, visiting Plato, and then returned to Cyzicus, where he died 355 B. C. The fame of the academy of Plato is to a large extent due to Eudoxus's pupils of the school at Cyzicus, among whom are Menæchmus, Dinostratus, Athenæus, and Helicon. Diogenes Laertius describes Eudoxus as astronomer, physician, legislator, as well as geometer. The *Eudemian Summary* says that Eudoxus "first increased the number of general theorems, added to the three proportions three more, and raised to a considerable quantity the learning, begun by Plato, on the subject of the section, to which he applied the analytical method." By this "section" is meant, no doubt, the "golden section" (*sectio aurea*), which cuts a line in extreme and mean ratio. The first five propositions in Euclid XIII relate to lines cut by this section, and are generally attributed to Eudoxus. Eudoxus added much to the knowledge of solid geometry. He proved, says Archimedes, that a pyramid is exactly one-third of a prism, and a cone one-third of a cylinder, having equal base and altitude. The proof that spheres are to each other as the cubes of their radii is probably due to him. He made frequent and skilful use of the method of exhaustion, of which he was in all probability the inventor. A scholiast on Euclid, thought to be Proclus, says further that Eudoxus practically invented the whole of Euclid's fifth book. Eudoxus also found two mean proportionals between two given lines, but the method of solution is not known.

Plato has been called a maker of mathematicians. Besides the pupils already named, the *Eudemian Summary* mentions the following: **Theætetus** of Athens, a man of great natural gifts, to whom, no doubt, Euclid was greatly indebted in the composition of the 10th book,[1] treating of incommensurables and of the 13th book; **Leodamas** of Thasos; **Neocleides** and his pupil **Leon,** who added much to the work of their predecessors, for Leon wrote an *Elements* carefully designed, both in number and utility of its proofs; **Theudius of Magnesia,** who composed a very good book of *Elements* and generalised propositions, which had been confined to particular cases; **Hermotimus of Colophon,** who discovered many propositions of the *Elements* and com-

[1] G. J. Allman, *op. cit.*, p. 212.

posed some on *loci;* and, finally, the names of **Amyclas of Heraclea, Cyzicenus of Athens,** and **Philippus of Mende.**

A skilful mathematician of whose life and works we have no details is **Aristæus,** the elder, probably a senior contemporary of Euclid. The fact that he wrote a work on conic sections tends to show that much progress had been made in their study during the time of Menæchmus. Aristæus wrote also on regular solids and cultivated the analytic method. His works contained probably a summary of the researches of the Platonic school.[1]

Aristotle (384–322 B. C.), the systematiser of deductive logic, though not a professed mathematician, promoted the science of geometry by improving some of the most difficult definitions. His *Physics* contains passages with suggestive hints of the principle of virtual velocities. He gave the best discussion of continuity and of Zeno's arguments against motion, found in antiquity. About his time there appeared a work called *Mechanica*, of which he is regarded by some as the author. Mechanics was totally neglected by the Platonic school.

The First Alexandrian School

In the previous pages we have seen the birth of geometry in Egypt, its transference to the Ionian Islands, thence to Lower Italy and to Athens. We have witnessed its growth in Greece from feeble childhood to vigorous manhood, and now we shall see it return to the land of its birth and there derive new vigor.

During her declining years, immediately following the Peloponnesian War, Athens produced the greatest scientists and philosophers of antiquity. It was the time of Plato and Aristotle. In 338 B. C., at the battle of Chæronea, Athens was beaten by Philip of Macedon, and her power was broken forever. Soon after, Alexander the Great, the son of Philip, started out to conquer the world. In eleven years he built up a great empire which broke to pieces in a day. Egypt fell to the lot of Ptolemy Soter. Alexander had founded the seaport of Alexandria, which soon became the "noblest of all cities." Ptolemy made Alexandria the capital. The history of Egypt during the next three centuries is mainly the history of Alexandria. Literature, philosophy, and art were diligently cultivated. Ptolemy created the university of Alexandria. He founded the great Library and built laboratories, museums, a zoölogical garden, and promenades. Alexandria soon became the great centre of learning.

Demetrius Phalereus was invited from Athens to take charge of the Library, and it is probable, says Gow, that **Euclid** was invited with him to open the mathematical school. According to the studies of H. Vogt,[2] Euclid was born about 365 B. C. and wrote his *Elements*

[1] G. J. Allman, *op. cit.*, p. 205.
[2] *Bibliotheca mathematica*, 3 S., Vol. 13, 1913, pp. 193–202.

between 330 and 320 B. C. Of the life of Euclid, little is known, except what is added by Proclus to the *Eudemian Summary*. Euclid, says Proclus, was younger than Plato and older than Eratosthenes and Archimedes, the latter of whom mentions him. He was of the Platonic sect, and well read in its doctrines. He collected the *Elements*, put in order much that Eudoxus had prepared, completed many things of Theætetus, and was the first who reduced to unobjectionable demonstration the imperfect attempts of his predecessors. When Ptolemy once asked him if geometry could not be mastered by an easier process than by studying the *Elements*, Euclid returned the answer, "There is no royal road to geometry." Pappus states that Euclid was distinguished by the fairness and kindness of his disposition, particularly toward those who could do anything to advance the mathematical sciences. Pappus is evidently making a contrast to Apollonius, of whom he more than insinuates the opposite character.[1] A pretty little story is related by Stobæus:[2] "A youth who had begun to read geometry with Euclid, when he had learnt the first proposition, inquired, 'What do I get by learning these things?' So Euclid called his slave and said, 'Give him threepence, since he must make gain out of what he learns.'" These are about all the personal details preserved by Greek writers. Syrian and Arabian writers claim to know much more, but they are unreliable. At one time Euclid of Alexandria was universally confounded with Euclid of Megara, who lived a century earlier.

The fame of Euclid has at all times rested mainly upon his book on geometry, called the *Elements*. This book was so far superior to the *Elements* written by Hippocrates, Leon, and Theudius, that the latter works soon perished in the struggle for existence. The Greeks gave Euclid the special title of "the author of the *Elements*." It is a remarkable fact in the history of geometry, that the *Elements* of Euclid, written over two thousand years ago, are still regarded by some as the best introduction to the mathematical sciences. In England they were used until the present century extensively as a text-book in schools. Some editors of Euclid have, however, been inclined to credit him with more than is his due. They would have us believe that a finished and unassailable system of geometry sprang at once from the brain of Euclid, "an armed Minerva from the head of Jupiter." They fail to mention the earlier eminent mathematicians from whom Euclid got his material. Comparatively few of the propositions and proofs in the *Elements* are his own discoveries. In fact, the proof of the "Theorem of Pythagoras" is the only one directly ascribed to him. Allman conjectures that the substance of Books I, II, IV comes from the Pythagoreans, that the substance of Book VI is due to the Pytha-

[1] A. De Morgan, "Eucleides" in *Smith's Dictionary of Greek and Roman Biography and Mythology*.
[2] J. Gow, *op. cit.*, p. 195.

goreans and Eudoxus, the latter contributing the doctrine of proportion as applicable to incommensurables and also the Method of Exhaustions (Book XII), that Theætetus contributed much toward Books X and XIII, that the principal part of the original work of Euclid himself is to be found in Book X.[1] Euclid was the greatest systematiser of his time. By careful selection from the material before him, and by logical arrangement of the propositions selected, he built up, from a few definitions and axioms, a proud and lofty structure. It would be erroneous to believe that he incorporated into his *Elements* all the elementary theorems known at his time. Archimedes, Apollonius, and even he himself refer to theorems not included in his *Elements*, as being well-known truths.

The text of the *Elements* that was commonly used in schools was Theon's edition. Theon of Alexandria, the father of Hypatia, brought out an edition, about 700 years after Euclid, with some alterations in the text. As a consequence, later commentators, especially Robert Simson, who labored under the idea that Euclid must be absolutely perfect, made Theon the scapegoat for all the defects which they thought they could discover in the text as they knew it. But among the manuscripts sent by Napoleon I from the Vatican to Paris was found a copy of the *Elements* believed to be anterior to Theon's recension. Many variations from Theon's version were noticed therein, but they were not at all important, and showed that Theon generally made only verbal changes. The defects in the *Elements* for which Theon was blamed must, therefore, be due to Euclid himself. The *Elements* used to be considered as offering models of scrupulously rigorous demonstrations. It is certainly true that in point of rigor it compares favorably with its modern rivals; but when examined in the light of strict mathematical logic, it has been pronounced by C. S. Peirce to be "riddled with fallacies." The results are correct only because the writer's experience keeps him on his guard. In many proofs Euclid relies partly upon intuition.

At the beginning of our editions of the *Elements*, under the head of definitions, are given the assumptions of such notions as the point, line, etc., and some verbal explanations. Then follow three postulates or demands, and twelve axioms. The term "axiom" was used by Proclus, but not by Euclid. He speaks, instead, of "common notions"—common either to all men or to all sciences. There has been much controversy among ancient and modern critics on the postulates and axioms. An immense preponderance of manuscripts and the testimony of Proclus place the "axioms" about *right angles* and *parallels* among the postulates.[2] This is indeed their proper place,

[1] G. J. Allman, *op. cit.*, p. 211.

[2] A. De Morgan, *loc. cit.;* H. Hankel, *Theorie der Complexen Zahlensysteme*, Leipzig, 1867, p. 52. In the various editions of Euclid's *Elements* different numbers are assigned to the axioms. Thus the parallel axiom is called by Robert Simson the

for they are really *assumptions*, and not *common notions* or axioms. The postulate about *parallels* plays an important rôle in the history of non-Euclidean geometry. An important postulate which Euclid missed was the one of superposition, according to which figures can be moved about in space without any alteration in form or magnitude.

The *Elements* contains thirteen books by Euclid, and two, of which it is supposed that Hypsicles and Damascius are the authors. The first four books are on plane geometry. The fifth book treats of the theory of proportion as applied to magnitudes in general. It has been greatly admired because of its rigor of treatment. Beginners find the book difficult. Expressed in modern symbols, Euclid's definition of proportion is thus: Four magnitudes, a, b, c, d, are in proportion, when for any integers m and n, we have simultaneously $ma \gtreqless nb$, and $mc \gtreqless nd$. Says T. L. Heath,[1] "certain it is that there is an exact correspondence, almost coincidence, between Euclid's definition of equal ratios and the modern theory of irrationals due to Dedekind. H. G. Zeuthen finds a close resemblance between Euclid's definition and Weierstrass' definition of equal numbers. The sixth book develops the geometry of similar figures. Its 27th Proposition is the earliest maximum theorem known to history. The seventh, eighth, ninth books are on the theory of numbers, or on arithmetic. According to P. Tannery, the knowledge of the existence of irrationals must have greatly affected the mode of writing the *Elements*. The old naïve theory of proportion being recognized as untenable, proportions are not used at all in the first four books. The rigorous theory of Eudoxus was postponed as long as possible, because of its difficulty. The interpolation of the arithmetical books VII–IX is explained as a preparation for the fuller treatment of the irrational in book X. Book VII explains the G. C. D. of two numbers by the process of division (the so-called "Euclidean method"). The theory of proportion of (rational) numbers is then developed on the basis of the definition, "Numbers are proportional when the first is the same multiple, part, or parts of the second that the third is of the fourth." This is believed to be the older, Pythagorean theory of proportion.[2] The tenth treats of the theory of incommensurables. De Morgan considered this the most wonderful of all. We give a fuller account of it under the head of Greek Arithmetic. The next three books are on

12th, by Bolyai the 11th, by Clavius the 13th, by F. Peyrard the 5th. It is called the 5th *postulate* in old manuscripts, also by Heiberg and Menge in their annotated edition of Euclid's works, in Greek and Latin, Leipzig, 1883, and by T. L. Heath in his *Thirteen Books of Euclid's Elements*, Vols. I–III, Cambridge, 1908. Heath's is the most recent translation into English and is very fully and ably annotated.

[1] T. L. Heath, *op. cit.*, Vol. II, p. 124.

[2] Read H. B. Fine, "Ratio, Proportion and Measurement in the Elements of Euclid," *Annals of Mathematics*, Vol. XIX, 1917, pp. 70–76.

stereometry. The eleventh contains its more elementary theorems; the twelfth, the metrical relations of the pyramid, prism, cone, cylinder, and sphere. The thirteenth treats of the regular polygons, especially of the triangle and pentagon, and then uses them as faces of the five regular solids; namely, the tetraedron, octaedron, icosaedron, cube, and dodecaedron. The regular solids were studied so extensively by the Platonists that they received the name of "Platonic figures." The statement of Proclus that the whole aim of Euclid in writing the *Elements* was to arrive at the construction of the regular solids, is obviously wrong. The fourteenth and fifteenth books, treating of solid geometry, are apocryphal. It is interesting to see that to Euclid, and to Greek mathematicians in general, the existence of areas was evident from intuition. The notion of non-quadrable areas had not occurred to them.

A remarkable feature of Euclid's, and of all Greek geometry before Archimedes is that it eschews mensuration. Thus the theorem that the area of a triangle equals half the product of its base and its altitude is foreign to Euclid.

Another extant book of Euclid is the *Data*. It seems to have been written for those who, having completed the *Elements*, wish to acquire the power of solving new problems proposed to them. The *Data* is a course of practice in *analysis*. It contains little or nothing that an intelligent student could not pick up from the *Elements* itself. Hence it contributes little to the stock of scientific knowledge. The following are the other works with texts more or less complete and generally attributed to Euclid: *Phænomena*, a work on spherical geometry and astronomy; *Optics*, which develops the hypothesis that light proceeds from the eye, and not from the object seen; *Catoptrica*, containing propositions on reflections from mirrors: *De Divisionibus*, a treatise on the division of plane figures into parts having to one another a given ratio; [1] *Sectio Canonis*, a work on musical intervals. His treatise on *Porisms* is lost; but much learning has been expended by Robert Simson and M. Chasles in restoring it from numerous notes found in the writings of Pappus. The term "porism" is vague in meaning. According to Proclus, the aim of a porism is not to state some property or truth, like a theorem, nor to effect a construction, like a problem, but to find and bring to view a thing which necessarily exists with given numbers or a given construction, as, to find the centre of a given circle, or to find the G. C. D. of two given numbers. Porisms, according to Chasles, are incomplete theorems, "expressing certain relations between things variable according to a common law." Euclid's other lost works are *Fallacies*, containing exercises in detection of fallacies; *Conic Sections*, in four books, which are the foundation of a work on the same subject by Apollonius; and *Loci on a Surface*,

[1] A careful restoration was brought out in 1915 by R. C. Archibald of Brown University.

the meaning of which title is not understood. Heiberg believes it to mean "loci which are surfaces."

The immediate successors of Euclid in the mathematical school at Alexandria were probably **Conon, Dositheus,** and **Zeuxippus,** but little is known of them.

Archimedes (287?–212 B. C.), the greatest mathematician of antiquity, was born in Syracuse. Plutarch calls him a relation of King Hieron; but more reliable is the statement of Cicero, who tells us he was of low birth. Diodorus says he visited Egypt, and, since he was a great friend of Conon and Eratosthenes, it is highly probable that he studied in Alexandria. This belief is strengthened by the fact that he had the most thorough acquaintance with all the work previously done in mathematics. He returned, however, to Syracuse, where he made himself useful to his admiring friend and patron, King Hieron, by applying his extraordinary inventive genius to the construction of various war-engines, by which he inflicted much loss on the Romans during the siege of Marcellus. The story that, by the use of mirrors reflecting the sun's rays, he set on fire the Roman ships, when they came within bow-shot of the walls, is probably a fiction. The city was taken at length by the Romans, and Archimedes perished in the indiscriminate slaughter which followed. According to tradition, he was, at the time, studying the diagram to some problem drawn in the sand. As a Roman soldier approached him, he called out, "Don't spoil my circles." The soldier, feeling insulted, rushed upon him and killed him. No blame attaches to the Roman general Marcellus, who admired his genius, and raised in his honor a tomb bearing the figure of a sphere inscribed in a cylinder. When Cicero was in Syracuse, he found the tomb buried under rubbish.

Archimedes was admired by his fellow-citizens chiefly for his mechanical inventions; he himself prized far more highly his discoveries in pure science. He declared that "every kind of art which was connected with daily needs was ignoble and vulgar." Some of his works have been lost. The following are the extant books, arranged approximately in chronological order: 1. Two books on *Equiponderance of Planes* or *Centres of Plane Gravities*, between which is inserted his treatise on the *Quadrature of the Parabola;* 2. *The Method;* 3. Two books on the *Sphere* and *Cylinder;* 4. The *Measurement of the Circle;* 5. *On Spirals;* 6. *Conoids* and *Spheroids;* 7. The *Sand-Counter;* 8. Two books on *Floating Bodies;* 9. Fifteen *Lemmas*.

In the book on the *Measurement of the Circle*, Archimedes proves first that the area of a circle is equal to that of a right triangle having the length of the circumference for its base, and the radius for its altitude. In this he assumes that there exists a straight line equal in length to the circumference—an assumption objected to by some ancient critics, on the ground that it is not evident that a straight line can equal a curved one. The finding of such a line was the next problem. He first finds an upper limit to the ratio of the circumference to the diameter, or π. To do this, he starts with an equilateral triangle of which the base is a tangent and the vertex is the centre of the circle. By successively bisecting the angle at the centre, by comparing ratios, and by taking the irrational square roots always a little too small, he finally arrived at the conclusion that $\pi < 3\frac{1}{7}$. Next he finds a lower limit by inscribing in the circle regular polygons of 6, 12, 24, 48, 96 sides, finding for each successive polygon its perimeter, which is, of course, always less than the circumference. Thus he finally concludes that "the circumference of a circle exceeds three times its diameter by a part which is less than $\frac{1}{7}$ but more than $\frac{10}{71}$ of the diameter." This approximation is exact enough for most purposes.

The *Quadrature of the Parabola* contains two solutions to the problem—one mechanical, the other geometrical. The method of exhaustion is used in both.

It is noteworthy that, perhaps through the influence of Zeno, infinitesimals (infinitely small constants) were not used in rigorous demonstration. In fact, the great geometers of the period now under consideration resorted to the radical measure of excluding them from demonstrative geometry by a postulate. This was done by Eudoxus, Euclid, and Archimedes. In the preface to the *Quadrature of the Parabola*, occurs the so-called "Archimedean postulate," which Archimedes himself attributes to Eudoxus: "When two spaces are unequal, it is possible to add to itself the difference by which the lesser is surpassed by the greater, so often that every finite space will be exceeded." Euclid (*Elements* V, 4) gives the postulate in the form of a definition: "Magnitudes are said to have a ratio to one another, when the less can be multiplied so as to exceed the other." Nevertheless, infinitesimals may have been used in tentative researches. That such was the case with Archimedes is evident from his book, *The Method*, formerly thought to be irretrievably lost, but fortunately discovered by Heiberg in 1906 in Constantinople. The contents of this book shows that he considered infinitesimals sufficiently scientific to suggest the truths of theorems, but not to furnish rigorous proofs. In finding the areas of parabolic segments, the volumes of spherical segments and other solids of revolution, he uses a mechanical process, consisting of the weighing of infinitesimal elements, which he calls straight lines or plane areas, but which are really infinitely narrow strips or infinitely thin plane laminæ.[1] The breadth or thickness is regarded as being the same in the elements weighed at any one time. The Archimedean postulate did not command the interest of mathematicians until the modern arithmetic continuum was created. It was O. Stolz that showed that it was a consequence of Dedekind's postulate relating to "sections."

[1] T. L. Heath, *Method of Archimedes*, Cambridge, 1912, p. 8.

It would seem that, in his great researches, Archimedes' mode of procedure was, to start with mechanics (centre of mass of surfaces and solids) and by his infinitesimal-mechanical method to discover new results for which later he deduced and published the rigorous proofs. Archimedes knew the integral[1] $\int x^3 dx$.

Archimedes studied also the ellipse and accomplished its quadrature, but to the hyperbola he seems to have paid less attention. It is believed that he wrote a book on conic sections.

Of all his discoveries Archimedes prized most highly those in his *Sphere and Cylinder*. In it are proved the new theorems, that the surface of a sphere is equal to four times a great circle; that the surface segment of a sphere is equal to a circle whose radius is the straight line drawn from the vertex of the segment to the circumference of its basal circle; that the volume and the surface of a sphere are $\frac{2}{3}$ of the volume and surface, respectively, of the cylinder circumscribed about the sphere. Archimedes desired that the figure to the last proposition be inscribed on his tomb. This was ordered done by Marcellus.

The spiral now called the "spiral of Archimedes," and described in the book *On Spirals*, was discovered by Archimedes, and not, as some believe, by his friend Conon.[2] His treatise thereon is, perhaps, the most wonderful of all his works. Nowadays, subjects of this kind are made easy by the use of the infinitesimal calculus. In its stead the ancients used the method of exhaustion. Nowhere is the fertility of his genius more grandly displayed than in his masterly use of this method. With Euclid and his predecessors the method of exhaustion was only the means of proving propositions which must have been seen and believed before they were proved. But in the hands of Archimedes this method, perhaps combined with his infinitesimal-mechanical method, became an instrument of discovery.

By the word "conoid," in his book on *Conoids and Spheroids*, is meant the solid produced by the revolution of a parabola or a hyperbola about its axis. Spheroids are produced by the revolution of an ellipse, and are long or flat, according as the ellipse revolves around the major or minor axis. The book leads up to the cubature of these solids. A few constructions of geometric figures were given by Archimedes and Apolonius which were effected by "insertions." In the following trisection of an angle, attributed by the Arabs to Archimedes, the "insertion" is achieved by the aid of a *graduated* ruler.[3] To trisect the angle CAB, draw the arc BCD. Then "insert" the

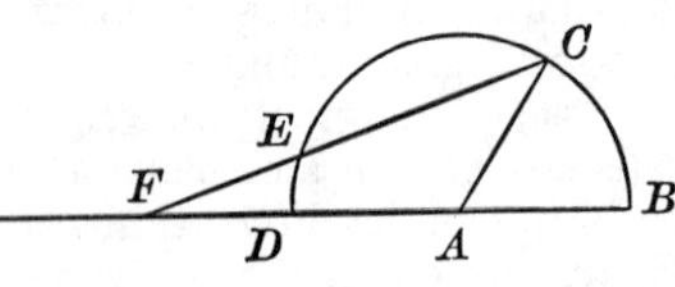

[1] H. G. Zeuthen in *Bibliotheca mathematica*, 3 S., Vol. 7, 1906–7, p. 347.

[2] M. Cantor, *op. cit.*, Vol. I, 3 Aufl., 1907, p. 306.

[3] F. Enriques, *Fragen der Elementargeometrie*, deutsche Ausg. v. H. Fleischer, II, Leipzig, 1907, p. 234.

distance FE, equal to AB, marked on an edge passing through C and moved until the points E and F are located as shown in the figure. The required angle is EFD.

His arithmetical treatise and problems will be considered later. We shall now notice his works on mechanics. Archimedes is the author of the first sound knowledge on this subject. Archytas, Aristotle, and others attempted to form the known mechanical truths into a science, but failed. Aristotle knew the property of the lever, but could not establish its true mathematical theory. The radical and fatal defect in the speculations of the Greeks, in the opinion of Whewell, was "that though they had in their possession facts and ideas, *the ideas were not distinct and appropriate to the facts.*" For instance, Aristotle asserted that when a body at the end of a lever is moving, it may be considered as having two motions; one in the direction of the tangent and one in the direction of the radius; the former motion is, he says, *according to nature*, the latter *contrary to nature.* These inappropriate notions of "natural" and "unnatural" motions, together with the habits of thought which dictated these speculations, made the perception of the true grounds of mechanical properties impossible.[1] It seems strange that even after Archimedes had entered upon the right path, this science should have remained absolutely stationary till the time of Galileo—a period of nearly two thousand years.

The proof of the property of the lever, given in his *Equiponderance of Planes*, holds its place in many text-books to this day. Mach [2] criticizes it. "From the mere assumption of the equilibrium of equal weights at equal distances is derived the inverse proportionality of weight and lever arm! How is that possible?" Archimedes' estimate of the efficiency of the lever is expressed in the saying attributed to him, "Give me a fulcrum on which to rest, and I will move the earth."

While the *Equiponderance* treats of solids, or the equilibrium of solids, the book on *Floating Bodies* treats of hydrostatics. His attention was first drawn to the subject of specific gravity when King Hieron asked him to test whether a crown, professed by the maker to be pure gold, was not alloyed with silver. The story goes that our philosopher was in a bath when the true method of solution flashed on his mind. He immediately ran home, naked, shouting, "I have found it!" To solve the problem, he took a piece of gold and a piece of silver, each weighing the same as the crown. According to one author, he determined the volume of water displaced by the gold, silver, and crown respectively, and calculated from that the amount of gold and silver

[1] William Whewell, *History of the Inductive Sciences*, 3rd Ed., New York, 1858, Vol. I, p. 87. William Whewell (1794–1866) was Master of Trinity College, Cambridge.

[2] E. Mach, *The Science of Mechanics*, tr. by T. McCormack, Chicago, 1907, p. 14. Ernst Mach (1838–1916) was professor of the history and theory of the inductive sciences at the university of Vienna.

in the crown. According to another writer, he weighed separately the gold, silver, and crown, while immersed in water, thereby determining their loss of weight in water. From these data he easily found the solution. It is possible that Archimedes solved the problem by both methods.

After examining the writings of Archimedes, one can well understand how, in ancient times, an "Archimedean problem" came to mean a problem too deep for ordinary minds to solve, and how an "Archimedean proof" came to be the synonym for unquestionable certainty. Archimedes wrote on a very wide range of subjects, and displayed great profundity in each. He is the Newton of antiquity.

Eratosthenes, eleven years younger than Archimedes, was a native of Cyrene. He was educated in Alexandria under Callimachus the poet, whom he succeeded as custodian of the Alexandrian Library. His many-sided activity may be inferred from his works. He wrote on *Good and Evil*, *Measurement of the Earth*, *Comedy*, *Geography*, *Chronology*, *Constellations*, and the *Duplication of the Cube*. He was also a philologian and a poet. He measured the obliquity of the ecliptic and invented a device for finding prime numbers, to be described later. Of his geometrical writings we possess only a letter to Ptolemy Euergetes, giving a history of the duplication problem and also the description of a very ingenious mechanical contrivance of his own to solve it. In his old age he lost his eyesight, and on that account is said to have committed suicide by voluntary starvation.

About forty years after Archimedes flourished **Apollonius of Perga,** whose genius nearly equalled that of his great predecessor. He incontestably occupies the second place in distinction among ancient mathematicians. Apollonius was born in the reign of Ptolemy Euergetes and died under Ptolemy Philopator, who reigned 222–205 B. C. He studied at Alexandria under the successors of Euclid, and for some time, also, at Pergamum, where he made the acquaintance of that Eudemus to whom he dedicated the first three books of his *Conic Sections*. The brilliancy of his great work brought him the title of the "Great Geometer." This is all that is known of his life.

His *Conic Sections* were in eight books, of which the first four only have come down to us in the original Greek. The next three books were unknown in Europe till the middle of the seventeenth century, when an Arabic translation, made about 1250, was discovered. The eighth book has never been found. In 1710 E. Halley of Oxford published the Greek text of the first four books and a Latin translation of the remaining three, together with his conjectural restoration of the eighth book, founded on the introductory lemmas of Pappus. The first four books contain little more than the substance of what earlier geometers had done. Eutocius tells us that Heraclides, in his life of Archimedes, accused Appolonius of having appropriated, in his *Conic Sections*, the unpublished discoveries of that great mathematician.

It is difficult to believe that this charge rests upon good foundation. Eutocius quotes Geminus as replying that neither Archimedes nor Apollonius claimed to have invented the conic sections, but that Apollonius had introduced a real improvement. While the first three or four books were founded on the works of Menæchmus, Aristæus, Euclid, and Archimedes, the remaining ones consisted almost entirely of new matter. The first three books were sent to Eudemus at intervals, the other books (after Eudemus's death) to one Attalus. The preface of the second book is interesting as showing the mode in which Greek books were "published" at this time. It reads thus: "I have sent my son Apollonius to bring you (Eudemus) the second book of my Conics. Read it carefully and communicate it to such others as are worthy of it. If Philonides, the geometer, whom I introduced to you at Ephesus, comes into the neighbourhood of Pergamum, give it to him also." [1]

The first book, says Apollonius in his preface to it, "contains the mode of producing the three sections and the conjugate hyperbolas and their principal characteristics, more fully and generally worked out than in the writings of other authors." We remember that Menæchmus, and all his successors down to Apollonius, considered only sections of *right* cones by a plane perpendicular to their elements, and that the three sections were obtained each from a different cone. Apollonius introduced an important generalisation. He produced all the sections from one and the same cone, whether right or scalene, and by sections which may or may not be perpendicular to its sides. The old names for the three curves were now no longer applicable. Instead of calling the three curves, sections of the "acute-angled," "right-angled," and "obtuse-angled" cone, he called them *ellipse*, *parabola*, and *hyperbola*, respectively. To be sure, we find the words "parabola" and "ellipse" in the works of Archimedes, but they are probably only interpolations. The word "ellipse" was applied because $y^2<px$, p being the parameter; the word "parabola" was introduced because $y^2=px$, and the term "hyperbola" because $y^2>px$.

The treatise of Apollonius rests on a unique property of conic sections, which is derived directly from the nature of the cone in which these sections are found. How this property forms the key to the system of the ancients is told in a masterly way by M. Chasles.[2] "Conceive," says he, "an oblique cone on a circular base; the straight line drawn from its summit to the centre of the circle forming its base is called the *axis* of the cone. The plane passing through the axis, perpendicular to its base, cuts the cone along two lines and determines in the circle a diameter; the triangle having this diameter for its base

[1] H. G. Zeuthen, *Die Lehre von den Kegelschnitten im Alterthum*, Kopenhagen, 1886, p. 502.

[2] M. Chasles, *Geschichte der Geometrie.* Aus dem Französischen übertragen durch, Dr. L. A. Sohncke, Halle, 1839, p. 15.

and the two lines for its sides, is called *the triangle through the axis.* In the formation of his conic sections, Apollonius supposed the cutting plane to be perpendicular to the plane of the triangle through the axis. The points in which this plane meets the two sides of this triangle are the *vertices* of the curve; and the straight line which joins these two points is a diameter of it. Apollonius called this diameter *latus transversum.* At one of the two vertices of the curve erect a perpendicular (*latus rectum*) to the plane of the triangle through the axis, of a certain length, to be determined as we shall specify later, and from the extremity of this perpendicular draw a straight line to the other vertex of the curve; now, through any point whatever of the diameter of the curve, draw at right angles an *ordinate:* the square of this ordinate, comprehended between the diameter and the curve, will be equal to the rectangle constructed on the portion of the ordinate comprised between the diameter and the straight line, and the part of the diameter comprised between the first vertex and the foot of the ordinate. Such is the characteristic property which Apollonius recognises in his conic sections and which he uses for the purpose of inferring from it, by adroit transformations and deductions, nearly all the rest. It plays, as we shall see, in his hands, almost the same rôle as the equation of the second degree with two variables (abscissa and ordinate) in the system of analytic geometry of Descartes." Apollonius made use of co-ordinates as did Menæchmus before him.[1] Chasles continues:

"It will be observed from this that the diameter of the curve and the perpendicular erected at one of its extremities suffice to construct the curve. These are the two elements which the ancients used, with which to establish their theory of conics. The perpendicular in question was called by them *latus erectum;* the moderns changed this name first to that of *latus rectum*, and afterwards to that of *parameter.*"

The first book of the *Conic Sections* of Apollonius is almost wholly devoted to the generation of the three principal conic sections.

The second book treats mainly of asymptotes, axes, and diameters.

The third book treats of the equality or proportionality of triangles, rectangles, or squares, of which the component parts are determined by portions of transversals, chords, asymptotes, or tangents, which are frequently subject to a great number of conditions. It also touches the subject of foci of the ellipse and hyperbola.

In the fourth book, Apollonius discusses the harmonic division of straight lines. He also examines a system of two conics, and shows that they cannot cut each other in more than four points. He investigates the various possible relative positions of two conics, as, for instance, when they have one or two points of contact with each other.

The fifth book reveals better than any other the giant intellect of its author. Difficult questions of *maxima and minima*, of which few

[1] T. L. Heath, *Apollonius of Perga*, Cambridge, 1896, p. CXV.

examples are found in earlier works, are here treated most exhaustively. The subject investigated is, to find the longest and shortest lines that can be drawn from a given point to a conic. Here are also found the germs of the subject of *evolutes* and *centres of osculation.*

The sixth book is on the similarity of conics.

The seventh book is on conjugate diameters.

The eighth book, as restored by Halley, continues the subject of conjugate diameters.

It is worthy of notice that Apollonius nowhere introduces the notion of *directrix* for a conic, and that, though he incidentally discovered the *focus* of an ellipse and hyperbola, he did not discover the focus of a parabola.[1] Conspicuous in his geometry is also the absence of technical terms and symbols, which renders the proofs long and cumbrous. R. C. Archibald claims that Apollonius was familiar with the centres of similitude of circles, usually attributed to Monge. T. L. Heath[2] comments thus: "The principal machinery used by Apollonius as well as by the earlier geometers comes under the head of what has been not inappropriately called a *geometrical algebra.*"

The discoveries of Archimedes and Apollonius, says M. Chasles, marked the most brilliant epoch of ancient geometry. Two questions which have occupied geometers of all periods may be regarded as having originated with them. The first of these is the quadrature of curvilinear figures, which gave birth to the infinitesimal calculus. The second is the theory of conic sections, which was the prelude to the theory of geometrical curves of all degrees, and to that portion of geometry which considers only the forms and situations of figures and uses only the intersection of lines and surfaces and the ratios of rectilineal distances. These two great divisions of geometry may be designated by the names of *Geometry of Measurements* and *Geometry of Forms and Situations*, or, Geometry of Archimedes and of Apollonius.

Besides the *Conic Sections*, Pappus ascribes to Apollonius the following works: *On Contacts*, *Plane Loci*, *Inclinations*, *Section of an Area*, *Determinate Section*, and gives lemmas from which attempts have been made to restore the lost originals. Two books on *De Sectione Rationis* have been found in the Arabic. The book on *Contacts*, as restored by F. Vieta, contains the so-called "Apollonian Problem": Given three circles, to find a fourth which shall touch the three.

Euclid, Archimedes, and Apollonius brought geometry to as high a state of perfection as it perhaps could be brought without first introducing some more general and more powerful method than the old method of exhaustion. A briefer symbolism, a Cartesian geometry, an infinitesimal calculus, were needed. The Greek mind was not

[1] J. Gow, *op. cit.*, p. 252.

[2] T. L. Heath, *Apollonius of Perga*, edited in modern notation. Cambridge, 1896, p. ci.

adapted to the invention of general methods. Instead of a climb to still loftier heights we observe, therefore, on the part of later Greek geometers, a descent, during which they paused here and there to look around for details which had been passed by in the hasty ascent.[1]

Among the earliest successors of Apollonius was **Nicomedes.** Nothing definite is known of him, except that he invented the *conchoid* ("mussel-like"), a curve of the fourth order. He devised a little machine by which the curve could be easily described. With aid of the conchoid he duplicated the cube. The curve can also be used for trisecting angles in a manner resembling that in the eighth lemma of Archimedes. Proclus ascribes this mode of trisection to Nicomedes, but Pappus, on the other hand, claims it as his own. The conchoid was used by Newton in constructing curves of the third degree.

About the time of Nicomedes (say, 180 B. C.), flourished also **Diocles,** the inventor of the *cissoid* ("ivy-like"). This curve he used for finding two mean proportionals between two given straight lines. The Greeks did not consider the companion-curve to the cissoid; in fact, they considered only the part of the cissoid proper which lies inside the circle used in constructing the curve. The part of the area of the circle left over when the two circular areas on the concave sides of the branches of the curve are removed, looks somewhat like an ivy-leaf. Hence, probably, the name of the curve. That the two branches extend to infinity appears to have been noticed first by G. P. de Roberal in 1640 and then by R. de Sluse.[2]

About the life of **Perseus** we know as little as about that of Nicomedes and Diocles. He lived some time between 200 and 100 B. C. From Heron and Geminus we learn that he wrote a work on the *spire*, a sort of anchor-ring surface described by Heron as being produced by the revolution of a circle around one of its chords as an axis. The sections of this surface yield peculiar curves called *spiral sections*, which, according to Geminus, were thought out by Perseus. These curves appear to be the same as the *Hippopede* of Eudoxus.

Probably somewhat later than Perseus lived **Zenodorus.** He wrote an interesting treatise on a new subject; namely, *isoperimetrical figures*. Fourteen propositions are preserved by Pappus and Theon. Here are a few of them: Of isoperimetrical, regular polygons, the one having the largest number of angles has the greatest area; the circle has a greater area than any regular polygon of equal periphery; of all isoperimentrical polygons of n sides, the regular is the greatest; of all solids having surfaces equal in area, the sphere has the greatest volume.

Hypsicles (between 200 and 100 B. C.) was supposed to be the author of both the fourteenth and fifteenth books of Euclid, but recent critics are of opinion that the fifteenth book was written by an author

[1] M. Cantor, *op. cit.*, Vol. I, 3 Aufl., 1907, p. 350.
[2] G. Loria, *Ebene Curven*, transl. by F. Schütte, I, 1910, p. 37.

who lived several centuries after Christ. The fourteenth book contains seven elegant theorems on *regular solids*. A treatise of Hypsicles on *Risings* is of interest because it gives the division of the circumference into 360 degrees after the fashion of the Babylonians.

Hipparchus of Nicæa in Bithynia was the greatest astronomer of antiquity. He took astronomical observations between 161 and 127 B. C. He established inductively the famous theory of epicycles and eccentrics. As might be expected, he was interested in mathematics, not *per se*, but only as an aid to astronomical inquiry. No mathematical writings of his are extant, but Theon of Alexandria informs us that Hipparchus originated the science of *trigonometry*, and that he calculated a "table of chords" in twelve books. Such calculations must have required a ready knowledge of arithmetical and algebraical operations. He possessed arithmetical and also graphical devices for solving geometrical problems in a plane and on a sphere. He gives indication of having seized the idea of co-ordinate representation, found earlier in Apollonius.

About 100 B. C. flourished **Heron the Elder** of Alexandria. He was the pupil of Ctesibius, who was celebrated for his ingenious mechanical inventions, such as the hydraulic organ, the water-clock, and catapult. It is believed by some that Heron was a son of Ctesibius. He exhibited talent of the same order as did his master by the invention of the eolipile and a curious mechanism known as "Heron's fountain." Great uncertainty exists concerning his writings. Most authorities believe him to be the author of an important *Treatise on the Dioptra*, of which there exist three manuscript copies, quite dissimilar. But M. Marie [1] thinks that the *Dioptra* is the work of *Heron the Younger*, who lived in the seventh or eighth century after Christ, and that *Geodesy*, another book supposed to be by Heron, is only a corrupt and defective copy of the former work. *Dioptra* contains the important formula for finding the area of a triangle expressed in terms of its sides; its derivation is quite laborious and yet exceedingly ingenious. "It seems to me difficult to believe," says Chasles, "that so beautiful a theorem should be found in a work so ancient as that of Heron the Elder, without that some Greek geometer should have thought to cite it." Marie lays great stress on this silence of the ancient writers, and argues from it that the true author must be Heron the Younger or some writer much more recent than Heron the Elder. But no reliable evidence has been found that there actually existed a second mathematician by the name of Heron. P. Tannery has shown that, in applying this formula, Heron used the approximation $\sqrt{A} \sim \frac{1}{2}(a + \frac{A}{a})$ for the irrational square roots where a^2 is the square nearest to

[1] Maximilien Marie, *Histoire des sciences mathématiques et physiques*. Paris, Tome I, 1883, p. 178.

A. When a more accurate value was wanted, Heron put $\frac{1}{2}(a+\frac{A}{a})$ in place of a in the above formula. Apparently, Heron sometimes found square and cube roots also by the method of "double false position."

"Dioptra," says Venturi, were instruments which had great resemblance to our modern theodolites. The book *Dioptra* is a treatise on geodesy containing solutions, with aid of these instruments, of a large number of questions in geometry, such as to find the distance between two points, of which one only is accessible, or between two points which are visible but both inaccessible; from a given point to draw a perpendicular to a line which cannot be approached; to find the difference of level between two points; to measure the area of a field without entering it.

Heron was a practical surveyor. This may account for the fact that his writings bear so little resemblance to those of the Greek authors, who considered it degrading the science to apply geometry to surveying. The character of his geometry is not Grecian, but decidedly Egyptian. This fact is the more surprising when we consider that Heron demonstrated his familiarity with Euclid by writing a commentary on the *Elements*. Some of Heron's formulas point to an old Egyptian origin. Thus, besides the above exact formula for the area of a triangle in terms of its sides, Heron gives the formula $\frac{a_1+a_2}{2}\times\frac{b}{2}$, which bears a striking likeness to the formula $\frac{a_1+a_2}{2}\times\frac{b_1+b_2}{2}$ for finding the area of a quadrangle, found in the Edfu inscriptions. There are, moreover, points of resemblance between Heron's writings and the ancient Ahmes papyrus. Thus Ahmes used unit-fractions exclusively (except the fraction $\frac{2}{3}$); Heron uses them oftener than other fractions. Like Ahmes and the priests at Edfu, Heron divides complicated figures into simpler ones by drawing auxiliary lines; like them, he shows, throughout, a special fondness for the isosceles trapezoid.

The writings of Heron satisfied a practical want, and for that reason were borrowed extensively by other peoples. We find traces of them in Rome, in the Occident during the Middle Ages, and even in India.

The works attributed to Heron, including the newly discovered *Metrica* published in 1903, have been edited by J. H. Heiberg, H. Schöne and W. Schmidt.

Geminus of Rhodes (about 70 B. C.) published an astronomical work still extant. He wrote also a book, now lost, on the *Arrangement of Mathematics*, which contained many valuable notices of the early history of Greek mathematics. Proclus and Eutocius quote it frequently. Theodosius is the author of a book of little merit on the geometry of the sphere. Investigations due to P. Tannery and A. A. Björnbo[1] seem to indicate that the mathematician Theodosius was not Theodosius of Tripolis, as formerly supposed, but was a resident of Bithynia and contemporary of Hipparchus. **Dionysodorus** of Amisus in Pontus applied the intersection of a parabola and hyperbola to the solution of a problem which Archimedes, in his *Sphere and Cylinder*, had left incomplete. The problem is "to cut a sphere so that its segments shall be in a given ratio."

We have now sketched the progress of geometry down to the time of Christ. Unfortunately, very little is known of the history of geometry between the time of Apollonius and the beginning of the Christian era. The names of quite a number of geometers have been mentioned, but very few of their works are now extant. It is certain, however, that there were no mathematicians of real genius from Apollonius to Ptolemy, excepting Hipparchus and perhaps Heron.

The Second Alexandrian School

The close of the dynasty of the Lagides which ruled Egypt from the time of Ptolemy Soter, the builder of Alexandria, for 300 years; the absorption of Egypt into the Roman Empire; the closer commercial relations between peoples of the East and of the West; the gradual decline of paganism and spread of Christianity,—these events were of far-reaching influence on the progress of the sciences, which then had their home in Alexandria. Alexandria became a commercial and intellectual emporium. Traders of all nations met in her busy streets, and in her magnificent Library, museums, lecture-halls, scholars from the East mingled with those of the West; Greeks began to study older literatures and to compare them with their own. In consequence of this interchange of ideas the Greek philosophy became fused with Oriental philosophy. Neo-Pythagoreanism and Neo-Platonism were the names of the modified systems. These stood, for a time, in opposition to Christianity. The study of Platonism and Pythagorean mysticism led to the revival of the theory of numbers. Perhaps the dispersion of the Jews and their introduction to Greek learning helped in bringing about this revival. The theory of numbers became a favorite study. This new line of mathematical inquiry ushered in what we may call a new school. There is no doubt that even now geometry continued to be one of the most important studies in the Alexandrian course. This Second Alexandrian School may be said to begin with the Christian era. It was made famous by the names of Claudius Ptolemæus, Diophantus, Pappus, Theon of Smyrna, Theon of Alexandria, Iamblichus, Porphyrius, and others.

By the side of these we may place **Serenus** of Antinœia, as having

[1] Axel Anthon Björnbo (1874–1911) of Copenhagen was a historian of mathematics. See *Bibliotheca mathematica*, 3 S., Vol. 12, 1911–12, pp. 337–344.

been connected more or less with this new school. He wrote on sections of the cone and cylinder, in two books, one of which treated only of the triangular section of the cone through the apex. He solved the problem, "given a cone (cylinder), to find a cylinder (cone), so that the section of both by the same plane gives similar ellipses." Of particular interest is the following theorem, which is the foundation of the modern theory of harmonics: If from D we draw DF, cutting the triangle ABC, and choose H on it, so that $DE : DF = EH : HF$, and if we draw the line AH, then every transversal through D, such as DG, will be divided by AH so that $DK : DG = KJ : JG$. **Menelaus** of Alexandria (about 98 A. D.) was the author of *Sphærica*, a work extant in Hebrew and Arabic, but not in Greek. In it he proves the theorems on the congruence of spherical triangles, and describes their properties in much the same way as Euclid treats plane triangles. In it are also found the theorems that the sum of the three sides of a spherical triangle is less than a great circle, and that the sum of the three angles exceeds two right angles. Celebrated are two theorems of his on plane and spherical triangles. The one on plane triangles is that, "if the three sides be cut by a straight line, the product of the three segments which have no common extremity is equal to the product of the other three." L. N. M. Carnot makes this proposition, known as the "lemma of Menelaus," the base of his theory of transversals. The corresponding theorem for spherical triangles, the so-called "regula sex quantitatum," is obtained from the above by reading "chords of three segments doubled," in place of "three segments."

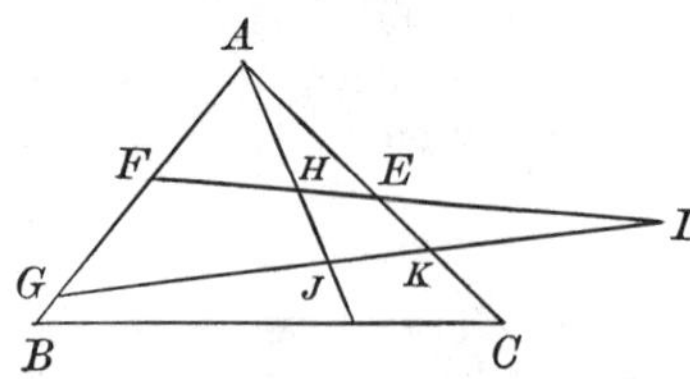

Claudius Ptolemy, a celebrated astronomer, was a native of Egypt. Nothing is known of his personal history except that he flourished in Alexandria in 139 A. D. and that he made the earliest astronomical observations recorded in his works, in 125 A. D., the latest in 151 A. D. The chief of his works are the *Syntaxis Mathematica* (or the *Almagest*, as the Arabs call it) and the *Geographica*, both of which are extant. The former work is based partly on his own researches, but mainly on those of Hipparchus. Ptolemy seems to have been not so much of an independent investigator, as a corrector and improver of the work of his great predecessors. The *Almagest* [1] forms the foundation of all astronomical science down to N. Copernicus. The fundamental idea of his system, the "Ptolemaic System," is that the earth is in the centre of the universe, and that the sun and planets revolve around the earth. Ptolemy did considerable for mathematics. He created,

[1] On the importance of the Almagest in the history of astronomy, consult P. Tannery, *Recherches sur l'histoire de l'astronomie*, Paris, 1893.

for astronomical use, a *trigonometry* remarkably perfect in form. The foundation of this science was laid by the illustrious Hipparchus.

The *Almagest* is in 13 books. Chapter 9 of the first book shows how to calculate tables of chords. The circle is divided into 360 degrees, each of which is halved. The diameter is divided into 120 divisions; each of these into 60 parts, which are again subdivided into 60 smaller parts. In Latin, these parts were called *partes minutæ primæ* and *partes minutæ secundæ*. Hence our names, "minutes" and "seconds." The sexagesimal method of dividing the circle is of Babylonian origin, and was known to Geminus and Hipparchus. But Ptolemy's method of calculating chords seems original with him. He first proved the proposition, now appended to Euclid VI (*D*), that "the rectangle contained by the diagonals of a quadrilateral figure inscribed in a circle is equal to both the rectangles contained by its opposite sides." He then shows how to find from the chords of two arcs the chords of their sum and difference, and from the chord of any arc that of its half. These theorems he applied to the calculation of his tables of chords. The proofs of these theorems are very pretty. Ptolemy's construction of sides of a regular inscribed pentagon and decagon was given later by C. Clavius and L. Mascheroni, and now is used much by engineers. Let the radius BD be $\perp$ to AC, $DE = EC$. Make $EF = EB$, then BF is the side of the pentagon and DF is the side of the decagon.

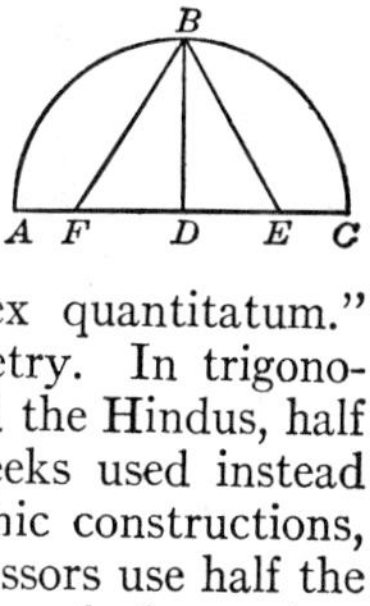

Another chapter of the first book in the *Almagest* is devoted to *trigonometry*, and to *spherical* trigonometry in particular. Ptolemy proved the "lemma of Menelaus," and also the "regula sex quantitatum." Upon these propositions he built up his trigonometry. In trigonometric computations, the Greeks did not use, as did the Hindus, half the chord of twice the arc (the "sine"); the Greeks used instead the whole chord of double the arc. Only in graphic constructions, referred to again later, did Ptolemy and his predecessors use half the chord of double the arc. The fundamental theorem of plane trigonometry, that two sides of a triangle are to each other as the chords of double the arcs measuring the angles opposite the two sides, was not stated explicitly by Ptolemy, but was contained implicitly in other theorems. More complete are the propositions in spherical trigonometry.

The fact that trigonometry was cultivated not for its own sake, but to aid astronomical inquiry, explains the rather startling fact that spherical trigonometry came to exist in a developed state earlier than plane trigonometry.

The remaining books of the *Almagest* are on astronomy. Ptolemy has written other works which have little or no bearing on mathematics, except one on geometry. Extracts from this book, made by Proclus, indicate that Ptolemy did not regard the parallel-axiom of

Euclid as self-evident, and that Ptolemy was the first of the long line of geometers from ancient time down to our own who toiled in the vain attempt to prove it. The untenable part of his demonstration is the assertion that, in case of parallelism, the sum of the interior angles on one side of a transversal must be the same as their sum on the other side of the transversal. Before Ptolemy an attempt to improve the theory of parallels was made by **Posidonius** (first cent. B. C.) who defined parallel lines as lines that are coplanar and equidistant. From an Arabic writer, *Al-Nirizi* (ninth cent.) it appears that Simplicius brought forward a proof of the 5th postulate, based upon this definition, and due to his friend Aganis (Geminus?).[1]

In the making of maps of the earth's surface and of the celestial sphere, Ptolemy (following Hipparchus) used stereographic projection. The eye is imagined to be at one of the poles, the projection being thrown upon the equatorial plane. He devised an instrument, a form of astrolabe planisphere, which is a stereographic projection of the celestial sphere.[2] Ptolemy wrote a monograph on the analemma which was a figure involving orthographic projections of the celestial sphere upon three mutually perpendicular planes (the horizontal, meridian and vertical circles). The analemma was used in determining positions of the sun, the rising and setting of the stars. The procedure was probably known to Hipparchus and the older astronomers. It furnished a graphic method for the solution of spherical triangles and was used subsequently by the Hindus, the Arabs, and Europeans as late as the seventeenth century.[3]

Two prominent mathematicians of this time were Nicomachus and Theon of Smyrna. Their favorite study was theory of numbers. The investigations in this science culminated later in the algebra of Diophantus. But no important geometer appeared after Ptolemy for 150 years. An occupant of this long gap was **Sextus Julius Africanus,** who wrote an unimportant work on geometry applied to the art of war, entitled *Cestes*. Another was the sceptic, **Sextus Empiricus** (200 A. D.); he endeavored to elucidate Zeno's "Arrow" by stating another argument equally paradoxical and therefore far from illuminating: Men never die, for if a man die, it must either be at a time when he is alive, or at a time when he is not alive; hence he never dies. Sextus Empiricus advanced also the paradox, that, when a line rotating in a plane about one of its ends describes a circle with each of its points, these concentric circles are of unequal area, yet each circle must be equal to the neighbouring circle which it touches.[1]

[1] R. Bonola, *Non-Euclidean Geometry*, trans. by H. S. Carslaw, Chicago, 1912, pp. 3–8. Robert Bonola (1875–1911) was professor in Rome.

[2] See M. Latham, "The Astrolabe," *Am. Math. Monthly*, Vol. 24, 1917, p. 162.

[3] See A. v. Braunmühl, *Geschichte der Trigonometrie*, Leipzig, I, 1900, p. 11. Alexander von Braunmühl (1853–1908) was professor at the technical high school in Munich.

Pappus, probably born about 340 A. D., in Alexandria, was the last great mathematician of the Alexandrian school. His genius was inferior to that of Archimedes, Apollonius, and Euclid, who flourished over 500 years earlier. But living, as he did, at a period when interest in geometry was declining, he towered above his contemporaries "like the peak of Teneriffa above the Atlantic." He is the author of a *Commentary on the Almagest*, a *Commentary on Euclid's Elements*, a *Commentary on the Analemma of Diodorus*,—a writer of whom nothing is known. All these works are lost. Proclus, probably quoting from the *Commentary on Euclid*, says that Pappus objected to the statement that an angle equal to a right angle is always itself a right angle.

The only work of Pappus still extant is his *Mathematical Collections*. This was originally in eight books, but the first and portions of the second are now missing. The *Mathematical Collections* seems to have been written by Pappus to supply the geometers of his time with a succinct analysis of the most difficult mathematical works and to facilitate the study of them by explanatory lemmas. But these lemmas are selected very freely, and frequently have little or no connection with the subject on hand. However, he gives very accurate summaries of the works of which he treats. The *Mathematical Collections* is invaluable to us on account of the rich information it gives on various treatises by the foremost Greek mathematicians, which are now lost. Mathematicians of the last century considered it possible to restore lost work from the *rêsumê* by Pappus alone.

We shall now cite the more important of those theorems in the *Mathematical Collections* which are supposed to be original with Pappus. First of all ranks the elegant theorem re-discovered by *P. Guldin*, over 1000 years later, that the volume generated by the revolution of the area under a plane curve which lies wholly on one side of the axis equals that area multiplied by the circumference described by its center of gravity. Pappus proved also that the centre of gravity of a triangle is that of another triangle whose vertices lie upon the sides of the first and divide its three sides in the same ratio. In the fourth book are new and brilliant propositions on the quadratrix which indicate an intimate acquaintance with curved surfaces. He generates the quadratrix as follows: Let a spiral line be drawn upon a right circular cylinder; then the perpendiculars to the axis of the cylinder drawn from each point of the spiral line form the surface of a screw. A plane passed through one of these perpendiculars, making any convenient angle with the base of the cylinder, cuts the screw-surface in a curve, the orthogonal projection of which upon the base is the *quadratrix*. A second mode of generation is no less admirable: If we make the spiral of Archimedes the base of a right

[1] See K. Lasswitz, *Geschichte der Atomistik*, I, Hamburg und Leipzig, 1890, p. 148.

cylinder, and imagine a cone of revolution having for its axis the side of the cylinder passing through the initial point of the spiral, then this cone cuts the cylinder in a curve of double curvature. The perpendiculars to the axis drawn through every point in this curve form the surface of a screw which Pappus here calls the *plectoidal surface*. A plane passed through one of the perpendiculars at any convenient angle cuts that surface in a curve whose orthogonal projection upon the plane of the spiral is the required *quadratrix*. Pappus considers curves of double curvature still further. He produces a *spherical spiral* by a point moving uniformly along the circumference of a great circle of a sphere, while the great circle itself revolves uniformly around its diameter. He then finds the area of that portion of the surface of the sphere determined by the spherical spiral, "a complanation which claims the more lively admiration, if we consider that, although the entire surface of the sphere was known since Archimedes' time, to measure portions thereof, such as spherical triangles, was then and for a long time afterwards an unsolved problem."[1] A question which was brought into prominence by Descartes and Newton is the "problem of Pappus." Given several straight lines in a plane, to find the locus of a point such that when perpendiculars (or, more generally, straight lines at given angles) are drawn from it to the given lines, the product of certain ones of them shall be in a given ratio to the product of the remaining ones. It is worth noticing that it was Pappus who first found the focus of the parabola and propounded the theory of the involution of points. He used the directrix and was the first to put in definite form the definition of the conic sections as loci of those points whose distances from a fixed point and from a fixed line are in a constant ratio. He solved the problem to draw through three points lying in the same straight line, three straight lines which shall form a triangle inscribed in a given circle. From the *Mathematical Collections* many more equally difficult theorems might be quoted which are original with Pappus as far as we know. It ought to be remarked, however, that he has been charged in three instances with copying theorems without giving due credit, and that he may have done the same thing in other cases in which we have no data by which to ascertain the real discoverer.[2]

About the time of Pappus lived **Theon** of Alexandria. He brought out an edition of Euclid's *Elements* with notes, which he probably used as a text-book in his classes. His commentary on the *Almagest* is valuable for the many historical notices, and especially for the specimens of Greek arithmetic which it contains. Theon's daughter **Hypatia,** a woman celebrated for her beauty and modesty, was the last Alexandrian teacher of reputation, and is said to have been an

[1] M. Cantor, *op. cit.*, Vol. I, 3 Aufl., 1907, p. 451.

[2] For a defence of Pappus against these charges, see J. H. Weaver in *Bull. Am Math. Soc.*, Vol. 23, 1916, pp. 131–133.

abler philosopher and mathematician than her father. Her notes on the works of Diophantus and Apollonius have been lost. Her tragic death in 415 A. D. is vividly described in Kingsley's *Hypatia*.

From now on, mathematics ceased to be cultivated in Alexandria. The leading subject of men's thoughts was Christian theology. Paganism disappeared, and with it pagan learning. The Neo-Platonic school at Athens struggled on a century longer. Proclus, Isidorus, and others kept up the "golden chain of Platonic succession." **Proclus,** the successor of Syrianus, at the Athenian school, wrote a commentary on Euclid's *Elements*. We possess only that on the first book, which is valuable for the information it contains on the history of geometry. **Damascius** of Damascus, the pupil of Isidorus, is now believed to be the author of the fifteenth book of Euclid. Another pupil of Isidorus was **Eutocius** of Ascalon, the commentator of Apollonius and Archimedes. **Simplicius** wrote a commentary on Aristotle's *De Cœlo*. Simplicius reports Zeno as saying: "That which, being added to another, does not make it greater, and being taken away from another does not make it less, is nothing." According to this, the denial of the existence of the infinitesimal goes back to Zeno. This momentous question presented itself centuries later to Leibniz, who gave different answers. The report made by Simplicius of the quadratures of Antiphon and Hippocrates of Chios is one of the best sources of historical information on this point.[1] In the year 529, Justinian, disapproving heathen learning, finally closed by imperial edict the schools at Athens.

As a rule, the geometers of the last 500 years showed a lack of creative power. They were commentators rather than discoverers.

The principal characteristics of ancient geometry are:—

(1) A wonderful clearness and definiteness of its concepts and an almost perfect logical rigor of its conclusions.

(2) A complete want of general principles and methods. Ancient geometry is decidedly *special*. Thus the Greeks possessed no general method of drawing tangents. "The determination of the tangents to the three conic sections did not furnish any rational assistance for drawing the tangent to any other new curve, such as the conchoid, the cissoid, etc." In the demonstration of a theorem, there were, for the ancient geometers, as many different cases requiring separate proof as there were different positions for the lines. The greatest geometers considered it necessary to treat all possible cases independently of each other, and to prove each with equal fulness. To devise methods by which the various cases could all be disposed of by one stroke, was beyond the power of the ancients. "If we compare a mathematical problem with a huge rock, into the interior of which we desire to penetrate, then the work of the Greek mathe-

[1] See F. Rudio in *Bibliotheca mathematica*, 3 S., Vol. 3, 1902, pp. 7–62.

maticians appears to us like that of a vigorous stonecutter who, with chisel and hammer, begins with indefatigable perseverance, from without, to crumble the rock slowly into fragments; the modern mathematician appears like an excellent miner, who first bores through the rock some few passages, from which he then bursts it into pieces with one powerful blast, and brings to light the treasures within." [1]

Greek Arithmetic and Algebra

Greek mathematicians were in the habit of discriminating between the *science* of numbers and the *art* of calculation. The former they called *arithmetica*, the latter *logistica*. The drawing of this distinction between the two was very natural and proper. The difference between them is as marked as that between theory and practice. Among the Sophists the art of calculation was a favorite study. Plato, on the other hand, gave considerable attention to philosophical arithmetic, but pronounced calculation a vulgar and childish art.

In sketching the history of Greek calculation, we shall first give a brief account of the Greek mode of counting and of writing numbers. Like the Egyptians and Eastern nations, the earliest Greeks counted on their fingers or with pebbles. In case of large numbers, the pebbles were probably arranged in parallel vertical lines. Pebbles on the first line represented units, those on the second tens, those on the third hundreds, and so on. Later, frames came into use, in which strings or wires took the place of lines. According to tradition, Pythagoras, who travelled in Egypt and, perhaps, in India, first introduced this valuable instrument into Greece. The *abacus*, as it is called, existed among different peoples and at different times, in various stages of perfection. An abacus is still employed by the Chinese under the name of *Swan-pan*. We possess no specific information as to how the Greek abacus looked or how it was used. Boethius says that the Pythagoreans used with the abacus certain nine signs called *apices*, which resembled in form the nine "Arabic numerals." But the correctness of this assertion is subject to grave doubts.

The oldest Grecian numerical symbols were the so-called *Herodianic signs* (after Herodianus, a Byzantine grammarian of about 200 A. D., who describes them). These signs occur frequently in Athenian inscriptions and are, on that account, now generally called *Attic*. For some unknown reason these symbols were afterwards replaced by the *alphabetic numerals*, in which the letters of the Greek alphabet were used, together with three strange and antique letters ϛ, ϙ, and ϡ, and the symbol M. This change was decidedly for the worse, for the old Attic numerals were less burdensome on the memory, inasmuch

[1] H. Hankel, *Die Entwickelung der Mathematik in den letzten Jahrhunderten*. Tübingen, 1884, p. 16.

as they contained fewer symbols and were better adapted to show forth analogies in numerical operations. The following table shows the Greek alphabetic numerals and their respective values:—

α	β	γ	δ	ε	ϛ	ζ	η	θ	ι	κ	λ	μ	ν	ξ	ο	π	ϙ
1	2	3	4	5	6	7	8	9	10	20	30	40	50	60	70	80	90

ρ	σ	τ	υ	φ	χ	ψ	ω	ϡ	͵α	͵β	͵γ etc.
100	200	300	400	500	600	700	800	900	1000	2000	3000

M	$\overset{\beta}{\mathrm{M}}$	$\overset{\gamma}{\mathrm{M}}$ etc.
10,000	20,000	30,000

It will be noticed that at 1000, the alphabet is begun over again, but, to prevent confusion, a stroke is now placed before the letter and generally somewhat below it. A horizontal line drawn over a number served to distinguish it more readily from words. The coefficient for M was sometimes placed before or behind instead of over the M. Thus 43,678 was written δM͵γχοη. It is to be observed that the Greeks had no zero.

Fractions were denoted by first writing the numerator marked with an accent, then the denominator marked with two accents and written twice. Thus, $\iota\gamma'\kappa\theta''\kappa\theta''=\frac{13}{29}$. In case of fractions having unity for the numerator, the α' was omitted and the denominator was written only once. Thus $\mu\delta''=\frac{1}{44}$.

The Greeks had the name epimorion for the ratio $\frac{n}{n+1}$. Archytas proved the theorem that if an epimorion $\frac{\alpha}{\beta}$ is reduced to its lowest terms $\frac{\mu}{\nu}$, then $\nu=\mu+1$. This theorem is found later in the musical writings of Euclid and of the Roman Boethius. The Euclidean form of arithmetic, without perhaps the representation of numbers by lines, existed as early as the time of Archytas.[1]

Greek writers seldom refer to calculation with alphabetic numerals. Addition, subtraction, and even multiplication were probably performed on the abacus. Expert mathematicians may have used the symbols. Thus Eutocius, a commentator of the sixth century after Christ, gives a great many multiplications of which the following is a specimen: [2]—

[1] P. Tannery in *Bibliotheca mathematica*, 3 S., Vol. VI, 1905, p. 228.
[2] J. Gow, *op. cit.*, p. 50.

Greek	Modern
$\overline{\sigma\xi\epsilon}$	265
$\overline{\sigma\xi\epsilon}$	265
$\overset{\delta}{\mathrm{M}}\overset{\alpha}{\mathrm{M}},\beta,\alpha$	40000, 12000, 1000
$\overset{\alpha}{\mathrm{M}},\beta,\overline{\gamma\chi\tau}$	12000, 3600, 300
$,\alpha\ \tau\ \kappa\epsilon$	1000, 300, 25
$\overset{\zeta}{\mathrm{M}}\ \overline{\sigma\kappa\epsilon}$	70225

The operation is explained sufficiently by the modern numerals appended. In case of mixed numbers, the process was still more clumsy. Divisions are found in Theon of Alexandria's commentary on the *Almagest*. As might be expected, the process is long and tedious.

We have seen in geometry that the more advanced mathematicians frequently had occasion to extract the square root. Thus Archimedes in his *Mensuration of the Circle* gives a large number of square roots. He states, for instance, that $\sqrt{3}<\frac{1351}{780}$ and $\sqrt{3}>\frac{265}{153}$, but he gives no clue to the method by which he obtained these approximations. It is not improbable that the earlier Greek mathematicians found the square root by trial only. Eutocius says that the method of extracting it was given by Heron, Pappus, Theon, and other commentators on the *Almagest*. Theon's is the only one of these methods known to us. It is the same as the one used nowadays, except that sexagesimal fractions are employed in place of our decimals. What the mode of procedure actually was when sexagesimal fractions were not used, has been the subject of conjecture on the part of numerous modern writers.

Of interest, in connection with arithmetical symbolism, is the *Sand-Counter* (Arenarius), an essay addressed by **Archimedes** to Gelon, king of Syracuse. In it Archimedes shows that people are in error who think the sand cannot be counted, or that if it can be counted, the number cannot be expressed by arithmetical symbols. He shows that the number of grains in a heap of sand not only as large as the whole earth, but as large as the entire universe, can be arithmetically expressed. Assuming that 10,000 grains of sand suffice to make a little solid of the magnitude of a poppy-seed, and that the diameter of a poppy-seed be not smaller than $\frac{1}{40}$ part of a finger's breadth; assuming further, that the diameter of the universe (supposed to extend to the sun) be less than 10,000 diameters of the earth, and that the latter be less than 1,000,000 stadia, Archimedes finds a number which would exceed the number of grains of sand in the sphere of the universe. He goes on even further. Supposing the universe to reach out to the fixed stars, he finds that the sphere, having the distance from the earth's centre to the fixed stars for its radius, would contain a number of grains of sand less than 1000 myriads of the eighth octad. In our notation, this number would be 10^{63} or 1 with 63 ciphers after it. It can hardly be doubted that one object which Archimedes had in view in making this calculation was the improvement of the Greek symbolism. It is not known whether he invented some short notation by which to represent the above number or not.

We judge from fragments in the second book of Pappus that Apollonius proposed an improvement in the Greek method of writing numbers, but its nature we do not know. Thus we see that the Greeks never possessed the boon of a clear, comprehensive symbolism. The honor of giving such to the world was reserved by the irony of fate for a nameless Indian of an unknown time, and we know not whom to thank for an invention of such importance to the general progress of intelligence.[1]

Passing from the subject of *logistica* to that of *arithmetica*, our attention is first drawn to the science of numbers of **Pythagoras.** Before founding his school, Pythagoras studied for many years under the Egyptian priests and familiarised himself with Egyptian mathematics and mysticism. If he ever was in Babylon, as some authorities claim, he may have learned the sexagesimal notation in use there; he may have picked up considerable knowledge on the theory of proportion, and may have found a large number of interesting astronomical observations. Saturated with that speculative spirit then pervading the Greek mind, he endeavored to discover some principle of homogeneity in the universe. Before him, the philosophers of the Ionic school had sought it in the matter of things; Pythagoras looked for it in the structure of things. He observed various numerical relations or analogies between numbers and the phenomena of the universe. Being convinced that it was in numbers and their relations that he was to find the foundation to true philosophy, he proceeded to trace the origin of all things to numbers. Thus he observed that musical strings of equal length stretched by weights having the proportion of $\frac{1}{2}$, $\frac{2}{3}$, $\frac{3}{4}$, produced intervals which were an octave, a fifth, and a fourth. Harmony, therefore, depends on musical proportion; it is nothing but a mysterious numerical relation. Where harmony is, there are numbers. Hence the order and beauty of the universe have their origin in numbers. There are seven intervals in the musical scale, and also seven planets crossing the heavens. The same numerical relations which underlie the former must underlie the latter. But where numbers are, there is harmony. Hence his spiritual ear discerned in the planetary motions a wonderful "harmony of the spheres." The Pythagoreans invested particular numbers with extraordinary attributes. Thus *one* is the essence of things; it is an absolute number; hence the origin of all numbers and so of all things. *Four* is the most perfect number, and was in some mystic way conceived to correspond to the human soul. Philolaus believed that 5 is the cause of color, 6 of cold, 7 of mind and health and light, 8 of love and friendship.[2] In Plato's works are evidences of a similar belief in religious relations of numbers. Even Aristotle referred the virtues to numbers.

Enough has been said about these mystic speculations to show what lively interest in mathematics they must have created and

[1] J. Gow, *op. cit.*, p. 63.

[2] J. Gow, *op. cit.*, p. 69.

maintained. Avenues of mathematical inquiry were opened up by them which otherwise would probably have remained closed at that time.

The Pythagoreans classified numbers into odd and even. They observed that the sum of the series of odd numbers from 1 to $2n+1$ was always a complete square, and that by addition of the even numbers arises the series 2, 6, 12, 20, in which every number can be decomposed into two factors differing from each other by unity. Thus, $6=2.3$, $12=3.4$, etc. These latter numbers were considered of sufficient importance to receive the separate name of *heteromecic* (not equilateral). Numbers of the form $\frac{n(n+1)}{2}$ were called *triangular*, because they could always be arranged thus, ∴∴ Numbers which were equal to the sum of all their possible factors, such as 6, 28, 496, were called[1] *perfect;* those exceeding that sum, *excessive;* and those which were less, *defective. Amicable* numbers were those of which each was the sum of the factors in the other. Much attention was paid by the Pythagoreans to the subject of proportion. The quantities a, b, c, d were said to be in *arithmetical* proportion when $a-b=c-d$; in *geometrical* proportion, when $a:b=c:d$; in *harmonic* proportion, when $a-b:b-c=a:c$. It is probable that the Pythagoreans were also familiar with the *musical* proportion $a:\frac{a+b}{2}=\frac{2ab}{a+b}:b$. Iamblichus says that Pythagoras introduced it from Babylon.

In connection with arithmetic, Pythagoras made extensive investigations into geometry. He believed that an arithmetical fact had its analogue in geometry, and *vice versa.* In connection with his theorem on the right triangle he devised a rule by which integral numbers could be found, such that the sum of the squares of two of them equalled the square of the third. Thus, take for one side an odd number $(2n+1)$; then $\frac{(2n+1)^2-1}{2}=2n^2+2n=$ the other side, and $(2n^2+2n+1)=$ hypotenuse. If $2n+1=9$, then the other two numbers are 40 and 41. But this rule only applies to cases in which the hypotenuse differs from one of the sides by 1. In the study of the right triangle there doubtless arose questions of puzzling subtlety. Thus, given a number equal to the side of an isosceles right triangle, to find the number which the hypotenuse is equal to. The side may have been taken equal to 1, 2, $\frac{3}{2}$, $\frac{6}{5}$, or any other number, yet in every instance all efforts to find a number exactly equal to the hypotenuse must have remained fruitless. The problem may have been attacked again and again, until finally "some rare genius, to whom it is granted, during some happy moments, to soar with eagle's flight above the level of human thinking," grasped the happy thought that this prob-

[1] By the later Greeks and Euclid.

lem cannot be solved. In some such manner probably arose the theory of *irrational quantities*, which is attributed by Eudemus to the Pythagoreans. It was indeed a thought of extraordinary boldness, to assume that straight lines could exist, differing from one another not only in length,—that is, in quantity,—but also in a quality, which, though real, was absolutely invisible.[1] Need we wonder that the Pythagoreans saw in irrationals a deep mystery, a symbol of the unspeakable? We are told that the one who first divulged the theory of irrationals, which the Pythagoreans kept secret, perished in consequence in a shipwreck, "for the unspeakable and invisible should always be kept secret." Its discovery is ascribed to Pythagoras, but we must remember that all important Pythagorean discoveries were, according to Pythagorean custom, referred back to him. The first incommensurable ratio known seems to have been that of the side of a square to its diagonal, as $1 : \sqrt{2}$. **Theodorus of Cyrene** added to this the fact that the sides of squares represented in length by $\sqrt{3}$, $\sqrt{5}$, etc., up to $\sqrt{17}$, and Theætetus, that the sides of any square, represented by a surd, are incommensurable with the linear unit. **Euclid** (about 300 B. C.), in his *Elements*, X, 9, generalised still further: Two magnitudes whose squares are (or are not) to one another as a square number to a square number are commensurable (or incommensurable), and conversely. In the tenth book, he treats of incommensurable quantities at length. He investigates every possible variety of lines which can be represented by $\sqrt{\sqrt{a} \pm \sqrt{b}}$, a and b representing two commensurable lines, and obtains 25 species. Every individual of every species is incommensurable with all the individuals of every other species. "This book," says De Morgan, "has a completeness which none of the others (not even the fifth) can boast of; and we could almost suspect that Euclid, having arranged his materials in his own mind, and having completely elaborated the tenth book, wrote the preceding books after it, and did not live to revise them thoroughly."[2] The theory of incommensurables remained where Euclid left it, till the fifteenth century.

If it be recalled that the early Egyptians had some familiarity with quadratic equations, it is not surprising if similar knowledge is displayed by Greek writers in the time of Pythagoras. Hippocrates, in the fifth century B. C., when working on the areas of lunes, assumes the geometrical equivalent of the solution of the quadratic equation $x^2 + \sqrt{\frac{3}{2}}\, ax = a^2$. The complete geometrical solution was given by Euclid in his *Elements*, VI, 27–29. He solves certain types of quadratic equations geometrically in Book II, 5, 6, 11.

[1] H. Hankel, *Zur Geschichte der Mathematik in Alterthum und Mittelalter*, 1874, p. 102.

[2] A. De Morgan, "Eucleides" in *Smith's Dictionary of Greek and Roman Biog. and Myth.*

Euclid devotes the seventh, eighth, and ninth books of his *Elements* to arithmetic. Exactly how much contained in these books is Euclid's own invention, and how much is borrowed from his predecessors, we have no means of knowing. Without doubt, much is original with Euclid. The *seventh book* begins with twenty-one definitions. All but that for "prime" numbers and for perfect numbers had been given by the Pythagoreans. Next follows a process for finding the G. C. D. of two or more numbers. The *eighth book* deals with numbers in continued proportion, and with the mutual relations of squares, cubes, and plane numbers. Thus, XXII, if three numbers are in continued proportion, and the first is a square, so is the third. In the *ninth book*, the same subject is continued. It contains the proposition that the number of primes is greater than any given number.

After the death of Euclid, the theory of numbers remained almost stationary for 400 years. Geometry monopolised the attention of all Greek mathematicians. Only two are known to have done work in arithmetic worthy of mention. **Eratosthenes** (275–194 B. C.) invented a "sieve" for finding prime numbers. All composite numbers are "sifted" out in the following manner: Write down the odd numbers from 3 up, in succession. By striking out every third number after the 3, we remove all multiples of 3. By striking out every fifth number after the 5, we remove all multiples of 5. In this way, by rejecting multiples of 7, 11, 13, etc., we have left prime numbers only. **Hypsicles** (between 200 and 100 B. C.) worked at the subjects of polygonal numbers and arithmetical progressions, which Euclid entirely neglected. In his work on "risings of the stars," he showed (1) that in an arithmetical series of $2n$ terms, the sum of the last n terms exceeds the sum of the first n by a multiple of n^2; (2) that in such a series of $2n+1$ terms, the sum of the series is the number of terms multiplied by the middle term; (3) that in such a series of $2n$ terms, the sum is half the number of terms multiplied by the two middle terms.[1]

For two centuries after the time of Hypsicles, arithmetic disappears from history. It is brought to light again about 100 A. D. by **Nicomachus,** a Neo-Pythagorean, who inaugurated the final era of Greek mathematics. From now on, arithmetic was a favorite study, while geometry was neglected. Nicomachus wrote a work entitled *Introductio Arithmetica*, which was very famous in its day. The great number of commentators it has received vouch for its popularity. Boethius translated it into Latin. Lucian could pay no higher compliment to a calculator than this: "You reckon like Nicomachus of Gerasa." The *Introductio Arithmetica* was the first exhaustive work in which arithmetic was treated quite independently of geometry. Instead of drawing lines, like Euclid, he illustrates things by real numbers. To be sure, in his book the old geometrical nomenclature is retained, but the method is inductive instead of deductive. "Its sole

[1] J. Gow, *op. cit.*, p. 87.

business is classification, and all its classes are derived from, and exhibited by, actual numbers." The work contains few results that are really original. We mention one important proposition which is probably the author's own. He states that cubical numbers are always equal to the sum of successive odd numbers. Thus, $8=2^3=3+5$, $27=3^3=7+9+11$, $64=4^3=13+15+17+19$, and so on. This theorem was used later for finding the sum of the cubical numbers themselves. **Theon** of Smyrna is the author of a treatise on "the mathematical rules necessary for the study of Plato." The work is ill arranged and of little merit. Of interest is the theorem, that every square number, or that number minus 1, is divisible by 3 or 4 or both. A remarkable discovery is a proposition given by **Iamblichus** in his treatise on Pythagorean philosophy. It is founded on the observation that the Pythagoreans called 1, 10, 100, 1000, units of the first, second, third, fourth "course" respectively. The theorem is this: If we add any three consecutive numbers, of which the highest is divisible by 3, then add the digits of that sum, then, again, the digits of *that* sum, and so on, the final sum will be 6. Thus, $61+62+63=186$, $1+8+6=15$, $1+5=6$. This discovery was the more remarkable, because the ordinary Greek numerical symbolism was much less likely to suggest any such property of numbers than our "Arabic" notation would have been.

Hippolytus, who appears to have been bishop at Portus Romae in Italy in the early part of the third century, must be mentioned for the giving of "proofs" by casting out the 9's and the 7's.

The works of Nicomachus, Theon of Smyrna, Thymaridas, and others contain at times investigations of subjects which are really algebraic in their nature. Thymaridas in one place uses the Greek, word meaning "unknown quantity" in a way which would lead one to believe that algebra was not far distant. Of interest in tracing the invention of algebra are the arithmetical epigrams in the *Palatine Anthology*, which contain about fifty problems leading to linear equations. Before the introduction of algebra these problems were propounded as puzzles. A riddle attributed to Euclid and contained in the *Anthology* is to this effect: A mule and a donkey were walking along, laden with corn. The mule says to the donkey, "If you gave me one measure, I should carry twice as much as you. If I gave you one, we should both carry equal burdens. Tell me their burdens, O most learned master of geometry." [1]

It will be allowed, says Gow, that this problem, if authentic, was not beyond Euclid, and the appeal to geometry smacks of antiquity. A far more difficult puzzle was the famous "cattle-problem," which Archimedes propounded to the Alexandrian mathematicians. The problem is indeterminate, for from only seven equations, eight unknown quantities in integral numbers are to be found. It may be

[1] J. Gow, *op. cit.*, p. 99.

stated thus: The sun had a herd of bulls and cows, of different colors. (1) Of Bulls, the white (W) were, in number, $(\frac{1}{2}+\frac{1}{3})$ of the blue (B) and yellow (Y): the B were $(\frac{1}{4}+\frac{1}{5})$ of the Y and piebald (P): the P were $(\frac{1}{6}+\frac{1}{7})$ of the W and Y. (2) Of Cows, which had the same colors (w, b, y, p),

$$w=(\tfrac{1}{3}+\tfrac{1}{4})\ (B+b):\ b=(\tfrac{1}{4}+\tfrac{1}{5})\ (P+p):\ p=(\tfrac{1}{5}+\tfrac{1}{6})\ (Y+y):\ y=(\tfrac{1}{6}+\tfrac{1}{7}).\ (W+w).$$

Find the number of bulls and cows.[1] This leads to high numbers, but, to add to its complexity, the conditions are superadded that W+B=a square, and P+Y=a triangular number, leading to an indeterminate equation of the second degree. Another problem in the *Anthology* is quite familiar to school-boys: "Of four pipes, one fills the cistern in one day, the next in two days, the third in three days, the fourth in four days: if all run together, how soon will they fill the cistern?" A great many of these problems, puzzling to an arithmetician, would have been solved easily by an algebraist. They became very popular about the time of Diophantus, and doubtless acted as a powerful stimulus on his mind.

Diophantus was one of the last and most fertile mathematicians of the second Alexandrian school. He flourished about 250 A. D. His age[2] was eighty-four, as is known from an epitaph to this effect: Diophantus passed $\frac{1}{6}$ of his life in childhood, $\frac{1}{12}$ in youth, and $\frac{1}{7}$ more as a bachelor; five years after his marriage was born a son who died four years before his father, at half his father's age.[2] The place of nativity and parentage of Diophantus are unknown. If his works were not written in Greek, no one would think for a moment that they were the product of Greek mind. There is nothing in his works that reminds us of the classic period of Greek mathematics. His were almost entirely new ideas on a new subject. In the circle of Greek mathematicians he stands alone in his specialty. Except for him, we should be constrained to say that among the Greeks *algebra* was almost an unknown science.

Of his works we have lost the *Porisms*, but possess a fragment of *Polygonal Numbers*, and seven books of his great work on *Arithmetica*, said to have been written in 13 books. Recent editions of the *Arithmetica* were brought out by the indefatigable historians, P. Tannery and T. L. Heath, and by G. Wertheim.

If we except the Ahmes papyrus, which contains the first suggestions of algebraic notation, and of the solution of equations, then his *Arithmetica* is the earliest treatise on algebra now extant. In this work is introduced the idea of an algebraic equation expressed in algebraic symbols. His treatment is purely analytical and completely divorced from geometrical methods. He states that "a number to be sub-

[1] J. Gow, *op. cit.*, p. 63. [2] At death.

tracted, multiplied by a number to be subtracted, gives a number to be added." This is applied to the multiplication of differences, such as $(x-1)$ $(x-2)$. It must be remarked, that Diophantus had no notion whatever of negative numbers standing by themselves. All he knew were differences, such as $(2x-10)$, in which $2x$ could not be smaller than 10 without leading to an absurdity. He appears to be the first who could perform such operations as $(x-1)\times(x-2)$ without reference to geometry. Such identities as $(a+b)^2=a^2+2ab+b^2$, which with Euclid appear in the elevated rank of geometric theorems, are with Diophantus the simplest consequences of the algebraic laws of operation. His sign for subtraction was ⋔, for equality ι. For unknown quantities he had only one symbol, ς. He had no sign for addition except juxtaposition. Diophantus used but few symbols, and sometimes ignored even these by describing an operation in words when the symbol would have answered just as well.

In the solution of simultaneous equations Diophantus adroitly managed with only one symbol for the unknown quantities and arrived at answers, most commonly, by the method of *tentative assumption*, which consists in assigning to some of the unknown quantities preliminary values, that satisfy only one or two of the conditions. These values lead to expressions palpably wrong, but which generally suggest some stratagem by which values can be secured satisfying all the conditions of the problem.

Diophantus also solved determinate equations of the second degree. Such equations were solved geometrically by Euclid and Hippocrates. Algebraic solutions appear to have been found by Heron of Alexandria, who gives $8\frac{1}{2}$ as an approximate answer to the equation $144x(14-x)=6720$. In the *Geometry*, doubtfully attributed to Heron, the solution of the equation $\frac{11}{14}x^2+\frac{27}{7}x=212$ is practically stated in the form $x=\frac{\sqrt{(154\times 212+841)}-29}{11}$. Diophantus nowhere goes through with the whole process of solving quadratic equations; he merely states the result. Thus, "$84x^2+7x=7$, whence x is found $=\frac{1}{4}$." From partial explanations found here and there it appears that the quadratic equation was so written that all terms were positive. Hence, from the point of view of Diophantus, there were three cases of equations with a positive root: $ax^2+bx=c$, $ax^2=bx+c$, $ax^2+c=bx$, each case requiring a rule slightly different from the other two. Notice he gives only one root. His failure to observe that a quadratic equation has two roots, even when both roots are positive, rather surprises us. It must be remembered, however, that this same inability to perceive more than one out of the several solutions to which a problem may point is common to all Greek mathematicians. Another point to be observed is that he never accepts as an answer a quantity which is negative or irrational.

Diophantus devotes only the first book of his *Arithmetica* to the solution of determinate equations. The remaining books extant treat mainly of *indeterminate quadratic equations* of the form $Ax^2 + Bx + C = y^2$, or of two simultaneous equations of the same form. He considers several but not all the possible cases which may arise in these equations. The opinion of Nesselmann on the method of Diophantus, as stated by Gow, is as follows: "(1) Indeterminate equations of the second degree are treated completely only when the quadratic or the absolute term is wanting: his solution of the equations $Ax^2 + C = y^2$ and $Ax^2 + Bx + C = y^2$ is in many respects cramped. (2) For the 'double equation' of the second degree he has a definite rule only when the quadratic term is wanting in both expressions: even then his solution is not general. More complicated expressions occur only under specially favourable circumstances." Thus, he solves $Bx + C^2 = y^2$, $B_1x + C_1{}^2 = y_1{}^2$.

The extraordinary ability of Diophantus lies rather in another direction, namely, in his wonderful ingenuity to reduce all sorts of equations to particular forms which he knows how to solve. Very great is the variety of problems considered. The 130 problems found in the great work of Diophantus contain over 50 different classes of problems, which are strung together without any attempt at classification. But still more multifarious than the problems are the solutions. General methods are almost unknown to Dipohantus. Each problem has its own distinct method, which is often useless for the most closely related problems. "It is, therefore, difficult for a modern, after studying 100 Diophantine solutions, to solve the 101st." This statement, due to Hankel, is somewhat overdrawn, as is shown by Heath.[1]

That which robs his work of much of its scientific value is the fact that he always feels satisfied with one solution, though his equation may admit of an indefinite number of values. Another great defect is the absence of general methods. Modern mathematicians, such as L. Euler, J. Lagrange, K. F. Gauss, had to begin the study of indeterminate analysis anew and received no direct aid from Diophantus in the formulation of methods. In spite of these defects we cannot fail to admire the work for the wonderful ingenuity exhibited therein in the solution of particular equations.

[1] T. L. Heath, *Diophantus of Alexandria*, 2 Ed., Cambridge, 1910, pp. 54–97.

THE ROMANS

Nowhere is the contrast between the Greek and Roman minds shown forth more distinctly than in their attitude toward the mathematical science. The sway of the Greek was a flowering time for mathematics, but that of the Roman a period of sterility. In philosophy, poetry, and art the Roman was an imitator. But in mathematics he did not even rise to the desire for imitation. The mathematical fruits of Greek genius lay before him untasted. In him a science which had no direct bearing on practical life could awake no interest. As a consequence, not only the higher geometry of Archimedes and Apollonius, but even the *Elements* of Euclid, were neglected. What little mathematics the Romans possessed did not come altogether from the Greeks, but came in part from more ancient sources. Exactly where and how some of it originated is a matter of doubt. It seems most probable that the "Roman notation," as well as the early practical geometry of the Romans, came from the old Etruscans, who, at the earliest period to which our knowledge of them extends, inhabited the district between the Arno and Tiber.

Livy tells us that the Etruscans were in the habit of representing the number of years elapsed, by driving yearly a nail into the sanctuary of Minerva, and that the Romans continued this practice. A less primitive mode of designating numbers, presumably of Etruscan origin, was a notation resembling the present "Roman notation." This system is noteworthy from the fact that a principle is involved in it which is rarely met with in others, namely, the principle of subtraction. If a letter be placed before another of greater value, its value is not to be added to, but subtracted from, that of the greater. In the designation of large numbers a horizontal bar placed over a letter was made to increase its value one thousand fold. In fractions the Romans used the duodecimal system.

Of arithmetical calculations, the Romans employed three different kinds: Reckoning on the fingers, upon the abacus, and by tables prepared for the purpose.[1] Finger-symbolism was known as early as the time of King Numa, for he had erected, says Pliny, a statue of the double-faced Janus, of which the fingers indicated 365 (355?), the number of days in a year. Many other passages from Roman authors point out the use of the fingers as aids to calculation. In fact, a finger-symbolism of practically the same form was in use not only in Rome, but also in Greece and throughout the East, certainly as early as the beginning of the Christian era, and continued to be used in Europe

[1] M. Cantor, *op. cit.*, Vol. I, 3 Aufl., 1907, p. 526.

during the Middle Ages. We possess no knowledge as to where or when it was invented. The second mode of calculation, by the abacus, was a subject of elementary instruction in Rome. Passages in Roman writers indicate that the kind of abacus most commonly used was covered with dust and then divided into columns by drawing straight lines. Each column was supplied with pebbles (calculi, whence "calculare" and "calculate") which served for calculation.

The Romans used also another kind of abacus, consisting of a metallic plate having grooves with movable buttons. By its use all integers between 1 and 9,999,999, as well as some fractions, could be represented. In the two adjoining figures [1] the lines represent grooves

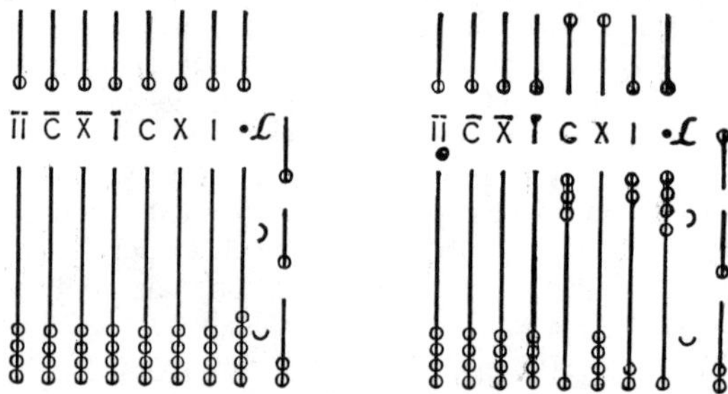

and the circles buttons. The Roman numerals indicate the value of each button in the corresponding groove below, the button in the shorter groove above having a fivefold value. Thus ⅂Γ = 1,000,000; hence each button in the long left-hand groove, when in use, stands for 1,000,000, and the button in the short upper groove stands for 5,000,000. The same holds for the other grooves labelled by Roman numerals. The eighth long groove from the left (having 5 buttons) represents duodecimal fractions, each button indicating $\frac{1}{12}$, while the button above the dot means $\frac{6}{12}$. In the ninth column the upper button represents $\frac{1}{24}$, the middle $\frac{1}{48}$, and two lower each $\frac{1}{72}$. Our first figure represents the positions of the buttons before the operation begins; our second figure stands for the number $852\,\frac{1}{3}\,\frac{1}{24}$. The eye has here to distinguish the buttons in use and those left idle. Those counted are one button above c (=500), and three buttons below c (=300); one button above x (=50); two buttons below I (=2); four buttons indicating duodecimals $(=\frac{1}{3})$; and the button for $\frac{1}{24}$.

Suppose now that $10{,}318\,\frac{1}{4}\,\frac{1}{8}\,\frac{1}{48}$ is to be added to $852\,\frac{1}{3}\,\frac{1}{24}$. The operator could begin with the highest units, or the lowest units, as he pleased. Naturally the hardest part is the addition of the fractions.

[1] G. Friedlein, *Die Zahlzeichen und das elementare Rechnen der Griechen und Römer*, Erlangen, 1869, Fig. 21. Gottfried Friedlein (1828–1875) was "Rektor der Kgl. Studienanstalt zu Hof" in Bavaria.

In this case the button for $\frac{1}{48}$, the button above the dot and three buttons below the dot were used to indicate the sum $\frac{3}{4}\,\frac{1}{48}$. The addition of 8 would bring all the buttons above and below I into play, making 10 units. Hence, move them all back and move up one button in the groove below x. Add 10 by moving up another of the buttons below x; add 300 to 800 by moving back all buttons above and below c, except one button below, and moving up one button below $\bar{\text{I}}$; add 10,000 by moving up one button below $\bar{\text{x}}$. In subtraction the operation was similar.

Multiplication could be carried out in several ways. In case of $38\,\frac{1}{2}\,\frac{1}{14}$ times $25\,\frac{1}{3}$, the abacus may have shown successively the following values: $600\ (=30.20)$, $760\ (=600+20.8)$, $770\ (=760+\frac{1}{2}.20)$, $770\frac{10}{12}\ (=770+\frac{1}{24}.20)$, $920\frac{10}{12}\ (=770\frac{10}{12}+30.5)$, $960\,\frac{10}{12}\ (=920\,\frac{10}{12}+8.5)$, $963\,\frac{1}{3}\ (=960\,\frac{10}{12}+\frac{1}{2}.5)$, $963\,\frac{1}{2}\,\frac{1}{24}\ (=963\,\frac{1}{3}+\frac{1}{24}.5)$, $973\,\frac{1}{2}\,\frac{1}{24}\ (=963\,\frac{1}{2}\,\frac{1}{24}+\frac{1}{3}.30)$, $976\,\frac{2}{12}\,\frac{1}{24}\ (=973.\frac{1}{2}\,\frac{1}{24}+8.\frac{1}{3})$, $976\,\frac{1}{3}\,\frac{1}{24}\ (=976\,\frac{2}{12}\,\frac{1}{24}+\frac{1}{2}.\frac{1}{3})$, $976\,\frac{1}{3}\,\frac{1}{24}\,\frac{1}{72}\ (=976\,\frac{1}{3}\,\frac{1}{24}+\frac{1}{3}.\frac{1}{24})$.[1]

In division the abacus was used to represent the remainder resulting from the subtraction from the dividend of the divisor or of a convenient multiple of the divisor. The process was complicated and difficult. These methods of abacal computation show clearly how multiplication or division can be carried out by a series of successive additions or subtractions. In this connection we suspect that recourse was had to mental operations and to the multiplication table. Possibly finger-multiplication may also have been used. But the multiplication of large numbers must, by either method, have been beyond the power of the ordinary arithmetician. To obviate this difficulty, the arithmetical tables mentioned above were used, from which the desired products could be copied at once. Tables of this kind were prepared by *Victorius* of Aquitania. His tables contain a peculiar notation for fractions, which continued in use throughout the Middle Ages. Victorius is best known for his *canon paschalis*, a rule for finding the correct date for Easter, which he published in 457 A. D.

Payments of interest and problems in interest were very old among the Romans. The Roman laws of inheritance gave rise to numerous arithmetical examples. Especially unique is the following: A dying man wills that, if his wife, being with child, gives birth to a son, the son shall receive $\frac{2}{3}$ and she $\frac{1}{3}$ of his estates; but if a daughter is born, she shall receive $\frac{1}{3}$ and his wife $\frac{2}{3}$. It happens that twins are born, a boy and a girl. How shall the estates be divided so as to satisfy the will? The celebrated Roman jurist, Salvianus Julianus, decided that the estates shall be divided into seven equal parts of which the son receives four, the wife two, the daughter one.

We next consider Roman geometry. He who expects to find in

[1] Friedlein, *op. cit.*, p. 89.

Rome a science of geometry, with definitions, axioms, theorems, and proofs arranged in logical order, will be disappointed. The only geometry known was a *practical* geometry, which, like the old Egyptian, consisted only of empirical rules. This practical geometry was employed in surveying. Treatises thereon have come down to us, compiled by the Roman surveyors, called *agrimensores* or *gromatici.* One would naturally expect rules to be clearly formulated. But no; they are left to be abstracted by the reader from a mass of numerical examples. "The total impression is as though the Roman gromatic were thousands of years older than Greek geometry, and as though a deluge were lying between the two." Some of their rules were probably inherited from the Etruscans, but others are identical with those of Heron. Among the latter is that for finding the area of a triangle from its sides and the approximate formula, $\frac{13}{30}a^2$, for the area of equilateral triangles (a being one of the sides). But the latter area was also calculated by the formulas $\frac{1}{2}(a^2+a)$ and $\frac{1}{2}a^2$, the first of which was unknown to Heron. Probably the expression $\frac{1}{2}a^2$ was derived from the Egyptian formula $\frac{a+b}{2}\cdot\frac{c+d}{2}$ for the determination of the surface of a quadrilateral. This Egyptian formula was used by the Romans for finding the area, not only of rectangles, but of any quadrilaterals whatever. Indeed, the gromatici considered it even sufficiently accurate to determine the areas of cities, laid out irregularly, simply by measuring their circumferences.[1] Whatever Egyptian geometry the Romans possessed was transplanted across the Mediterranean at the time of *Julius Cæsar*, who ordered a survey of the whole empire to secure an equitable mode of taxation. Cæsar also reformed the calendar, and, for that purpose, drew from Egyptian learning. He secured the services of the Alexandrian astronomer, *Sosigenes.*

Two Roman philosophical writers deserve our attention. The philosophical poet, *Titus Lucretius* (96?–55 B. C.), in his *De rerum natura*, entertains conceptions of an infinite multitude and of an infinite magnitude which accord with the modern definitions of those terms as being not variables but constants. However, the Lucretian infinites are not composed of abstract things, but of material particles. His infinite multitude is of the denumerable variety; he made use of the whole-part property of infinite multitudes.[2]

Cognate topics are discussed several centuries later by the celebrated father of the Latin church, *St. Augustine* (354–430 A. D.), in his references to Zeno of Elea. In a dialogue on the question, whether or not the mind of man moves when the body moves, and travels with the body, he is led to a definition of motion, in which he displays some levity. It has been said of scholasticism that it has no sense of humor.

[1] H. Hankel, *op. cit.*, p. 297.
[2] C. J. Keyser in *Bull. Am. Math. Soc.*, Vol. 24, 1918, p. 268, 321.

Hardly does this apply to St. Augustine. He says: "When this discourse was concluded, a boy came running from the house to call us to dinner. I then remarked that this boy compels us not only to define motion, but to see it before our very eyes. So let us go, and pass from this place to another; for that is, if I am not mistaken, nothing else than motion." St. Augustine deserves the credit of having accepted the existence of the actually infinite and to have recognized it as being, not a variable, but a constant. He recognized all finite positive integers as an infinity of that type. On this point he occupied a radically different position than his forerunner, the Greek father of the church, Origen of Alexandria. Origen's arguments against the actually infinite have been pronounced by Georg Cantor the profoundest ever advanced against the actually infinite.

In the fifth century, the Western Roman Empire was fast falling to pieces. Three great branches—Spain, Gaul, and the province of Africa—broke off from the decaying trunk. In 476, the Western Empire passed away, and the Visigothic chief, Odoacer, became king. Soon after, Italy was conquered by the Ostrogoths under Theodoric. It is remarkable that this very period of political humiliation should be the one during which Greek science was studied in Italy most zealously. School-books began to be compiled from the elements of Greek authors. These compilations are very deficient, but are of absorbing interest, from the fact that, down to the twelfth century, they were the only sources of mathematical knowledge in the Occident. Foremost among these writers is **Boethius** (died 524). At first he was a great favorite of King Theodoric, but later, being charged by envious courtiers with treason, he was imprisoned, and at last decapitated. While in prison he wrote *On the Consolations of Philosophy*. As a mathematician, Boethius was a Brobdingnagian among Roman Scholars, but a Liliputian by the side of Greek masters. He wrote an *Institutis Arithmetica*, which is essentially a translation of the arithmetic of Nicomachus, and a *Geometry* in several books. Some of the most beautiful results of Nicomachus are omitted in Boethius' arithmetic. The first book on geometry is an extract from Euclid's *Elements*, which contains, in addition to definitions, postulates, and axioms, the theorems in the first three books, without proofs. How can this omission of proofs be accounted for? It has been argued by some that Boethius possessed an incomplete Greek copy of the *Elements;* by others, that he had Theon's edition before him, and believed that only the theorems came from Euclid, while the proofs were supplied by Theon. The second book, as also other books on geometry attributed to Boethius, teaches, from numerical examples, the mensuration of plane figures after the fashion of the agrimensores.

A celebrated portion in the geometry of Boethius is that pertaining to an abacus, which he attributes to the Pythagoreans. A considerable improvement on the old abacus is there introduced. Pebbles are discarded, and *apices* (probably small cones) are used. Upon each of these apices is drawn a numeral giving it some value below 10. The names of these numerals are pure Arabic, or nearly so, but are added, apparently, by a later hand. The o is not mentioned by Boethius in the text. These numerals bear striking resemblance to the Gubar-numerals of the West-Arabs, which are admittedly of Indian origin. These facts have given rise to an endless controversy. Some contended that Pythagoras was in India, and from there brought the nine numerals to Greece, where the Pythagoreans used them secretly. This hypothesis has been generally abandoned, for it is not certain that Pythagoras or any disciple of his ever was in India, nor is there any evidence in any Greek author, that the apices were known to the Greeks, or that numeral signs of any sort were used by them with the abacus. It is improbable, moreover, that the Indian signs, from which the apices are derived, are so old as the time of Pythagoras. A second theory is that the *Geometry* attributed to Boethius is a forgery; that it is not older than the tenth, or possibly the ninth, century, and that the apices are derived from the Arabs. But there is an Encyclopædia written by *Cassiodorius* (died about 585) in which both the arithmetic and geometry of Boethius are mentioned. Some doubt exists as to the proper interpretation of this passage in the Encyclopædia. At present the weight of evidence is that the geometry of Boethius, or at least the part mentioning the numerals, is spurious.[1] A third theory (Woepcke's) is that the Alexandrians either directly or indirectly obtained the nine numerals from the Hindus, about the second century A. D., and gave them to the Romans on the one hand, and to the Western Arabs on the other. This explanation is the most plausible.

It is worthy of note that Cassiodorius was the first writer to use the terms "rational" and "irrational" in the sense now current in arithmetic and algebra.[2]

[1] A good discussion of this so-called "Boethius question," which has been debated for two centuries, is given by D. E. Smith and L. C. Karpinski in their *Hindu-Arabic Numerals*, 1911, Chap. V.

[2] *Encyclopédie des sciences mathématiques*, Tome I, Vol. 2, 1907, p. 2. An illuminating article on ancient finger-symbolism is L. J. Richardson's "Digital Reckoning Among the Ancients" in the *Am. Math. Monthly*, Vol 23, 1916, pp. 7–13.

THE MAYA

The Maya of Central America and Southern Mexico developed hieroglyphic writing, as found in inscriptions and codices dating apparently from about the beginning of the Christian era, that ranks "probably as the foremost intellectual achievement of pre-Columbian times in the New World." Maya number systems and chronology are remarkable for the extent of their early development. Perhaps five or six centuries before the Hindus gave a systematic exposition of their *decimal* number system with its zero and principle of local value, the Maya in the flatlands of Central America had evolved systematically a *vigesimal* number system employing a zero and the principle of local value. In the Maya number system found in the codices the ratio of increase of successive units was not 10, as in the Hindu system; it was 20 in all positions except the third. That is, 20 units of the lowest order (*kins*, or days) make one unit of the next higher order (*uinals*, or 20 days), 18 uinals make one unit of the third order (*tun*, or 360 days), 20 tuns make one unit of the fourth order (*katun*, or 7200 days), 20 katuns make one unit of the fifth order (*cycle*, or 144,000 days) and finally, 20 cycles make 1 *great cycle* of 2,880,000 days. In Maya codices we find symbols for 1 to 19, expressed by bars and dots. Each bar stands for 5 units, each dot for 1 unit. For instance,

·	··	∷	—	·· —	· =	···· ≡
1	2	4	5	7	11	19

The zero is represented by a symbol that looks roughly like a half-closed eye. In writing 20 the principle of local value enters. It is expressed by a dot placed over the symbol for zero. The numbers are written vertically, the lowest order being assigned the lowest position. Accordingly, 37 was expressed by the symbols for 17 (three bars and two dots) in the kin place, and one dot representing 20, placed above 17 in the uinal place. To write 360 the Maya drew two zeros, one above the other, with one dot higher up, in third place ($1 \times 18 \times 20 + 0 + 0 = 360$). The highest number found in the codices is in our decimal notation 12,489,781.

A second numeral system is found on Maya inscriptions. It employs the zero, but not the principle of local value. Special symbols are employed to designate the different units. It is as if we were to write 203 as "2 hundreds, 0 tens, 3 ones." [1]

[1] For an account of the Maya number-systems and chronology, see S. G. Morley *An Introduction to the Study of the Maya Hieroglyphs*, Government Printing Office, Washington, 1915.

The Maya had a sacred year of 260 days, an official year of 360 days and a solar year of 365+ days. The fact that $18 \times 20 = 360$ seems to account for the break in the vigesimal system, making 18 (instead of 20) uinals equal to 1 tun. The lowest common multiple of 260 and 365, or 18980, was taken by the Maya as the "calendar round," a period of 52 years, which is "the most important period in Maya chronology."

We may add here that the number systems of Indian tribes in North America, while disclosing no use of the zero nor of the principle of local value, are of interest as exhibiting not only quinary, decimal, and vigesimal systems, but also ternary, quarternary, and octonary systems.[1]

[1] See W. C. Eells, "Number Systems of the North American Indians" in *American Math. Monthly*, Vol. 20, 1913, pp. 263–272, 293–299; also *Bibliotheca mathematica*, 3 S., Vol. 13, 1913, pp. 218–222.

THE CHINESE [1]

The oldest extant Chinese work of mathematical interest is an anonymous publication, called *Chou-pei* and written before the second century, A. D., perhaps long before. In one of the dialogues the Chou-pei is believed to reveal the state of mathematics and astronomy in China as early as 1100 B. C. The Pythagorean theorem of the right triangle appears to have been known at that early date.

Next to the Chou-pei in age is the *Chiu-chang Suan-shu* ("Arithmetic in Nine Sections"), commonly called the *Chiu-chang*, the most celebrated Chinese Text on arithmetic. Neither its authorship nor the time of its composition is known definitely. By an edict of the despotic emperor Shih Hoang-ti of the Ch'in Dynasty "all books were burned and all scholars were buried in the year 213 B. C." After the death of this emperor, learning revived again. We are told that a scholar named CHANG T'SANG found some old writings, upon which he based this famous treatise, the Chiu-chang. About a century later a revision of it was made by Ching Ch'ou-ch'ang; commentaries on this classic text were made by Liu Hui in 263 A. D. and by Li Ch'un-fêng in the seventh century. How much of the "Arithmetic in Nine Sections," as it exists to-day, is due to the old records ante-dating 213 B. C., how much to Chang T'sang and how much to Ching Ch'ou-ch'ang, it has not yet been found possible to determine.

The "Arithmetic in Nine Sections" begins with mensuration; it gives the area of a triangle as $\frac{1}{2}\ b\ h$, of a trapezoid as $\frac{1}{2}\ (b+b')h$, of a circle variously as $\frac{1}{2}c\,.\,\frac{1}{2}d$, $\frac{1}{4}cd$, $\frac{3}{4}d^2$ and $\frac{1}{12}c^2$, where c is the circumference and d is the diameter. Here π is taken equal to 3. The area of a segment of a circle is given as $\frac{1}{2}(ca+a^2)$, where c is the chord and a the altitude. Then follow fractions, commercial arithmetic including percentage and proportion, partnership, and square and cube root of numbers. Certain parts exhibit a partiality for unit-fractions. Division by a fraction is effected by inverting the fraction and multiplying. The rules of operation are usually stated in obscure language. There are given rules for finding the volumes of the prism, cylinder, pyramid, truncated pyramid and cone, tetrahedron and wedge. Then follow problems in alligation. There are indications of the use of positive and negative numbers. Of interest is the following problem because centuries later it is found in a work of the Hindu Brahmagupta:

[1] All our information on Chinese mathematics is drawn from Yoshio Mikami's *The Development of Mathematics in China and Japan*, Leipzig, 1912, and from David Eugene Smith and Yoshio Mikami's *History of Japanese Mathematics*, Chicago, 1914.

There is a bamboo 10 ft. high, the upper end of which is broken and reaches to the ground 3 ft. from the stem. What is the height of the break? In the solution the height of the break is taken $=\frac{10}{2}-\frac{3^2}{2\times 10}$. Here is another: A square town has a gate at the mid-point of each side. Twenty paces north of the north gate there is a tree which is visible from a point reached by walking from the south gate 14 paces south and then 1775 paces west. Find the side of the square. The problem leads to the quadratic equation $x^2+(20+14)x-2\times 20\times 1775=0$. The derivation and solution of this equation are not made clear in the text. There is an obscure statement to the effect that the answer is obtained by evolving the root of an expression which is not monomial but has an additional term [the term of the first degree $(20+14)x$]. It has been surmised that the process here referred to was evolved more fully later and led to the method closely resembling Horner's process of approximating to the roots, and that the process was carried out by the use of calculating boards. Another problem leads to a quadratic equation, the rule for the solution of which fits the solution of literal quadratic equations.

We come next to the *Sun-Tsu Suan-ching* ("Arithmetical Classic of Sun-Tsu"), which belongs to the first century, A. D. The author, SUN-TSU, says: "In making calculations we must first know positions of numbers. Unity is vertical and ten horizontal; the hundred stands while the thousand lies; and the thousand and the ten look equally, and so also the ten thousand and the hundred." This is evidently a reference to abacal computation, practiced from time immemorial in China, and carried on by the use of computing rods. These rods, made of small bamboo or of wood, were in Sun-Tsu's time much longer. The later rods were about $1\frac{1}{2}$ inches long, red and black in color, representing respectively positive and negative numbers. According to Sun-Tsu, units are represented by vertical rods, tens by horizontal rods, hundreds by vertical, and so on; for 5 a single rod suffices. The numbers 1–9 are represented by rods thus: 𝍠, 𝍡, 𝍢, 𝍣, 𝍤, 𝍥, 𝍦, 𝍧, 𝍨; the numbers in the tens column, 10, 20, . . ., 90 are written thus: 𝍩, 𝍪, 𝍫, 𝍬, 𝍭, 𝍮, 𝍯, 𝍰, 𝍱. The number 6728 is designated by 𝍮𝍦𝍪𝍧. The rods were placed on a board ruled in columns, and were rearranged as the computation advanced. The successive steps in the multiplication of 321 by 46 must have been about as follows:

321	321	321
138	1472	14766
46	46	46

The product was placed between the multiplicand and multiplier. The 46 is multiplied first by 3, then by 2, and last by 1, the 46 being moved to the right one place at each step. Sun-Tsu does not take up division, except when the divisor consists of one digit. Square root is explained more clearly than in the "Arithmetic in Nine Sections." Algebra is involved in the problem suggested by the reply made by a woman washing dishes at a river: "I don't know how many guests there were; but every two used a dish for rice between them; every three a dish for broth; every four a dish for meat; and there were 65 dishes in all.—Rule: Arrange the 65 dishes, and multiply by 12, when we get 780. Divide by 13, and thus we obtain the answer."

An indeterminate equation is involved in the following: "There are certain things whose number is unknown. Repeatedly divide by 3, the remainder is 2; by 5 the remainder is 3; and by 7 the remainder is 2. What will be the number?" Only one solution is given, viz. 23.

The *Hai-tao Suan-ching* ("Sea-island Arithmetical Classic") was written by Liu Hui, the commentator on the "Arithmetic in Nine Sections," during the war-period in the third century, A. D. He gives complicated problems indicating marked proficiency in algebraic manipulation. The first problem calls for the determination of the distance of an island and the height of a peak on the island, when two rods 30′ high and 1000′ apart are in line with the peak, the top of the peak being in line with the top of the nearer (more remote) rod, when seen from a point on the level ground 123′ (127′) behind this nearer (more remote) rod. The rules given for solving the problem are equivalent to the expressions obtained from proportions arising from the similar triangles.

Of the treatises brought forth during the next centuries only a few are extant. We mention the "Arithmetical Classic of Chang Ch'iu-chien" of the sixth century which gives problems on proportion, arithmetical progression and mensuration. He proposes the "problem of 100 hens" which is given again by later Chinese authors: "A cock costs 5 pieces of money, a hen 3 pieces, and 3 chickens 1 piece. If then we buy with 100 pieces 100 of them, what will be their respective numbers?"

The early values of π used in China were 3 and $\sqrt{10}$. Liu Hui calculated the perimeters of regular inscribed polygons of 12, 24, 48, 96, 192 sides and arrived at $\pi = 3.14+$. Tsu Ch'ung-chih in the fifth century took the diameter 10^8 and obtained as upper and lower limits for π 3.1415927 and 3.1415926, and from these the "accurate" and "inaccurate" values 355/113, 22/7. The value 22/7 is the upper limit given by Archimedes and is found here for the first time in Chinese history. The ratio 355/113 became known to the Japanese, but in the West it was not known until *Adriaen Anthonisz*, the father of *Adriaen Metius*, derived it anew, sometime between 1585 and 1625. However, M. Curtze's researches would seem to show that it was known to Valentin Otto as early as 1573.[1]

[1] *Bibliotheca mathematica*, 3 S., Vol. 13, 1913, p. 264. A neat geometric construc-

In the first half of the seventh century WANG HS' IAO-T'UNG brought forth a work, the *Ch'i-ku Suan-ching*, in which numerical cubic equations appear for the first time in Chinese mathematics. This took place seven or eight centuries after the first Chinese treatment of quadratics. Wang Hs'iao-t'ung gives several problems leading to cubics: "There is a right triangle, the product of whose two sides is 706 $\frac{1}{50}$, and whose hypotenuse is greater than the first side by 30 $\frac{9}{60}$. It is required to know the lengths of the three sides." He gives the answer 14 $\frac{7}{10}$, 49 $\frac{1}{5}$, 51 $\frac{1}{4}$, and the rule: "The Product (P) being squared and being divided by twice the Surplus (S), make the result *shih* or the constant class. Halve the surplus and make it the *lien-fa* or the second degree class. And carry out the operation of evolution according to the extraction of cube root. The result gives the first side. Adding the surplus to it, one gets the hypotenuse. Divide the product with the first side and the quotient is the second side." This rule leads to the cubic equation $x^3+S/2x^2-\frac{P^2}{2S}=0$. The mode of solution is similar to the process of extracting cube roots, but details of the process are not revealed.

In 1247 CH'IN CHIU-SHAO wrote the *Su-shu Chiu-chang* ("Nine Sections of Mathematics") which makes a decided advance on the solution of numerical equations. At first Ch'in Chiu-shao led a military life; he lived at the time of the Mongolian invasion. For ten years stricken with disease, he recovered and then devoted himself to study. The following problem led him to an equation of the tenth degree: There is a circular castle of unknown diameter, having 4 gates. Three miles north of the north gate is a tree which is visible from a point 9 miles east of the south gate. The unknown diameter is found to be 9. He passes beyond Sun-Tsu in his ability to solve indeterminate equations arising for a number which will give the residues $r_1, r_2, \ldots, r_n$ when divided by $m_1, m_2, \ldots, m_n$, respectively.

Ch'in Chiu-shao solves the equation $-x^4+763200x^2-40642560000=0$ by a process almost identical with Horner's method. However, the computations were very probably carried out on a computing board, divided into columns, and by the use of computing rods. Hence the arrangement of the work must have been different from that of Horner. But the operations performed were the same. The first digit in the root being 8, (8 hundreds), a transformation is effected which yields $x^4-3200x^3-3076800x^2-826880000x+38205440000=0$, the same equation that is obtained by Horner's process. Then, taking 4 as the second figure in the root, the absolute term vanishes in the operation, giving the root 840. Thus the Chinese had

tion of the fraction $\frac{355}{113}=3+4^2\div(7^2+8^2)$ is given anonymously in *Grunert's Archiv*, Vol. 12, 1849, p. 98. Using $\frac{355}{113}$, T. M. P. Hughes gives in *Nature*, Vol. 93, 1914, p. 110, a method of constructing a triangle that gives the area of a given circle with great accuracy.

invented Horner's method of solving numerical equations more than five centuries before Ruffini and Horner. This solution of higher numerical equations is given later in the writings of *Li Yeh* and others. Ch'in Chiu-shao marks an advance over Sun-Tsu in the use of o as a symbol for zero. Most likely this symbol is an importation from India. Positive and negative numbers were distinguished by the use of red and black computing rods. This author gives for the first time a problem which later became a favorite one among the Chinese; it involved the trisection of a trapezoidal field under certain restrictions in the mode of selection of boundaries.

We have already mentioned a contemporary of Ch'in Chiu-shao, namely, LI YEH; he lived far apart in a rival monarchy and worked independently. He was the author of *T'sê-yüan Hai-ching* ("Sea-Mirror of the Circle-Measurements"), 1248, and of the *I-ku Yen-tuan*, 1259. He used the symbol o for zero. On account of the inconvenience of writing and printing positive and negative numbers in different colors, he designated negative numbers by drawing a cancellation mark across the symbol. Thus ⊥o stood for 60, ⊥̸o stood for −60. The unknown quantity was represented by unity which was probably represented on the counting board by a rod easily distinguished from the other rods. The terms of an equation were written, not in a horizontal, but in a vertical line. In Li Yeh's work of 1259, as also in the work of Ch'in Chiu-shao, the absolute term is put in the top line; in Li Yeh's work of 1248 the order of the terms is reversed, so that the absolute term is in the bottom line and the highest power of the unknown in the top line. In the thirteenth century Chinese algebra reached a much higher development than formerly. This science, with its remarkable method (our Horner's) of solving numerical equations, was designated by the Chinese "the celestial element method."

A third prominent thirteenth century mathematician was YANG HUI, of whom several books are still extant. They deal with the summation of arithmetical progressions, of the series $1+3+6+\dots+(1+2+\dots+n)=n(n+1)(n+2)\div 6$, $1^2+2^2+\dots+n^2=\frac{1}{3}n(n+\frac{1}{2})(n+1)$, also with proportion, simultaneous linear equations, quadratic and quartic equations.

Half a century later, Chinese algebra reached its height in the treatise *Suan-hsiao Chi-mêng* ("Introduction to Mathematical Studies"), 1299, and the *Szu-yuen Yü-chien* ("The Precious Mirror of the Four Elements"), 1303, which came from the pen of CHU SHIH-CHIEH. The first work contains no new results, but exerted a great stimulus on Japanese mathematics in the seventeenth century. At one time the book was lost in China, but in 1839 it was restored by the discovery of a copy of a Korean reprint, made in 1660. The "Precious Mirror" is a more original work. It treats fully of the "celestial element method." He gives as an "ancient method" a triangle (known in the West as Pascal's arithmetical triangle), displaying the binomial coefficients, which were known to the Arabs in the eleventh century and were probably imported into China. Chu shih-Chieh's algebraic notation was altogether different from our modern notation. Thus, $a+b+c+d$ was written

```
                                    I
    I                             2 O 2
  I * I                        I O *2 O I
    I                            2² O 2
                                    I
```

as shown on the left, except that, in the central position, we employ an asterisk in place of the Chinese character *t'ai* (great extreme, absolute term) and that we use the modern numerals in place of the *sangi* forms. The square of $a+b+c+d$, namely, $a^2+b^2+c^2+d^2+2ab+2ac+2ad+2bc+2bd+2cd$, is represented as shown on the right. In further illustration of the Chinese notation, at the time of Chu Shih-Chieh, we give [1]

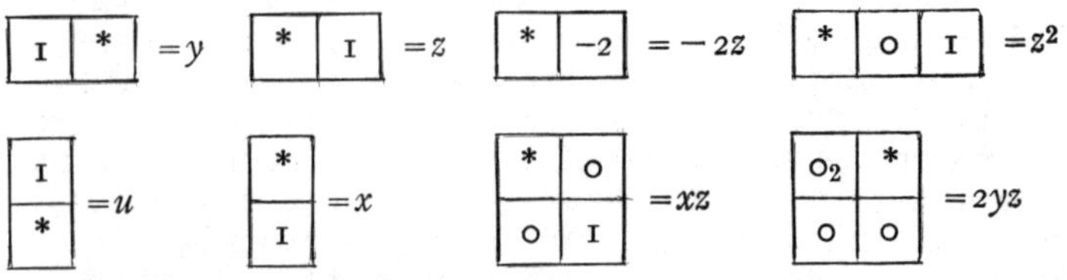

In the fourteenth century astronomy and the calendar were studied. They involved the rudiments of geometry and spherical trigonometry. In this field importations from the Arabs are disclosed.

After the noteworthy achievements of the thirteenth century, Chinese mathematics for several centuries was in a period of decline. The famous "celestial element method" in the solution of higher equations was abandoned and forgotten. Mention must be made, however, of CH'ÊNG TAI-WEI, who in 1593 issued his *Suan-fa T'ung-tsung* ("A Systematised Treatise on Arithmetic"), which is the oldest work now extant that contains a diagram of the form of the abacus, called *suan-pan*, and the explanation of its use. The instrument was known in China in the twelfth century. Resembling the old Roman abacus, it contained balls, movable along rods held by a wooden frame. The suan-pan replaced the old computing rods. The "Systematised Treatise on Arithmetic" is famous also for containing some magic squares and magic circles. Little is known of the early history

[1] In the symbol for "xz" notice that the "1" is one space down (x) and one space to the right (z) of *, and is made to stand for the product xz. In the symbol for "$2yz$" the three o's indicate the absence of the terms y, x, xy; the small "2" means twice the product of the two letters in the same row, respectively one space to the right and to the left of *, i. e., $2yz$. The limitations of this notation are obvious.

of magic squares. Myth tells us that, in early times, the sage *Yü*, the enlightened emperor, saw on the calamitous Yellow River a divine tortoise, whose back was decorated with the figure made up of the numbers from 1 to 9, arranged in form of a magic square or *lo-shu.*

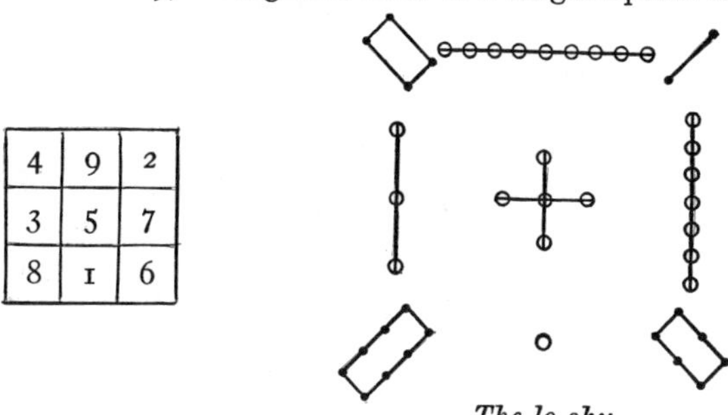

4	9	2
3	5	7
8	1	6

The lo-shu.

The numerals are indicated by knots in strings: black knots represent even numbers (symbolizing imperfection), white knots represent odd numbers (perfection).

Christian missionaries entered China in the sixteenth century. The Italian Jesuit *Matteo Ricci* (1552–1610) introduced European astronomy and mathematics. With the aid of a Chinese scholar named *Hsü*, he brought out in 1607 a translation of the first six books of Euclid. Soon after followed a sequel to Euclid and a treatise on surveying. The missionary *Mu Ni-ko* sometime before 1660 introduced logarithms. In 1713 Adrian Vlack's logarithmic tables to 11 places were reprinted. *Ferdinand Verbiest*[1] of West Flanders, a noted Jesuit missionary and astronomer, was in 1669 made vice-president of the Chinese astronomical board and in 1673 its president. European algebra found its way into China. *Mei Ku-ch'êng* noticed that the European algebra was essentially of the same principles as the Chinese "celestial element method" of former days which had been forgotten. Through him there came a revival of their own algebraic method, without, however, displacing European science. Later Chinese studies touched mainly three subjects: The determination of π by geometry and by infinite series, the solution of numerical equations, and the theory of logarithms.

We shall see later that Chinese mathematics stimulated the growth of mathematics in Japan and India. We have seen that, in a small way, there was a taking as well as a giving. Before the influx of recent European science, China was influenced somewhat by Hindu and Arabic mathematics. The Chinese achievements which stand out most conspicuously are the solution of numerical equations and the origination of magic squares and magic circles.

[1] Consult H. Bosmans, *Ferdinand Verbiest*, Louvain, 1912. Extract from *Revue des Questions scientifiques*, January–April, 1912.

THE JAPANESE [1]

According to tradition, there existed in Japan in remote times a system of numeration which extended to high powers of ten and resembled somewhat the sand counter of Archimedes. About 552 A. D. Buddhism was introduced into Japan. This new movement was fostered by Prince Shōtoku Taishi who was deeply interested in all learning. Mathematics engaged his attention to such a degree that he came to be called the father of Japanese mathematics. A little later the Chinese system of weights and measures was adopted. In 701 a university system was established in which mathematics figured prominently. Chinese science was imported, special mention being made in the official Japanese records of nine Chinese texts on mathematics, which include the *Chou-pei*, the *Suan-ching* written by Sun-Tsu and the great arithmetical work, the *Chiu-chang*. But this eighth century interest in mathematics was of short duration; the Chiu-chang was forgotten and the dark ages returned. Calendar reckoning and the rudiments of computation are the only signs of mathematical activity until about the seventeenth century of our era. On account of the crude numeral systems, devoid of the principal of local value and of a symbol for zero, mechanical aids of computation became a necessity. These consisted in Japan, as in China, of some forms of the abacus. In China there came to be developed an instrument, called the *suan-pan*, in Japan it was called the *soroban*. The importation of the suan-pan into Japan is usually supposed to have occurred before the close of the sixteenth century. Bamboo computing rods were used in Japan in the seventh century. These round pieces were replaced later by the square prisms (*sangi* pieces). Numbers were represented by these rods in the manner practiced by the Chinese. The numerals were placed inside the squares of a surface ruled like a chess board. The *soroban* was simply a more highly developed form of abacal instrument.

The years 1600 to 1675 mark a period of great mathematical activity. It was inaugurated by Mōri Kambei Shigeyoshi, who popularized the use of the *soroban*. His pupil, Yoshida Shichibei Kōyū, is the author of *Jinkō-ki*, 1627, which attained wide popularity and is the oldest Japanese mathematical work now extant. It explains operations on the *soroban*, including square and cube root. In one of

[1] This account is compiled from David Eugene Smith and Yoshio Mikami's *History of Japanese Mathematics*, Chicago, 1914, from Yoshio Mikami's *Development of Mathematics in China and Japan*, Leipzig, 1912, and from T. Hayashi's *A Brief History of the Japanese Mathematics*, Overgedrukt uit het *Nieuw Archief voor Wiskunde* VI, pp. 296–361; VII, pp. 105–161.

his later editions Yoshida appended a number of advanced problems to be solved by competitors. This procedure started among the Japanese the practice of issuing problems, which was kept up until 1813 and helped to stimulate mathematical activity.

Another pupil of Mōri was IMAMURA CHISHŌ who, in 1639, published a treatise entitled *Jugairoku*, written in classical Chinese. He took up the mensuration of the circle, sphere and cone. Another author, ISOMURA KITTOKU, in his *Ketsugishō*, 1660 (second edition 1684), when considering problems on mensuration, makes a crude approach to integration. He gives magic squares, both odd and even celled, and also magic circles. Such squares and circles became favorite topics among the Japanese. In the 1684 edition, Isomura gives also magic wheels. TANAKA KISSHIN arranges the integers 1–96 in six 4^2-celled magic squares, such that the sum in each row and column are 194; placing the six squares upon a cube, he obtains his "magic cube." Tanaka formed also "magic rectangles."[1] MURAMATSU in 1663 gives a magic square containing as many as 19^2 cells and a magic circle involving 129 numbers. Muramatsu gives also the famous "Josephus Problem" in the following form: 'Once upon a time there lived a wealthy farmer who had thirty children, half being of his first wife and half of his second one. The latter wished a favorite son to inherit all the property, and accordingly she asked him one day, saying: Would it not be well to arrange our 30 children on a circle, calling one of them the first and counting out every tenth one until there should remain only one, who should be called the heir. The husband assenting, the wife arranged the children . . ; the counting . . resulted in the elimination of 14 step-children at once, leaving only one. Thereupon the wife, feeling confident of her success, said, . . let us reverse the order. . The husband agreed again, and the counting proceeded in the reverse order, with the unexpected result that all of the second wife's children were stricken out and there remained only the step-child, and accordingly he inherited the property." The origin of this problem is not known. It is found much earlier in the Codex Einsidelensis (Einsideln, Switzerland) of the tenth century, while a Latin work of Roman times attributes it to Flavius Josephus. It commonly appears as a problem relating to Turks and Christians, half of whom must be sacrificed to save a sinking ship. It was very common in early printed European books on arithmetic and in books on mathematical recreations.

In 1666 SATŌ SEIKŌ wrote his *Kongenki* which, in common with other works of his day, considers the computation of $\pi(=3.14)$. He is the first Japanese to take up the Chinese "celestial element method" in algebra. He applies it to equations of as high a degree as the sixth. His successor, SAWAGUCHI, and a contemporary NOZAWA, give a crude calculus resembling that of Cavalieri. Sawaguchi rises

[1] Y. Mikami in *Archiv der Mathematik u. Physik*, Vol. 20, pp. 183–186.

above the Chinese masters in recognising the plurality of roots, but he declares problems which yield them to be erroneous in their nature. Another evidence of a continued Chinese influence is seen in the Chinese value of π, $\frac{355}{113}$, which was made known in Japan by IKEDA.

We come now to SEKI KŌWA (1642–1708) whom the Japanese consider the greatest mathematician that their country has produced. The year of his birth was the year in which Galileo died and Newton was born. Seki was a great teacher who attracted many gifted pupils. Like Pythagoras, he discouraged divulgence of mathematical discoveries made by himself and his school. For that reason it is difficult to determine with certainty the exact origin and nature of some of the discoveries attributed to him. He is said to have left hundreds of manuscripts; the transcripts of a few of them still remain. He published only one book, the *Hatsubi Sampō*, 1674, in which he solved 15 problems issued by a contemporary writer. Seki's explanations are quite incomplete and obscure. Takebe, one of his pupils, lays stress upon Seki's clearness. The inference is that Seki gave his explanations orally, probably using the computing rods or *sangi*, as he proceeded. Noteworthy among his mathematical achievements are the *tenzan* method and the *yendan* method. Both of these refer to improvements in algebra. The *tenzan* method is an improvement of the Chinese "celestial element" method, and has reference to notation, while the *yendan* refers to explanations or method of analysis. The exact nature and value of these two methods are not altogether clear. By the Chinese "celestial element" method the roots of equations were computed one digit at a time. Seki removed this limitation. Building on results of his predecessors, Seki gives also rules for writing down magic squares of $(2n+1)^2$ cells. In the case of the more troublesome even celled squares, Seki first gives a rule for the construction of a magic square of 4^2 cells, then of $4(n+1)^2$ and $16\ n^2$ cells. He simplified also the treatment of magic circles. Perhaps the most original and important work of Seki is the invention of determinants, sometime before 1683. Leibniz, to whom the first idea of determinants is usually assigned, made his discovery known in 1693 when he stated that three linear equations in x and y can be consistent only when the determinant arising from the coefficients vanishes. Seki took n equations and gave a more general treatment. Seki knew that a determinant of the nth order, when expanded, has $n!$ terms and that rows and columns are interchangeable.[1] Usually attributed to Seki is the invention of the *yenri* or "circle-principle" which, it is claimed, accomplishes somewhat the same things as the differential and integral calculus. Neither the exact nature nor the origin of the *yenri* is well understood. Doubt exists whether Seki was its discoverer. TAKEBE, a pupil of Seki, used the *yenri* and may be the chief originator

[1] For details consult Y. Mikami, "On the Japanese theory of determinants" in *Isis*, Vol. II, 1914, pp. 9–36.

of it, but his explanations are incomplete and obscure. Seki, Takebe and their co-workers dealt with infinite series, especially in the study of the circle and of π. Probably some knowledge of European mathematics found its way into Japan in the seventeenth century. A Japanese, under the name of *Petrus Hartsingius*, is known to have studied at Leyden under Van Schooten, but there is no clear evidence that he returned to Japan. In 1650 a Portuguese astronomer, whose real name is not known and whose adopted name was *Sawano Chūan*, translated a European astronomical work into Japanese.[1]

In the eighteenth century the followers of Seki were in control. Their efforts were expended upon problem-solving, the mensuration of the circle and the study of infinite series. Of KURUSHIMA GITA, who died in 1757, fragmentary manuscripts remain, which show a "magic cube" composed of four 4^2-celled magic squares in which the sums of rows and columns is 130, and the sums of corresponding cells of the four squares is likewise 130. This "magic cube" is evidently a different thing from Tanaka's "Magic cube." Near the close of the eighteenth century, during the waning days of the Seki school, there arose a bitter controversy between FUJITA SADASUKE, then the head of the Seki school, and AIDA AMMEI. Of the two, Aida was the younger and more gifted—an insurgent against the old and involved methods of exposition. Aida worked on the approximate solution of numerical equations.[2] The most noted work of the time was done by a man living in peaceful seclusion, AJIMA CHOKUYEN of Yedo, who died in 1798. He worked on Diophantine analysis and on a problem known in the West as "the Malfatti problem," to inscribe three circles in a triangle, each tangent to the other two. This problem appeared in Japan in 1781. Malfatti's publication on it appeared in 1803, but the special case of the isosceles triangle had been considered by Jakob Bernoulli before 1744. Ajima treats special cases of the problem to determine the number of figures in the repetend in circulating decimals. He improved the *yenri* and placed mathematics on the highest plane that it reached in Japan during the eighteenth century.

In the early part of the nineteenth century there was greater infiltration of European mathematics. There was considerable activity, but no noteworthy names appeared, except WADA NEI (1787–1840) who perfected the *yenri* still further, developing an integral calculus that served the ordinary purposes of mensuration, and giving reasons where his predecessors ordinarily gave only rules. He worked particularly on maxima and minima, and on roulettes. Japanese researches of his day relate to groups of ellipses and other figures which

[1] Y. Mikami in *Annaes da academia polyt. do Porto*, Vol. VIII, 1913.

[2] Y. Mikami, "On Aida Ammei's solution of an equation" in *Annaes da Academia Polyt. do Porto*, Vol. VIII, 1913. This article gives details on the solution of equations in China and Japan. See also Mikami's article on Miyai Antai in the *Tohoku Math. Journal*, Vol. 5, Nos. 3, 4, 1914.

can be drawn upon a folding fan. Here mathematics finds application to artistic design.

After the middle of the nineteenth century the native mathematics of Japan yielded to a strong influx of Western mathematics. The movement in Japan became a part of the great international advance. In 1911 there was started the *Tôhoku Mathematical Journal*, under the editorship of T. Hayashi. It is devoted to advanced mathematics, contains articles in many of our leading modern languages and is quite international in character.[1]

Looking back we see that Japan produced some able mathematicians, but on account of her isolation, geographically and socially, her scientific output did not affect or contribute to the progress of mathematics in the West. The Babylonians, Hindus, Arabs, and to some extent even the Chinese through their influence on the Hindus, contributed to the onward march of mathematics in the West. But the Japanese stand out in complete isolation.

[1] G. A. Miller, *Historical Introduction to Mathematical Literature*, 1916, p. 24.

THE HINDUS

After the time of the ancient Greeks, the first people whose researches wielded a wide influence in the world march of mathematics, belonged, like the Greeks, to the Aryan race. It was, however, not a European, but an Asiatic nation, and had its seat in far-off India.

Unlike the Greek, Indian society was fixed into castes. The only castes enjoying the privilege and leisure for advanced study and thinking were the *Brahmins*, whose prime business was religion and philosophy, and the *Kshatriyas*, who attended to war and government.

Of the development of Hindu mathematics we know but little. A few manuscripts bear testimony that the Indians had climbed to a lofty height, but their path of ascent is no longer traceable. It would seem that Greek mathematics grew up under more favorable conditions than the Hindu, for in Greece it attained an independent existence, and was studied for its own sake, while Hindu mathematics always remained merely a servant to astronomy. Furthermore, in Greece mathematics was a science of the people, free to be cultivated by all who had a liking for it; in India, as in Egypt, it was in the hands chiefly of the priests. Again, the Indians were in the habit of putting into verse all mathematical results they obtained, and of clothing them in obscure and mystic language, which, though well adapted to aid the memory of him who already understood the subject, was often unintelligible to the uninitiated. Although the great Hindu mathematicians doubtless reasoned out most or all of their discoveries, yet they were not in the habit of preserving the proofs, so that the naked theorems and processes of operation are all that have come down to our time. Very different in these respects were the Greeks. Obscurity of language was generally avoided, and proofs belonged to the stock of knowledge quite as much as the theorems themselves. Very striking was the difference in the bent of mind of the Hindu and Greek; for, while the Greek mind was pre-eminently *geometrical*, the Indian was first of all *arithmetical*. The Hindu dealt with number, the Greek with form. Numerical symbolism, the science of numbers, and algebra attained in India far greater perfection than they had previously reached in Greece. On the other hand, Hindu geometry was merely mensuration, unaccompanied by demonstration. Hindu trigonometry is meritorious, but rests on arithmetic more than on geometry.

An interesting but difficult task is the tracing of the relation between Hindu and Greek mathematics. It is well known that more or less trade was carried on between Greece and India from early times. After Egypt had become a Roman province, a more lively commercial intercourse sprang up between Rome and India, by way of Alexandria. *A priori*, it does not seem improbable, that with the traffic of merchandise there should also be an interchange of ideas. That communications of thought from the Hindus to the Alexandrians actually did take place, is evident from the fact that certain philosophic and theologic teachings of the Manicheans, Neo-Platonists, Gnostics, show unmistakable likeness to Indian tenets. Scientific facts passed also from Alexandria to India. This is shown plainly by the Greek origin of some of the technical terms used by the Hindus. Hindu astronomy was influenced by Greek astronomy. A part of the geometrical knowledge which they possessed is traceable to Alexandria, and to the writings of Heron in particular. In algebra there was, probably, a mutual giving and receiving.

There is evidence also of an intimate connection between Indian and Chinese mathematics. In the fourth and succeeding centuries of our era Indian embassies to China and Chinese visits to India are recorded by Chinese authorities.[1] We shall see that undoubtedly there was an influx of Chinese mathematics into India.

The history of Hindu mathematics may be resolved into two periods: First the *S'ulvasūtra period* which terminates not later than 200 A. D., second, *the astronomical and mathematical period*, extending from about 400 to 1200 A. D.

The term S'ulvasūtra means "the rules of the cord"; it is the name given to the supplements of the Kalpasūtras which explain the construction of sacrificial altars.[2] The S'ulvasūtras were composed sometime between 800 B. C. and 200 A. D. They are known to modern scholars through three quite modern manuscripts. Their aim is primarily not mathematical, but religious. The mathematical parts relate to the construction of squares and rectangles. Strange to say, none of these geometrical constructions occur in later Hindu works; later Indian mathematics ignores the S'ulvasūtras!

The second period of Hindu mathematics probably originated with an influx from Alexandria of western astronomy. To the fifth century of our era belongs an anonymous Hindu astronomical work, called the *Sūrya Siddhānta* ("Knowledge from the sun") which came to be regarded a standard work. In the sixth century A. D., **Varāha Mihira** wrote his *Pañcha Siddhāntikā* which gives a summary of the *Sūrya Siddhānta* and four other astronomical works then in use; it contains matters of mathematical interest.

In 1881 there was found at *Bakhshālī*, in northwest India, buried in the earth, an anonymous arithmetic, supposed, from the peculiar-

[1] G. R. Kaye, *Indian Mathematics*, Calcutta & Simla, 1915, p. 38. We are drawing heavily upon this book which embodies the results of recent studies of Hindu mathematics.

[2] G. R. Kaye, *op. cit.*, p. 3.

ities of its verses, to date from the third or fourth century after Christ. The document that was found is of birch bark, and is an incomplete copy, prepared probably about the eighth century, of an older manuscript.[1] It contains arithmetical computation.

The noted Hindu astronomer **Āryabhata** was born 476 A. D. at Pāṭaliputra, on the upper Ganges. His celebrity rests on a work entitled *Āryabhaṭiya*, of which the third chapter is devoted to mathematics. About one hundred years later, mathematics in India reached the highest mark. At that time flourished **Brahmagupta** (born 598). In 628 he wrote his *Brahma-sphuta-siddhānta* ("The Revised System of Brahma"), of which the twelfth and eighteenth chapters belong to mathematics.

Probably to the ninth century belongs **Mahāvīra,** a Hindu author on elementary mathematics, whose writings have only recently been brought to the attention of historians. He is the author of the *Gaṇita-Sāra-Sangraha* which throws light upon Hindu geometry and arithmetic. The following centuries produced only two names of importance; namely, **S'rīdhara,** who wrote a *Ganita-sara* ("Quintessence of Calculation"), and **Padmanābha,** the author of an algebra. The science seems to have made but little progress at this time; for a work entitled *Siddhānta S'iromani* ("Diadem of an Astronomical System"), written by **Bhāskara** in 1150, stands little higher than that of Brahmagupta, written over 500 years earlier. The two most important mathematical chapters in this work are the *Līlāvatī* (="the beautiful," *i. e.* the noble science) and *Vīja-gaṇita* (="root-extraction"), devoted to arithmetic and algebra. From now on, the Hindus in the Brahmin schools seemed to content themselves with studying the masterpieces of their predecessors. Scientific intelligence decreases continually, and in more modern times a very deficient Arabic work of the sixteenth century has been held in great authority.

The mathematical chapters of the *Brahma-siddhānta* and *Siddhānta S'iromani* were translated into English by H. T. Colebrooke, London, 1817. The *Sūrya-siddhānta* was translated by E. Burgess, and annotated by W. D. Whitney, New Haven, Conn., 1860. Mahāvīra's *Gaṇita-Sāra-Sangraha* was published in 1912 in Madras by M. Rangācārya.

We begin with geometry, the field in which the Hindus were least proficient. The *S'ulvasūtras* indicate that the Hindus, perhaps as early as 800 B. C., applied geometry in the construction of altars. Kaye[2] states that the mathematical rules found in the *S'ulvasūtras* "relate to (1) the construction of squares and rectangles, (2) the relation of the diagonal to the sides, (3) equivalent rectangles and squares, (4) equivalent circles and squares." A knowledge of the Pythagorean

[1] The *Bakhshālī Manuscript*, edited by Rudolf Hoernly in the *Indian Antiquary*, xvii, 33–48 and 275–279, Bombay, 1888.

[2] G. R. Kaye, *op. cit.*, p. 4.

theorem is disclosed in such relations as $3^2+4^2=5^2$, $12^2+16^2=20^2$, $15^2+36^2=39^2$. There is no evidence that these expressions were obtained from any general rule. It will be remembered that special cases of the Pythagorean theorem were known as early as 1000 B. C. in China and as early as 2000 B. C. in Egypt. A curious expression for the ratio of the diagonal to the side of a square, namely,

$$\sqrt{2}=1+\frac{1}{3}+\frac{1}{3\cdot 4}-\frac{1}{3\cdot 4\cdot 34},$$

is explained by Kaye as being "an expression of a direct measurement" which may be obtained by the use of a scale of the kind named in one of the S'ulvasūtra manuscripts, and based upon the change ratios 3, 4, 34. It is noteworthy that the fractions used are all unit fractions and that the expression yields a result correct to five decimal places. The S'ulvasūtra rules yield, by the aid of the Pythagorean theorem, constructions for finding a square equal to the sum or difference of two squares; they yield a rectangle equal to a given square, with $a\sqrt{2}$ and $\frac{1}{2}a\sqrt{2}$ as the sides of the rectangle; they yield by geometrical construction a square equal to a given rectangle, and satisfying the relation $ab=(b+[a-b]/2)^2-\frac{1}{4}(a-b)^2$, corresponding to Euclid II, 5. In the S'ulvasūtras the altar building ritual explains the construction of a square equal to a circle. Let a be the side of a square and d the diameter of an equivalent circle, then the given rules may be expressed thus:[1] $d=a+(a\sqrt{2}-a)/3$, $a=d-2d/15$, $a=d(1-\frac{1}{8}+\frac{1}{8\cdot 29}-\frac{1}{8\cdot 29\cdot 6}+\frac{1}{8\cdot 29\cdot 6\cdot 8})$. This third expression may be obtained from the first by the aid of the approximation for $\sqrt{2}$, given above. Strange to say, none of these geometrical constructions occur in later Hindu works; the latter completely ignore the mathematical contents of the S'ulvasūtras.

During the six centuries from the time of Āryabhaṭa to that of Bhāskara, Hindu geometry deals mainly with mensuration. The Hindu gave no definitions, no postulates, no axioms, no logical chain of reasoning. His knowledge of mensuration was largely borrowed from the Mediterranean and from China, through imperfect channels of communication. Āryabhaṭa gives a rule for the area of triangle which holds only for the isosceles triangle. Brahmagupta distinguishes between approximate and exact areas, and gives Heron of Alexandria's famous formula for the triangular area, $\sqrt{s(s-a)(s-b)(s-c)}$. Heron's formula is given also by Mahāvīra[2] who advanced beyond his predecessors in giving the area of an equilateral triangle as $a^2\sqrt{3}/4$. Brahmagupta and Mahāvīra make a remarkable extension of Heron's formula by giving $\sqrt{(s-a)(s-b)(s-c)(s-d)}$ as the area of a quadrilateral whose sides are a, b, c, d, and whose semiperimeter is s. That this formula is true only for quadrilaterals that can be in-

[1] G. R. Kaye, *op. cit.*, p. 7.
[2] D. E. Smith, in *Isis*, Vol. 1, 1913, pp. 199, 200.

scribed in a circle was recognized by Brahmagupta, according to Cantor's [1] and Kaye's [2] interpretation of Brahmagupta's obscure exposition, but Hindu commentators did not understand the limitation and Bhāskara finally pronounced the formula unsound. Remarkable is "Brahmagupta's theorem" on cyclic quadrilaterals, $x^2=(ad+bc)\cdot(ac+bd)/(ab+cd)$ and $y^2=(ab+cd)\ (ac+bd)/(ad+bc)$, where x and y are the diagonals and a, b, c, d, the lengths of the sides; also the theorem that, if $a^2+b^2=c^2$ and $A^2+B^2=C^2$, then the quadrilateral ("Brahmagupta's trapezium"), (aC, cB, bC, cA) is cyclic and has its diagonals at right angles. Kaye says: From the triangles (3, 4, 5) and (5, 12, 13) a commentator obtains the quadrilateral (39, 60, 52, 25), with diagonals 63 and 56, etc. Brahmagupta (says Kaye) also introduces a proof of Ptolemy's theorem and in doing this follows Diophantus (III, 19) in constructing from right triangles (a, b, c) and (α, β, γ) new right triangles $(a\gamma, b\gamma, c\gamma)$ and $(\alpha c, \beta c, \gamma c)$ and uses the actual examples given by Diophantus, namely (39, 52, 65) and (25, 60, 65). Parallelisms of this sort show unmistakably that the Hindus drew from Greek sources.

In the mensuration of solids remarkable inaccuracies occur in Āryabhaṭa. He gives the volume of a pyramid as *half* the product of the base and the height; the volume of a sphere as $\pi^{\frac{3}{2}} r^3$. Āryabhaṭa gives in one place an extremely accurate value for π, viz. $3\frac{177}{1250}$ $(=3.1416)$, but he himself never utilized it, nor did any other Hindu mathematician before the twelfth century. A frequent Indian practice was to take $\pi=3$, or $\sqrt{10}$. Bhāskara gives two values,—the above mentioned 'accurate,' $\frac{3927}{1250}$, and the 'inaccurate,' Archimedean value, $\frac{22}{7}$. A commentator on *Līlāvatī* says that these values were calculated by beginning with a regular inscribed hexagon, and applying repeatedly the formula $AD=\sqrt{2-\sqrt{4-AB^2}}$, wherein AB is the side of the given polygon, and AD that of one with double the number of sides. In this way were obtained the perimeters of the inscribed polygons of 12, 24, 48, 96, 192, 384 sides. Taking the radius $=100$, the perimeter of the last one gives the value which Āryabhaṭa used for 100π. The empirical nature of Hindu geometry is illustrated by Bhāskara's proof of the Pythagorean theorem. He draws the right triangle four times in the square of the hypotenuse, so that in the middle there remains a square whose side equals the difference between the two sides

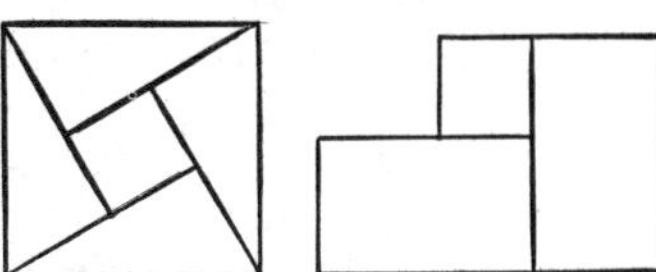

of the right triangle. Arranging this square and the four triangles in a different way, they are seen, together, to make up the sum of the square of the two sides. "Behold!" says Bhāskara, without adding

[1] Cantor, *op. cit.*, Vol. I, 3rd Ed., 1907, pp. 649–653.

[2] G. R. Kaye, *op. cit.*, pp. 20–22.

another word of explanation. Bretschneider conjectures that the Pythagorean proof was substantially the same as this. Recently it has been shown that this interesting proof is not of Hindu origin, but was given much earlier (early in the Christian era) by the Chinese writer *Chang Chun-Ch'ing*, in his commentary upon the ancient treatise, the Chou-pei.[1] In another place, Bhāskara gives a second demonstration of this theorem by drawing from the vertex of the right angle a perpendicular to the hypotenuse, and comparing the two triangles thus obtained with the given triangle to which they are similar. This proof was unknown in Europe till Wallis rediscovered it. The only Indian work that touches the subject of the conic sections is Mahāvīra's book, which gives an inaccurate treatment of the ellipse. It is readily seen that the Hindus cared little for geometry. Brahmagupta's cyclic quadrilaterals constitute the only gem in their geometry.

The grandest achievement of the Hindus and the one which, of all mathematical inventions, has contributed most to the progress of intelligence, is the perfecting of the so-called "Arabic Notation." That this notation did not originate with the Arabs is now admitted by every one. Until recently the preponderance of authority favored the hypothesis that our numeral system, with its concept of local value and our symbol for zero, was wholly of Hindu origin. Now it appears that the principal of local value was used in the sexagesimal system found on Babylonian tablets dating from 1600 to 2300 B. C. and that Babylonian records from the centuries immediately preceding the Christian era contain a symbol for zero which, however, was not used in computation. These sexagesimal fractions appear in Ptolemy's *Almagest* in 130 A. D., where the omicron **o** is made to designate blanks in the sexagesimal numbers, but was not used as a regular zero. The Babylonian origin of the sexagesimal fractions used by Hindu astronomers is denied by no one. The earliest form of the Indian symbol for zero was that of a dot which, according to Bühler,[2] was "commonly used in inscriptions and manuscripts in order to mark a blank." This restricted early use of the symbol for zero resembles somewhat the still earlier use made of it by the Babylonians and by Ptolemy. It is therefore probable that an imperfect notation involving the principle of local value and the use of the zero was imported into India, that it was there transferred from the sexagesimal to the decimal scale and then, in the course of centuries, brought to final perfection. If these views are found by further research to be correct, then the name "Babylonic-Hindu" notation will be more appropriate than either "Arabic" or "Hindu Arabic." It appears

[1] Yoshio Mikami, "The Pythagorean Theorem" in *Archiv d. Math. u. Physik*, 3. S., Vol. 22, 1912, pp. 1–4.

[2] Quoted by D. E. Smith and L. C. Karpinski in their *Hindu-Arabic Numerals*, Boston and London, 1911, p. 53.

that in India various numeral forms were used long before the principle of local value and the zero came to be used. Early Hindu numerals have been classified under three great groups. Numeral forms of one of these groups date from the third century B. C.[1] and are believed to be the forms from which our present system developed. That the nine figures were introduced quite early and that the principle of local value and the zero were incorporated later is a belief which receives support from the fact that on the island of Ceylon a notation resembling the Hindu, but without the zero has been preserved. We know that Buddhism and Indian culture were transplanted to Ceylon about the third century after Christ, and that this culture remained stationary there, while it made progress on the continent. It seems highly probable, then, that the numerals of Ceylon are the old, imperfect numerals of India. In Ceylon, nine figures were used for the units, nine others for the tens, one for 100, and also one for 1000. These 20 characters enabled them to write all the numbers up to 9999. Thus, 8725 would have been written with six signs, representing the following numbers: 8, 1000, 7, 100, 20, 5. These Singhalesian signs, like the old Hindu numerals, are supposed originally to have been the initial letters of the corresponding numeral adjectives. There is a marked resemblance between the notation of Ceylon and the one used by Āryabhaṭa in the first chapter of his work, and there only. Although the zero and the principle of position were unknown to the scholars of Ceylon, they were probably known to Āryabhaṭa; for, in the second chapter, he gives directions for extracting the square and cube roots, which seem to indicate a knowledge of them. The symbol for zero was called *sunya* (the void). It is found in the form of a dot in the Bakhshālī arithmetic, the date of which is uncertain. The earliest undoubted occurrence of our zero in India is in 876 A. D.[2] The earliest known mention of Hindu numerals outside of India was made in 662 A. D. by the Syrian writer, Severus Sebokht. He speaks of Hindu computations "which excel the spoken word and . . . are done with nine symbols." [3]

There appear to have been several notations in use in different parts of India, which differed, not in principle, but merely in the forms of the signs employed. Of interest is also a *symbolical system of position*, in which the figures generally were not expressed by numerical adjectives, but by objects suggesting the particular numbers in question. Thus, for 1 were used the words *moon*, *Brahma*, *Creator*, or *form; for* 4, the words *Veda*, (because it is divided into four parts) or *ocean*, etc. The following example, taken from the *Sūrya Siddhānta*, illustrates the idea. The number 1,577,917,828 is expressed from right to left as

[1] D. E. Smith and L. C. Karpinski, *op. cit.*, p. 22.
[2] *Ibid*, p. 52.
[3] M. F. Nau in *Journal Asiatique*, S. 10, Vol. 16, 1910; D. E. Smith in *Bull. Am. Math. Soc.*, Vol. 23, 1917, p. 366.

follows: Vasu (a class of 8 gods)+two+eight+mountains (the 7 mountain-chains)+form+digits (the 9 digits)+seven+mountains+lunar days (half of which equal 15). The use of such notations made it possible to represent a number in several different ways. This greatly facilitated the framing of verses containing arithmetical rules or scientific constants, which could thus be more easily remembered.

At an early period the Hindus exhibited great skill in calculating, even with large numbers. Thus, they tell us of an examination to which Buddha, the reformer of the Indian religion, had to submit, when a youth, in order to win the maiden he loved. In arithmetic, after having astonished his examiners by naming all the periods of numbers up to the 53d, he was asked whether he could determine the number of primary atoms which, when placed one against the other, would form a line one mile in length. Buddha found the required answer in this way: 7 primary atoms make a very minute grain of dust, 7 of these make a minute grain of dust, 7 of *these* a grain of dust whirled up by the wind, and so on. Thus he proceeded, step by step, until he finally reached the length of a mile. The multiplication of all the factors gave for the multitude of primary atoms in a mile a number consisting of 15 digits. This problem reminds one of the "Sand-Counter" of Archimedes.

After the numerical symbolism had been perfected, figuring was made much easier. Many of the Indian modes of operation differ from ours. The Hindus were generally inclined to follow the motion from left to right, as in writing. Thus, they *added* the left-hand columns first, and made the necessary corrections as they proceeded. For instance, they would have added 254 and 663 thus: $2+6=8$, $5+6=11$, which changes 8 into 9, $4+3=7$. Hence the sum 917. In *subtraction* they had two methods. Thus in $821-348$ they would say, 8 from $11=3$, 4 from $11=7$, 3 from $7=4$. Or they would say, 8 from $11=3$, 5 from $12=7$, 4 from $8=4$. In *multiplication* of a number by another of only one digit, say 569 by 5, they generally said, $5.5=25$, $5.6=30$, which changes 25 into 28, $5.9=45$, hence the 0 must be increased by 4. The product is 2845. In the multiplication with each other of many-figured numbers, they first multiplied, in the manner just indicated, with the left-hand digit of the multiplier, which was written above the multiplicand, and placed the product above the multiplier. On multiplying with the next digit of the multiplier, the product was not placed in a new row, as with us, but the first product obtained was corrected, as the process continued, by erasing, whenever necessary, the old digits, and replacing them by new ones, until finally the whole product was obtained. We who possess the modern luxuries of pencil and paper, would not be likely to fall in love with this Hindu method. But the Indians wrote "with a cane-pen upon a small blackboard with a white, thinly liquid paint which made marks that could be easily erased, or upon a white tablet, less than a foot square, strewn with red flour, on which they wrote the figures with a small stick, so that the figures appeared white on a red ground."[1] Since the digits had to be quite large to be distinctly legible, and since the boards were small, it was desirable to have a method which would not require much space. Such a one was the above method of multiplication. Figures could be easily erased and replaced by others without sacrificing neatness. But the Hindus had also other ways of multiplying, of which we mention the following: The tablet was divided into squares like a chess-board. Diagonals were also drawn, as seen in the figure. The multiplication of $12 \times 735 = 8820$ is exhibited in the adjoining diagram.[2] According to Kaye,[3] this mode of multiplying was not of Hindu origin and was known earlier to the Arabs. The manuscripts extant give no information of how *divisions* were executed.

	7	3	5
1	⁄7	⁄3	⁄5
2	1⁄4	⁄6	1⁄0
8	8	2	0

Hindu mathematicians of the twelfth century test the correctness of arithmetical computations by "casting out nines," but this process is not of Hindu origin; it was known to the Roman bishop Hippolytos in the third century.

In the *Bakhshālī* arithmetic a knowledge of the processes of computation is presupposed. In fractions, the numerator is written above the denominator without a dividing line. Integers are written as fractions with the denominator 1. In mixed expressions the integral part is written above the fraction. Thus, $\begin{smallmatrix}1\\1\\3\end{smallmatrix} = 1\frac{1}{3}$. In place of our = they used the word *phalam*, abbreviated into *pha*. Addition was indicated by *yu*, abbreviated from *yuta*. Numbers to be combined were often enclosed in a rectangle. Thus, *pha* 12 $\boxed{\begin{smallmatrix}5\\1\end{smallmatrix}\ \begin{smallmatrix}7\\1\end{smallmatrix}\ yu}$ means $\frac{5}{1}+\frac{7}{1}=12$. An unknown quantity is *sunya*, and is designated thus **.** by a heavy dot. The word *sunya* means "empty," and is the word for zero, which is here likewise represented by a dot. This double use of the word and dot rested upon the idea that a position is "empty" if not filled out. It is also to be considered "empty" so long as the number to be placed there has not been ascertained.[4]

The Bakhshālī arithmetic contains problems of which some are solved by reduction to unity or by a sort of *false position*. Example: B gives twice as much as A, C three times as much as B, D four times as much as C; together they give 132; how much did A give? Take 1 for the unknown (*sunya*), then A = 1, B = 2, C = 6, D = 24, their sum = 33. Divide 132 by 33, and the quotient 4 is what A gave.

The method of *false position* we have encountered before among

[1] H. Hankel, *op. cit.*, 1874, p. 186.
[2] M. Cantor, *op. cit.*, Vol. I, 3 Aufl., 1907, p. 611.
[3] G. R. Kaye, *op. cit.*, p. 34.
[4] Cantor, I, 3 Ed., 1907, pp. 613–618.

the early Egyptians. With them it was an instinctive procedure; with the Hindus it had risen to a conscious method. Bhāskara uses it, but while the Bakhshālī document preferably assumes 1 as the unknown, Bhāskara is partial to 3. Thus, if a certain number is taken five-fold, $\frac{1}{3}$ of the product be subtracted, the remainder divided by 10, and $\frac{1}{3}$, $\frac{1}{2}$ and $\frac{1}{4}$ of the original number added, then 68 is obtained. What is the number? Choose 3, then you get 15, 10, 1, and $1+\frac{3}{3}+\frac{3}{2}+\frac{3}{4}=\frac{17}{4}$. Then $(68\div\frac{17}{4})3=48$, the answer.

We shall now proceed to the consideration of some arithmetical problems and the Indian modes of solution. A favorite method was that of *inversion*. With laconic brevity, Āryabhaṭa describes it thus: "Multiplication becomes division, division becomes multiplication; what was gain becomes loss, what loss, gain; inversion." Quite different from this quotation in style is the following problem from Āryabhaṭa, which illustrates the method: "Beautiful maiden with beaming eyes, tell me, as thou understandst the right method of inversion, which is the number which multiplied by 3, then increased by $\frac{3}{4}$ of the product, divided by 7, diminished by $\frac{1}{3}$ of the quotient, multiplied by itself, diminished by 52, the square root extracted, addition of 8, and division by 10, gives the number 2?" The process consists in beginning with 2 and working backwards. Thus, $(2.10-8)^2+52=196$, $\sqrt{196}=14$, and $14.\frac{3}{2}.7.\frac{4}{7}\div3=28$, the answer.

Here is another example taken from *Līlāvatī*, a chapter in Bhāskara's great work: "The square root of half the number of bees in a swarm has flown out upon a jessamine-bush, $\frac{8}{9}$ of the whole swarm has remained behind; one female bee flies about a male that is buzzing within a lotus-flower into which he was allured in the night by its sweet odor, but is now imprisoned in it. Tell me the number of bees." Answer, 72. The pleasing poetic garb in which all arithmetical problems are clothed is due to the Indian practice of writing all school-books in verse, and especially to the fact that these problems, propounded as puzzles, were a favorite social amusement. Says Brahmagupta: "These problems are proposed simply for pleasure; the wise man can invent a thousand others, or he can solve the problems of others by the rules given here. As the sun eclipses the stars by his brilliancy, so the man of knowledge will eclipse the fame of others in assemblies of the people if he proposes algebraic problems, and still more if he solves them."

The Hindus solved problems in interest, discount, partnership, alligation, summation of arithmetical and geometric series, and devised rules for determining the numbers of combinations and permutations. It may here be added that chess, the profoundest of all games, had its origin in India. The invention of magic squares is sometimes erroneously attributed to the Hindus. Among the Chinese and Arabs magic squares appear much earlier. The first occurrence of a magic square among the Hindus is at Dudhai, Jhansi, in northern India. It is engraved upon a stone found in the ruins of a temple assigned to the eleventh century, A. D.[1] After the time of Bhāskara magic squares are mentioned by Hindu writers.

The Hindus made frequent use of the "rule of three." Their method of "false position," is almost identical with that of the "tentative assumption" of Diophantus. These and other rules were applied to a large number of problems.

Passing now to *algebra*, we shall first take up the symbols of operation. Addition was indicated simply by juxtaposition as in Diophantine algebra; subtraction, by placing a dot over the subtrahend; multiplication, by putting after the factors, *bha*, the abbreviation of the word *bhavita*, "the product"; division, by placing the divisor beneath the dividend; square-root, by writing *ka*, from the word *karana* (irrational), before the quantity. The unknown quantity was called by Brahmagupta *yâvattâvat* (*quantum tantum*). When several unknown quantities occurred, he gave, unlike Diophantus, to each a distinct name and symbol. The first unknown was designated by the general term "unknown quantity." The rest were distinguished by names of colors, as the black, blue, yellow, red, or green unknown. The initial syllable of each word constituted the symbol for the respective unknown quantity. Thus *yâ* meant x; *kâ* (from *kâlaka* = black) meant y; *yâ kâ bha*, "x times y"; *ka* 15 *ka* 10, "$\sqrt{15}-\sqrt{10}$."

The Indians were the first to recognize the existence of absolutely negative quantities. They brought out the difference between positive and negative quantities by attaching to the one the idea of "possession," to the other that of "debts." The conception also of opposite directions on a line, as an interpretation of + and − quantities, was not foreign to them. They advanced beyond Diophantus in observing that a quadratic (having real roots) has always two roots. Thus Bhāskara gives $x=50$ and $x=-5$ for the roots of $x^2-45x=250$. "But," says he, "the second value is in this case not to be taken, for it is inadequate; people do not approve of negative roots." Commentators speak of this as if negative roots were seen, but not admitted.

Another important generalization, says Hankel, was this, that since the time of Bhāskara the Hindus never confined their arithmetical operations to rational numbers. For instance, Bhāskara showed how, by the formula $\sqrt{a+\sqrt{b}}=\sqrt{\frac{a+\sqrt{a^2-b}}{2}}+\sqrt{\frac{a-\sqrt{a^2-b}}{2}}$ the square root of the sum of rational and irrational numbers could be found. The Hindus never discerned the dividing line between numbers and magnitudes, set up by the Greeks, which, though the product of a scientific spirit, greatly retarded the progress of mathematics. They passed from magnitudes to numbers and from numbers to magnitudes without anticipating that gap which to a sharply discriminating mind

[1] *Bull. Am. Math. Soc.*, Vol. 24, 1917, p. 106.

exists between the continuous and discontinuous. Yet by doing so the Indians greatly aided the general progress of mathematics. "Indeed, if one understands by algebra the application of arithmetical operations to complex magnitudes of all sorts, whether rational or irrational numbers or space-magnitudes, then the learned Brahmins of Hindostan are the real inventors of algebra." [1]

Let us now examine more closely the Indian algebra. In extracting the square and cube roots they used the formulas $(a+b)^2=a^2+2ab+b^2$ and $(a+b)^3=a^3+3a^2b+3ab^2+b^3$. In this connection Āryabhaṭa speaks of dividing a number into periods of two and three digits. From this we infer that the principle of position and the zero in the numerical notation were already known to him. In figuring with zeros, a statement of Bhāskara is interesting. A fraction whose denominator is zero, says he, admits of no alteration, though much be added or subtracted. Indeed, in the same way, no change takes place in the infinite and immutable Deity when worlds are destroyed or created, even though numerous orders of beings be taken up or brought forth. Though in this he apparently evinces clear mathematical notions, yet in other places he makes a complete failure in figuring with fractions of zero denominator.

In the Hindu solutions of determinate equations, Cantor thinks he can see traces of Diophantine methods. Some technical terms betray their Greek origin. Even if it be true that the Indians borrowed from the Greeks, they deserve credit for improving the solutions of linear and quadratic equations. Recognizing the existence of negative numbers, Brahmagupta was able to unify the treatment of the three forms of quadratic equations considered by Diophantus, viz., $ax^2+bx=c$, $bx+c=ax^2$, $ax^2+c=bx$, (a, b and c being positive numbers), by bringing the three under the one general case, $px^2+qx+r=0$. To S'rīdhara is attributed the "Hindu method" of completing the square which begins by multiplying both sides of the equation by $4p$. Bhāskara advances beyond the Greeks and even beyond Brahmagupta when he says that "the square of a positive, as also of a negative number, is positive; that the square root of a positive number is twofold, positive and negative. There is no square root of a negative number, for it is not a square." Kaye points out, however, that the Hindus were not the first to give double solutions of quadratic equations.[2] The Arab Al-Khowarizmi of the ninth century gave both solutions of $x^2+21=10x$. Of equations of higher degrees, the Indians succeeded in solving only some special cases in which both sides of the equation could be made perfect powers by the addition of certain terms to each.

Incomparably greater progress than in the solution of determinate equations was made by the Hindus in the treatment of *indeterminate equations*. Indeterminate analysis was a subject to which the Hindu

[1] H. Hankel, *op. cit.*, p. 195. [2] G. R. Kaye, *op. cit.*, p. 34.

mind showed a happy adaptation. We have seen that this very subject was a favorite with Diophantus, and that his ingenuity was almost inexhaustible in devising solutions for particular cases. But the glory of having invented *general* methods in this most subtle branch of mathematics belongs to the Indians. The Hindu indeterminate analysis differs from the Greek not only in method, but also in aim. The object of the former was to find all possible integral solutions. Greek analysis, on the other hand, demanded not necessarily integral, but simply rational answers. Diophantus was content with a single solution; the Hindus endeavored to find all solutions possible. Āryabhata gives solutions in integers to linear equations of the form $ax \pm by = c$, where a, b, c are integers. The rule employed is called the *pulverizer*. For this, as for most other rules, the Indians give no proof. Their solution is essentially the same as the one of Euler. Euler's process of reducing $\frac{a}{b}$ to a continued fraction amounts to the same as the Hindu process of finding the greatest common divisor of a and b by division. This is frequently called the Diophantine method. Hankel protests against this name, on the ground that Diophantus not only never knew the method, but did not even aim at solutions purely integral.[1] These equations probably grew out of problems in astronomy. They were applied, for instance, to determine the time when a certain constellation of the planets would occur in the heavens.

Passing by the subject of linear equations with more than two unknown quantities, we come to indeterminate quadratic equations. In the solution of $xy = ax + by + c$, they applied the method re-invented later by Euler, of decomposing $(ab + c)$ into the product of two integers $m.n$ and of placing $x = m + b$ and $y = n + a$.

Remarkable is the Hindu solution of the quadratic equation $cy^2 = ax^2 + b$. With great keenness of intellect they recognized in the special case $y^2 = ax^2 + 1$ a fundamental problem in indeterminate quadratics. They solved it by the *cyclic method*. "It consists," says De Morgan, "in a rule for finding an indefinite number of solutions of $y^2 = ax^2 + 1$ (a being an integer which is not a square), by means of one solution given or found, and of feeling for one solution by making a solution of $y^2 = ax^2 + b$ give a solution of $y^2 = ax^2 + b^2$. It amounts to the following theorem: If p and q be one set of values of x and y in $y^2 = ax^2 + b$ and p' and q' the same or another set, then $qp' + pq'$ and $app' + qq'$ are values of x and y in $y^2 = ax^2 + b^2$. From this it is obvious that one solution of $y^2 = ax^2 + 1$ may be made to give any number, and that if, taking b at pleasure, $y^2 = ax^2 + b^2$ can be solved so that x and y are divisible by b, then one preliminary solution of $y^2 = ax^2 + 1$ can be found. Another mode of trying for solutions is a combination of the preceding with the *cuttaca* (pulverizer)." These calculations were used in astronomy.

[1] H. Hankel, *op. cit.*, p. 196.

Doubtless this "cyclic method" constitutes the greatest invention in the theory of numbers before the time of Lagrange. The perversity of fate has willed it, that the equation $y^2=ax^2+1$ should now be called *Pell's equation;* the first incisive work on it is due to Brahmin scholarship, reinforced, perhaps, by Greek research. It is a problem that has exercised the highest faculties of some of our greatest modern analysts. By them the work of the Greeks and Hindus was done over again; for, unfortunately, only a small portion of the Hindu algebra and the Hindu manuscripts, which we now possess, were known in the Occident. Hankel attributed the invention of the "cyclic method" entirely to the Hindus, but later historians, P. Tannery, M. Cantor, T. Heath, G. R. Kaye favor the hypothesis of ultimate Greek origin. If the missing parts of Diophantus are ever found, light will probably be thrown upon this question.

Greater taste than for geometry was shown by the Hindus for trigonometry. Interesting passages are found in Varāha Mihira's *Pañcha Siddhāntikā* of the sixth century A. D.,[1] which, in our notation for unit radius, gives $\pi=\sqrt{10}$, $\sin 30^\circ=\frac{1}{2}$, $\sin 60^\circ=\sqrt{1-\frac{1}{4}}$, $\sin^2\gamma=[\{\sin 2\gamma\}^2+\{1-\sin(90^\circ-2\gamma)\}^2]/4$. This is followed by a table of 24 sines, the angles increasing by intervals of $3^\circ 45'$ (the eighth part of 30°), obviously taken from Ptolemy's table of chords. However, instead of dividing the radius into 60 parts in the manner of Ptolemy, the Hindu astronomer divides it into 120 parts, which device enabled him to convert Ptolemy's table of chords into a table of sines without changing the numerical values. Āryabhaṭa took a still different value for the radius, namely, 3438, obtained apparently from the relation $2\times 3.1416r=21{,}600$. The Hindus followed the Greeks and Babylonians in the practice of dividing the circle into quadrants, each quadrant into 90 degrees and 5400 minutes—thus dividing the whole circle into 21,600 equal parts. Each quadrant was divided also into 24 equal parts, so that each part embraced 225 units of the whole circumference, and corresponded to $3\frac{3}{4}$ degrees. Notable is the fact that the Indians never reckoned, like the Greeks, with the whole chord of double the arc, but always with the *sine* (*joa*) and *versed sine.* Their mode of calculating tables was theoretically very simple. The sine of 90° was equal to the radius, or 3438; the sine of 30° was evidently half that, or 1719. Applying the formula $\sin^2 a+\cos^2 a=r^2$, they obtained $\sin 45^\circ=\sqrt{\frac{r^2}{2}}=2431$. Substituting for $\cos a$ its equal $\sin(90-a)$, and making $a=60^\circ$, they obtained $\sin 60^\circ=\frac{\sqrt{3r^2}}{2}=2978$. With the sines of 90, 60, 45, and 30 as starting-points, they reckoned the sines of half the angles by the formula $\text{ver}\sin 2a=2\sin^2 a$, thus obtaining the sines of $22^\circ 30'$, 15°,

[1] G. R. Kaye, *op. cit.*, p. 10.

$11^\circ 15'$, $7^\circ 30'$, $3^\circ 45'$. They now figured out the sines of the complements of these angles, namely, the sines of $86^\circ 15'$, $82^\circ 30'$, $78^\circ 45'$, 75°, $67^\circ 30'$; then they calculated the sines of half these angles; then of their complements; then, again, of half their complements; and so on. By this very simple process they got the sines of angles at intervals of $3^\circ 45'$. In this table they discovered the unique law that if a, b, c be three successive arcs such that $a-b=b-c=3^\circ 45'$, then $\sin a - \sin b = (\sin b - \sin c) - \frac{\sin b}{225}$. This formula was afterwards used whenever a re-calculation of tables had to be made. No Indian trigonometrical treatise on the triangle is extant. In astronomy they solved plane and spherical triangles.[1]

Now that we have a fairly complete history of Chinese mathematics, Kaye has been able to point out parallelisms between Hindu and Chinese mathematics which indicate that India is indebted to China. The [2] Chiu-chang Suan-shū ("Arithmetic in Nine Sections") was composed in China at least as early as 200 B. C.; the Chinese writer *Chang T'sang* wrote a commentary on it in 263 A. D. The "Nine Sections" gives the approximate area of a segment of a circle $= \frac{1}{2}(c+a)a$, where c is the chord and a is the perpendicular. This rule occurs in the work of the later Hindu author Mahāvīra. Again, the Chinese problem of the bamboo 10 ft. high, the upper end of which being broken reaches the ground three feet from the stem; to determine the height of the break—occurs in all Hindu books after the sixth century. The Chinese arithmetical treatise, Sun-Tsū Suan-ching, of about the first century A. D. has an example asking for a number which, divided by 3 yields the remainder 2, by 5 the remainder 3, and by 7 the remainder 2. Examples of this type occur in Indian works of the seventh and ninth centuries, particularly in Brahmagupta and Mahāvīra. On a preceding page we called attention to the fact that Bhāskara's dissection proof of the Pythagorean theorem is found much earlier in China. Kaye gives several other examples of Chinese origin that are found later in Hindu books.

Notwithstanding the Hindu indebtedness to other nations, it is remarkable to what extent Indian mathematics enters into the science of our time. Both the form and the spirit of the arithmetic and algebra of modern times are essentially Indian. Think of our notation of numbers, brought to perfection by the Hindus, think of the Indian arithmetical operations nearly as perfect as our own, think of their elegant algebraical methods, and then judge whether the Brahmins on the banks of the Ganges are not entitled to some credit. Unfortunately, some of the most brilliant results in indeterminate analysis, found in Hindu works, reached Europe too late to exert the in-

[1] A. Arneth, *Geschichte der reinen Mathematik*. Stuttgart, 1852, p. 174.
[2] G. R. Kaye, *op. cit.*, pp. 38–41.

fluence they would have exerted, had they come two or three centuries earlier.

At the beginning of the twentieth century, mathematical activity along modern lines sprang up in India. In the year 1907 there was founded the *Indian Mathematical Society;* in 1909 there was started at Madras the *Journal of the Indian Mathematical Society.*[1]

[1] (Three recent writers have advanced arguments tending to disprove the Hindu origin of our numerals. We refer (1) to G. R. Kaye's articles in *Scientia,* Vol. 24, 1918, pp. 53–55; in *Journal Asiatic Soc. Bengal*, III, 1907, pp. 475–508, also VII, 1911, pp. 801–816: in *Indian Antiquary*, 1911, pp. 50–56; (2) to Carra de Vaux's article in *Scientia*, Vol. 21, 1917, pp. 273–282; (3) to a Russian book brought out by Nikol. Bubnow in 1908 and translated into German in 1914 by Jos. Lezius. Kaye claims to show that the proofs of the Hindu origin of our numerals are largely legendary, that the question has been clouded by a confusion between the words *hindi* (Indian) and *hindasi* (measure geometrical), that the symbols are not modified letters of the alphabet. We must hold our minds in suspense on this difficult question and await further evidence.)

THE ARABS

After the flight of Mohammed from Mecca to Medina in 622 A. D., an obscure people of Semitic race began to play an important part in the drama of history. Before the lapse of ten years, the scattered tribes of the Arabian peninsula were fused by the furnace blast of religious enthusiasm into a powerful nation. With sword in hand the united Arabs subdued Syria and Mesopotamia. Distant Persia and the lands beyond, even unto India, were added to the dominions of the Saracens. They conquered Northern Africa, and nearly the whole Spanish peninsula, but were finally checked from further progress in Western Europe by the firm hand of Charles Martel (732 A. D.). The Moslem dominion extended now from India to Spain; but a war of succession to the caliphate ensued, and in 755 the Mohammedan empire was divided,—one caliph reigning at Bagdad, the other at Cordova in Spain. Astounding as was the grand march of conquest by the Arabs, still more so was the ease with which they put aside their former nomadic life, adopted a higher civilization, and assumed the sovereignty over cultivated peoples. Arabic was made the written language throughout the conquered lands. With the rule of the Abbasides in the East began a new period in the history of learning. The capital, Bagdad, situated on the Tigris, lay half-way between two old centres of scientific thought,—India in the East, and Greece in the West. The Arabs were destined to be the custodians of the torch of Greek science, to keep it ablaze during the period of confusion and chaos in the Occident, and afterwards to pass it over to the Europeans. This remark applies in part also to Hindu science. Thus science passed from Aryan to Semitic races, and then back again to the Aryan. Formerly it was held that the Arabs added but little to the knowledge of mathematics; recent studies indicate that they must be credited with novelties once thought to be of later origin.

The Abbasides at Bagdad encouraged the introduction of the sciences by inviting able specialists to their court, irrespective of nationality or religious belief. Medicine and astronomy were their favorite sciences. Thus Harun-al-Rashid, the most distinguished Saracen ruler, drew Indian physicians to Bagdad. In the year 772 there came to the court of Caliph Almansur a Hindu astronomer with astronomical tables which were ordered to be translated into Arabic. These tables, known by the Arabs as the *Sindhind*, and probably taken from the *Brahma-sphuta-siddhānta* of Brahmagupta, stood in great authority. They contained the important Hindu table of sines.

Doubtless at this time, and along with these astronomical tables, the Hindu numerals, with the zero and the principle of position, were introduced among the Saracens. Before the time of Mohammed the Arabs had no numerals. Numbers were written out in words. Later, the numerous computations connected with the financial administration over the conquered lands made a short symbolism indispensable. In some localities, the numerals of the more civilized conquered nations were used for a time. Thus, in Syria, the Greek notation was retained; in Egypt, the Coptic. In some cases, the numeral adjectives may have been abbreviated in writing. The *Diwani-numerals*, found in an Arabic-Persian dictionary, are supposed to be such abbreviations. Gradually it became the practice to employ the 28 Arabic letters of the alphabet for numerals, in analogy to the Greek system. This notation was in turn superseded by the Hindu notation, which quite early was adopted by merchants, and also by writers on arithmetic. Its superiority was generally recognized, except in astronomy, where the alphabetic notation continued to be used. Here the alphabetic notation offered no great disadvantage, since in the sexagesimal arithmetic, taken from the *Almagest*, numbers of generally only one or two places had to be written.[1]

As regards the form of the so-called Arabic numerals, the statement of the Arabic writer *Al-Biruni* (died 1039), who spent many years in India, is of interest. He says that the shape of the numerals, as also of the letters in India, differed in different localities, and that the Arabs selected from the various forms the most suitable. An Arabian astronomer says there was among people much difference in the use of symbols, especially of those for 5, 6, 7, and 8. The symbols used by the Arabs can be traced back to the tenth century. We find material differences between those used by the Saracens in the East and those used in the West. But most surprising is the fact that the symbols of both the East and of the West Arabs deviate so extraordinarily from the Hindu *Devanagari* numerals (=divine numerals) of to-day, and that they resemble much more closely the apices of the Roman writer Boethius. This strange similarity on the one hand, and dissimilarity on the other, is difficult to explain. The most plausible theory is the one of Woepcke: (1) that about the second century after Christ, before the zero had been invented, the Indian numerals were brought to Alexandria, whence they spread to Rome and also to West Africa; (2) that in the eighth century, after the notation in India had been already much modified and perfected by the invention of the zero, the Arabs at Bagdad got it from the Hindus; (3) that the Arabs of the West borrowed the Columbus-egg, the zero, from those in the East, but retained the old forms of the nine numerals, if for no other reason, simply to be contrary to their political enemies of the East; (4) that the old forms were remembered by the West-Arabs to be of Indian origin, and were hence called *Gubar-numerals*

[1] H. Hankel, *op. cit.*, p. 255.

(=dust-numerals, in memory of the Brahmin practice of reckoning on tablets strewn with dust or sand; (5) that, since the eighth century, the numerals in India underwent further changes, and assumed the greatly modified forms of the modern Devanagari-numerals.[1] This is rather a bold theory, but, whether true or not, it explains better than any other yet propounded, the relations between the apices, the Gubar, the East-Arabic, and Devanagari numerals.

It has been mentioned that in 772 the Indian *Siddhānta* was brought to Bagdad and there translated into Arabic. There is no evidence that any intercourse existed between Arabic and Indian astronomers either before or after this time, excepting the travels of Al-Biruni. But we should be very slow to deny the probability that more extended communications actually did take place.

Better informed are we regarding the way in which Greek science, in successive waves, dashed upon and penetrated Arabic soil. In Syria the sciences, especially philosophy and medicine, were cultivated by Greek Christians. Celebrated were the schools at Antioch and Emesa, and, first of all, the flourishing Nestorian school at Edessa. From Syria, Greek physicians and scholars were called to Bagdad. Translations of works from the Greek began to be made. A large number of Greek manuscripts were secured by Caliph *Al-Mamun* (813–833)✡ from the emperor in Constantinople and were turned over to Syria. The successors of Al-Mamun continued the work so auspiciously begun, until, at the beginning of the tenth century, the more important philosophic, medical, mathematical, and astronomical works of the Greeks could all be read in the Arabic tongue. The translations of mathematical works must have been very deficient at first, as it was evidently difficult to secure translators who were masters of both the Greek and Arabic and at the same time proficient in mathematics. The translations had to be revised again and again before they were satisfactory. The first Greek authors made to speak in Arabic were Euclid and Ptolemy. This was accomplished during the reign of the famous Harun-al-Rashid. A revised translation of Euclid's *Elements* was ordered by Al-Mamun. As this revision still contained numerous errors, a new translation was made, either by the learned Hunain ibn Ishak, or by his son, Ishak ibn Hunain. The word "ibn" means "son." To the thirteen books of the *Elements* were added the fourteenth, written by Hypsicles, and the fifteenth attributed by some to Damascius. But it remained for Tabit ibn Korra to bring forth an Arabic Euclid satisfying every need. Still greater difficulty was experienced in securing an intelligent translation of the *Almagest*. Among other important translations into Arabic were the works of Apollonius, Archimedes, Heron, and Diophantus. Thus we see that in the course of one century the Arabs gained access to the vast treasures of Greek science.

[1] M. Cantor, *op. cit.*, Vol. I, 1907, p. 711.

In astronomy great activity in original research existed as early as the ninth century. The religious observances demanded by Mohammedanism presented to astronomers several practical problems. The Moslem dominions being of such enormous extent, it remained in some localities for the astronomer to determine which way the "Believer" must turn during prayer that he may be facing Mecca. The prayers and ablutions had to take place at definite hours during the day and night. This led to more accurate determinations of time. To fix the exact date for the Mohammedan feasts it became necessary to observe more closely the motions of the moon. In addition to all this, the old Oriental superstition that extraordinary occurrences in the heavens in some mysterious way affect the progress of human affairs added increased interest to the prediction of eclipses.[1]

For these reasons considerable progress was made. Astronomical tables and instruments were perfected, observatories erected, and a connected series of observations instituted. This intense love for astronomy and astrology continued during the whole Arabic scientific period. As in India, so here, we hardly ever find a man exclusively devoted to pure mathematics. Most of the so-called mathematicians were first of all astronomers.

The first notable author of mathematical books was **Mohammed ibn Musa Al-Khowarizmi,** who lived during the reign of Caliph Al-Mamun (813–833). Our chief source of information about Al-Khowrizmi is the book of chronicles, entitled *Kitab Al-Fihrist,* written by *Al-Nadim,* about 987 A. D., and containing biographies of learned men. Al-Khowarizmi was engaged by the caliph in making extracts from the *Sindhind,* in revising the tablets of Ptolemy, in taking observations at Bagdad and Damascus, and in measuring a degree of the earth's meridian. Important to us is his work on algebra and arithmetic. The portion on arithmetic is not extant in the original, and it was not till 1857 that a Latin translation of it was found. It begins thus: "Spoken has Algoritmi. Let us give deserved praise to God, our leader and defender." Here the name of the author, *Al-Khowarizmi* has passed into *Algoritmi,* from which come our modern word *algorithm,* signifying the art of computing in any particular way, and the obsolete form *augrim,* used by Chaucer.[2] The arithmetic of Khowarizmi, being based on the principle of position and the Hindu method of calculation, "excels," says an Arabic writer, "all others in brevity and easiness, and exhibits the Hindu intellect and sagacity in the grandest inventions." This book was followed by a large number of arithmetics by later authors, which differed from the earlier ones chiefly in the greater variety of methods. Arabian arithmetics generally contained the four operations with inte-

[1] H. Hankel, *op. cit.*, pp. 226–228.

[2] See L. C. Karpinski, "Augrimstones" in *Modern Language Notes,* Vol. 27, 1912, pp. 206–209.

gers and fractions, modelled after the Indian processes. They explained the operation of *casting out the 9's*, also the *regula falsa* and the *regula duorum falsorum*, sometimes called the rules of "false position" and of "double position" or "double false position," by which algebraical examples could be solved without algebra. The *regula falsa* or *falsa positio* was the assigning of an assumed value to the unknown quantity, which value, if wrong, was corrected by some process like the "rule of three." It was known to the Hindus and to the Egyptian Ahmes. Diophantus used a method almost identical with this. The *regula duorum falsorum* was as follows:[1] To solve an equation $f(x)=V$, assume, for the moment, two values for x; namely, $x=a$ and $x=b$. Then form $f(a)=A$ and $f(b)=B$, and determine the errors $V-A=E_a$ and $V-B=E_b$, then the required $x=\frac{bE_a-aE_b}{E_a-E_b}$ is generally a close approximation, but is absolutely accurate whenever $f(x)$ is a linear function of x.

We now return to Khowarizmi, and consider the other part of his work,—the *algebra*. This is the first book known to contain this word itself as title. Really the title consists of two words, *al-jebr w'almuqabala*, the nearest English translation of which is "restoration and reduction." By "restoration" was meant the transposing of negative terms to the other side of the equation; by "reduction," the uniting of similar terms. Thus, $x^2-2x=5x+6$ passes by al-jebr into $x^2=5x+2x+6$; and this, by almuqabala, into $x^2=7x+6$. The work on algebra, like the arithmetic, by the same author, contains little that is original. It explains the elementary operations and the solutions of linear and quadratic equations. From whom did the author borrow his knowledge of algebra? That it came entirely from Indian sources is impossible, for the Hindus had no rules like the "restoration" and "reduction." They were, for instance, never in the habit of making all terms in an equation positive, as is done by the process of "restoration." Diophantus gives two rules which resemble somewhat those of our Arabic author, but the probability that the Arab got all his algebra from Diophantus is lessened by the considerations that he recognized both roots of a quadratic, while Diophantus noticed only one; and that the Greek algebraist, unlike the Arab, habitually rejected irrational solutions. It would seem, therefore, that the algebra of Al-Khowarizmi was neither purely Indian nor purely Greek. Al-Khowarizimi's fame among the Arabs was great. He gave the examples $x^2+10x=39$, $x^2+21=10x$, $3x+4=x^2$ which are used by later authors, for instance, by the poet and mathematician Omar Khayyam. "The equation $x^2+10x=39$ runs like a thread of gold through the algebras of several centuries" (L. C. Karpinski). It appears in the algebra of *Abu Kamil* who drew extensively upon the work of Al-

[1] H. Hankel, *op. cit.*, p. 259.

Khowarizmi. Abu Kamil, in turn, was the source largely drawn upon by the Italian, Leonardo of Pisa, in his book of 1202.

The algebra of Al-Khowarizmi contains also a few meagre fragments on *geometry*. He gives the theorem of the right triangle, but proves it after Hindu fashion and only for the simplest case, when the right triangle is isosceles. He then calculates the areas of the triangle, parallelogram, and circle. For π he uses the value $3\frac{1}{7}$, and also the two Indian, $\pi=\sqrt{10}$ and $\pi=\frac{62832}{20000}$. Strange to say, the last value was afterwards forgotten by the Arabs, and replaced by others less accurate. Al-Khowarizmi prepared astronomical tables, which, about 1000 A. D., were revised by *Maslama al-Majrītī* and are of importance as containing not only the *sine* function, but also the *tangent* function.[1] The former is evidently of Hindu origin, the latter may be an addition made by Maslama and was formerly attributed to Abu'l Wefa.

Next to be noticed are the three sons of **Musa Sakir,** who lived in Bagdad at the court of the Caliph Al-Mamun. They wrote several works, of which we mention a geometry containing the well-known formula for the area of a triangle expressed in terms of its sides. We are told that one of the sons travelled to Greece, probably to collect astronomical and mathematical manuscripts, and that on his way back he made acquaintance with Tabit ibn Korra. Recognizing in him a talented and learned astronomer, Mohammed procured for him a place among the astronomers at the court in Bagdad. **Tabit ibn Korra** (836–901) was born at Harran in Mesopotamia. He was proficient not only in astronomy and mathematics, but also in the Greek, Arabic, and Syrian languages. His translations of Apollonius, Archimedes, Euclid, Ptolemy, Theodosius, rank among the best. His dissertation on *amicable numbers* (of which each is the sum of the factors of the other) is the first known specimen of original work in mathematics on Arabic soil. It shows that he was familiar with the Pythagorean theory of numbers. Tabit invented the following rule for finding amicable numbers, which is related to Euclid's rule for perfect numbers: If $p=3.2^n-1$, $q=3.2^{n-1}-1$, $r=9.2^{2n-1}-1$ (n being a whole number) are three primes, then $a=2^npq$, $b=2^nr$ are a pair of amicable numbers. Thus, if $n=2$, then $p=11$, $q=5$, $r=71$, and $a=220$, $b=284$. Tabit also trisected an angle.

Tabit ibn Korra is the earliest writer outside of China to discuss magic squares. Other Arabic tracts on this subject are due to Ibn Al-Haitam and later writers.[2]

Foremost among the astronomers of the ninth century ranked **Al-**

[1] See H. Suter, "Die astronomischen Tafeln des Muḥammed ibn Mūsā Al-Khwārizmī in der Bearbeitung des Maslama ibn Aḥmed Al-Madjrıtī und der Latein. Uebersetzung des Athelhard von Bath," in *Mémoires de l'Académie R. des Sciences et des Lettres de Danemark*, Copenhague, 7me S., Section des Lettres, t. III, no. 1, 1914.

[2] See H. Suter, *Die Mathematiker u. Astronomen der Araber u. ihre Werke*, 1900, pp. 36, 93, 136, 139, 140, 146, 218.

Battani, called *Albategnius* by the Latins. Battan in Syria was his birthplace. His observations were celebrated for great precission. His work, *De scientia stellarum*, was translated into Latin by Plato Tiburtinus, in the twelfth century. Out of this translation sprang the word "sinus," as the name of a trigonometric function. The Arabic word for "sine," *jiba* was derived from the Sanskrit *jiva,* and resembled the Arabic word *jaib*, meaning an indentation or gulf. Hence the Latin "sinus."[1] Al-Battani was a close student of Ptolemy, but did not follow him altogether. He took an important step for the better, when he introduced the Indian "sine" or *half* the chord, in place of the *whole* chord of Ptolemy. He was the first to prepare a table of *cotangents*. He dealt with horizontal and also vertical sun dials, and accordingly considered a horizontal shadow (*umbra extensa* in Latin translation) and vertical shadow (*umbra versa*). These denoted, respectively, the "cotangent" and "tangent"; the former came to be called *umbra recta* by Latin writers. Al-Battani probably knew the law of sines; that this law was known to Al-Biruni is certain. Another improvement on Greek trigonometry made by the Arabs points likewise to Indian influences. Propositions and operations which were treated by the Greeks geometrically are expressed by the Arabs algebraically. Thus, *Al-Battani* at once gets from an equation $\frac{\sin\theta}{\cos\theta}=D$, the value of θ by means of $\sin\theta=\frac{D}{\sqrt{(1+D^2)}}$,—a process unknown to the ancients. He knows all the formulas for spherical triangles given in the *Almagest*, but goes further, and adds an important one of his own for oblique-angled triangles; namely, $\cos a=\cos b.\cos c+\sin b\sin c\cos A$.

At the beginning of the tenth century political troubles arose in the East, and as a result the house of the Abbasides lost power. One province after another was taken, till, in 945, all possessions were wrested from them. Fortunately, the new rulers at Bagdad, the Persian Buyides, were as much interested in astronomy as their predecessors. The progress of the sciences was not only unchecked, but the conditions for it became even more favorable. The Emir *Adud-ed-daula* (978–983) gloried in having studied astronomy himself. His son *Saraf-ed-daula* erected an observatory in the garden of his palace, and called thither a whole group of scholars.[2] Among them were *Abu'l-Wefa, Al-Kuhi, Al-Sagani.*

Abu'l Wefa (940–998) was born at Buzshan in Chorassan, a region among the Persian mountains, which has brought forth many Arabic astronomers. He made the brilliant discovery of the *variation* of the moon, an inequality usually supposed to have been first discovered by Tycho Brahe. Abu'l-Wefa translated Diophantus. He is one of the

[1] M. Cantor, *op. cit.*, Vol. I, 3 Aufl., 1907, p. 737.
[2] H. Hankel, *op. cit.*, p. 242.

last Arabic translators and commentators of Greek authors. The fact that he esteemed the algebra of Mohammed ibn Musa Al-Khowarizimi worthy of his commentary indicates that thus far algebra had made little or no progress on Arabic soil. Abu'l-Wefa invented a method for computing tables of sines which gives the sine of half a degree correct to nine decimal places. He used the *tangent* and calculated a table of tangents. In considering the shadow-triangle of sun-dials he introduced also the *secant* and *cosecant*. Unfortunately, these new trigonometric functions and the discovery of the moon's variation excited apparently no notice among his contemporaries and followers. A treatise by Abu'l-Wefa on "geometric constructions" indicates that efforts were being made at that time to improve draughting. It contains a neat construction of the corners of the regular polyedrons on the circumscribed sphere. Here, for the first time, appears the condition which afterwards became very famous in the Occident, that the construction be effected with a single opening of the compasses.

Al-Kuhi, the second astronomer at the observatory of the emir at Bagdad, was a close student of Archimedes and Apollonius. He solved the problem, to construct a segment of a sphere equal in volume to a given segment and having a curved surface equal in area to that of another given segment. He, **Al-Sagani,** and **Al-Biruni** made a study of the trisection of angles. **Abu'l Jud,** an able geometer, solved the problem by the intersection of a parabola with an equilateral hyperbola.

The Arabs had already observed the theorem that the sum of two cubes can never be a cube. This is a special case of the "last theorem of Fermat." **Abu Mohammed Al-Khojandi** of Chorassan thought he had proved this. His proof, now lost, is said to have been defective. Several centuries later Beha-Eddin declared the impossibility of $x^3+y^3=z^3$. Creditable work in theory of numbers and algebra was done by **Al-Karkhi** of Bagdad, who lived at the beginning of the eleventh century. His treatise on algebra is the greatest algebraic work of the Arabs. In it he appears as a disciple of Diophantus. He was the first to operate with higher roots and to solve equations of the form $x^{2n}+ax^n=b$. For the solution of quadratic equations he gives both arithmetical and geometrical proofs. He was the first Arabic author to give and prove the theorems on the summation of the series:—

$$1^2+2^2+3^2+\ldots+n^2=(1+2+\ldots+n)\frac{2n+1}{3},$$

$$1^3+2^3+3^3+\ldots+n^3=(1+2+\ldots+n)^2.$$

Al-Karkhi also busied himself with indeterminate analysis. He showed skill in handling the methods of Diophantus, but added nothing whatever to the stock of knowledge already on hand. Rather surprising is the fact that Al-Karkhi's algebra shows no traces whatever of Hindu indeterminate analysis. But most astonishing it is, that an arithmetic by the same author completely excludes the Hindu numerals. It is constructed wholly after Greek pattern. Abu'l-Wefa, also, in the second half of the tenth century, wrote an arithmetic in which Hindu numerals find no place. This practice is the very opposite to that of other Arabian authors. The question, why the Hindu numerals were ignored by so eminent authors, is certainly a puzzle. Cantor suggests that at one time there may have been rival schools, of which one followed almost exclusively Greek mathematics, the other Indian.

The Arabs were familiar with geometric solutions of quadratic equations. Attempts were now made to solve cubic equations geometrically. They were led to such solutions by the study of questions like the Archimedean problem, demanding the section of a sphere by a plane so that the two segments shall be in a prescribed ratio. The first to state this problem in form of a cubic equation was **Al-Mahani** of Bagdad, while **Abu Ja'far Alchazin** was the first Arab to solve the equation by conic sections. Solutions were given also by Al-Kuhi, Al-Hasan ibn Al-Haitam, and others. Another difficult problem, to determine the side of a regular heptagon, required the construction of the side from the equation $x^3-x^2-2x+1=0$. It was attempted by many and at last solved by Abu'l Jud.

The one who did most to elevate to a *method* the solution of algebraic equations by intersecting conics, was the poet **Omar Khayyam** of Chorassan (about 1045–1123). He divides cubics into two classes, the trinomial and quadrinomial, and each class into families and species. Each species is treated separately but according to a general plan. He believed that cubics could not be solved by calculation, nor bi-quadratics by geometry. He rejected negative roots and often failed to discover all the positive ones. Attempts at bi-quadratic equations were made by Abu'l-Wefa,[1] who solved geometrically $x^4=a$ and $x^4+ax^3=b$.

The solution of cubic equations by intersecting conics was the greatest achievement of the Arabs in algebra. The foundation to this work had been laid by the Greeks, for it was Menæchmus who first constructed the root of $x^3-a=0$ or $x^3-2a^3=0$. It was not his aim to find the number corresponding to x, but simply to determine the side x of a cube double another cube of side a. The Arabs, on the other hand, had another object in view: to find the roots of given numerical equations. In the Occident, the Arabic solutions of cubics remained unknown until quite recently. Descartes and Thomas Baker invented these constructions anew. The works of Al-Khayyam, Al-Karkhi, Abu'l Jud, show how the Arabs departed further and further from

[1] L. Matthiessen, *Grundzüge der Antiken und Modernen Algebra der Litteralen Gleichungen*, Leipzig, 1878, p. 923. Ludwig Matthiessen (1830–1906) was professor of physics at Rostock.

the Indian methods, and placed themselves more immediately under Greek influences.

With Al-Karkhi and Omar Khayyam, mathematics among the Arabs of the East reached flood-mark, and now it begins to ebb. Between 1100 and 1300 A. D. come the crusades with war and bloodshed, during which European Christians profited much by their contact with Arabian culture, then far superior to their own. The crusaders were not the only adversaries of the Arabs. During the first half of the thirteenth century, they had to encounter the wild Mongolian hordes, and, in 1256, were conquered by them under the leadership of *Hulagu*. The caliphate at Bagdad now ceased to exist. At the close of the fourteenth century still another empire was formed by Timur or *Tamerlane*, the Tartar. During such sweeping turmoil, it is not surprising that science declined. Indeed, it is a marvel that it existed at all. During the supremacy of Hulagu, lived **Nasir-Eddin** (1201–1274), a man of broad culture and an able astronomer. He persuaded Hulagu to build him and his associates a large observatory at Maraga. Treatises on algebra, geometry, arithmetic, and a translation of Euclid's *Elements*, were prepared by him. He for the first time elaborated trigonometry independently of astronomy and to such great perfection that, had his work been known, Europeans of the fifteenth century might have spared their labors.[1] He tried his skill at a proof of the parallel-postulate. His proof assumes that if AB is perpendicular to CD at C, and if another straight line EDF makes an angle EDC acute, then the perpendiculars to AB, comprehended between AB and EF, and drawn on the side of CD toward E, are shorter and shorter, the further they are from CD. His proof, in Latin translation, was published by Wallis in 1651.[2] Even at the court of Tamerlane in Samarkand, the sciences were by no means neglected. A group of astronomers was drawn to this court. **Uleg Beg** (1393–1449), a grandson of Tamerlane, was himself an astronomer. Most prominent at this time was **Al-Kashi,** the author of an arithmetic. Thus, during intervals of peace, science continued to be cultivated in the East for several centuries. The last prominent Oriental writer was *Beha-Eddin* (1547–1622). His *Essence of Arithmetic* stands on about the same level as the work of Mohammed ibn Musa Khowarizmi, written nearly 800 years before.

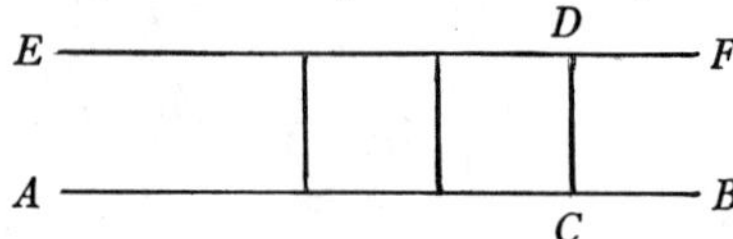

"Wonderful is the expansive power of Oriental peoples, with which upon the wings of the wind they conquer half the world, but more wonderful the energy with which, in less than two generations, they raise themselves from the lowest stages of cultivation to scientific

[1] *Bibliotheca mathematica* (2), 7, 1893, p. 6.

[2] R. Bonola, *Non-Euclidean Geometry*, transl. by H. S. Carslaw, Chicago, 1917, pp. 10–12.

efforts." During all these centuries, astronomy and mathematics in the Orient greatly excel these sciences in the Occident.

Thus far we have spoken only of the Arabs in the East. Between the Arabs of the East and of the West, which were under separate governments, there generally existed considerable political animosity. In consequence of this, and of the enormous distance between the two great centres of learning, Bagdad and Cordova, there was less scientific intercourse among them than might be expected to exist between peoples having the same religion and written language. Thus the course of science in Spain was quite independent of that in Persia. While wending our way westward to Cordova, we must stop in Egypt long enough to observe that there, too, scientific activity was rekindled. Not Alexandria, but Cairo with its library and observatory, was now the home of learning. Foremost among her scientists ranked **Ibn Junos** (died 1008), a contemporary of Abu'l-Wefa. He solved some difficult problems in spherical trigonometry. Another Egyptian astronomer was **Ibn Al-Haitam** (died 1038), who computed the volumes of paraboloids formed by revolving a parabola about any diameter or any ordinate; he used the method of exhaustion and gave the four summation formulas for the first four powers of the natural numbers.[1] Travelling westward, we meet in Morocco **Abu'l Hasan Ali,** whose treatise "on astronomical instruments" discloses a thorough knowledge of the *Conics* of Apollonius. Arriving finally in Spain at the capital, Cordova, we are struck by the magnificent splendor of her architecture. At this renowned seat of learning, schools and libraries were founded during the tenth century.

Little is known of the progress of mathematics in Spain. The earliest name that has come down to us is **Al-Majriti** (died 1007), the author of a mystic paper on "amicable numbers." His pupils founded schools at Cordova, Dania, and Granada. But the only great astronomer among the Saracens in Spain is **Jabir ibn Aflah** of Sevilla, frequently called *Geber*. He lived in the second half of the eleventh century. It was formerly believed that he was the inventor of algebra, and that the word *algebra* came from "Jabir" or "Geber." He ranks among the most eminent astronomers of this time, but, like so many of his contemporaries, his writings contain a great deal of mysticism. His chief work is an astronomy in nine books, of which the first is devoted to trigonometry. In his treatment of spherical trigonometry, he exercises great independence of thought. He makes war against the time-honored procedure adopted by Ptolemy of applying "the rule of six quantities," and gives a new way of his own, based on the "rule of four quantities." This is: If PP_1 and QQ_1 be two arcs of great circles intersecting in A, and if PQ and P_1Q_1 be arcs of great circles drawn perpendicular to QQ_1, then we have the proportion

$$\sin AP : \sin PQ = \sin AP_1 : \sin P_1Q_1.$$

[1] H. Suter in *Bibliotheca mathematica*, 3. S., Vol. 12, 1911–12, pp. 320–322.

From this he derives the formulas for spherical right triangles. This sine-formula was probably known before this to Tabit ibn Korra and others.[1] To the four fundamental formulas already given by Ptolemy, he added a fifth, discovered by himself. If a, b, c, be the sides, and A, B, C, the angles of a spherical triangle, right-angled at A, then $\cos B = \cos b \sin C$. This is frequently called "Geber's Theorem." Radical and bold as were his innovations in spherical trigonometry, in plane trigonometry he followed slavishly the old beaten path of the Greeks. Not even did he adopt the Indian "sine" and "cosine," but still used the Greek "chord of double the angle." So painful was the departure from old ideas, even to an independent Arab!

It is a remarkable fact that among the early Arabs no trace whatever of the use of the abacus can be discovered. At the close of the thirteenth century, for the first time, do we find an Arabic writer, **Ibn Albanna,** who uses processes which are a mixture of abacal and Hindu computation. Ibn Albanna lived in Bugia, an African seaport, and it is plain that he came under European influences and thence got a knowledge of the abacus. Ibn Albanna and Abraham ibn Esra before him, solved equations of the first degree by the rule of "double false position." After Ibn Albanna we find it used by *Al-Kalsadi* and *Beha-Eddin* (1547–1622).[2] If $ax+b=0$, let m and n be any two numbers ("double false position"), let also $am+b=M$, $an+b=N$, then $x=(nM-mN)\div(M-N)$.

Of interest is the approximate solution of the cubic $x^3+Q=Px$, which grew out of the computation of $x=\sin 1°$. The method is shown only in this one numerical example. It is given in *Miram Chelebi* in 1498, in his annotations of certain Arabic astronomical tables. The solution is attributed to *Atabeddin Jamshid*.[3] Write $x=(Q+x^3)\div \text{P}$. If $Q\div P=a+R\div P$, then a is the first approximation, x being snall. We have $Q=aP+R$, and consequently $x=a+(R+a^3)\div P=a+b+S\div P$, say. Then $a+b$ is the second approximation. We have $R=bP+S-a^3$ and $Q=(a+b)P+S-a^3$. Hence $x=a+b+(S-a^3+(a+b)^3\div P=a+b+c+T\div P$, say. Here $a+b+c$ is the third approximation, and so on. In general, the amount of computation is considerable, though for finding $x=\sin 1°$ the method answered very well. This example is the only known approximate arithmetical solution of an affected equation due to Arabic writers. Nearly three centuries before this, the Italian, Leonardo of Pisa, carried the solution of a cubic to a high degree of approximation, but without disclosing his method.

The latest prominent Spanish-Arabic scholar was **Al-Kalsadi** of Granada, who died in 1486. He wrote the *Raising of the Veil of the Science of Gubar*. The word "gubar" meant originally "dust" and

[1] See *Bibliotheca mathematica*, 2 S., Vol. 7, 1893, p. 7.

[2] L. Matthiessen, *Grundzüge d. Antiken u. modernen Algebra*, Leipzig, 1878, p. 275.

[3] See Cantor, *op. cit.* Vol. I, 3rd Ed., 1907, p. 782.

stands here for written arithmetic with numerals, in contrast to mental arithmetic. In addition, subtraction and multiplication, the result is written *above* the other figures. The square root was indicated by the initial Arabic letter of the word "jidre," meaning "root," particularly "square root." He had symbols for the unknown and had, in fact, a considerable amount of algebraic symbolism. His approximation for the square root $\sqrt{a^2+b}$, namely $(4a^3+3ab)/(4a^2+b)$, is believed by S. Günther✡ to disclose a method of continued fractions, without our modern notation, since $(4a^3+3ab)/(4a^2+b)=a+b/(2a+b/2a)$. Al-Kalsadi's work excels other Arabic works in the amount of algebraic symbolism used. Arabic algebra before him contained much less symbolism then Hindu algebra. With Nesselmann[1], we divide algebras, with respect to notation, into three classes: (1) *Rhetorical algebras*, in which no symbols are used, everything being written out in words, (2) *Syncopated algebras*, in which, as in the first class, everything is written out in words, except that abbreviations are used for certain frequently recurring operations and ideas, (3) *Symbolic algebras*, in which all forms and operations are represented by a fully developed algebraic symbolism, as for example, $x^2+10x+7$. According to this classification, Arabic works (excepting those of the later western Arabs), the Greek works of Iamblichus and Thymaridas, and the works of the early Italian writers and of Regiomontanus are *rhetorical* in form; the works of the later western Arabs, of Diophantus and of the later European writers down to about the middle of the seventeenth century (excepting Vieta's and Oughtred's) are *syncopated* in form; the Hindu works and those of Vieta and Oughtred, and of the Europeans since the middle of the seventeenth century, are *symbolic* in form. It is thus seen that the western Arabs took an advanced position in matters of algebraic notation, and were inferior to none of their predecessors or contemporaries, except the Hindus.

In the year in which Columbus discovered America, the Moors lost their last foot-hold on Spanish soil; the productive period of Arabic science was passed.

We have witnessed a laudable intellectual activity among the Arabs. They had the good fortune to possess rulers who, by their munificence, furthered scientific research. At the courts of the caliphs, scientists were supplied with libraries and observatories. A large number of astronomical and mathematical works were written by Arabic authors. It has been said that the Arabs were learned, but not original, With our present knowledge of their work, this dictum needs revision; they have to their credit several substantial accomplishments. They solved cubic equations by geometric construction, perfected trigonometry to a marked degree and made nu-

[1] G. H. F. Nesselmann, *Die Algebra der Griechen*, Berlin, 1842, pp. 301–306.

merous smaller advances all along the line of mathematics, physics and astronomy. Not least of their services to science consists in this, that they adopted the learning of Greece and India, and kept what they received with care. When the love for science began to grow in the Occident, they transmitted to the Europeans the valuable treasures of antiquity. Thus a Semitic race was, during the Dark Ages, the custodian of the Aryan intellectual possessions.

EUROPE DURING THE MIDDLE AGES

With the third century after Christ begins an era of migration of nations in Europe. The powerful Goths quit their swamps and forests in the North and sweep onward in steady southwestern current, dislodging the Vandals, Sueves, and Burgundians, crossing the Roman territory, and stopping and recoiling only when reaching the shores of the Mediterranean. From the Ural Mountains wild hordes sweep down on the Danube. The Roman Empire falls to pieces, and the Dark Ages begin. But dark though they seem, they are the germinating season of the institutions and nations of modern Europe. The Teutonic element, partly pure, partly intermixed with the Celtic and Latin, produces that strong and luxuriant growth, the modern civilization of Europe. Almost all the various nations of Europe belong to the Aryan stock. As the Greeks and the Hindus—both Aryan races—were the great thinkers of antiquity, so the nations north of the Alps and Italy became the great intellectual leaders of modern times.

Introduction of Roman Mathematics

We shall now consider how these as yet barbaric nations of the North gradually came in possession of the intellectual treasures of antiquity. With the spread of Christianity the Latin language was introduced not only in ecclesiastical but also in scientific and all important worldly transactions. Naturally the science of the Middle Ages was drawn largely from Latin sources. In fact, during the earlier of these ages Roman authors were the only ones read in the Occident. Though Greek was not wholly unknown, yet before the thirteenth century not a single Greek scientific work had been read or translated into Latin. Meagre indeed was the science which could be gotten from Roman writers, and we must wait several centuries before any substantial progress is made in mathematics.

After the time of Boethius and Cassiodorius mathematical activity in Italy died out. The first slender blossom of science among tribes that came from the North was an encyclopædia entitled *Origenes*, written by **Isidorus** (died 636 as bishop of Seville). This work is modelled after the Roman encyclopædias of Martianus Capella of Carthage and of Cassiodorius. Part of it is devoted to the quadrivium, arithmetic, music, geometry, and astronomy. He gives definitions and grammatical explications of technical terms, but does not describe the modes of computation then in vogue. After Isidorus there follows a century of darkness which is at last dissipated by the appearance of **Bede the Venerable** (672–735), the most learned man of his time. He was a native of Wearmouth, in England. His works contain treatises on the *Computus*, or the computation of Easter-time, and on finger-reckoning. It appears that a finger-symbolism was then widely used for calculation. The correct determination of the time of Easter was a problem which in those days greatly agitated the Church. It became desirable to have at least one monk at each monastery who could determine the day of religious festivals and could compute the calendar. Such determinations required some knowledge of arithmetic. Hence we find that the art of calculating always found some little corner in the curriculum for the education of monks.

The year in which Bede died is also the year in which **Alcuin** (735–804) was born. Alcuin was educated in York, and was called to the court of Charlemagne to direct the progress of education in the great Frankish Empire. Charlemagne was a great patron of learning and of learned men. In the great sees and monasteries he founded schools in which were taught the psalms, writing, singing, computation (*computus*), and grammar. By *computus* was here meant, probably, not merely the determination of Easter-time, but the art of computation in general. Exactly what modes of reckoning were then employed we have no means of knowing. It is not likely that Alcuin was familiar with the apices of Boethius or with the Roman method of reckoning on the abacus. He belongs to that long list of scholars who dragged the theory of numbers into theology. Thus the number of beings created by God, who created all things well, is 6, because 6 is a perfect number (the sum of its divisors being $1+2+3=6$); 8, on the other hand, is an imperfect number ($1+2+4<8$); hence the second origin of mankind emanated from the number 8, which is the number of souls said to have been in Noah's ark.

There is a collection of "Problems for Quickening the Mind" (*propositiones ad acuendos iuvenes*), which are certainly as old as 1000 A. D. and possibly older. Cantor is of the opinion that they were written much earlier and by Alcuin. The following is a specimen of these "Problems": A dog chasing a rabbit, which has a start of 150 feet, jumps 9 feet every time the rabbit jumps 7. In order to determine in how many leaps the dog overtakes the rabbit, 150 is to be divided by 2. In this collection of problems, the areas of triangular and quadrangular pieces of land are found by the same formulas of approximation as those used by the Egyptians and given by Boethius in his geometry. An old problem is the "cistern-problem" (given the time in which several pipes can fill a cistern singly, to find the time in which they fill it jointly), which has been found previously in Heron, in the Greek *Anthology*, and in Hindu works. Many of the problems show that the collection was compiled chiefly from Roman sources. The problem which, on account of its uniqueness, gives the most positive testimony regarding the Roman origin is that on the interpretation of a will in a case where twins are born. The problem is identical with the Roman, except that different ratios are chosen. Of the exercises for recreation, we mention the one of the wolf, goat, and cabbage, to be rowed across a river in a boat holding only one besides the ferry-man. Query: How must he carry them across so that the goat shall not eat the cabbage, nor the wolf the goat? [1] The solutions of the "problems for quickening the mind" require no further knowledge than the recollection of some few formulas used in surveying, the ability to solve linear equations and to perform the four fundamental operations with integers. Extraction of roots was nowhere demanded; fractions hardly ever occur.[2]

The great empire of Charlemagne tottered and fell almost immediately after his death. War and confusion ensued. Scientific pursuits were abandoned, not to be resumed until the close of the tenth century, when under Saxon rule in Germany and Capetian in France, more peaceful times began. The thick gloom of ignorance commenced to disappear. The zeal with which the study of mathematics was now taken up by the monks is due principally to the energy and influence of one man,—**Gerbert.** He was born in Aurillac in Auvergne. After receiving a monastic education, he engaged in study, chiefly of mathematics, in Spain. On his return he taught school at Rheims for ten years and became distinguished for his profound scholarship. By King Otto I, and his successors Gerbert was held in highest esteem. He was elected bishop of Rheims, then of Ravenna, and finally was made Pope under the name of Sylvester II, by his former Emperor Otho III. He died in 1003, after a life intricately involved in many political and ecclesiastical quarrels. Such was the career of the greatest mathematician of the tenth century in Europe. By his contemporaries his mathematical knowledge was considered wonderful. Many even accused him of criminal intercourse with evil spirits.

Gerbert enlarged the stock of his knowledge by procuring copies of rare books. Thus in Mantua he found the geometry of Boethius. Though this is of small scientific value, yet it is of great importance in history. It was at that time the principal book from which European scholars could learn the elements of geometry. Gerbert studied it with zeal, and is generally believed himself to be the author of a geometry. H. Weissenborn denied his authorship, and claimed that the book in question consists of three parts which cannot come from one and the same author. More recent study favors the conclusion that Gerbert is the author and that he compiled it from different sources.[3] This geometry contains little more than the one of Boethius, but the fact that occasional errors in the latter are herein corrected shows that

[1] S. Günther, *Geschichte des mathem. Unterrichts im deutschen Mittelalter.* Berlin, 1887, p. 32.

[2] M. Cantor, *op. cit.*, Vol. I, 3. Aufl., 1907, p. 839.

[3] S. Günther, *Geschichte der Mathematik*, 1. Teil, Leipzig, 1908, p. 249.

the author had mastered the subject. "The first mathematical paper of the Middle Ages which deserves this name," says Hankel, "is a letter of Gerbert to Adalbold, bishop of Utrecht," in which is explained the reason why the area of a triangle, obtained "geometrically" by taking the product of the base by half its altitude, differs from the area calculated "arithmetically," according to the formula $\frac{1}{2}a(a+1)$, used by surveyors, where a stands for a side of an equilateral triangle. He gives the correct explanation that in the latter formula all the small squares, in which the triangle is supposed to be divided, are counted in wholly, even though parts of them project beyond it. D. E. Smith [1] calls attention to a great medieval number game, called rithmomachia, claimed by some to be of Greek origin. It was played as late as the sixteenth century. It called for considerable arithmetical ability, and was known to Gerbert, Oronce Fine, Thomas Bradwardine and others. A board resembling a chess board was used. Relations like $81 = 72 + \frac{1}{8}$ of 72, $42 = 36 + \frac{1}{6}$ of 36 were involved.

Gerbert made a careful study of the arithmetical works of Boethius. He himself published the first, perhaps both, of the following two works,—*A Small Book on the Division of Numbers*, and *Rule of Computation on the Abacus*. They give an insight into the methods of calculation practised in Europe before the introduction of the Hindu numerals. Gerbert used the abacus, which was probably unknown to Alcuin. **Bernelinus,** a pupil of Gerbert, describes it as consisting of a smooth board upon which geometricians were accustomed to strew blue sand, and then to draw their diagrams. For arithmetical purposes the board was divided into 30 columns, of which 3 were reserved for fractions, while the remaining 27 were divided into groups with 3 columns in each. In every group the columns were marked respectively by the letters C (*centum*), D (*decem*), and S (*singularis*) or M (*monas*). Bernelinus gives the nine numerals used, which are the apices of Boethius, and then remarks that the Greek letters may be used in their place. By the use of these columns any number can be written without introducing a zero, and all operations in arithmetic can be performed in the same way as we execute ours without the columns, but with the symbol for zero. Indeed, the methods of adding, subtracting, and multiplying in vogue among the abacists agree substantially with those of to-day. But in division there is a very great difference. The early rules for division appear to have been framed to satisfy the following three conditions: (1) The use of the multiplication table shall be restricted as far as possible; at least, it shall never be required to multiply mentally a figure of two digits by another of one digit. (2) Subtractions shall be avoided as much as possible and replaced by additions. (3) The operation shall proceed in a purely mechanical way, without requiring trials.[2] That it should be necessary to make such conditions seems strange to us; but it must be re-

[1] *Am. Math. Monthly*, Vol. 28, 1911, pp. 73–80. [2] H. Hankel, *op. cit.*, p. 318.

remembered that the monks of the Middle Ages did not attend school during childhood and learn the multiplication table while the memory was fresh. Gerbert's rules for division are the oldest extant. They are so brief as to be very obscure to the uninitiated. They were probably intended simply to aid the memory by calling to mind the successive steps in the work. In later manuscripts they are stated more fully. In dividing any number by another of one digit, say 668 by 6, the divisor was first increased to 10 by adding 4. The process is exhibited in an adjoining figure.[1] As it continues, we must imagine the digits which are crossed out, to be erased and then replaced by the ones beneath. It is as follows: $600 \div 10 = 60$, but, to rectify the error, 4×60, or 240, must be added; $200 \div 10 = 20$, but 4×20, or 80, must be added. We now write for $60 + 40 + 80$, its sum 180, and continue thus: $100 \div 10 = 10$; the correction necessary is 4×10, or 40, which, added to 80, gives 120. Now $100 \div 10 = 10$, and the correction 4×10, together with the 20, gives 60. Proceeding as before, $60 \div 10 = 6$; the correction is $4 \times 6 = 24$. Now $20 \div 10 = 2$, the correction being $4 \times 2 = 8$. In the column of units we have now $8 + 4 + 8$, or 20. As before, $20 \div 10 = 2$; the correction is $2 \times 4 = 8$, which is not divisible by 10, but only by 6, giving the quotient 1 and the remainder 2. All the partial quotients taken together give $60 + 20 + 10 + 10 + 6 + 2 + 2 + 1 = 111$, and the remainder 2.

Similar but more complicated, is the process when the divisor contains two or more digits. Were the divisor 27, then the next higher multiple of 10, or 30, would be taken for the divisor, but corrections would be required for the 3. He who has the patience to carry such a division through to the end, will understand why it has been said of Gerbert that "Regulas dedit, quæ a sudantibus abacistis vix intelliguntur." He will also perceive why the Arabic method of division, when first introduced, was called the *divisio aurea*, but the one on the abacus, the *divisio ferrea*.

In his book on the abacus, Bernelinus devotes a chapter to fractions. These are, of course, the *duodecimals*, first used by the Romans. For want of a suitable notation, calculation with them was exceedingly difficult. It would be so even to us, were we accustomed, like the early abacists, to express them, not by a numerator or denominator, but by the application of names, such as *uncia* for $\frac{1}{12}$, *quincunx* for $\frac{5}{12}$, *dodrans* for $\frac{9}{12}$.

In the tenth century, Gerbert was the central figure among the learned. In his time the Occident came into secure possession of all mathematical knowledge of the Romans. During the eleventh century it was studied assiduously. Though numerous works were written on arithmetic and geometry, mathematical knowledge in the Occident was still very insignificant. Scanty indeed were the mathematical treasures obtained from Roman sources.

[1] M. Cantor, *op. cit.*, Vol. I, 3. Aufl., 1907, p. 882.

Translation of Arabic Manuscripts

By his great erudition and phenomenal activity, Gerbert infused new life into the study not only of mathematics, but also of philosophy. Pupils from France, Germany, and Italy gathered at Rheims to enjoy his instruction. When they themselves became teachers, they taught of course not only the use of the abacus and geometry, but also what they had learned of the philosophy of Aristotle. His philosophy was known, at first, only through the writings of Boethius. But the growing enthusiasm for it created a demand for his complete works. Greek texts were wanting. But the Latins heard that the Arabs, too, were great admirers of Peripatetism, and that they possessed translations of Aristotle's works and commentaries thereon. This led them finally to search for and translate Arabic manuscripts. During this search, mathematical works also came to their notice, and were translated into Latin. Though some few unimportant works may have been translated earlier, yet the period of greatest activity began about 1100. The zeal displayed in acquiring the Mohammedan treasures of knowledge excelled even that of the Arabs themselves, when, in the eighth century, they plundered the rich coffers of Greek and Hindu science.

Among the earliest scholars engaged in translating manuscripts into Latin was **Athelard of Bath.** The period of his activity is the first quarter of the twelfth century. He travelled extensively in Asia Minor, Egypt, perhaps also in Spain, and braved a thousand perils, that he might acquire the language and science of the Mohammedans. He made one of the earliest translations, from the Arabic, of Euclid's *Elements*. He translated the astronomical tables of Al-Khowarizmi. In 1857, a manuscript was found in the library at Cambridge, which proved to be the arithmetic by Al-Khowarizmi in Latin. This translation also is very probably due to Athelard.

At about the same time flourished *Plato of Tivoli* or *Plato Tiburtinus*. He effected a translation of the astronomy of Al-Battani and of the *Sphærica* of Theodosius.

About the middle of the twelfth century there was a group of Christian scholars busily at work at Toledo, under the leadership of Raymond, then archbishop of Toledo. Among those who worked under his direction, **John of Seville** was most prominent. He translated works chiefly on Aristotelian philosophy. Of importance to us is a *liber alghoarismi*, compiled by him from Arabic authors. The rule for the division of one fraction by another is proved as follows: $\frac{a}{d}\div\frac{c}{d}=\frac{ad}{bd}\div\frac{bc}{bd}=\frac{ad}{bc}$. This same explanation is given by the thirteenth century German writer, Jordanus Nemorarius. On comparing works like this with those of the abacists, we notice at once the most striking difference, which shows that the two parties drew from independent sources. It is argued by some that Gerbert got his apices and his arithmetical knowledge, not from Boethius, but from the Arabs in Spain, and that part or the whole of the geometry of Boethius is a forgery, dating from the time of Gerbert. If this were the case, then the writings of Gerbert would betray Arabic sources, as do those of John of Seville. But no points of resemblance are found. Gerbert could not have learned from the Arabs the use of the abacus, because all evidence we have goes to show that they did not employ it. Nor is it probable that he borrowed from the Arabs the apices, because they were never used in Europe except on the abacus. In illustrating an example in division, mathematicians of the tenth and eleventh centuries state an example in Roman numerals, then draw an abacus and insert in it the necessary numbers with the apices. Hence it seems probable that the abacus and apices were borrowed from the same source. The contrast between authors like John of Seville, drawing from Arabic works, and the abacists, consists in this, that, unlike the latter, the former mention the Hindus, use the term *algorism*, calculate with the zero, and do not employ the abacus. The former teach the extraction of roots, the abacists do not; they teach the sexagesimal fractions used by the Arabs, while the abacists employ the duodecimals of the Romans.[1]

A little later than John of Seville flourished **Gerard of Cremona** in Lombardy. Being desirous to gain possession of the *Almagest*, he went to Toledo, and there, in 1175, translated this great work of Ptolemy. Inspired by the richness of Mohammedan literature, he gave himself up to its study. He translated into Latin over 70 Arabic works. Of mathematical treatises, there were among these, besides the *Almagest*, the 15 books of Euclid, the *Sphærica* of Theodosius, a work of Menelaus, the algebra of Al-Khowarizmi, the astronomy of Jabir ibn Aflah, and others less important. Through Gerard of Cremona the term *sinus* was introduced into trigonometry. Al-Khawarizmi's algebra was translated also by Robert of Chester; his translation probably antedated Cremona's.

In the thirteenth century, the zeal for the acquisition of Arabic learning continued. Foremost among the patrons of science at this time ranked Emperor Frederick II of Hohenstaufen (died 1250). Through frequent contact with Mohammedan scholars, he became familiar with Arabic science. He employed a number of scholars in translating Arabic manuscripts, and it was through him that we came in possession of a new translation of the *Almagest*. Another royal head deserving mention as a zealous promoter of Arabic science was Alfonso X of Castile (died 1284). He gathered around him a number of Jewish and Christian scholars, who translated and compiled astronomical works from Arabic sources. Astronomical tables prepared by two Jews spread rapidly in the Occident, and constituted the basis

[1] M. Cantor, *op. cit.*, Vol. I, 3. Aufl., 1907, p. 879, chapter 40.

of all astronomical calculation till the sixteenth century. The number of scholars who aided in transplanting Arabic science upon Christian soil was large. But we mention only one, **Giovanni Campano** of Novara (about 1260), who brought out a new translation of Euclid, which drove the earlier ones from the field, and which formed the basis of the printed editions.[1]

At the middle of the twelfth century, the Occident was in possession of the so-called Arabic notation. At the close of the century, the Hindu methods of calculation began to supersede the cumbrous methods inherited from Rome. Algebra, with its rules for solving linear and quadratic equations, had been made accessible to the Latins. The geometry of Euclid, the *Sphærica* of Theodosius, the astronomy of Ptolemy, and other works were now accessible in the Latin tongue. Thus a great amount of new scientific material had come into the hands of the Christians. The talent necessary to digest this heterogeneous mass of knowledge was not wanting. The figure of Leonardo of Pisa adorns the vestibule of the thirteenth century.

It is important to notice that no work either on mathematics or astronomy was translated directly from the Greek previous to the fifteenth century.

The First Awakening and its Sequel

Thus far, France and the British Isles have been the headquarters of mathematics in Christian Europe. But at the beginning of the thirteenth century the talent and activity of one man was sufficient to assign the mathematical science a new home in Italy. This man was not a monk, like Bede, Alcuin, or Gerbert, but a layman who found time for scientific study. **Leonardo of Pisa** is the man to whom we owe the first renaissance of mathematics on Christian soil. He is also called *Fibonacci, i.e.* son of Bonaccio. His father was secretary at one of the numerous factories erected on the south and east coast of the Mediterranean by the enterprising merchants of Pisa. He made Leonardo, when a boy, learn the use of the abacus. The boy acquired a strong taste for mathematics, and, in later years, during extensive travels in Egypt, Syria, Greece, and Sicily, collected from the various peoples all the knowledge he could get on this subject. Of all the methods of calculation, he found the Hindu to be unquestionably the best. Returning to Pisa, he published, in 1202, his great work, the *Liber Abaci.* A revised edition of this appeared in 1228. This work contains the knowledge the Arabs possessed in arithmetic and algebra, and treats the subject in a free and independent way. This, together with the other books of Leonardo, shows that he was not merely a compiler, nor, like other writers of the Middle Ages, a slavish imitator of the form in which the subject had been previously presented. The extent

[1] H. Hankel, *op. cit.*, pp. 338, 339.

of his originality is not definitely known, since the sources from which he drew have not all been ascertained. Karpinski has shown that Leonardo drew extensively from Abu Kamil's algebra. Leonardo's *Practica geometriæ* is partly drawn from the *Liber embadorum* of Savasorda, a learned Jew of Barcelona and a co-worker of Plato of Tivoli.

Leonardo was the first great mathematician to advocate the adoption of the "Arabic notation." The calculation with the zero was the portion of Arabic mathematics earliest adopted by the Christians. The minds of men had been prepared for the reception of this by the use of the abacus and the apices. The reckoning with columns was gradually abandoned, and the very word *abacus* changed its meaning and became a synonym for *algorism*. For the zero, the Latins adopted the name *zephirum*, from the Arabic *sifr* (*sifra*=empty); hence our English word *cipher*. The new notation was accepted readily by the enlightened masses, but, at first, rejected by the learned circles. The merchants of Italy used it as early as the thirteenth century, while the monks in the monasteries adhered to the old forms. In 1299, nearly 100 years after the publication of Leonardo's *Liber Abaci*, the Florentine merchants were forbidden the use of the Arabic numeral in book-keeping, and ordered either to employ the Roman numerals or to write the numeral adjectives out in full. This decree is probably due to the variety of forms of certain digits and the consequent ambiguity, misunderstanding and fraud. Some interest attaches to the earliest dates indicating the use of Hindu-Arabic numerals in the Occident. Many erroneous or doubtful early dates have been given by writers inexperienced in the reading of manuscripts and inscriptions. The numerals are first found in manuscripts of the tenth century, but they were not well known until the beginning of the thirteenth century.[1] About 1275 they began to be widely used. The earliest Arabic manuscripts containing the numerals are of 874 and 888 A. D. They appear in a work written at Shiraz in Persia in 970 A. D. A church-pillar not far from the Jeremias Monastery in Egypt has the date 349 A. H. (=961 A. D.) The oldest definitely dated European manuscript known to contain the numerals is the Codex Vigilanus, written in the Albelda Cloister in Spain in 976 A. D. The nine characters without the zero are given, as an addition, in a Spanish copy of the *Origines* by Isidorus of Seville, 992 A. D. A tenth century manuscript with forms differing materially from those in the Codex Vigilanus was found in the St. Gall manuscript now in the University Library at Zürich. The numerals are contained in a Vatican manuscript of 1077, a Sicilian coin of 1138, a Regensburg (Bavaria) chronicle of 1197. The earliest manuscript in French giving the numerals dates about 1275. In the

[1] G. F. Hill, *The Development of Arabic Numerals in Europe*, Oxford, 1915, p. 11. Our dates are taken from this book and from D. E. Smith and L. C. Karpinski's *Hindu-Arabic Numerals*, Boston and London, 1911, pp. 133–146.

British Museum one English manuscript is of about 1230–50, another is of 1246. The earliest undoubted Arabic numerals on a gravestone are at Pforzheim in Baden of 1371 and one at Ulm of 1388. The earliest coins dated in the Arabic numerals are as follows: Swiss 1424, Austrian 1484, French 1485, German 1489, Scotch 1539, English 1551. The earliest calendar with Arabic figures is that of Köbel, 1518. The forms of the numerals varied considerably. The 5 was the most freakish. An upright 7 was rare in the earlier centuries.

In the fifteenth century the abacus with its counters ceased to be used in Spain and Italy. In France it was used later, and it did not disappear in England and Germany before the middle of the seventeenth century.[1] The method of abacal computation is found in the English exchequer for the last time in 1676. In the reign of Henry I the exchequer was distinctly organized as a court of law, but the financial business of the crown was also carried on there. The term "exchequer" is derived from the chequered cloth which covered the table at which the accounts were made up. Suppose the sheriff was summoned to answer for the full annual dues "in money or in tallies." "The liabilities and the actual payments of the sheriff were balanced by means of counters placed upon the squares of the chequered table, those on the one side of the table representing the value of the tallies, warrants and specie presented by the sheriff, and those on the other the amount for which he was liable," so that it was easy to see whether the sheriff had met his obligations or not. In Tudor times "pen and ink dots" took the place of counters. These dots were used as late as 1676.[2] The "tally" upon which accounts were kept was a peeled wooden rod split in such a way as to divide certain notches previously cut in it. One piece of the tally was given to the payer; the other piece was kept by the exchequer. The transaction could be verified easily by fitting the two halves together and noticing whether the notches "tallied" or not. Such tallies remained in use as late as 1783.

In the *Winter's Tale* (IV. 3), Shakespeare lets the clown be embarrassed by a problem which he could not do without counters. Iago (in *Othello*, i, 1) expresses his contempt for Michael Cassio, "forsooth a great mathematician," by calling him a "counter-caster." [3] So general, indeed, says Peacock, appears to have been the practice of this species of arithmetic, that its rules and principles form an essential part of the arithmetical treatises of that day. The real fact seems to be that the old methods were used long after the Hindu numerals were

[1] George Peacock, "Arithmetic" in the *Encyclopædia of Pure Mathematics*, London, 1847, p. 408.

[2] Article "Exchequer" in Palgrave's *Dictionary of Political Economy*, London, 1894.

[3] For additional information, consult F. P. Barnard, *The Casting-Counter and the Counting-Board*, Oxford, 1916. He gives a list of 159 extracts from English inventories referring to counting boards and also photographs of reckoning tables at Basel and Nürnberg, of reckoning cloths at Munich, etc.

in common and general use. With such dogged persistency does man cling to the old!

The *Liber Abaci* was, for centuries, one of the storehouses from which authors got material for works on arithmetic and algebra. In it are set forth the most perfect methods of calculation with integers and fractions, known at that time; the square and cube root are explained, cube root not having been considered in the Christian occident before; equations of the first and second degree leading to problems, either determinate or indeterminate, are solved by the methods of "single" or "double position," and also by real algebra. He recognized that the quadratic $x^2+c=bx$ may be satisfied by two values of x. He took no cognizance of negative and imaginary roots. The book contains a large number of problems. The following was proposed to Leonardo of Pisa by a magister in Constantinople, as a difficult problem: If A gets from B 7 denare, then A's sum is five-fold B's; if B gets from A 5 denare, then B's sum is seven-fold A's. How much has each? The *Liber Abaci* contains another problem, which is of historical interest, because it was given with some variations by Ahmes, 3000 years earlier: 7 old women go to Rome; each woman has 7 mules, each mule carries 7 sacks, each sack contains 7 loaves, with each loaf are 7 knives, each knife is put up in 7 sheaths. What is the sum total of all named? *Ans.* 137,256.[1] Following the practice of Arabic and of Greek and Egyptian writers, Leonardo frequently uses unit fractions. This was done also by other European writers of the Middle Ages. He explained how to resolve a fraction into the sum of unit fractions. He was one of the first to separate the numerator from the denominator by a fractional line. Before his time, when fractions were written in Hindu-Arabic numerals, the denominator was written beneath the numerator, without any sign of separation.

In 1220, Leonardo of Pisa published his *Practica Geometriæ*, which contains all the knowledge of geometry and trigonometry transmitted to him. The writings of Euclid and of some other Greek masters were known to him, either from Arabic manuscripts directly or from the translations made by his countrymen, Gerard of Cremona and Plato of Tivoli. As previously stated, a principal source of his geometrical knowledge was Plato of Tivoli's translation in 1116, from the Hebrew into Latin, of the *Liber embadorum* of Abraham Savasorda.[2] Leonardo's *Geometry* contains an elegant geometrical demonstration of Heron's formula for the area of a triangle, as a function of its three sides; the proof resembles Heron's. Leonardo treats the rich material before him with skill, some originality and Euclidean rigor.

Of still greater interest than the preceding works are those contain-

[1] M. Cantor, *op. cit.*, Vol. II, 2. Aufl., 1900, p. 26. See a problem in the Ahmes papyrus believed to be of the same type as this.

[2] See M. Curtze, *Urkunden zur Geschichte der Mathematik*, I Theil, Leipzig, 1902, p. 5.

ing Fibonacci's more original investigations. We must here preface that after the publication of the *Liber Abaci*, Leonardo was presented by the astronomer Dominicus to Emperor Frederick II of Hohenstaufen. On that occasion, John of Palermo, an imperial notary, proposed several problems, which Leonardo solved promptly. The first (probably an old familiar problem to him) was to find a number x, such that x^2+5 and x^2-5 are each square numbers. The answer is $x=3\frac{5}{12}$; for $(3\frac{5}{12})^2+5=(4\frac{1}{12})^2$, $(3\frac{5}{12})^2-5=(2\frac{7}{12})^2$. His masterly solution of this is given in his *liber quadratorum*, a manuscript which was not printed, but to which reference is made in the second edition of his *Liber Abaci*. The problem was not original with John of Palermo, since the Arabs had already solved similar ones. Some parts of Leonardo's solution may have been borrowed from the Arabs, but the method which he employed of building squares by the summation of odd numbers is original with him.

The second problem proposed to Leonardo at the famous scientific tournament which accompanied the presentation of this celebrated algebraist to that great patron of learning, Emperor Frederick II, was the solving of the equation $x^3+2x^2+10x=20$. As yet cubic equations had not been solved algebraically. Instead of brooding stubbornly over this knotty problem, and after many failures still entertaining new hopes of success, he changed his method of inquiry and showed by clear and rigorous demonstration that the roots of this equation could not be represented by the Euclidean irrational quantities, or, in other words, that they could not be constructed with the ruler and compass only. He contented himself with finding a very close approximation to the required root. His work on this cubic is found in the *Flos*, together with the solution of the following third problem given him by John of Palermo: Three men possess in common an unknown sum of money t; the share of the first is $\frac{t}{2}$; that of the second, $\frac{t}{3}$; that of the third, $\frac{t}{6}$. Desirous of depositing the sum at a safer place, each takes at hazard a certain amount; the first takes x, but deposits only $\frac{x}{2}$; the second carries y, but deposits only $\frac{y}{3}$; the third takes z, and deposits $\frac{z}{6}$. Of the amount deposited each one must receive exactly $\frac{1}{3}$, in order to possess his share of the whole sum. Find x, y, z. Leonardo shows the problem to be indeterminate. Assuming 7 for the sum drawn by each from the deposit, he finds $t=47$, $x=33$, $y=13$, $z=1$.

One would have thought that after so brilliant a beginning, the sciences transplanted from Mohammedan to Christian soil would have enjoyed a steady and vigorous development. But this was not the case. During the fourteenth and fifteenth centuries, the mathematical science was almost stationary. Long wars absorbed the energies of the people and thereby kept back the growth of the sciences. The death of Frederick II in 1254 was followed by a period of confusion in Germany. The German emperors and the popes were continually quarrelling, and Italy was inevitably drawn into the struggles between the Guelphs and the Ghibellines. France and England were engaged in the Hundred Years' War (1338–1453). Then followed in England the Wars of the Roses. The growth of science was retarded not only by war, but also by the injurious influence of scholastic philosophy. The intellectual leaders of those times quarrelled over subtle subjects in metaphysics and theology. Frivolous questions, such as "How many angels can stand on the point of a needle?" were discussed with great interest. Indistinctness and confusion of ideas characterized the reasoning during this period. The writers on mathematics during this period were not few in number, but their scientific efforts were vitiated by the method of scholastic thinking. Though they possessed the *Elements* of Euclid, yet the true nature of a mathematical proof was so little understood, that Hankel believes it no exaggeration to say that "since Fibonacci, not a single proof, not borrowed from Euclid, can be found in the whole literature of these ages, which fulfils all necessary conditions."

The only noticeable advance is a simplification of numerical operations and a more extended application of them. Among the Italians are evidences of an early maturity of arithmetic. Peacock [1] says: The Tuscans generally, and the Florentines in particular, whose city was the cradle of the literature and arts of the thirteenth and fourteenth centuries, were celebrated for their knowledge of arithmetic and book-keeping, which were so necessary for their extensive commerce; the Italians were in familiar possession of commercial arithmetic long before the other nations of Europe; to them we are indebted for the formal introduction into books of arithmetic, under distinct heads, of questions in the single and double rule of three, loss and gain, fellowship, exchange, simple and compound interest, discount, and so on.

There was also a slow improvement in the algebraic notation. The Hindu algebra possessed a tolerable symbolic notation, which was, however, completely ignored by the Mohammedans. In this respect, Arabic algebra approached much more closely to that of Diophantus, which can scarcely be said to employ symbols in a systematic way. Leonardo of Pisa possessed no algebraic symbolism. Like the early Arabs, he expressed the relations of magnitudes to each other by lines or in words. But in the mathematical writings of Chuquet (1484), of Widmann (1489) and of the monk *Luca Pacioli* (also called Lucas de Burgo sepulchri) symbols began to appear. Pacioli's consisted merely

[1] G. Peacock, *op. cit.*, 1847, p. 429.

in abbreviations of Italian words, such as *p* for *piu* (more), *m* for *meno* (less), *co* for *cosa* (the unknown x), *ce* for *censo* (x^2), *cece* for *censocenso* (x^3), "Our present notation has arisen by almost insensible degrees as convenience suggested different marks of abbreviation to different authors; and that perfect symbolic language which addresses itself solely to the eye, and enables us to take in at a glance the most complicated relations of quantity, is the result of a large series of small improvements." [1]

We shall now mention a few authors who lived during the thirteenth and fourteenth and the first half of the fifteenth centuries.

We begin with the philosophic writings of **Thomas Aquinas** (1225–1274), the great Italian philosopher of the Middle Ages, who gave in the completest form the ideas of Origen on infinity. Aquinas' notion of a continuum, particularly a linear continuum, made it *potentially* divisible to infinity, since practically the divisions could not be carried out to infinity. There was, therefore, no minimum line. On the other hand, the point is not a constituent part of the line, since it does not possess the property of infinite divisibility that parts of a line possess, nor can the continuum be constructed out of points. However, a point by its motion has the capacity of generating a line.[2] This continuum held a firm ascendancy over the ancient atomistic doctrine which assumed matter to be composed of very small, indivisible particles. No continuum superior to this was created before the nineteenth century. Aquinas explains Zeno's arguments against motion, as they are given by Aristotle, but hardly presents any new point of view. The Englishman, **Roger Bacon** [1214(?)–1294] likewise argued against a continuum of indivisible parts different from points. Renewing arguments presented by the Greeks and early Arabs, he held that the doctrine of indivisible parts of uniform size would make the diagonal of a square commensurable with a side. Likewise, if through the ends of an indivisible arc of a circle radii are drawn, these radii intercept an arc on a concentric circle of smaller radius; from this it would follow that the inner circle is of the same length as the outer circle, which is impossible. Bacon argued against infinity. If time were infinite, the absurdity would follow that the part is equal to the whole. Bacon's views were made known more widely through **Duns Scotus** (1265–1308), the theological and philosophical opponent of Thomas Aquinas. However, both argued against the existence of indivisible parts (points). Duns Scotus wrote on Zeno's paradoxies, but without reaching new points of view. His commentaries were annotated later by the Italian theologian, **Franciscus de Pitigianis,** who expressed himself in favor of the admission of the actual infinity to explain the "Dichotomy" and the "Achilles," but fails to adequately elaborate the subject. Scholastic ideas on infinity and the

[1] J. F. W. Herschel, "Mathematics" in *Edinburgh Encyclopædia.*

[2] C. R. Wallner, in *Bibliotheca mathematica,* 3. F., Bd. IV, 1903, pp. 29, 30.

continuum find expression in the writings of Bradwardine, the English *doctor profundus*.[1]

About the time of Leonardo of Pisa (1200 A. D.), lived the German monk **Jordanus Nemorarius** (?–1237), who wrote a once famous work on the properties of numbers, printed in 1496 and modelled after the arithmetic of Boethius. The most trifling numeral properties are treated with nauseating pedantry and prolixity. A practical arithmetic based on the Hindu notation was also written by him. **John Halifax** (Sacro Bosco, died 1256) taught in Paris and made an extract from the *Almagest* containing only the most elementary parts of that work. This extract was for nearly 400 years a work of great popularity and standard authority, as was also his arithmetical work, the *Tractatus de arte numerandi*. Other prominent writers are **Albertus Magnus** (1193?–1280) and **Georg Peurbach** (1423–1461) in Germany. It appears that here and there some of our modern ideas were anticipated by writers of the Middle Ages. Thus, **Nicole Oresme** (about 1323–1382), a bishop in Normandy, first conceived the notion of fractional powers, afterwards rediscovered by Stevin, and suggested a notation. Since $4^3=64$, and $64^{1/2}=8$, Oresme concluded that $4^{1 1/2}=8$. In his notation, $4^{1 1/2}$ is expressed, $\boxed{1p.\frac{1}{2}}4$, or $\boxed{\frac{p.1}{1.2}}4$. Some of the mathematicians of the Middle Ages possessed some idea of a function. Oresme even attempted a graphic representation. But of a numeric dependance of one quantity upon another, as found in Descartes, there is no trace among them.[2]

In an unpublished manuscript Oresme found the sum of the infinite series $\frac{1}{2}+\frac{2}{4}+\frac{3}{8}+\frac{4}{16}+\frac{5}{32}+$ .. *in inf*. Such recurrent infinite series were formerly supposed to have made their first appearance in the eighteenth century. The use of infinite series is explained also in the *Liber de triplici motu*, by the Portuguese mathematician *Alvarus Thomas*,[3] in 1509. He gives the division of a line-segment into parts representing the terms of a convergent geometric series; that is, a segment AB is divided into parts such that $AB : P_1B = P_1B : P_2B =$ .. $= P_iB : P_{i+1}B =$ .. Such a division of a line-segment occurs later in Napier's kinematical discussion of logarithms.

Thomas Bradwardine (about 1290–1349), archbishop of Canterbury, studied star-polygons. The first appearance of such polygons was with Pythagoras and his school. We next meet with such polygons in the geometry of Boethius and also in the translation of Euclid from the Arabic by Athelard of Bath. To England falls the honor of having produced the earliest European writers on trigonometry. The

[1] F. Cajori, *Americ. Math. Monthly*, Vol. 22, 1915, pp. 45–47.

[2] H. Wieleitner in *Bibliotheca mathematica*, 3. S., Vol. 13, 1913, pp. 115–145.

[3] See *Études sur Léonard da Vinci*, Vol. III, Paris, 1913, pp. 393, 540, 541, by Pierre Duhem (1861–1916) of the University of Bordeaux; see also Wieleitner in *Bibliotheca mathematica*, Vol. 14, 1914, pp. 150–168.

writings of Bradwardine, of Richard of Wallingford, and John Maudith, both professors at Oxford, and of Simon Bredon of Winchecombe, contain trigonometry drawn from Arabic sources.

The works of the Greek monk **Maximus Planudes** (about 1260–1310), are of interest only as showing that the Hindu numerals were then known in Greece. A writer belonging, like Planudes, to the Byzantine school, was **Manuel Moschopulus** who lived in Constantinople in the early part of the fourteenth century. To him appears to be due the introduction into Europe of magic squares. He wrote a treatise on this subject. Magic squares were known before this to the Arabs and Japanese; they originated with the Chinese. Mediæval astrologers and physicians believed them to possess mystical properties and to be a charm against plague, when engraved on silver plate.

Recently there has been printed a Hebrew arithmetical work by the French Jew, *Levi ben Gerson*, written in 1321,[1] and handed down in several manuscripts. It contains formulas for the number of permutations and combinations of n things taken k at a time. It is worthy of note that the earliest practical arithmetic known to have been brought out in print appeared anonymously in Treviso, Italy, in 1478, and is referred to as the "Treviso arithmetic." Four years later, in 1482, came out at Bamberg the first printed German arithmetic. It is by *Ulrich Wagner*, a teacher of arithmetic at Nürnberg. It was printed on parchment, but only fragments of one copy are now extant.[2]

According to Eneström, Ph. Calandri's *De arithmetrica opusculum*, Florence, 1491, is the first printed treatise containing the word "zero"; it is found in some fourteenth century manuscripts.

In 1494 was *printed* the *Summa de Arithmetica, Geometria, Proportione et Proportionalita*, written by the Tuscan monk **Luca Pacioli** (1445–1517), who, as we remarked, introduced several symbols in algebra. This contains all the knowledge of his day on arithmetic, algebra, and trigonometry, and is the first comprehensive work which appeared after the *Liber Abaci* of Fibonacci. It contains little of importance which cannot be found in Fibonacci's great work, published three centuries earlier. Pacioli came in personal touch with two artists who were also mathematicians, *Leonardo da Vinci*[3] (1452–1519) and *Pier della Francesca* (1416–1492). Da Vinci inscribed regular polygons in circles, but did not distinguish between accurate and approximate constructions. It is interesting to note that da Vinci was familiar with the Greek text of Archimedes on the measurement of the circle. Pier della Francesca advanced the theory of perspective, and left a manuscript on regular solids which was published by

[1] *Bibliotheca mathematica*, 3. S., Vol. 14, 1916, p. 261.

[2] See D. E. Smith, *Rara arithmetica*, Boston and London, 1908, pp. 3, 12, 15; F. Unger, *Methodik der Praktischen Arithmetik in Historischer Entwickelung*, Leipzig, 1888, p. 39.

[3] Consult P. Duhem's *Études sur Léonard de Vinci*, Paris, 1909.

Pacioli in 1509 as his own work, in a book entitled, *Divina proportione.*

Perhaps the greatest result of the influx of Arabic learning was the establishment of universities. What was their attitude toward mathematics? The *University of Paris,* so famous at the beginning of the twelfth century under the teachings of Abelard paid but little attention to this science during the Middle Ages. Geometry was neglected, and Aristotle's logic was the favorite study. In 1336, a rule was introduced that no student should take a degree without attending lectures on mathematics, and from a commentary on the first six books of Euclid, dated 1536, it appears that candidates for the degree of A. M. had to give an oath that they had attended lectures on these books.[1] Examinations, when held at all, probably did not extend beyond the first book, as is shown by the nickname "magister matheseos," applied to the Theorem of Pythagoras, the last in the first book. More attention was paid to mathematics at the *University of Prague,* founded 1384. For the Baccalaureate degree, students were required to take lectures on Sacro Bosco's famous work on astronomy. Of candidates for the A.M. were required not only the six books of Euclid, but an additional knowledge of applied mathematics. Lectures were given on the *Almagest.* At the *University of Leipzig,* the daughter of Prague, and at *Cologne,* less work was required, and, as late as the sixteenth century, the same requirements were made at these as at Prague in the fourteenth. The universities of Bologna, Padua, Pisa, occupied similar positions to the ones in Germany, only that purely astrological lectures were given in place of lectures on the *Almagest.* At Oxford, in the middle of the fifteenth century, the first two books of Euclid were read.[2]

Thus it will be seen that the study of mathematics was maintained at the universities only in a half-hearted manner. No great mathematician and teacher appeared, to inspire the students. The best energies of the schoolmen were expended upon the stupid subtleties of their philosophy. The genius of Leonardo of Pisa left no permanent impress upon the age, and another Renaissance of mathematics was wanted.

[1] H. Hankel, *op. cit.*, p. 355.

[2] J. Gow, *op. cit.*, p. 207.

EUROPE DURING THE SIXTEENTH, SEVENTEENTH AND EIGHTEENTH CENTURIES

We find it convenient to choose the time of the capture of Constantinople by the Turks as the date at which the Middle Ages ended and Modern Times began. In 1453, the Turks battered the walls of this celebrated metropolis with cannon, and finally captured the city; the Byzantine Empire fell, to rise no more. Calamitous as was this event to the East, it acted favorably upon the progress of learning in the West. A great number of learned Greeks fled into Italy, bringing with them precious manuscripts of Greek literature. This contributed vastly to the reviving of classic learning. Up to this time, Greek masters were known only through the often very corrupt Arabic manuscripts, but now they began to be studied from original sources and in their own language. The first English translation of Euclid was made in 1570 from the Greek by *Sir Henry Billingsley*, assisted by *John Dee*.[1] About the middle of the fifteenth century, printing was invented; books became cheap and plentiful; the printing-press transformed Europe into an audience-room. Near the close of the fifteenth century, America was discovered, and, soon after, the earth was circumnavigated. The pulse and pace of the world began to quicken. Men's minds became less servile; they became clearer and stronger. The indistinctness of thought, which was the characteristic feature of mediæval learning, began to be remedied chiefly by the steady cultivation of Pure Mathematics and Astronomy. Dogmatism was attacked; there arose a long struggle with the authority of the Church and the established schools of philosophy. The Copernican System was set up in opposition to the time-honored Ptolemaic System. The long and eager contest between the two culminated in a crisis at the time of Galileo, and resulted in the victory of the new system. Thus, by slow degrees, the minds of men were cut adrift from their old scholastic moorings and sent forth on the wide sea of scientific inquiry, to discover new islands and continents of truth.

The Renaissance

With the sixteenth century began a period of increased intellectual activity. The human mind made a vast effort to achieve its freedom. Attempts at its emancipation from Church authority had been made before, but they were stifled and rendered abortive. The first great and successful revolt against ecclesiastical authority was made in

[1] G. B. Halsted in *Am. Jour. of Math.*, Vol. II, 1879.

Germany. The new desire for judging freely and independently in matters of religion was preceded and accompanied by a growing spirit of scientific inquiry. Thus it was that, for a time, Germany led the van in science. She produced *Regiomontanus*, *Copernicus*, *Rhæticus* and *Kepler*, at a period when France and England had, as yet, brought forth hardly any great scientific thinkers. This remarkable scientific productiveness was no doubt due, to a great extent, to the commercial prosperity of Germany. Material prosperity is an essential condition for the progress of knowledge. As long as every individual is obliged to collect the necessaries for his subsistence, there can be no leisure for higher pursuits. At this time, Germany had accumulated considerable wealth. The Hanseatic League commanded the trade of the North. Close commercial relations existed between Germany and Italy. Italy, too, excelled in commercial activity and enterprise. We need only mention Venice, whose glory began with the crusades, and Florence, with her bankers and her manufacturers of silk and wool. These two cities became great intellectual centres. Thus, Italy, too, produced men in art, literature, and science, who shone forth in fullest splendor. In fact, Italy was the fatherland of what is termed the Renaissance.

For the first great contributions to the mathematical sciences we must, therefore, look to Italy and Germany. In Italy brilliant accessions were made to algebra, in Germany progress was made in astronomy and trigonometry.

On the threshold of this new era we meet in Germany with the figure of John Mueller, more generally called **Regiomontanus** (1436–1476). Chiefly to him we owe the revival of trigonometry. He studied astronomy and trigonometry at Vienna under the celebrated George Peurbach. The latter perceived that the existing Latin translations of the *Almagest* were full of errors, and that Arabic authors had not remained true to the Greek original. Peurbach therefore began to make a translation directly from the Greek. But he did not live to finish it. His work was continued by Regiomontanus, who went beyond his master. Regiomontanus learned the Greek language from Cardinal Bessarion, whom he followed to Italy, where he remained eight years collecting manuscripts from Greeks who had fled thither from the Turks. In addition to the translation of and the commentary on the *Almagest*, he prepared translations of the *Conics* of Apollonius, of Archimedes, and of the mechanical works of Heron. Regiomontanus and Peurbach adopted the Hindu *sine* in place of the Greek *chord of double the arc*. The Greeks and afterwards the Arabs divided the radius into 60 equal parts, and each of these again into 60 smaller ones. The Hindu expressed the length of the radius by parts of the circumference, saying that of the 21,600 equal divisions of the latter, it took 3438 to measure the radius. Regiomontanus, to secure greater precision, constructed one table of sines on a radius divided into

600,000 parts, and another on a radius divided decimally into 10,000,000 divisions. He emphasized the use of the *tangent* in trigonometry. Following out some ideas of his master, he calculated a table of tangents. German mathematicians were not the first Europeans to use this function. In England it was known a century earlier to Bradwardine, who speaks of tangent (*umbra versa*) and cotangent (*umbra recta*), and to John Maudith. Even earlier, in the twelfth century, the *umbra versa* and *umbra recta* are used in a translation from Arabic into Latin, effected by Gerard of Cremona, of the Toledian Tables of Al-Zarkali, who lived in Toledo about 1080. Regiomontanus was the author of an arithmetic and also of a complete treatise on trigonometry, containing solutions of both plane and spherical triangles. Some innovations in trigonometry, formerly attributed to Regiomontanus, are now known to have been introduced by the Arabs before him. Nevertheless, much credit is due to him. His complete mastery of astronomy and mathematics, and his enthusiasm for them, were of far-reaching influence throughout Germany. So great was his reputation, that Pope Sixtus IV called him to Italy to improve the calendar. Regiomontanus left his beloved city of Nürnberg for Rome, where he died in the following year.

After the time of Peurbach and Regiomontanus, trigonometry and especially the calculation of tables continued to occupy German scholars. More refined astronomical instruments were made, which gave observations of greater precision; but these would have been useless without trigonometrical tables of corresponding accuracy. Of the several tables calculated, that by *Georg Joachim* of Feldkirch in Tyrol, generally called **Rhæticus** (1514–1567) deserves special mention. He calculated a table of sines with the radius = 10,000,000,000 and from 10″ to 10″; and, later on, another with the radius = 1,000,000,000,000,000, and proceeding from 10″ to 10″. He began also the construction of tables of tangents and secants, to be carried to the same degree of accuracy; but he died before finishing them. For twelve years he had had in continual employment several calculators. The work was completed in 1596 by his pupil, **Valentine Otho** (1550?–1605). This was indeed a gigantic work,—a monument of German diligence and indefatigable perseverance. The tables were republished in 1613 by **Bartholomäus Pitiscus** (1561–1613) of Heidelberg, who spared no pains to free them of errors. Pitiscus was perhaps the first to use the word "trigonometry." Astronomical tables of so great a degree of accuracy had never been dreamed of by the Greeks, Hindus, or Arabs. That Rhæticus was not a ready calculator only, is indicated by his views on trigonometrical lines. Up to his time, the trigonometric functions had been considered always with relation to the arc; he was the first to construct the right triangle and to make them depend directly upon its angles. It was from the right triangle that Rhæticus got his idea of calculating the hypotenuse; *i. e.* he was the first to plan a table of secants. Good work in trigonometry was done also by **Vieta** and Romanus.

We shall now leave the subject of trigonometry to witness the progress in the solution of algebraical equations. To do so, we must quit Germany for Italy. The first comprehensive algebra printed was that of Luca Pacioli. He closes his book by saying that the solution of the equations $x^3+mx=n$, $x^3+n=mx$ is as impossible at the present state of science as the quadrature of the circle. This remark doubtless stimulated thought. The first step in the algebraic solution of cubics was taken by **Scipione del Ferro** (1465–1526), a professor of mathematics at Bologna, who solved the equation $x^3+mx=n$. He imparted it to his pupil, *Floridas*, in 1505, but did not publish it. It was the practice in those days and for two centuries afterwards to keep discoveries secret, in order to secure by that means an advantage over rivals by proposing problems beyond their reach. This practice gave rise to numberless disputes regarding the priority of inventions. A second solution of cubics was given by *Nicolo* of Brescia [1499(?)–1557]. When a boy of six, Nicolo was so badly cut by a French soldier that he never again gained the free use of his tongue. Hence he was called **Tartaglia,** *i. e.* the stammerer. His widowed mother being too poor to pay his tuition in school, he learned to read and picked up a knowledge of Latin, Greek, and mathematics by himself. Possessing a mind of extraordinary power, he was able to appear as teacher of mathematics at an early age. He taught in Venice, then in Brescia, and later again in Venice. In 1530, one Colla proposed him several problems, one leading to the equation $x^3+px^2=q$. Tartaglia found an imperfect method for solving this, but kept it secret. He spoke about his secret in public and thus caused Del Ferro's pupil, Floridas, to proclaim his own knowledge of the form $x^3+mx=n$. Tartaglia, believing him to be a mediocrist and braggart, challenged him to a public discussion, to take place on the 22d of February, 1535. Hearing, meanwhile, that his rival had gotten the method from a deceased master, and fearing that he would be beaten in the contest, Tartaglia put in all the zeal, industry, and skill to find the rule for the equations, and he succeeded in it ten days before the appointed date, as he himself modestly says.[1] The most difficult step was, no doubt, the passing from quadratic irrationals, used in operating from time of old, to cubic irrationals. Placing $x=\sqrt[3]{t}-\sqrt[3]{u}$, Tartaglia perceived that the irrationals disappeared from the equation $x^3+mx=n$, making $n=t-u$. But this last equality, together with $(\frac{1}{3}m)^3=tu$, gives at once

$$t=\sqrt{\left(\frac{n}{2}\right)^2+\left(\frac{m}{3}\right)^3}+\frac{n}{2},\quad u=\sqrt{\left(\frac{n}{2}\right)^2+\left(\frac{m}{3}\right)^3}-\frac{n}{2}.$$

This is Tartaglia's solution of $x^3+mx=n$. On the 13th of February, he found a similar solution for $x^3=mx+n$. The contest began on the

[1] H. Hankel, *op. cit.*, p. 362.

22d. Each contestant proposed thirty problems. The one who could solve the greatest number within fifty days should be the victor. Tartaglia solved the thirty problems proposed by Floridas in two hours; Floridas could not solve any of Tartaglia's. From now on, Tartaglia studied cubic equations with a will. In 1541 he discovered a general solution for the cubic $x^3 \pm px^2 = \pm q$, by transforming it into the form $x^3 \pm mx = \pm n$. The news of Tartaglia's victory spread all over Italy. Tartaglia was entreated to make known his method, but he declined to do so, saying that after his completion of the translation from the Greek of Euclid and Archimedes, he would publish a large algebra containing his method. But a scholar from Milan, named **Hieronimo Cardano** (1501–1576), after many solicitations, and after giving the most solemn and sacred promises of secrecy, succeeded in obtaining from Tartaglia a knowledge of his rules. Cardan was a singular mixture of genius, folly, self-conceit and mysticism. He was successively professor of mathematics and medicine at Milan, Pavia and Bologna, In 1570 he was imprisoned for debt. Later he went to Rome, was admitted to the college of physicians and was pensioned by the pope.

At this time Cardan was writing his *Ars Magna*, and he knew no better way to crown his work than by inserting the much sought for rules for solving cubics. Thus Cardan broke his most solemn vows, and published in 1545 in his *Ars Magna* Tartaglia's solution of cubics. However, Cardan did credit "his friend Tartaglia" with the discovery of the rule. Nevertheless, Tartaglia became desperate. His most cherished hope, of giving to the world an immortal work which should be the monument of his deep learning and power for original research, was suddenly destroyed; for the crown intended for his work had been snatched away. His first step was to write a history of his invention; but, to completely annihilate his enemies, he challenged Cardan and his pupil Lodovico Ferrari to a contest: each party should propose thirty-one questions to be solved by the other within fifteen days. Tartaglia solved most questions in seven days, but the other party did not send in their solutions before the expiration of the fifth month; moreover, all their solutions except one were wrong. A replication and a rejoinder followed. Endless were the problems proposed and solved on both sides. The dispute produced much chagrin and heart-burnings to the parties, and to Tartaglia especially, who met with many other disappointments. After having recovered himself again, Tartaglia began, in 1556, the publication of the work which he had had in his mind for so long; but he died before he reached the consideration of cubic equations. Thus the fondest wish of his life remained unfulfilled. How much credit for the algebraic solution of the general cubic is due to Tartaglia and how much to Del Ferro it is now impossible to ascertain definitely. Del Ferro's researches were never published and were lost. We know of them only through the remarks of Cardan and his pupil L. Ferrari who say that Del Ferro's and Tartaglia's methods were alike. Certain it is that the customary designation, "Cardan's solution of the cubic" ascribes to Cardan what belongs to one or the other of his predecessors.

Remarkable is the great interest that the solution of cubics excited throughout Italy. It is but natural that after this great conquest mathematicians should attack bi-quadratic equations. As in the case of cubics, so here, the first impulse was given by Colla, who, in 1540, proposed for solution the equation $x^4+6x^2+36=60x$. To be sure, Cardan had studied particular cases as early as 1539. Thus he solved the equation $13x^2=x^4+2x^3+2x+1$ by a process similar to that employed by Diophantus and the Hindus; namely, by adding to both sides $3x^2$ and thereby rendering both numbers complete squares. But Cardan failed to find a general solution; it remained for his pupil **Lodovico Ferrari** (1522–1565) of Bologna to make the brilliant discovery of the general solution of bi-quadratic equations. Ferrari reduced Colla's equation to the form $(x^2+6)^2=60x+6x^2$. In order to give also the right member the form of a complete square he added to both members the expression $2(x^2+6)y+y^2$, containing a new unknown quantity y. This gave him $(x^2+6+y)^2=(6+2y)x^2+60x+(12y+y^2)$. The condition that the right member be a complete square is expressed by the cubic equation $(2y+6)(12y+y^2)=900$. Extracting the square root of the bi-quadratic, he got $x^2+6+y=x\sqrt{2y+6}+\frac{900}{\sqrt{2y+6}}$. Solving the cubic for y and substituting, it remained only to determine x from the resulting quadratic. L. Ferrari pursued a similar method with other numerical bi-quadratic equations.[1] Cardan had the pleasure of publishing this discovery in his *Ars Magna* in 1545. Ferrari's solution is sometimes ascribed to *R. Bombelli*, but he is no more the discoverer of it than Cardan is of the solution called by his name.

To Cardan algebra is much indebted. In his *Ars Magna* he takes notice of negative roots of an equation, calling them *fictitious*, while the positive roots are called real. He paid some attention to computations involving the square root of negative numbers, but failed to recognize imaginary roots. Cardan also observed the difficulty in the irreducible case in the cubics, which, like the quadrature of the circle, has since "so much tormented the perverse ingenuity of mathematicians." But he did not understand its nature. It remained for **Raphael Bombelli** of Bologna, who published in 1572 an algebra of great merit, to point out the reality of the apparently imaginary expression which a root assumes, also to assign its value, when rational, and thus to lay the foundation of a more intimate knowledge of imaginary quantities. Cardan was an inveterate gambler. In 1663 there was published posthumously his gambler's manual, *De ludo aleæ*,

[1] H. Hankel, *op. cit.*, p. 368.

which contains discussions relating to the chances favorable for throwing a particular number with two dice and also with three dice. Cardan considered another problem in probabilities. Stated in general terms, the problem is: What is the proper division of a stake between two players, if the game is interrupted and one player has taken s_1 points, the other s_2 points, s points being required to win.[1] Cardan gives the ratio $(1+2+\ldots+[s-s_2])/(1+2+\ldots+[s-s_1])$, Tartaglia gives $(s+s_1-s_2)/(s+s_2-s_1)$. Both of these answers are wrong. Cardan considered also what later became known as the "Petersburg problem."

After the brilliant success in solving equations of the third and fourth degrees, there was probably no one who doubted, that with aid of irrationals of higher degrees, general equations of any degree whatever could be solved. But all attempts at the algebraic solution of the quintic were fruitless, and, finally, Abel demonstrated that all hopes of finding algebraic solutions to general equations of higher than the fourth degree were purely Utopian.

Since no solution by radicals of equations of higher degrees could be found, it seemed nothing remained to be done but the devising of processes by which the real roots of numerical equations could be found by approximation. The Chinese method used by them as early as the thirteenth century was unknown in the Occident. We have seen that in the early part of the thirteenth century Leonardo of Pisa solved a cubic to a high degree of approximation, but we are ignorant of his method. The earliest known process in the Occident of approaching to a root of an affected numerical equation was invented by Nicolas Chuquet, who, in 1484 at Lyons, wrote a work of high rank, entitled *Le triparty en la science des nombres*. It was not printed until 1880.[2] If $\frac{a}{c}>x<\frac{b}{d}$, then Chuquet takes the intermediate value $\frac{a+b}{c+d}$ as a closer approximation to the root x. He finds a series of successive intermediate values. We stated earlier that in 1498 the Arabic writer Miram Chelebi gave a method of solving $x^3+Q=Px$ which he attributes to Atabeddin Jamshid. This cubic arose in the computation of $x=\sin 1^\circ$.

The earliest printed method of approximation to the roots of affected equations is that of Cardan, who gave it in the *Ars Magna*, 1545, under the title of *regula aurea*. It is a skilful application of the rule of "false position," and is applicable to equations of any degree. This mode of approximation was exceedingly rough, yet this fact hardly explains why Clavius, Stevin and Vieta did not refer to it.

[1] M. Cantor, II, 2 Aufl., 1900, pp. 501, 520, 537.

[2] Printed in the *Bulletino Boncompagni*, T xiii, 1880; see pp. 653–654. See also F. Cajori, "A History of the Arithmetical Methods of Approximation to the Roots of Numerical Equations of one Unknown Quantity" in *Colorado College Publication*, General Series Nos. 51 and 52, 1910.

Processes of approximation were given by the Frenchman J. Peletier (1554), the Italian R. Bombelli (1572), the German R. Ursus (1601), the Swiss Joost Bürgi, the German Pitiscus (1612), and the Belgian Simon Stevin. But far more important than the processes of these men was that of the Frenchman, **Francis Vieta** (1540–1603), which initiates a new era. It is contained in a work published at Paris in 1600 by Marino Ghetaldi as editor, with Vieta's consent, under the title: *De numerosa protestatum purarum atque adfectarum ad exegesin resolutione tractatus*. His method is not of the nature of the rule of "double false position," used by Cardan and Bürgi, but resembles the method of ordinary root-extraction. Taking $f(x)=k$, where k is taken positive, Vieta separates the required root from the rest, then substitutes an approximate value for it and shows that another figure of the root can be obtained by division. A repetition of this process gives the next figure, and so on. Thus, in $x^5-5x^3+500x=7905504$, he takes $r=20$, then computes $7905504-r^5+5r^3-500r$ and divides the result by a value which in our modern notation takes the form $|(f(r+s^1)-f(r))|-s_1^n$, where n is the degree of the equation and s_1 is a unit of the denomination of the digit next to be found. Thus, if the required root is 243, and r has been taken to be 200, then s_1 is 10; but if r is taken as 240, then s_1 is 1. In our example, where $r=20$, the divisor is 878295, and the quotient yields the next digit of the root equal to 4. We obtain $x=20+4=24$, the required root. Vieta's procedure was greatly admired by his contemporaries, particularly the Englishmen, T. Harriot, W. Oughtred and J. Wallis, each of whom introduced some minor improvements.

We pause a moment to sketch the life of Vieta, the most eminent French mathematician of the sixteenth century. He was born in Poitou and died at Paris. He was employed throughout life in the service of the state, under Henry III. and Henry IV. He was, therefore, not a mathematician by profession, but his love for the science was so great that he remained in his chamber studying, sometimes several days in succession, without eating and sleeping more than was necessary to sustain himself. So great devotion to abstract science is the more remarkable, because he lived at a time of incessant political and religious turmoil. During the war against Spain, Vieta rendered service to Henry IV by deciphering intercepted letters written in a species of cipher, and addressed by the Spanish Court to their governor of Netherlands. The Spaniards attributed the discovery of the key to magic.

In 1579 Vieta published his *Canon mathematicus seu ad triangula cum appendicibus*, which contains very remarkable contributions to trigonometry. It gives the first systematic elaboration in the Occident of the methods of computing plane and spherical triangles by the aid of the six trigonometric functions.[1] He paid special attention

[1] A. v. Braunmühl, *Geschichte der Trigonometrie*, I, Leipzig, 1900, p. 160.

also to goniometry, developing such relations as $\sin \alpha = \sin (60° + \alpha) - \sin (60° - \alpha)$, $csc\, \alpha + ctn\, \alpha = \text{ctn}\frac{\alpha}{2}$, $-ctn\, \alpha + csc\, \alpha = \tan\frac{\alpha}{2}$, with the aid of which he could compute from the functions of angles below 30° or 45°, the functions of the remaining angles below 90°, essentially by addition and subtraction alone. Vieta is the first to apply algebraic transformation to trigonometry, particularly to the multisection of angles. Letting $2 \cos \alpha = x$, he expresses $\cos n\alpha$ as a function of x for all integers $n < 11$; letting $2 \sin \alpha = x$ and $2 \sin 2\alpha = y$, he expresses $2x^{n-2} \sin n\alpha$ in terms of x and y. Vieta exclaims: "Thus the analysis of angular sections involves geometric and arithmetic secrets which hitherto have been penetrated by no one."

An ambassador from Netherlands once told Henry IV that France did not possess a single geometer capable of solving a problem propounded to geometers by a Belgian mathematician, Adrianus Romanus. It was the solution of the equation of the forty-fifth degree:—

$$45y - 3795y^3 + 95634y^5 - \ldots + 945y^{41} - 45y^{43} + y^{45} = C.$$

Henry IV called Vieta, who, having already pursued similar investigations, saw at once that this awe-inspiring problem was simply the equation by which $C = 2 \sin \phi$ was expressed in terms of $y = 2 \sin \frac{1}{45} \phi$; that, since $45 = 3.3.5$, it was necessary only to divide an angle once into 5 equal parts, and then twice into 3,—a division which could be effected by corresponding equations of the fifth and third degrees. Brilliant was the discovery by Vieta of 23 roots to this equation, instead of only one. The reason why he did not find 45 solutions, is that the remaining ones involve negative sines, which were unintelligible to him. Detailed investigations on the famous old problem of the section of an angle into an odd number of equal parts, led Vieta to the discovery of a trigonometrical solution of Cardan's irreducible case in cubics. He applied the equation $(2 \cos \frac{1}{3} \phi)^3 - 3\left(2 \cos \frac{1}{3}\phi\right) = 2 \cos\phi$ to the solution of $x^3 - 3a^2x = a^2b$, when $a > \frac{1}{2}b$, by placing $x = 2a \cos\frac{1}{3}\phi$, and determining ϕ from $b = 2a \cos\phi$.

The main principle employed by him in the solution of equations is that of *reduction*. He solves the quadratic by making a suitable substitution which will remove the term containing x to the first degree. Like Cardan, he reduces the general expression of the cubic to the form $x^3 + mx + n = 0$; then, assuming $x = (\frac{1}{3}a - z^2) \div z$ and substituting, he gets $z^6 - bz^3 - \frac{1}{27}a^3 = 0$. Putting $z^3 = y$, he has a quadratic. In the solution of bi-quadratics, Vieta still remains true to his principle of reduction. This gives him the well-known cubic resolvent. He thus adheres throughout to his favorite principle, and thereby introduces into algebra a uniformity of method which claims our lively admiration. In Vieta's algebra we discover a partial knowledge of the relations existing between the coefficients and the roots of an equation. He shows that if the coefficient of the second term in an equation of the second degree is minus the sum of two numbers whose product is the third term, then the two numbers are roots of the equation. Vieta rejected all except positive roots; hence it was impossible for him to fully perceive the relations in question.

The most epoch-making innovation in algebra due to Vieta is the denoting of general or indefinite quantities by letters of the alphabet. To be sure, Regiomontanus and Stifel in Germany, and Cardan in Italy, used letters before him, but Vieta extended the idea and first made it an essential part of algebra. The new algebra was called by him *logistica speciosa* in distinction to the old *logistica numerosa*. Vieta's formalism differed considerably from that of to-day. The equation $a^3+3a^2b+3ab^2+b^3=(a+b)^3$ was written by him "A cubus $+B$ in A quadr. $3+A$ in B quadr. $3+B$ cubo æqualia $\overline{A+B}$ cubo." In numerical equations the unknown quantity was denoted by N, its square by Q, and its cube by C. Thus the equation $x^3-8x^2+16x=40$ was written $1\ C-8Q+16\ N$ *æqual.* 40. Vieta used the term "coefficient," but it was little used before the close of the seventeenth century.[1] Sometimes he uses also the term "polynomial." Observe that exponents and our symbol (=) for equality were not yet in use; but that Vieta employed the Maltese cross (+) as the short-hand symbol for addition, and the (−) for subtraction. These two characters had not been in very general use before the time of Vieta. "It is very singular," says Hallam, "that discoveries of the greatest convenience, and, apparently, not above the ingenuity of a village schoolmaster, should have been overlooked by men of extraordinary acuteness like Tartaglia, Cardan, and L. Ferrari; and, hardly less so that, by dint of that acuteness, they dispensed with the aid of these contrivances in which we suppose that so much of the utility of algebraic expression consists." Even after improvements in notation were once proposed, it was with extreme slowness that they were admitted into general use. They were made oftener by accident than design, and their authors had little notion of the effect of the change which they were making. The introduction of the + and − symbols seems to be due to the Germans, who, although they did not enrich algebra during the Renaissance with great inventions, as did the Italians, still cultivated it with great zeal. The arithmetic of **John Widmann,** brought out in 1489 in Leipzig, is the earliest printed book in which the + and −symbols have been found. The + sign is not restricted by him to ordinary addition; it has the more general meaning "et" or "and" as in the heading, "regula augmenti + decrementi." The − sign is used to indicate subtraction, but not regularly so. The word "plus" does not occur in Widmann's text; the word "minus" is used only two or three times. The symbols + and − are used regularly for addi-

[1] *Encyclopédie des sciences mathématiques*, Tome I, Vol. 2, 1907, p. 2.

tion and subtraction, in 1521,[1] in the arithmetic of *Grammateus*, (Heinrich Schreiber, died 1525) a teacher at the University of Vienna. His pupil, **Christoff Rudolff,** the writer of the first text-book on algebra in the German language (printed in 1525), employs these symbols also. So did Stifel, who brought out a second edition of Rudolff's *Coss* in 1553. Thus, by slow degrees, their adoption became universal. Several independent paleographic studies of Latin manuscripts of the fourteenth and fifteenth centuries make it almost certain that the sign + comes from the Latin *et*, as it was cursively written in manuscripts just before the time of the invention of printing.[2] The origin of the sign − is still uncertain. There is another short-hand symbol of which we owe the origin to the Germans. In a manuscript published sometime in the fifteenth century, a dot placed before a number is made to signify the extraction of a root of that number. This dot is the embryo of our present symbol for the square root. Christoff Rudolff, in his algebra, remarks that "the radix quadrata is, for brevity, designated in his algorithm with the character √, as √4." Here the dot has grown into a symbol much like our own. This same symbol was used by *Michael Stifel*. Our sign of equality is due to **Robert Recorde** (1510–1558), the author of *The Whetstone of Witte* (1557), which is the first English treatise on algebra. He selected this symbol because no two things could be more equal than two parallel lines =. The sign ÷ for division was first used by *Johann Heinrich Rahn*, a Swiss, in his *Teutsche Algebra*, Zurich, 1659, and was introduced in England through Thomas Brancker's translation of Rahn's book, London, 1668.

Michael Stifel (1486?–1567), the greatest German algebraist of the sixteenth century, was born in Esslingen, and died in Jena. He was educated in the monastery of his native place, and afterwards became Protestant minister. The study of the significance of mystic numbers in Revelation and in Daniel drew him to mathematics. He studied German and Italian works, and published in 1544, in Latin, a book entitled *Arithmetica integra*. Melanchthon wrote a preface to it. Its three parts treat respectively of rational numbers, irrational numbers, and algebra. Stifel gives a table containing the numerical values of the binomial coefficients for powers below the 18th. He observes an advantage in letting a geometric progression correspond to an arithmetical progression, and arrives at the designation of integral powers by numbers. Here are the germs of the theory of exponents and of logarithms. In 1545 Stifel published an arithmetic in German. His edition of Rudolff's *Coss* contains rules for solving cubic equations, derived from the writings of Cardan.

[1] G. Eneström in *Bibliotheca mathematica*, 3. S., Vol. 9, 1908–09, pp. 155–157; Vol. 14, 1914, p. 278.

[2] For references see M. Cantor, *op. cit.*, Vol. II, 2. Ed., 1900, p. 231; J. Tropfke, *op. cit.*, Vol. I, 1902, pp. 133, 134.

We remarked above that Vieta discarded negative roots of equations. Indeed, we find few algebraists before and during the Renaissance who understood the significance even of negative quantities. Fibonacci seldom uses them. Pacioli states the rule that "minus times minus gives plus," but applies it really only to the development of the product of $(a-b)$ $(c-d)$; purely negative quantities do not appear in his work. The German "Cossist" (algebraist), *Michael Stifel*, speaks as early as 1544 of numbers which are "absurd" or "fictitious below zero," and which arise when "real numbers above zero" are subtracted from zero. Cardan, at last, speaks of a "pure minus"; "but these ideas," says H. Hankel, "remained sparsely, and until the beginning of the seventeenth century, mathematicians dealt exclusively with absolute positive quantities." One of the first algebraists who occasionally place a purely negative quantity by itself on one side of an equation, is *T. Harriot* in England. As regards the recognition of negative roots, Cardan and Bombelli were far in advance of all writers of the Renaissance, including Vieta. Yet even they mentioned these so-called false or fictitious roots only in passing, and without grasping their real significance and importance. On this subject Cardan and Bombelli had advanced to about the same point as had the Hindu Bhāskara, who saw negative roots, but did not approve of them. The generalization of the conception of quantity so as to include the negative, was an exceedingly slow and difficult process in the development of algebra.

We shall now consider the history of geometry during the Renaissance. Unlike algebra, it made hardly any progress. The greatest gain was a more intimate knowledge of Greek geometry. No essential progress was made before the time of Descartes. Regiomontanus, Xylander (Wilhelm Holzmann, 1532–1576) of Augsburg, Tartaglia, Federigo Commandino (1509–1575) of Urbino in Italy, Maurolycus and others, made translations of geometrical works from the Greek. The description and instrumental construction of a new curve, the epicycloid, is explained by **Albrecht Dürer** (1471–1528), the celebrated painter and sculptor of Nürnberg, in a book, *Underweysung der Messung mit dem Zyrkel und rychtscheyd*, 1525. The idea of such a curve goes back at least as far as Hipparchus who used it in his astronomical theory of epicycles. The epicycloid does not again appear in history until the time of G. Desargues and P. La Hire. Dürer is the earliest writer in the Occident to call attention to magic squares. A simple magic square appears in his celebrated painting called "Melancholia."

Johannes Werner (1468–1528) of Nürnberg published in 1522 the first work on conics which appeared in Christian Europe.✡ Unlike the geometers of old, he studied the sections in relation with the cone, and derived their properties directly from it. This mode of studying the conics was followed by **Franciscus Maurolycus** (1494–1575) of Messina. The latter is, doubtless, the greatest geometer of the sixteenth century. From the notes of Pappus, he attempted to restore the missing fifth book of Apollonius on *maxima* and *minima*. His chief work is his masterly and original treatment of the conic sections, wherein he discusses tangents and asymptotes more fully than Apollonius had done, and applies them to various physical and astronomical problems. To Maurolycus has been ascribed also the discovery of the inference by mathematical induction.[1] It occurs in his introduction to his *Opuscula mathematica*, Venice, 1575. Later, mathematical induction was used by Pascal in his *Traité du triangle arithmétique* (1662). Processes akin to mathematical induction, some of which would yield the modern mathematical induction by introducing some slight change in the mode of presentation or in the point of view, were given before Maurolycus. *Giovanni Campano* (latinized form, Campanus) of Novara in Italy, in his edition of Euclid (1260), proves the irrationality of the golden section by a recurrent mode of inference resulting in a reductio ad absurdum. But he does not descend by a regular progression from n to n−1, n−2, etc., but leaps irregularly over, perhaps, several integers. Campano's process was used later by Fermat. A recurrent mode of inference is found in Bhāskara's "cyclic method" of solving indeterminate equations, in Theon of Smyrna (about 130 A. D.) and in Proclus's process for finding numbers representing the sides and diagonals of squares; it is found in Euclid's proof (*Elements* IX, 20) that the number of primes is infinite.

The foremost geometrician of Portugal was **Pedro Nunes**[2] (1502–1578) or *Nonius*. He showed that a ship sailing so as to make equal angles with the meridians does not travel in a straight line, nor usually along the arc of a great circle, but describes a path called the loxodromic curve. Nunes invented the "nonius" and described it in his *De crepusculis*, Lisbon, 1542. It consists in the juxtaposition of equal arcs, one arc divided into m equal parts and the other into $m+1$ equal parts. Nonius took $m=89$. The instrument is also called a "vernier," after the Frenchman *Pierre Vernier*, who re-invented it in 1631. The foremost French mathematician before Vieta was **Peter Ramus** (1515–1572), who perished in the massacre of St. Bartholomew. *Vieta* possessed great familiarity with ancient geometry. The new form which he gave to algebra, by representing general quantities by letters, enabled him to point out more easily how the construction of the roots of cubics depended upon the celebrated ancient problems of the duplication of the cube and the trisection of an angle. He reached the interesting conclusion that the former problem includes the solutions of all cubics in which the radical in Tartaglia's formula is real, but that the latter problem includes only those leading to the irreducible case.

[1] G. Vacca in *Bulletin Am. Math. Society*, 2. S., Vol. 16, 1909, p. 70. See also F. Cajori in Vol. 15, pp. 407–409.

[2] See R. Guimarães, *Pedro Nunes*, Coïmbre, 1915.

The problem of the quadrature of the circle was revived in this age, and was zealously studied even by men of eminence and mathematical ability. The army of circle-squarers became most formidable during the seventeenth century. Among the first to revive this problem was the German Cardinal **Nicolaus Cusanus** (1401–1464), who had the reputation of being a great logician. His fallacies were exposed to full view by Regiomontanus. As in this case, so in others, every quadrator of note raised up an opposing mathematician: Oronce Fine was met by Jean Buteo (c. 1492–1572) and P. Nunes; Joseph Scaliger by Vieta, Adrianus Romanus, and Clavius; a Quercu by Adriaen Anthonisz (1527–1607). Two mathematicians of Netherlands, **Adrianus Romanus** (1561–1615) and **Ludolph van Ceulen** (1540–1610), occupied themselves with approximating to the ratio between the circumference and the diameter. The former carried the value π to 15, the latter to 35, places. The value of π is therefore often named "Ludolph's number." His performance was considered so extraordinary, that the numbers were cut on his tomb-stone (now lost) in St. Peter's churchyard, at Leyden. These men had used the Archimedian method of in- and circum-scribed polygons, a method refined in 1621 by *Willebrord Snellius* (1580–1626) who showed how narrower limits may be obtained for π without increasing the number of sides of the polygons. Snellius used two theorems equivalent to $\frac{1}{3}$ $(2 \sin\theta \tan\theta) \angle \theta \angle 3/(2 \csc\theta+\cot\theta)$. The greatest refinements in the use of the geometrical method of Archimedes were reached by C. Huyghens in his *De circuli magnitudine inventa*, 1654, and by **James Gregory** (1638–1675), professor at St. Andrews and Edinburgh, in his *Exercitationes geometricæ*, 1668, and *Vera circuli et hyperbolae quadratura*, 1667. Gregory gave several formulas for approximating to π and in the second of these publications boldly attempted to prove by the Archimedean algorithm that the quadrature of the circle is impossible. Huyghens showed that Gregory's proof is not conclusive, although he himself believed that the quadrature is impossible. Other attempts to prove this impossibility were made by Thomas Fautat De Lagny (1660–1734) of Paris, in 1727, Joseph Saurin (1659–1737) in 1720, Isaac Newton in his *Principia* I, 6, lemma 28, E. Waring, L. Euler, 1771.

That these proofs would lack rigor was almost to be expected, as long as no distinction was made between algebraical and transcendental numbers.

The earliest explicit expression for π by an infinite number of operations was found by Vieta. Considering regular polygons of 4, 8, 16, . . . sides, inscribed in a circle of unit radius, he found that the area of the circle is

$$2 \frac{1}{\sqrt{\frac{1}{2}}\sqrt{\frac{1}{2}+\frac{1}{2}\sqrt{\frac{1}{2}}}\sqrt{\frac{1}{2}+\frac{1}{2}\sqrt{\frac{1}{2}+\frac{1}{2}\sqrt{2}\ldots}}},$$

from which we obtain

$$\frac{\pi}{2}=\frac{1}{\sqrt{\frac{1}{2}}\sqrt{\frac{1}{2}+\frac{1}{2}\sqrt{\frac{1}{2}\ldots}}}$$, which may be derived from Euler's formula [1]

$$\theta=\frac{\sin\theta}{\cos^{\theta}/2\,\cos^{\theta}/4\,\cos^{\theta}/8\ldots}$$, $(\theta\angle\pi)$, by taking $\theta=\pi/2$.

As mentioned earlier, it was **Adrianus Romanus** (1561–1615) of Louvain who propounded for solution that equation of the forty-fifth degree solved by Vieta. On receiving Vieta's solution, he at once departed for Paris, to make his acquaintance with so great a master. Vieta proposed to him the Apollonian problem, to draw a circle touching three given circles. "Adrianus Romanus solved the problem by the intersection of two hyperbolas; but this solution did not possess the rigor of the ancient geometry. Vieta caused him to see this, and then, in his turn, presented a solution which had all the rigor desirable." [2] Romanus did much toward simplifying spherical trigonometry by reducing, by means of certain projections, the 28 cases in triangles then considered to only six.

Mention must here be made of the improvements of the Julian calendar. The yearly determination of the movable feasts had for a long time been connected with an untold amount of confusion. The rapid progress of astronomy led to the consideration of this subject, and many new calendars were proposed. Pope Gregory XIII convoked a large number of mathematicians, astronomers, and prelates, who decided upon the adoption of the calendar proposed by the Jesuit **Christophorus Clavius** (1537–1612) of Rome. To rectify the errors of the Julian calendar it was agreed to write in the new calendar the 15th of October immediately after the 4th of October of the year 1582. The Gregorian calendar met with a great deal of opposition both among scientists and among Protestants. Clavius, who ranked high as a geometer, met the objections of the former most ably and effectively; the prejudices of the latter passed away with time.

The passion for the study of mystical properties of numbers descended from the ancients to the moderns. Much was written on numerical mysticism even by such eminent men as Pacioli and Stifel. The *Numerorum Mysteria* of Peter Bungus covered 700 quarto pages. He worked with great industry and satisfaction on 666, which is the number of the beast in Revelation (xiii, 18), the symbol of Antichrist. He reduced the name of the "impious" Martin Luther to a form which may express this formidable number. Placing $a=1$, $b=2$, etc., $k=10$, $l=20$, etc., he finds, after misspelling the name, that $M_{(30)}A_{(1)}R_{(80)}T_{(100)}I_{(9)}N_{(40)}L_{(20)}V_{(200)}T_{(100)}E_{(5)}R_{(80)}A_{(1)}$ constitutes the number required. These attacks on the great reformer were not unprovoked, for his

[1] E. W. Hobson, *Squaring the Circle*, Cambridge, 1913, pp. 26, 27, 31.

[2] A. Quetelet, *Histoire des Sciences mathématiques et physiques chez les Belges.* Bruxelles, 1864, p. 137.

friend, Michael Stifel, the most acute and original of the early mathematicians of Germany, exercised an equal ingenuity in showing that the above number referred to Pope Leo X,—a demonstration which gave Stifel unspeakable comfort.[1]

Astrology also was still a favorite study. It is well known that Cardan, Maurolycus, Regiomontanus, and many other eminent scientists who lived at a period even later than this, engaged in deep astrological study; but it is not so generally known that besides the occult sciences already named, men engaged in the mystic study of star-polygons and magic squares. "The pentagramma gives you pain," says Faust to Mephistopheles. It is of deep psychological interest to see scientists, like the great Kepler, demonstrate on one page a theorem on star-polygons, with strict geometric rigor, while on the next page, perhaps, he explains their use as amulets or in conjurations. Playfair, speaking of Cardan as an astrologer, calls him "a melancholy proof that there is no folly or weakness too great to be united to high intellectual attainments." [2] Let our judgment not be too harsh. The period under consideration is too near the Middle Ages to admit of complete emancipation from mysticism even among scientists. Scholars like Kepler, Napier, Albrecht Dürer, while in the van of progress and planting one foot upon the firm ground of truly scientific inquiry, were still resting with the other foot upon the scholastic ideas of preceding ages.

Vieta to Descartes

The ecclesiastical power, which in the ignorant ages was an unmixed benefit, in more enlightened ages became a serious evil. Thus, in France, during the reigns preceding that of Henry IV, the theological spirit predominated. This is painfully shown by the massacres of Vassy and of St. Bartholomew. Being engaged in religious disputes, people had no leisure for science and for secular literature. Hence, down to the time of Henry IV, the French "had not put forth a single work, the destruction of which would now be a loss to Europe." In England, on the other hand, no religious wars were waged. The people were comparatively indifferent about religious strifes; they concentrated their ability upon secular matters, and acquired, in the sixteenth century, a literature which is immortalized by the genius of Shakespeare and Spenser. This great literary age in England was followed by a great scientific age. At the close of the sixteenth century, the shackles of ecclesiastical authority were thrown off by France. The ascension of Henry IV to the throne was followed in 1598 by the Edict of Nantes, granting freedom of worship to the Huguenots, and thereby terminating religious wars. The genius of the French nation

[1] G. Peacock, *op. cit.*, p. 424.

[2] John Playfair, "Progress of the Mathematical and Physical Sciences" in *Encyclopædia Britannica*, 7th ed., continued in 8th Ed., by Sir John Leslie.

now began to blossom. Cardinal Richelieu, during the reign of Louis XIII, pursued the broad policy of not favoring the opinions of any sect, but of promoting the interests of the nation. His age was remarkable for the progress of knowledge. It produced that great secular literature, the counterpart of which was found in England in the sixteenth century. The seventeenth century was made illustrious also by the great French mathematicians, Roberval, Descartes, Desargues, Fermat, and Pascal.

More gloomy is the picture in Germany. The great changes which revolutionized the world in the sixteenth century, and which led England to national greatness, led Germany to degradation. The first effects of the Reformation there were salutary. At the close of the fifteenth and during the sixteenth century, Germany had been conspicuous for her scientific pursuits. She had been a leader in astronomy and trigonometry. Algebra also, excepting for the discoveries in cubic equations, was, before the time of Vieta, in a more advanced state there than elsewhere. But at the beginning of the seventeenth century, when the sun of science began to rise in France, it set in Germany. Theologic disputes and religious strife ensued. The Thirty Years' War (1618–1648) proved ruinous. The German empire was shattered, and became a mere lax confederation of petty despotisms. Commerce was destroyed; national feeling died out. Art disappeared, and in literature there was only a slavish imitation of French artificiality. Nor did Germany recover from this low state for 200 years; for in 1756 began another struggle, the Seven Years' War, which turned Prussia into a wasted land. Thus it followed that at the beginning of the seventeenth century, the great Kepler was the only German mathematician of eminence, and that in the interval of 200 years between Kepler and Gauss, there arose no great mathematician in Germany excepting Leibniz.

Up to the seventeenth century, mathematics was cultivated but little in Great Britain. During the sixteenth century, she brought forth no mathematician comparable with Vieta, Stifel, or Tartaglia. But with the time of Recorde, the English became conspicuous for numerical skill. The first important arithmetical work of English authorship was published in Latin in 1522 by **Cuthbert Tonstall** (1474–1559). He had studied at Oxford, Cambridge, and Padua, and drew freely from the works of Pacioli and Regiomontanus. Reprints of his arithmetic appeared in England and France. After Recorde the higher branches of mathematics began to be studied. Later, Scotland brought forth John Napier, the inventor of logarithms. The instantaneous appreciation of their value is doubtless the result of superiority in calculation. In Italy, and especially in France, geometry, which for a long time had been an almost stationary science, began to be studied with success. Galileo, Torricelli, Roberval, Fermat, Desargues, Pascal, Descartes, and the English Wallis are the great revolutioners of this science. Theoretical mechanics began to be studied. The foundations were laid by Fermat and Pascal for the theory of numbers and the theory of probability.

We shall first consider the improvements made in the art of calculating. The nations of antiquity experimented thousands of years upon numeral notations before they happened to strike upon the so-called "Arabic notation." In the simple expedient of the cipher, which was permanently introduced by the Hindus, mathematics received one of the most powerful impulses. It would seem that after the "Arabic notation" was once thoroughly understood, decimal fractions would occur at once as an obvious extension of it. But "it is curious to think how much science had attempted in physical research and how deeply numbers had been pondered, before it was perceived that the all-powerful simplicity of the 'Arabic notation' was as valuable and as manageable in an infinitely descending as in an infinitely ascending progression." [1] Simple as decimal fractions appear to us, the invention of them is not the result of one mind or even of one age. They came into use by almost imperceptible degrees. The first mathematicians identified with their history did not perceive their true nature and importance, and failed to invent a suitable notation. The idea of decimal fractions makes its first appearance in methods for approximating to the square roots of numbers. Thus John of Seville, presumably in imitation of Hindu rules, adds $2n$ ciphers to the number, then finds the square root, and takes this as the numerator of a fraction whose denominator is 1 followed by n ciphers. The same method was followed by *Cardan*, but it failed to be generally adopted even by his Italian contemporaries; for otherwise it would certainly have been at least mentioned by *Pietro Cataldi* (died 1626) in a work devoted exclusively to the extraction of roots. Cataldi, and before him Bombelli in 1572, find the square root by means of continued fractions—a method ingenious and novel, but for practical purposes inferior to Cardan's. *Oronce Fine* (1494–1555) in France (called also Orontius Finaeus), and **William Buckley** (died about 1550) in England extracted the square root in the same way as Cardan and John of Seville. The invention of decimals has been frequently attributed to Regiomontanus, on the ground that instead of placing the sinus totus, in trigonometry, equal to a multiple of 60, like the Greeks, he put it=100,000. But here the trigonometrical lines were expressed in *integers*, and not in fractions. Though he adopted a decimal division of the radius, he and his successors did not apply the idea outside of trigonometry and, indeed, had no notion whatever of decimal *fractions*. To **Simon Stevin** (1548–1620) of Bruges in Belgium, a man who did a great deal of work in most diverse fields of science, we owe the first systematic treatment of decimal fractions. In his *La Disme* (1585) he describes in very express

[1] Mark Napier, *Memoirs of John Napier of Merchiston*. Edinburgh, 1834.

terms the advantages, not only of decimal fractions, but also of the decimal division in systems of weights and measures. Stevin applied the new fractions "to all the operations of ordinary arithmetic."[1] What he lacked was a suitable notation. In place of our decimal point, he used a cipher; to each place in the fraction was attached the corresponding index. Thus, in his notation, the number 5.912 would be $\overset{0}{5}\overset{1}{9}\overset{2}{1}\overset{3}{2}$ or 5⓪9①1②2③. These indices, though cumbrous in practice, are of interest, because they embody the notion of powers of numbers. Stevin considered also fractional powers. He says that "$\frac{2}{3}$" placed within a circle would mean $x^{2/3}$, but he does not actually use his notation. This notion had been advanced much earlier by Oresme, but it had remained unnoticed. Stevin found the greatest common divisor of x^3+x^2 and x^2+7x+6 by the process of continual division, thereby applying to polynomials Euclid's mode of finding the greatest common divisor of numbers, as explained in Book VII of his *Elements*. Stevin was enthusiastic not only over decimal fractions, but also over the decimal division of weights and measures. He considered it the duty of governments to establish the latter. He advocated the decimal subdivision of the degree. No improvement was made in the notation of decimals till the beginning of the seventeenth century. After Stevin, decimals were used by *Joost Bürgi* (1552–1632), a Swiss by birth, who prepared a manuscript on arithmetic soon after 1592, and by **Johann Hartmann Beyer,** who assumes the invention as his own. In 1603, he published at Frankfurt on the Main a *Logistica Decimalis*. Historians of mathematics do not yet agree to whom the first introduction of the decimal point or comma should be ascribed. Among the candidates for the honor are Pellos (1492), Bürgi (1592), Pitiscus (1608, 1612), Kepler (1616), Napier (1616, 1617). This divergence of opinion is due mainly to different standards of judgment. If the requirement made of candidates is not only that the decimal point or comma was actually used by them, but that they must give evidence that the numbers used were actually decimal fractions, that the point or comma was with them not merely a general symbol to indicate a separation, that they must actually use the decimal point in operations including multiplication or division of decimal fractions, then it would seem that the honor falls to John Napier, who exhibits such use in his *Rabdologia*, 1617. Perhaps Napier received the suggestion for this notation from Pitiscus who, according to G. Eneström,[2] uses the point in his *Trigonometria* of 1608 and 1612, not as a regular decimal point, but as a more general sign of separation. Napier's decimal point did not meet with immediate adoption. *W. Oughtred* in 1631 designates the fraction .56 thus, 0|$\underline{56}$. *Albert Girard*, a pupil of Stevin, in 1629 uses the point on one occasion. John Wallis in 1657 writes

[1] A. Quetelet, *op. cit.*, p. 158.

[2] *Bibliotheca mathematica*, 3. S., Vol. 6, 1905, p. 109.

12|345, but afterwards in his algebra adopts the usual point. A. De Morgan says that "to the first quarter of the eighteenth century we must refer not only the complete and final victory of the decimal point, but also that of the now universal method of performing the operations of division and extraction of the square root." [1] We have dwelt at some length on the progress of the decimal notation, because "the history of language . . . is of the highest order of interest, as well as utility: its suggestions are the best lesson for the future which a reflecting mind can have."

The miraculous powers of modern calculation are due to three inventions: the Arabic Notation, Decimal Fractions, and Logarithms. The invention of logarithms in the first quarter of the seventeenth century was admirably timed, for Kepler was then examining planetary orbits, and Galileo had just turned the telescope to the stars. During the Renaissance German mathematicians had constructed trigonometrical tables of great accuracy, but its greater precision enormously increased the work of the calculator. It is no exaggeration to say that the invention of logarithms "by shortening the labors doubled the life of the astronomer." Logarithms were invented by **John Napier** (1550–1617), Baron of Merchiston, in Scotland. It is one of the greatest curiosities of the history of science that Napier constructed logarithms before exponents were used. To be sure, Stifel and Stevin made some attempts to denote powers by indices, but this notation was not generally known,—not even to *T. Harriot*, whose algebra appeared long after Napier's death. That logarithms flow naturally from the exponential symbol was not observed until much later. What, then, was Napier's line of thought?

Let AB be a definite line, DE a line extending from D indefinitely. Imagine two points starting at the same moment; the one moving

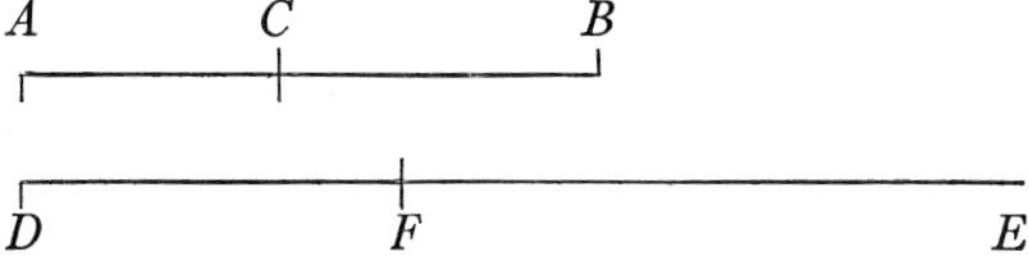

from A toward B, the other from D toward E. Let the velocity during the first moment be the same for both: let that of the point on line DE be uniform; but the velocity of the point on AB decreasing in such a way that when it arrives at any point C, its velocity is proportional to the remaining distance BC. While the first point moves over a distance AC, the second one moves over a distance DF. Napier calls DF the logarithm of BC.

He first sought the logarithms only of sines; the line AB was the sine of 90° and was taken $=10^7$; BC was the sine of the arc, and

[1] A. De Morgan, *Arithmetical Books from the Invention of Printing to the Present Time*, London, 1847, p. xxvii.

DF its logarithm. We notice that as the motion proceeds, BC decreases in geometrical progression, while DF increases in arithmetical progression. Let $AB=a=10^7$, let $x=DF$, $y=BC$, then $AC=a-y$. The velocity of the point C is $\frac{d(a-y)}{dt}=y$; this gives $-$nat. log $y=t+c$. When $t=0$, then $y=a$ and $c=-$nat. log a. Again, let $\frac{dx}{dt}=a$ be the velocity of the point F, then $x=at$. Substituting for t and c their values and remembering that $a=10^7$ and that by definition $x=$Nap. log y, we get

$$\text{Nap. log } y=10^7 \text{ nat. log } \frac{10^7}{y}.$$

It is evident from this formula that Napier's logarithms are not the same as the natural logarithms. Napier's logarithms increase as the number itself decreases. He took the logarithm of sin $90°=0$; *i. e.* the logarithm of $10^7=0$. The logarithm of sin α increased from zero as α decreased from $90°$. Napier's genesis of logarithms from the conception of two flowing points reminds us of Newton's doctrine of fluxions. The relation between geometric and arithmetical progressions, so skilfully utilized by Napier, had been observed by Archimedes, Stifel, and others. What was the base of Napier's system of logarithms? To this we reply that not only did the notion of a "base" never suggest itself to him, but it is inapplicable to his system. This notion demands that zero be the logarithm of 1; in Napier's system, zero is the logarithm of 10^7. Napier's great invention was given to the world in 1614 in a work entitled *Mirifici logarithmorum canonis descriptio*. In it he explained the nature of his logarithms, and gave a logarithmic table of the natural sines of a quadrant from minute to minute. In 1619 appeared Napier's *Mirifici logarithmorum canonis constructio*, as a posthumous work, in which his method of calculating logarithms is explained. An English translation of the *Constructio*, by W. R. Macdonald, appeared in Edinburgh, in 1889.

Henry Briggs (1556–1631), in Napier's time professor of geometry at Gresham College, London, and afterwards professor at Oxford, was so struck with admiration of Napier's book, that he left his studies in London to do homage to the Scottish philosopher. Briggs was delayed in his journey, and Napier complained to a common friend, "Ah, John, Mr. Briggs will not come." At that very moment knocks were heard at the gate, and Briggs was brought into the lord's chamber. Almost one-quarter of an hour was spent, each beholding the other without speaking a word. At last Briggs began: "My lord, I have undertaken this long journey purposely to see your person, and to know by what engine of wit or ingenuity you came first to think of this most excellent help in astronomy, viz. the logarithms; but, my lord, being by you found out, I wonder nobody found it out before, when now known it is so easy." Briggs suggested to Napier the advantage that would result from retaining zero for the logarithm of the whole sine, but choosing 10,000,000,000 for the logarithm of the 10th part of that same sine, *i. e.* of $5° 44' 22''$. Napier said that he had already thought of the change, and he pointed out a slight improvement on Briggs' idea; viz. that zero should be the logarithm of 1, and 10,000,000,000 that of the whole sine, thereby making the characteristic of numbers greater than unity positive and not negative, as suggested by Briggs. Briggs admitted this to be more convenient. The invention of "Briggian logarithms" occurred, therefore, to Briggs and Napier independently. The great practical advantage of the new system was that its fundamental progression was accommodated to the base, 10, of our numerical scale. Briggs devoted all his energies to the construction of tables upon the new plan. Napier died in 1617, with the satisfaction of having found in Briggs an able friend to bring to completion his unfinished plans. In 1624 Briggs published his *Arithmetica logarithmica*, containing the logarithms to 14 places of numbers, from 1 to 20,000 and from 90,000 to 100,000. The gap from 20,000 to 90,000 was filled up by that illustrious successor of Napier and Briggs, *Adrian Vlacq* (1600?–1667). He was born at Gouda in Holland and lived ten years in London as a bookseller and publisher. Being driven out by London bookdealers, he settled in Paris where he met opposition again, for selling foreign books. He died at The Hague. John Milton, in his *Defensio secunda*, published an abuse of him. Vlacq published in 1628 a table of logarithms from 1 to 100,000, of which 70,000 were calculated by himself. The first publication of Briggian logarithms of trigonometric functions was made in 1620 by **Edmund Gunter** (1581–1626) of London, a colleague of Briggs, who found the logarithmic sines and tangents for every minute to seven places. Gunter was the inventor of the words *cosine* and *cotangent* (1620).

The word *cosine* was an abbreviation of *complemental sine*. The invention of the words *tangent* and *secant* is due to the physician and mathematician, *Thomas Finck*, a native of Flensburg, who used them in his *Geometria rotundi*, Basel, 1583. Gunter is known to engineers for his "Gunter's chain." It is told of him that "When he was a student at Christ College, it fell to his lot to preach the Passion sermon, which some old divines that I knew did hear, but they said that it was said of him then in the University that our Savior never suffered so much since his passion as in that sermon, it was such a lamented one." [1] Briggs devoted the last years of his life to calculating more extensive Briggian logarithms of trigonometric functions, but he died in 1631, leaving his work unfinished. It was carried on by **Henry Gel-**

[1] Aubrey's *Brief Lives*, Edition A. Clark, 1898, Vol. I, p. 276.

librand (1597–1637) of Gresham College in London, and then published by Vlacq at his own expense. Briggs divided a degree into 100 parts, as was done also by N. Roe in 1633, W. Oughtred in 1657, John Newton in 1658, but owing to the publication by Vlacq of trigonometrical tables constructed on the old sexagesimal division, Briggs' innovation did not prevail. Briggs and Vlacq published four fundamental works, the results of which have not been superseded by any subsequent calculations until very recently.

The word "characteristic," as used in logarithms, first occurs in Briggs' *Arithmetica logarithmica*, 1624; the word "mantissa" was introduced by John Wallis in the Latin edition of his *Algebra*, 1693, p. 41, and was used by L. Euler in his *Introductio in analysin* in 1748, p. 85.

The only rival of John Napier in the invention of logarithms was the Swiss **Joost Bürgi** (1552–1632). He published a table of logarithms, *Arithmetische und Geometrische Progresstabulen*, Prague, 1620, but he conceived the idea and constructed his table independently of Napier. He neglected to have it published until Napier's logarithms were known and admired throughout Europe.

Among the various inventions of Napier to assist the memory of the student or calculator, is "Napier's rule of circular parts" for the solution of spherical right triangles. It is, perhaps, "the happiest example of artificial memory that is known." Napier gives in the *Descriptio* a proof of his rule; proofs were given later by Johann Heinrich Lambert (1765) and Leslie Ellis (1863).[1] Of the four formulas for oblique spherical triangles which are sometimes called "Napier's Analogies," only two are due to Napier himself; they are given in his *Constructio*. The other two were added by Briggs in his annotations to the *Constructio*.

A modification of Napier's logarithms was made by **John Speidell,** a teacher of mathematics in London, who published the *New Logarithmes*, London, 1619, containing the logarithms of sines, tangents and secants. Speidell did not advance a new theory. He simply aimed to improve on Napier's tables by making all logarithms positive. To achieve this end he subtracted Napier's logarithmic numbers from 10^8 and then discarded the last two digits. Napier gave *log sin* $30'=47413852$. Subtracting this from 10^8 leaves 52586148. Speidell wrote *log sin* $30'=525861$. It has been said that Speidell's logarithms of 1619 are logarithms to the natural base e. This is not quite true, on account of complications arising from the fact that the logarithms in Speidell's table appear as *integral* numbers and that the natural trigonometric values (not printed in Speidell's tables) are likewise written as integral numbers. If the last five figures in Speidell's logarithms are taken as decimals (mantissas), then the logarithms are the natural logarithms (with 10 added to every negative character-

[1] R. Mortiz, *Am. Math. Monthly*, Vol. 22, 1915, p. 221.

istic) of the trigonometric values, provided the latter are expressed decimally as ratios. For instance, Napier gives *sin* 30′=87265, the radius being 10^7. In reality, *sin* 30′=.0087265. The natural logarithm of this fraction is approximately $\bar{5}.25861$. Adding 10 gives 5.25861. As seen above, Speidell writes *log sin* 30′=525861. The relation between the natural logarithms and the logarithms in Speidell's trigonometric tables is shown by the formula, *Sp. log* $x=10^5\left(10+\log_e \frac{x}{10^5}\right)$. For secants and the latter half of the tangents the addition of 10 is omitted. In Speidell's table, *log tan* 89°=404812, the natural logarithm of *tan* 89° being 4.04812. In the 1622 edition of his *New Logarithmes*, Speidell included also a table of logarithms of the numbers 1–1000. Except for the omission of the decimal point, the logarithms in this table are genuinely natural logarithms. Thus, he gives *log* 10=2302584; in modern notation, $\log_e 10=2.302584$. J. W. L. Glaisher has pointed out [1] that these are not the earliest natural logarithms. The second (1618) edition of Edward Wright's translation of Napier's *Descriptio* contains an anonymous *Appendix*, very probably written by William Oughtred, describing a process of interpolation with the aid of a small table containing the logarithms of 72 sines. The latter are natural logarithms with the decimal point omitted. Thus, log 10=2302584, log 50=3911021. This *Appendix* is noteworthy also as containing the earliest account of the radix method of computing logarithms. After the time of Speidell no tables of natural logarithms were published until 1770, when J. H. Lambert inserted a seven place table of natural logarithms of the numbers 1–100 in his *Zusätze zu den Logarithmischen und Trigonometrischen Tabellen*. Most of the early methods of computing logarithms originated in England. Napier begins the computations of his logarithms of 1614 by forming a geometric progression of 101 terms, the first term being 10^7 and the common ratio $\left(1-\frac{1}{10^7}\right)$ and the last term 9,999,900.0004950. This progression constitutes the "First Table" given in his *Constructio*. Omitting the decimal part of the last term, he takes 9,999,900 as the second term of a new progression of 51 terms whose first term is 10^7, the common ratio being $\left(1-\frac{1}{10^5}\right)$ and the last term 9,995,001.222927 (should be 9,995001.224804). A third geometric progression of 21 terms has 10^7 as its first term, 9,995000 for its second term, the common ratio $\left(1-\frac{1}{2000}\right)$ and 9,900,473.57808 as its last term. This progression of 21 terms constitutes the first of 69 columns of numbers in Napier's

[1] *Quarterly Jour. of Pure & Appl. Math.*, Vol. 46, 1915, p. 145.

"Third Table." Each column is a geometric progression of 21 terms with $\left(1-\frac{1}{2000}\right)$ as the common ratio. The 69 first or top numbers in the 69 columns themselves constitute a geometric progression having the ratio $\left(1-\frac{1}{100}\right)$, the first top number being 10^7, the second 9900000, and so on. The last number in the 69th column is 4998609.4034. Thus this "Third Table" gives a series of numbers very nearly, but not exactly in geometrical progression, and lying between 10^7 and very nearly $\frac{1}{2}.10^7$. Says Hutton, these tables were "found in the most simple manner, by little more than easy subtractions." The numbers are taken as the sines of angles between 90° and 30°. Kinematical considerations yield him an upper and a lower limit for the logarithm of a given sine. By these limits he obtains the logarithm of each number in his "Third Table." To obtain the logarithms of sines between 0° and 30° Napier indicates two methods. By one of them he computes log sin θ, $15° < \theta < 30°$, by the aid of his "Third Table" and the formula $\sin 2\theta = 2 \sin\theta \sin(90° - \theta)$. A repetition of this process gives the logarithms of sines down to $\theta = 7° \, 30'$, and so on.

Bürgi's method of computation was more primitive than Napier's. In his table the logarithms were printed in red and were called "red numbers"; the antilogarithms were in black. The expressions $r_n = 10n$, $b_n = b_{n-1}\left(1+\frac{1}{10^4}\right)$, where $r_0 = 0$, $b_0 = 100{,}000{,}000$, and $n = 1, 2, 3, \ldots$, indicate the mode of computation. Any term b_n of the geometric series is obtained by *adding* to the preceding term b_{n-1}, the $\frac{1}{10^4}$th part of that term. Proceeding thus Bürgi arrives at $r = 230{,}270{,}022$ and $b = 1{,}000{,}000{,}000$, this last pair of numbers being obtained by interpolation.

In the Appendix to the *Constructio* there are described three methods of computing logarithms which are probably the result of the joint labors of Napier and Briggs. The first method rests on the successive extractions of fifth roots. The second calls for square roots only. Taking log 1=0 and log $10 = 10^{10}$, find the logarithm of the mean proportion between 1 and 10. There follows $\log\sqrt{1\times 10} = \log 3.16227766017 = \frac{1}{2}(10^{10})$; then $\log\sqrt{10\times 3.16227766017} = \log 5.62341325191 = \frac{3}{4}(10^{10})$, and so on. Substantially this method was used by Kepler in his book on logarithms of 1624 and by Vlacq. The third method in the Appendix to the *Constructio* lets log 1=0, log $10 = 10^{10}$, and takes 2 as a factor 10^{10} times, yielding a number composed of 301029996 figures; hence log 2=0,301029996.

A famous method of computing logarithms is the so-called "radix method." It requires the aid of a table of radices or numbers of the form $1 \pm \frac{r}{10^n}$, with their logarithms. The logarithm of a number is found by resolving the number into factors of the form $1 \pm \frac{r}{10^n}$ and then adding the logarithms of the factors. The earliest appearance of this method is in the anonymous "Appendix" (very probably due to Oughtred) to Edward Wright's 1618 edition of Napier's *Descriptio*.[1] It is fully developed by Briggs who, in his *Arithmetica logarithmica*, 1624, gives a table of radices. The method has been frequently rediscovered and given in various forms.[2] A slight simplification of Briggs' process was given as one of three methods by *Robert Flower* in a tract, *The Radix a new way of making Logarithms*, London, 1771. He divides a given number by a power of 10 and a single digit, so as to reduce the first figure to .9, and then multiplies by a procession of radices until all the digits become nines. The radix method was rediscovered in 1786 by *George Atwood* (1746–1807), the inventor of "Atwood's machine," in *An essay on the Arithmetic of Factors*, and again by *Zecchini Leonelli* in 1802, by Thomas Manning (1772–1840), scholar of Caius College, Cambridge, in 1806, by Thomas Weddle in 1845, Hearn in 1847 and Orchard in 1848. Extensions and variations of the radix method have been published by Peter Gray (1807?–1887), a writer on life contingencies, Thoman, A. J. Ellis (1814–1890), and others. The three distinct methods of its application are due to Briggs, Flower and Weddle.

Another method of computing common logarithms is by the repeated formation of geometric means. If $A=1$, $B=10$, then $C=\sqrt{AB}=3.162278$ has the logarithm .5, $D=\sqrt{BC}=5.623413$ has the logarithm .75, etc. Perhaps suggested by Napier's remarks in the *Constructio*, this method was developed by French writers, of whom Jacques Ozanam (1640–1717) in 1670 was perhaps the first.[3] Ozanam is best known for his *Récréations mathématiques et physiques*, 1694.

Still different devices for the computation of logarithms were invented by Brook Taylor (1717), John Long (1714), William Jones, Roger Cotes (1722), Andrew Reid (1767), James Dodson (1742), Abel Bürja (1786), and others.[4]

[1] J. W. L. Glaisher, in *Quarterly Jour. of Math's*, Vol. 46, 1915, p. 125.

[2] For the detailed history of this method consult also A. J. Ellis in *Proceedings of the Royal Society* (London), Vol. 31, 1881, pp. 398–413; S. Lupton, *Mathematical Gazette*, Vol. 7, 1913, pp. 147–150, 170–173; Ch. Hutton's Introduction to his *Mathematical Tables*.

[3] See J. W. L. Glaisher in *Quarterly Journal of Pure and Appl. Math's*, Vol. 47, 1916, pp. 249–301.

[4] For details see Ch. Hutton's Introduction to his *Mathematical Tables*, also the *Encyclopédie des sciences mathématiques*, 1908; I, 23, "Tables de logarithmes."

After the labor of computing logarithms was practically over, the facile methods of computing by infinite series came to be discovered. James Gregory, Lord William Brounker (1620–1684), Nicholas Mercator (1620–1687), John Wallis and Edmund Halley are the pioneer workers. Mercator in 1668 derived what amounts to the infinite series for log $(1+a)$. Transformations of this series yielded rapidly converging results. James Gregory in 1668 and Edmund Halley in 1695 obtained substantially the series $\frac{1}{2}\log(1+z)-\frac{1}{2}\log(1-z)$. G. Vega in his *Thesaurus* of 1794 lets $z=1/(2y^2-1)$.

The theoretic view-point of the logarithm was broadened somewhat during the seventeenth century by the graphic representation, both in rectangular and polar coördinates, of a variable and its variable logarithm. Thus were invented the logarithmic curve and the logarithmic spiral. It has been thought that the earliest reference to the logarithmic curve was made by the Italian Evangelista Torricelli in a letter of the year 1644, but Paul Tannery made it practically certain that Descartes knew the curve in 1639.[1] Descartes described the logarithmic spiral in 1638 in a letter to P. Mersenne, but does not give its equation, nor connect it with logarithms. He describes it as the curve which makes equal angles with all the radii drawn through the origin. The name "logarithmic spiral" was coined by Jakob Bernoulli in a paper published in the *Acta eruditorum* in 1691.

The most brilliant conquest in algebra during the sixteenth century had been the solution of cubic and biquadratic equations. All attempts at solving algebraically equations of higher degrees remaining fruitless, a new line of inquiry—the properties of equations and their roots—was gradually opened up. We have seen that Vieta had attained a partial knowledge of the relations between roots and coefficients. *Jacques Peletier* (1517–1582), a French man of letters, poet and mathematician, had observed as early as 1558, that the root of an equation is a divisor of the last term. In passing he writes equations with all terms on one side, and equated to zero. This was done also by Buteo and Harriot. One who extended the theory of equations somewhat further than Vieta, was **Albert Girard** (1590?–1633?), a mathematician of Lorraine. Like Vieta, this ingenious author applied algebra to geometry, and was the first who understood the use of negative roots in the solution of geometric problems. He spoke of imaginary quantities, inferred by induction that every equation has as many roots as there are units in the number expressing its degree, and first showed how to express the sums of their powers in terms of the coefficients. Another algebraist of considerable power was the English **Thomas Harriot** (1560–1621). He accompanied the first colony sent out by Sir Walter Raleigh to Virginia. After having sur-

[1] See G. Loria, *Bibliotheca math.*, 3. S., Vol. 1, 1900, p. 75; *L'intermédiaire des mathématiciens*, Vol. 7, 1900, p. 95.

[2] For details and references, see F. Cajori, "History of the Exponential and Logarithmic Concepts," *Am. Math. Monthly*, Vol. 20, 1913, pp. 10, 11.

veyed that country he returned to England. As a mathematician, he was the boast of his country. He brought the theory of equations under one comprehensive point of view by grasping that truth in its full extent to which Vieta and Girard only approximated; viz. that in such an equation in its simplest form, the coefficient of the second term with its sign changed is equal to the sum of the roots; the coefficient of the third is equal to the sum of the products of every two of the roots, etc. He was the first to decompose equations into their simple factors; but, since he failed to recognize imaginary and even negative roots, he failed also to prove that every equation could be thus decomposed. Harriot made some changes in algebraic notation, adopting small letters of the alphabet in place of the capitals used by Vieta. The symbols of inequality $>$ and $<$ were introduced by him. The signs $\geqq$ and $\leqq$ were first used about a century later by the Parisian hydrographer, Pierre Bouguer.[1] Harriot's work, *Artis Analyticæ praxis*, was published in 1631, ten years after his death. **William Oughtred** (1574–1660) contributed vastly to the propagation of mathematical knowledge in England by his treatises, the *Clavis mathematicæ*, 1631 (later Latin editions, 1648, 1652, 1667, 1693; English editions, 1647, 1694), *Circles of Proportion*, 1632, *Trigonometrie*, 1657.[2] Oughtred was an episcopal minister at Albury, near London, and gave private lessons, free of charge, to pupils interested in mathematics. Among his most noted pupils are the mathematician John Wallis and the astronomer Seth Ward. Oughtred laid extraordinary emphasis upon the use of mathematical symbols; altogether he used over 150 of them. Only three have come down to modern times, namely $\times$ as the symbol of multiplication, :: as that of proportion, and $\backsim$ as that for "difference." The symbol $\times$ occurs in the *Clavis*, but the letter X which closely resembles it, occurs as a sign of multiplication in the anonymous "Appendix to the Logarithmes" in Edward Wright's translation of Napier's *Descriptio*, published in 1618.[3] This appendix was most probably written by Oughtred. A proportion $A:B=C:D$ he wrote $A\cdot B::C\cdot D$. Oughtred's notation for ratio and proportion was widely used in England and on the Continent, but as early as 1651 the English astronomer Vincent Wing began to use (:) for ratio,[4] a notation which gained ground and freed the dot (.) for use as the symbol of separation in decimal fractions. It is interesting to note the attitude of Leibniz toward some of these symbols. On July 29, 1698, he wrote in a letter to John Bernoulli: "I do not like $\times$ as a symbol for multiplication, as it is easily confounded with x; . . . often I simply relate two quantities by an interposed dot and indicate multiplication

[1] P. H. Fuss, *Corresp. math. phys.*, I, 1843, p. 304; *Encyclopédie des sciences mathématiques*, T. I, Vol. I, 1904, p. 23.
[2] See F. Cajori, *William Oughtred*, Chicago and London, 1916.
[3] F. Cajori, in *Nature*, Vol. XCIV, 1914, p. 363.
[4] *Ibid.*, p. 477.

by ZC·LM. Hence in designating ratio I use not one point but two points, which I use, at the same time, for division; thus, for your *dy.x*::*dt.a* I write *dy*:*x*=*dt*:*a;* for, *y* is to *x* as *dt* is to *a*, is indeed the same as, *dy* divided by *x* is equal to *dt* divided by *a*. From this equation follow then all the rules of proportion." This conception of ratio and proportion was far in advance of that in contemporary arithmetics. Through the aid of *Christian Wolf* the dot was generally adopted in the eighteenth century as a symbol of multiplication. Presumably Leibniz had no knowledge that Harriot in his *Artis analyticæ praxis*, 1631, used a dot for multiplication, as in $aaa-3.bba=+2.ccc$. Harriot's dot received no attention, not even from Wallis.

Oughtred and some of his English contemporaries, Richard Norwood, John Speidell and others were prominent in introducing abbreviations for the trigonometric functions: *s*, *si*, or *sin* for *sine; s co* or *si co* for "sine complement" or *cosine; se* for *secant*, etc. Oughtred did not use parentheses. Terms to be aggregated were enclosed between double colons. He wrote $\sqrt{(A+E)}$ thus, $\sqrt{q}:A+E:$ The two dots at the end were sometimes omitted. Thus, $C:A+B-E$ meant $(A+B-E)$.[3] Before Oughtred the use of parentheses had been suggested by Clavius in 1608 and Girard in 1629. In fact, as early as 1556 Tartaglia wrote $\sqrt{\sqrt{28}-\sqrt{10}}$ thus ℞ v. (℞28 men ℞10), where ℞ v. means "radix universalis," but he did not use parentheses in indicating the product of two expressions.[1] Parentheses were used by I. Errard de Bar-le-Duc (1619), Jacobo de Billy (1643), Richard Norwood (1631), Samuel Foster (1659); nevertheless parentheses did not become popular in algebra before the time of Leibniz and the Bernoullis.

It is noteworthy that Oughtred denotes $3\frac{1}{7}$ and $\frac{355}{113}$, the approximate ratios of the circumference to the diameter, by the symbol $\frac{\pi}{\delta}$; it occurs in the 1647 edition and in the later editions of his *Clavis mathematicæ*. Oughtred's notation was adopted and used extensively by Isaac Barrow. It was the forerunner of the notation $\pi=3.14159\ldots$, first used by William Jones in 1706 in his *Synopsis palmariorum matheseos*, London, 1706, p. 263. L. Euler first used $\pi=3.14159\ldots$ in 1737. In his time, the symbol met with general adoption.

Oughtred stands out prominently as the inventor of the circular and the rectilinear slide rules. The circular slide rule was described in print in his book, the *Circles of Proportion*, 1632. His rectilinear slide rule was described in 1633 in an *Addition* to the above work. But Oughtred was not the first to describe the circular slide rule in print; this was done by one of his pupils, Richard Delamain, in 1630, in a booklet, entitled *Grammelogia*.[2] A bitter controversy arose be-

[1] G. Eneström in *Bibliotheca mathematica*, 3. S., Vol. 7, p. 296.
[2] See F. Cajori, *William Oughtred*, Chicago and London, 1916, p. 46.

tween Delamain and Oughtred. Each accused the other of having stolen the invention from him. Most probably each was an independent inventor. To the invention of the rectilinear slide rule Oughtred has a clear title. He states that he designed his slide rules as early as 1621. The slide rule was improved in England during the seventeenth and eighteenth centuries and was used quite extensively.[1]

Some of the stories told about Oughtred are doubtless apocryphal, as for instance, that his economical wife denied him the use of a candle for study in the evening, and that he died of joy at the Restoration, after drinking "a glass of sack" to his Majesty's health. De Morgan humorously remarks, "It should be added, by way of excuse, that he was eighty-six years old."

Algebra was now in a state of sufficient perfection to enable Descartes and others to take that important step which forms one of the grand epochs in the history of mathematics,—the application of algebraic analysis to define the nature and investigate the properties of algebraic curves.

In geometry, the determination of the areas of curvilinear figures was diligently studied at this period. **Paul Guldin** (1577–1643), a Swiss mathematician of considerable note, rediscovered the following theorem, published in his *Centrobaryca*, which has been named after him, though first found in the *Mathematical Collections* of Pappus: The volume of a solid of revolution is equal to the area of the generating figure, multiplied by the circumference described by the centre of gravity. We shall see that this method excels that of Kepler and Cavalieri in following a more exact and natural course; but it has the disadvantage of necessitating the determination of the centre of gravity, which in itself may be a more difficult problem than the original one of finding the volume. Guldin made some attempts to prove his theorem, but Cavalieri pointed out the weakness of his demonstration.

Johannes Kepler (1571–1630) was a native of Würtemberg and imbibed Copernican principles while at the University of Tübingen. His pursuit of science was repeatedly interrupted by war, religious persecution, pecuniary embarrassments, frequent changes of residence, and family troubles. In 1600 he became for one year assistant to the Danish astronomer, Tycho Brahe, in the observatory near Prague. The relation between the two great astronomers was not always of an agreeable character. Kepler's publications are voluminous. His first attempt to explain the solar system was made in 1596, when he thought he had discovered a curious relation between the five regular solids and the number and distance of the planets. The publication of this pseudo-discovery brought him much fame. At one time he tried to represent the orbit of Mars by the oval curve which we now write in polar coördinates, $\rho = 2r \cos^3 \theta$. Maturer reflection and intercourse with Tycho Brahe and Galileo led him to investigations and results

[1] See F. Cajori, *History of the Logarithmic Slide Rule*, New York, 1909.

worthy of his genius—"Kepler's laws." He enriched pure mathematics as well as astronomy. It is not strange that he was interested in the mathematical science which had done him so much service; for "if the Greeks had not cultivated conic sections, Kepler could not have superseded Ptolemy." [1] The Greeks never dreamed that these curves would ever be of practical use; Aristæus and Apollonius studied them merely to satisfy their intellectual cravings after the ideal; yet the conic sections assisted Kepler in tracing the march of the planets in their elliptic orbits. Kepler made also extended use of logarithms and decimal fractions, and was enthusiastic in diffusing a knowledge of them. At one time, while purchasing wine, he was struck by the inaccuracy of the ordinary modes of determining the contents of kegs. This led him to the study of the volumes of solids of revolution and to the publication of the *Stereometria Doliorum* in 1615. In it he deals first with the solids known to Archimedes and then takes up others. Kepler made wide application of an old but neglected idea, that of infinitely great and infinitely small quantities. Greek mathematicians usually shunned this notion, but with it modern mathematicians completely revolutionized the science. In comparing rectilinear figures, the method of superposition was employed by the ancients, but in comparing rectilinear and curvilinear figures with each other, this method failed because no addition or subtraction of rectilinear figures could ever produce curvilinear ones. To meet this case, they devised the Method of Exhaustion, which was long and difficult; it was purely synthetical, and in general required that the conclusion should be known at the outset. The new notion of infinity led gradually to the invention of methods immeasurably more powerful. Kepler conceived the circle to be composed of an infinite number of triangles having their common vertices at the centre, and their bases in the circumference; and the sphere to consist of an infinite number of pyramids. He applied conceptions of this kind to the determination of the areas and volumes of figures generated by curves revolving about any line as axis, but succeeded in solving only a few of the simplest out of the 84 problems which he proposed for investigation in his *Stereometria*.

Other points of mathematical interest in Kepler's works are (1) the assertion that the circumference of an ellipse, whose axes are $2a$ and $2b$, is nearly $\pi\ (a+b)$; (2) a passage from which it has been inferred that Kepler knew the variation of a function near its maximum value to disappear; (3) the assumption of the principle of continuity (which differentiates modern from ancient geometry), when he shows that a parabola has a focus at infinity, that lines radiating from this "cæcus focus" are parallel and have no other point at infinity.

The *Stereometria* led Cavalieri, an Italian Jesuit, to the consideration

[1] William Whewell, *History of the Inductive Sciences*, 3rd Ed., New York, 1858, Vol. I, p. 311.

of infinitely small quantities. **Bonaventura Cavalieri** (1598–1647), a pupil of Galileo and professor at Bologna, is celebrated for his *Geometria indivisibilibus continuorum nova quadam ratione promota*, 1635. This work expounds his method of Indivisibles, which occupies an intermediate place between the method of exhaustion of the Greeks and the methods of Newton and Leibniz. "Indivisibles" were discussed by Aristotle and the scholastic philosophers. They commanded the attention of Galileo. Cavalieri does not define the term. He borrows the concept from the scholastic philosophy of Bradwardine and Thomas Aquinas, in which a point is the indivisible of a line, a line the indivisible of a surface, etc. Each indivisible is capable of generating the next higher continuum by motion; a moving point generates a line, etc. The relative magnitude of two solids or surfaces could then be found simply by the summation of series of planes or lines. For example, Cavalieri finds the sum of the squares of all lines making up a triangle equal to one-third the sum of the squares of all lines of a parallelogram of equal base and altitude; for if in a triangle, the first line at the apex be 1, then the second is 2, the third is 3, and so on; and the sum of their squares is

$$1^2+2^2+3^2+\ldots+n^2=n(n+1)\,(2n+1)\div 6.$$

In the parallelogram, each of the lines is n and their number is n; hence the total sum of their squares is n^3. The ratio between the two sums is therefore

$$n(n+1)\,(2n+1)\div 6n^3=\tfrac{1}{3},$$

since n is infinite. From this he concludes that the pyramid or cone is respectively $\frac{1}{3}$ of a prism or cylinder of equal base and altitude, since the polygons or circles composing the former decrease from the base to the apex in the same way as the squares of the lines parallel to the base in a triangle decrease from base to apex. By the Method of Indivisibles, Cavalieri solved the majority of the problems proposed by Kepler. Though expeditious and yielding correct results, Cavalieri's method lacks a scientific foundation. If a line has absolutely no width, then the addition of no number, however great, of lines can ever yield an area; if a plane has no thickness whatever, then even an infinite number of planes cannot form a solid. Though unphilosophical, Cavalieri's method was used for fifty years as a sort of integral calculus. It yielded solutions to some difficult problems. Guldin made a severe attack on Cavalieri and his method. The latter published in 1647, after the death of Guldin, a treatise entitled *Exercitationes geometricæ sex*, in which he replied to the objections of his opponent and attempted to give a clearer explanation of his method. Guldin had never been able to demonstrate the theorem named after him, except by metaphysical reasoning, but Cavalieri proved it by the method of indivisibles. A revised edition of the *Geometria* appeared in 1653.

There is an important curve, not known to the ancients, which now began to be studied with great zeal. Roberval gave it the name of "trochoid," Pascal the name of "roulette," Galileo the name of "cycloid." The invention of this curve seems to be due to *Charles Bouvelles* who in a geometry published in Paris in 1501 refers to this curve in connection with the problem of the squaring of the circle. Galileo valued it for the graceful form it would give to arches in architecture. He ascertained its area by weighing paper figures of the cycloid against that of the generating circle, and found thereby the first area to be nearly but not exactly thrice the latter. A mathematical determination was made by his pupil, **Evangelista Torricelli** (1608–1647), who is more widely known as a physicist than as a mathematician.

By the Method of Indivisibles he demonstrated its area to be triple that of the revolving circle, and published his solution. This same quadrature had been effected a few years earlier (about 1636) by Roberval in France, but his solution was not known to the Italians. Roberval, being a man of irritable and violent disposition, unjustly accused the mild and amiable Torricelli of stealing the proof. This accusation of plagiarism created so much chagrin with Torricelli that it is considered to have been the cause of his early death. **Vincenzo Viviani** (1622–1703), another prominent pupil of Galileo, determined the tangent to the cycloid. This was accomplished in France by Descartes and Fermat.

In France, where geometry began to be cultivated with greatest success, Roberval, Fermat, Pascal, employed the Method of Indivisibles and made new improvements in it. **Giles Persone de Roberval** (1602–1675), for forty years professor of mathematics at the College of France in Paris, claimed for himself the invention of the Method of Indivisibles. Since his complete works were not published until after his death, it is difficult to settle questions of priority. Montucla and Chasles are of the opinion that he invented the method independently of and earlier than the Italian geometer, though the work of the latter was published much earlier than Roberval's. Marie finds it difficult to believe that the Frenchman borrowed nothing whatever from the Italian, for both could not have hit independently upon the word *Indivisibles,* which is applicable to infinitely small quantities, as conceived by Cavalieri, but not as conceived by Roberval. Roberval and Pascal improved the rational basis of the Method of Indivisibles, by considering an area as made up of an indefinite number of rectangles instead of lines, and a solid as composed of indefinitely small solids instead of surfaces. Roberval applied the method to the finding of areas, volumes, and centres of gravity. He effected the quadrature of a parabola of any degree $y^m = a^{m-1}x$, and also of a parabola $y^m = a^{m-n}x^n$. We have already mentioned his quadrature of the cycloid. Roberval is best known for his method of drawing tangents, which, however, was invented at the same time, if not earlier, by Torricelli.

Torricelli's appeared in 1644 under the title *Opera geometrica*. Roberval gives the fuller exposition of it. Some of his special applications were published at Paris as early as 1644 in Mersenne's *Cogitata physico-mathematica*. Roberval presented the full development of the subject to the French Academy of Sciences in 1668 which published it in its *Mémoires*. This academy had grown out of scientific meetings held with Mersenne at Paris. It was founded by Minister Richelieu in 1635 and reorganized by Minister Colbert in 1666. *Marin Mersenne* (1588–1648) rendered great services to science. His polite and engaging manners procured him many friends, including Descartes and Fermat. He encouraged scientific research, carried on an extensive correspondence, and thereby was the medium for the intercommunication of scientific intelligence.

Roberval's method of drawing tangents is allied to Newton's principle of fluxions. Archimedes conceived his spiral to be generated by a double motion. This idea Roberval extended to all curves. Plane curves, as for instance the conic sections, may be generated by a point acted upon by two forces, and are the resultant of two motions. If at any point of the curve the resultant be resolved into its components, then the diagonal of the parallelogram determined by them is the tangent to the curve at that point. The greatest difficulty connected with this ingenious method consisted in resolving the resultant into components having the proper lengths and directions. Roberval did not always succeed in doing this, yet his new idea was a great step in advance. He broke off from the ancient definition of a tangent as a straight line having only one point in common with a curve,—a definition which by the methods then available was not adapted to bring out the properties of tangents to curves of higher degrees, nor even of curves of the second degree and the parts they may be made to play in the generation of the curves. The subject of tangents received special attention also from Fermat, Descartes, and Barrow, and reached its highest development after the invention of the differential calculus. Fermat and Descartes defined tangents as secants whose two points of intersection with the curve coincide; Barrow considered a curve a polygon, and called one of its sides produced a tangent.

A profound scholar in all branches of learning and a mathematician of exceptional powers was **Pierre de Fermat** (1601–1665). He studied law at Toulouse, and in 1631 was made councillor for the parliament of Toulouse. His leisure time was mostly devoted to mathematics, which he studied with irresistible passion. Unlike Descartes and Pascal, he led a quiet and unaggressive life. Fermat has left the impress of his genius upon all branches of mathematics then known. A great contribution to geometry was his *De maximis et minimis*. About twenty years earlier, Kepler had first observed that the increment of a variable, as, for instance, the ordinate of a curve, is evanescent for values very near a maximum or a minimum value of the variable. Developing this idea, Fermat obtained his rule for maxima and minima. He substituted $x+e$ for x in the given function of x and then equated to each other the two consecutive values of the function and divided the equation by e. If e be taken o, then the roots of this equation are the values of x, making the function a maximum or a minimum. Fermat was in possession of this rule in 1629. The main difference between it and the rule of the differential calculus is that it introduces the indefinite quantity e instead of the infinitely small dx. Fermat made it the basis for his method of drawing tangents, which involved the determination of the length of the subtangent for a given point of a curve.

Owing to a want of explicitness in statement, Fermat's method of maxima and minima, and of tangents, was severely attacked by his great contemporary, Descartes, who could never be brought to render due justice to his merit. In the ensuing dispute, Fermat found two zealous defenders in Roberval and Pascal, the father; while C. Mydorge, G. Desargues, and Claude Hardy supported Descartes.

Since Fermat introduced the conception of infinitely small differences between consecutive values of a function and arrived at the principle for finding the maxima and minima, it was maintained by Lagrange, Laplace, and Fourier, that Fermat may be regarded as the first inventor of the differential calculus. This point is not well taken, as will be seen from the words of Poisson, himself a Frenchman, who rightly says that the differential calculus "consists in a system of rules proper for finding the differentials of all functions, rather than in the use which may be made of these infinitely small variations in the solution of one or two isolated problems.")

A contemporary mathematician, whose genius perhaps equalled that of the great Fermat, was **Blaise Pascal** (1623–1662). He was born at Clermont in Auvergne. In 1626 his father retired to Paris, where he devoted himself to teaching his son, for he would not trust his education to others. Blaise Pascal's genius for geometry showed itself when he was but twelve years old. His father was well skilled in mathematics, but did not wish his son to study it until he was perfectly acquainted with Latin and Greek. All mathematical books were hidden out of his sight. The boy once asked his father what mathematics treated of, and was answered, in general, "that it was the method of making figures with exactness, and of finding out what proportions they relatively had to one another." He was at the same time forbidden to talk any more about it, or ever to think of it. But his genius could not submit to be confined within these bounds. Starting with the bare fact that mathematics taught the means of making figures infallibly exact, he employed his thoughts about it and with a piece of charcoal drew figures upon the tiles of the pavement, trying the methods of drawing, for example, an exact circle or equilateral triangle. He gave names of his own to these figures and then formed axioms, and, in short, came to make demonstrations. In this way he is reported to have arrived unaided at the theorem that the sum of the three angles of a triangle is equal to two right angles. His father caught him in the act of studying this theorem, and was so astonished at the sublimity and force of his genius as to weep for joy. The father now gave him Euclid's *Elements*, which he, without assistance, mastered easily. His regular studies being languages, the boy employed only his hours of amusement on the study of geometry, yet he had so ready and lively a penetration that, at the age of sixteen, he wrote a treatise upon conics, which passed for such a surprising effort of genius, that it was said nothing equal to it in strength had been produced since the time of Archimedes. Descartes refused to believe that it was written by one so young as Pascal. This treatise was never published, and is now lost. Leibniz saw it in Paris and reported on a portion of its contents. The precocious youth made vast progress in all the sciences, but the constant application at so tender an age greatly impaired his health. Yet he continued working, and at nineteen invented his famous machine for performing arithmetical operations mechanically. This continued strain from overwork resulted in a permanent indisposition, and he would sometimes say that from the time he was eighteen, he never passed a day free from pain. At the age of twenty-four he resolved to lay aside the study of the human sciences and to consecrate his talents to religion. His Provincial Letters against the Jesuits are celebrated. But at times he returned to the favorite study of his youth. Being kept awake one night by a toothache, some thoughts undesignedly came into his head concerning the roulette or cycloid; one idea followed another; and he thus discovered properties of this curve even to demonstration. A correspondence between him and Fermat on certain problems was the beginning of the theory of probability. Pascal's illness increased, and he died at Paris at the early age of thirty-nine years. By him the answer to the objection to Cavalieri's Method of Indivisibles was put in clearer form. Like Roberval, he explained "the sum of right lines" to mean "the sum of infinitely small rectangles." Pascal greatly advanced the knowledge of the cycloid. He determined the area of a section produced by any line parallel to the base; the volume generated by it revolving around its base or around the axis; and, finally, the centres of gravity of these volumes, and also of half these volumes cut by planes of symmetry. Before publishing his results, he sent, in 1658, to all mathematicians that famous challenge offering prizes for the first two solutions of these problems. Only Wallis and A. La Louvére competed for them. The latter was quite unequal to the task; the former, being pressed for time, made numerous mistakes: neither got a prize. Pascal then published his own solutions, which produced a great sensation among scientific men. Wallis, too, published his, with the errors corrected. Though not competing for the prizes,

Huygens, Wren, and Fermat solved some of the questions. The chief discoveries of **Christopher Wren** (1632–1723), the celebrated architect of St. Paul's Cathedral in London, were the rectification of a cycloidal arc and the determination of its centre of gravity. Fermat found the area generated by an arc of the cycloid. Huygens invented the cycloidal pendulum.

The beginning of the seventeenth century witnessed also a revival of synthetic geometry. One who treated conics still by ancient methods, but who succeeded in greatly simplifying many prolix proofs of Apollonius, was **Claude Mydorge** (1585–1647), in Paris, a friend of Descartes. But it remained for **Girard Desargues** (1593–1662) of Lyons, and for Pascal, to leave the beaten track and cut out fresh paths. They introduced the important method of Perspective. All conics on a cone with circular base appear circular to an eye at the apex. Hence Desargues and Pascal conceived the treatment of the conic sections as projections of circles. Two important and beautiful theorems were given by Desargues: The one is on the "involution of the six points," in which a transversal meets a conic and an inscribed quadrangle; the other is that, if the vertices of two triangles, situated either in space or in a plane, lie on three lines meeting in a point, then their sides meet in three points lying on a line; and conversely. This last theorem has been employed in recent times by Brianchon, C. Sturm, Gergonne, and Poncelet. Poncelet made it the basis of his beautiful theory of homological figures. We owe to Desargues the theory of involution and of transversals; also the beautiful conception that the two extremities of a straight line may be considered as meeting at infinity, and that parallels differ from other pairs of lines only in having their points of intersection at infinity. He re-invented the epicycloid and showed its application to the construction of gear teeth, a subject elaborated more fully later by La Hire. Pascal greatly admired Desargues' results, saying (in his *Essais pour les Coniques*), "I wish to acknowledge that I owe the little that I have discovered on this subject, to his writings." Pascal's and Desargues' writings contained some of the fundamental ideas of modern synthetic geometry. In Pascal's wonderful work on conics, written at the age of sixteen and now lost, were given the theorem on the anharmonic ratio, first found in Pappus, and also that celebrated proposition on the mystic hexagon, known as "Pascal's theorem," viz. that the opposite sides of a hexagon inscribed in a conic intersect in three points which are collinear. This theorem formed the keystone to his theory. He himself said that from this alone he deduced over 400 corollaries, embracing the conics of Apollonius and many other results. Less gifted than Desargues and Pascal was **Philippe de la Hire** (1640–1718). At first active as a painter, he afterwards devoted himself to astronomy and mathematics, and became professor of the Collège de France in Paris. He wrote three works on conic sections, published in 1673, 1679 and

1685. The last of these, the *Sectiones Conicae*, was best known. La Hire gave the polar properties of circles, and, by projection, transferred his polar theory from the circle to the conic sections. In the construction of maps De la Hire used "globular" projection in which the eye is not at the pole of the sphere, as in the Ptolemaic stereographic projection, but on the radius produced through the pole at a distance $r \sin 45°$ outside the sphere. Globular projection has the advantage that everywhere on the map there is approximately the same degree of exaggeration of distances. This mode of projection was modified by his countryman A. Parent. De la Hire wrote on roulettes, on graphic methods, epicycloids, conchoids, and on magic squares. The labors of De la Hire, the genius of Desargues and Pascal, uncovered several of the rich treasures of modern synthetic geometry; but owing to the absorbing interest taken in the analytical geometry of Descartes and later in the differential calculus, the subject was almost entirely neglected until the nineteenth century.

In the theory of numbers no new results of scientific value had been reached for over 1000 years, extending from the times of Diophantus and the Hindus until the beginning of the seventeenth century. But the illustrious period we are now considering produced men who rescued this science from the realm of mysticism and superstition, in which it had been so long imprisoned; the properties of numbers began again to be studied scientifically. Not being in possession of the Hindu indeterminate analysis, many beautiful results of the Brahmins had to be re-discovered by the Europeans. Thus a solution in integers of linear indeterminate equations was re-discovered by the Frenchman **Bachet de Méziriac** (1581–1638), who was the earliest noteworthy European Diophantist. In 1612 he published *Problèmes plaisants et délectables qui se font par les nombres*, and in 1621 a Greek edition of *Diophantus* with notes. An interest in prime numbers is disclosed in the so-called "Mersenne's numbers," of the form $M_p = 2^p - 1$, with p prime. Marin Mersenne asserted in the preface to his *Cogitata Physico-Mathematica*, 1644, that the only values of p not greater than 257 which make M_p a prime are 1, 2, 3, 5, 7, 13, 17, 19, 31, 67, 127, and 257. Four mistakes have now been detected in Mersenne's classification, viz., M_{67} is composite; M_{61}, M_{89} and M_{107} are prime. M_{181} has been found to be composite. Mersenne gave in 1644 also the first eight perfect numbers 6, 28, 496, 8128, 33550336, 8589869056, 137438691328, 2305843008139952128. In Euclid's *Elements*, Bk. 9, Prop. 36, is given the formula for perfect numbers $2^{p-1}(2^p - 1)$, where $2^{p-1} - 1$ is prime. The above eight perfect numbers are reproduced by taking $p = 2, 3, 5, 7, 13, 17, 19, 31$. A ninth was found in 1885 by J. Pervušin and P. Seelhoff, for which $p = 61$, a tenth in 1912 by R. E. Powers, for which $p = 89$. The father of the modern theory of numbers is **Fermat.** He was so uncommunicative in disposition, that he generally concealed his methods and made known his results only. In some cases later analysts have been greatly puzzled in the attempt of supplying the proofs. Fermat owned a copy of Bachet's *Diophantus*, in which he entered numerous marginal notes. In 1670 these notes were incorporated in a new edition of *Diophantus*, brought out by his son. Other theorems on numbers, due to Fermat, were published in his *Opera varia* (edited by his son) and in Wallis's *Commercium epistolicum* of 1658. Of the following theorems, the first seven are found in the marginal notes: [1]—

(1) $x^n+y^n=z^n$ is impossible for integral values of x, y, and z, when $n>2$.

This famous theorem was appended by Fermat to the problem of Diophantus II, 8: "To divide a given square number into two squares." Fermat's marginal note is as follows: "On the other hand it is impossible to separate a cube into two cubes, or a biquadrate into two biquadrates, or generally any power except a square into two powers with the same exponent. I have discovered a truly marvelous proof of this, which however the margin is not large enough to contain." That Fermat actually possessed a proof is doubtful. No general proof has yet been published. Euler proved the theorem for $n=3$ and $n=4$; Dirichlet for $n=5$ and $n=14$, G. Lamé for $n=7$ and Kummer for many other values. Repeatedly was the theorem made the prize question of learned societies, by the Academy of Sciences in Paris in 1823 and 1850, by the Academy of Brussels in 1883. The recent history of the theorem follows later.

(2) A prime of the form $4n+1$ is only once the hypothenuse of a right triangle; its square is twice; its cube is three times, etc. Example: $5^2=3^2+4^2$; $25^2=15^2+20^2=7^2+24^2$; $125^2=75^2+100^2=35^2+120^2=44^2+117^2$.

(3) A prime of the form $4n+1$ can be expressed once, and only once, as the sum of two squares. Proved by Euler.

(4) A number composed of two cubes can be resolved into two other cubes in an infinite multiplicity of ways.

(5) Every number is either a triangular number or the sum of two or three triangular numbers; either a square or the sum of two, three, or four squares; either a pentagonal number or the sum of two, three, four, or five pentagonal numbers; similarly for polygonal numbers in general. The proof of this and other theorems is promised by Fermat in a future work which never appeared. This theorem is also given, with others, in a letter of 1637 (?) addressed to *Pater Mersenne*.

(6) As many numbers as you please may be found, such that the square of each remains a square on the addition to or subtraction from it of the sum of all the numbers.

[1] For a fuller historical account of Fermat's Diophantine theorems and problems, see T. L. Heath, *Diophantus of Alexandria*, 2. Ed., 1910, pp. 267–328. See also *Annals of Mathematics*, 2. S., Vol. 18, 1917, pp. 161–187.

(7) $x^4+y^4=z^2$ is impossible.

(8) In a letter of 1640 he gives the celebrated theorem generally known as "Fermat's theorem," which we state in Gauss's notation: If p is prime, and a is prime to p, then $a^{p-1}\equiv 1 \pmod{p}$. It was proved by Leibniz and by Euler.

(9) Fermat died with the belief that he had found a long-sought-for law of prime numbers in the formula $2^{2^n}+1=$ a prime, but he admitted that he was unable to prove it rigorously. The law is not true, as was pointed out by Euler in the example $2^{2^5}+1=4,294,967,297=6,700,417$ times 641. The American lightning calculator *Zerah Colburn*, when a boy, readily found the factors, but was unable to explain the method by which he made his marvellous mental computation.

(10) An odd prime number can be expressed as the difference of two squares in one, and only one, way. This theorem, given in the *Relation*, was used by Fermat for the decomposition of large numbers into prime factors.

(11) If the integers a, b, c represent the sides of a right triangle, then its area cannot be a square number. This was proved by Lagrange.

(12) Fermat's solution of $ax^2+1=y^2$, where a is integral but not a square, has come down in only the broadest outline, as given in the *Relation*. He proposed the problem to the Frenchman, *Bernhard Frenicle de Bessy*, and in 1657 to all living mathematicians. In England, Wallis and Lord Brouncker conjointly found a laborious solution, which was published in 1658, and also in 1668 in Thomas Brancker's translation of Rahn's *Algebra*, "altered and augmented" by *John Pell* (1610–1685). The first solution was given by the Hindus. Though Pell had no other connection with the problem, it went by the name of "Pell's problem." Pell held at one time the mathematical chair at Amsterdam. In a controversy with Longomontanus who claimed to have effected the quadrature of the circle, Pell first used the now familiar trigonometric formula $\tan 2A=2\tan A/(1-\tan^2 A)$.

We are not sure that Fermat subjected all his theorems to rigorous proof. His methods of proof were entirely lost until 1879, when a document was found buried among the manuscripts of Huygens in the library of Leyden, entitled *Relation des découvertes en la science des nombres*. It appears from it that he used an inductive method, called by him *la descente infinie ou indefinie*. He says that this was particularly applicable in proving the impossibility of certain relations, as, for instance, Theorem 11, given above, but that he succeeded in using the method also in proving affirmative statements. Thus he proved Theorem 3 by showing that if we suppose there be a prime $4n+1$ which does not possess this property, then there will be a smaller prime of the form $4n+1$ not possessing it; and a third one smaller than the second, not possessing it; and so on. Thus descending indefinitely, he arrives at the number 5, which is the smallest prime factor of the form $4n+1$. From the above supposition it would follow that 5 is not the sum of two squares—a conclusion contrary to fact. Hence the supposition is false, and the theorem is established. Fermat applied this method of descent with success in a large number of theorems. By this method L. Euler, A. M. Legendre, P. G. L. Dirichlet, proved several of his enunciations and many other numerical propositions.

Fermat was interested in magic squares. These squares, to which the Chinese and Arabs were so partial, reached the Occident not later than the fifteenth century. A magic square of 25 cells was found by M. Curtze in a German manuscript of that time. The artist, Albrecht Dürer, exhibits one of 16 cells in 1514 in his painting called "Melancholie." The above-named *Bernhard Frenicle de Bessy* (about 1602–1675) brought out the fact that the number of magic squares increased enormously with the order by writing down 880 magic squares of the order four. Fermat gave a general rule for finding the number of magic squares of the order n, such that, for $n=8$, this number was 1,004,144,995,344; but he seems to have recognized the falsity of his rule. Bachet de Méziriac, in his *Problèmes plaisants et délectables*, Lyon, 1612, gave a rule "des terrasses" for writing down magic squares of odd order. Frenicle de Bessy gave a process for those of even order. In the seventeenth century magic squares were studied [1] by Antoine Arnauld, Jean Prestet, J. Ozanam; in the eighteenth century by Poignard, De la Hire, J. Sauveur, L. L. Pajot, J. J. Rallier des Ourmes, L. Euler and Benjamin Franklin. In a letter B. Franklin said of his magic square of 16^2 cells, "I make no question, but you will readily allow the square of 16 to be the most magically magical of any magic square ever made by any magician."

A correspondence between *B. Pascal* and *P. Fermat* relating to a certain game of chance was the germ of the theory of probabilities, of which some anticipations are found in Cardan, Tartaglia, J. Kepler and Galileo. Chevalier de Méré proposed to B. Pascal the fundamental "Problem of Points," [2] to determine the probability which each player has, at any given stage of the game, of winning the game. Pascal and Fermat supposed that the players have equal chances of winning a single point.

The former communicated this problem to Fermat, who studied it with lively interest and solved it by the theory of combinations, a theory which was diligently studied both by him and Pascal. The calculus of probabilities engaged the attention also of C. Huygens. The most important theorem reached by him was that, if A has p chances of winning a sum a, and q chances of winning a sum b, then

[1] *Encyclopédie des sciences math's*, T. I, Vol. 3, 1906, p. 66.

[2] *Oeuvres complètes de Blaise Pascal*, T. I, Paris, 1866, pp. 220–237. See also I. Todhunter, *History of the Mathematical Theory of Probability*, Cambridge and London, 1865, Chapter II.

he may expect to win the sum $\frac{ap+bq}{p+q}$. Huygens gave his results in a treatise on probability (1657), which was the best account of the subject until the appearance of Jakob Bernoulli's *Ars conjectandi* which contained a reprint of Huygens' treatise. An absurd abuse of mathematics in connection with the probability of testimony was made by *John Craig* who in 1699 concluded that faith in the Gospel so far as it depended on oral tradition expired about the year 800, and so far as it depended on written tradition it would expire in the year 3150.

Connected with the theory of probability were the investigations on mortality and insurance. The use of tables of mortality does not seem to have been altogether unknown to the ancients, but the first name usually mentioned in this connection is Captain *John Graunt* who published at London in 1662 his *Natural and Political Observations . . . made upon the bills of mortality*, basing his deductions upon records of deaths which began to be kept in London in 1592 and were first intended to make known the progress of the plague. Graunt was careful to publish the actual figures on which he based his conclusions, comparing himself, when so doing, to a "silly schoolboy, coming to say his lessons to the world (that peevish and tetchie master), who brings a bundle of rods, wherewith to be whipped for every mistake he has committed." [1] Nothing of marked importance was done after Graunt until 1693 when Edmund Halley [1] published in the *Philosophical Transactions* (London) his celebrated memoir on the *Degrees of Mortality of Mankind . . . with an Attempt to ascertain the Price of Annuities upon Lives.* To find the value of an annuity, multiply the chance that the individual concerned will be alive after n years by the present value of the annual payment due at the end of n years; then sum the results thus obtained for all values of n from 1 to the extreme possible age for the life of that individual. Halley considers also annuities on joint lives.

Among the ancients, Archimedes was the only one who attained clear and correct notions on theoretical statics. He had acquired firm possession of the idea of pressure, which lies at the root of mechanical science. But his ideas slept nearly twenty centuries, until the time of **S. Stevin** and **Galileo Galilei** (1564–1642). Stevin determined accurately the force necessary to sustain a body on a plane inclined at any angle to the horizon. He was in possession of a complete doctrine of equilibrium. While Stevin investigated statics, Galileo pursued principally dynamics. Galileo was the first to abandon the idea usually attributed to Aristotle that bodies descend more quickly in proportion as they are heavier; he established the first law of motion; determined the laws of falling bodies; and, having obtained

[1] I. Todhunter, *History of the Theory of Probability*, pp. 38, 42.

a clear notion of acceleration and of the independence of different motions, was able to prove that projectiles move in parabolic curves. Up to his time it was believed that a cannon-ball moved forward at first in a straight line and then suddenly fell vertically to the ground. Galileo had an understanding of *centrifugal forces*, and gave a correct definition of *momentum*. Though he formulated the fundamental principle of statics, known as the *parallelogram of forces*, yet he did not fully recognize its scope. The principle of virtual velocities was partly conceived by **Guido Ubaldo** (died 1607), and afterwards more fully by Galileo.

Galileo is the founder of the science of dynamics. Among his contemporaries it was chiefly the novelties he detected in the sky that made him celebrated, but J. Lagrange claims that his astronomical discoveries required only a telescope and perseverance, while it took an extraordinary genius to discover laws from phenomena, which we see constantly and of which the true explanation escaped all earlier philosophers. Galileo's dialogues on mechanics, the *Discorsi e demostrazioni matematiche*, 1638, touch also the subject of infinite aggregates. The author displays a keenness of vision and an originality which was not equalled until the time of Dedekind and Georg Cantor. Salviati, who in general represents Galileo's own ideas in these dialogues, says,[1] "infinity and indivisibility are in their very nature incomprehensible to us." Simplicio, who is the spokesman of Aristotelian scholastic philosophy, remarks that "the infinity of points in the long line is greater than the infinity of points in the short line." Then come the remarkable words of Salviati: "This is one of the difficulties which arise when we attempt, with our finite minds, to discuss the infinite, assigning to it those properties which we give to the finite and unlimited; but this I think is wrong, for we cannot speak of infinite quantities as being the one greater or less than or equal to another. . . . We can only infer that the totality of all numbers is infinite, and that the number of squares is infinite, and that the number of the roots is infinite; neither is the number of squares less than the totality of all numbers, nor the latter greater than the former; and finally the attributes 'equal,' 'greater,' and 'less' are not applicable to infinite, but only to finite quantities. . . . One line does not contain more or less or just as many points as another, but . . . each line contains an infinite number." From the time of Galileo and Descartes to Sir William Hamilton, there was held the doctrine of the finitude of the human mind and its consequent inability to conceive the infinite. A. De Morgan ridiculed this, saying, the argument amounts to this, "who drives fat oxen should himself be fat."

Infinite series, which sprang into prominence at the time of the

[1] See Galileo's *Dialogues concerning two new Sciences*, translated by Henry Crew and Alfonso de Salvio, New York, 1914, "First Day," pp. 30–32.

invention of the differential and integral calculus, were used by a few writers before that time. *Pietro Mengoli* (1626–1686) of Bologna [1] treats them in a book, *Novæ quadraturæ arithmeticæ*, of 1650. He proves the divergence of the harmonic series by dividing its terms into an infinite number of groups, such that the sum of the terms in each group is greater than 1. The first proof of this was formerly attributed to Jakob Bernoulli, 1689. Mengoli showed the convergence of the reciprocals of the triangular numbers, a result formerly supposed to have been first reached by C. Huygens, G. W. Leibniz, or Jakob Bernoulli. Mengoli reached creditable results on the summation of infinite series.

Descartes to Newton

Among the earliest thinkers of the seventeenth and eighteenth centuries, who employed their mental powers toward the destruction of old ideas and the up-building of new ones, ranks **René Descartes** (1596–1650). Though he professed orthodoxy in faith all his life, yet in science he was a profound sceptic. He found that the world's brightest thinkers had been long exercised in metaphysics, yet they had discovered nothing certain; nay, had even flatly contradicted each other. This led him to the gigantic resolution of taking nothing whatever on authority, but of subjecting everything to scrutinous examination, according to new methods of inquiry. The certainty of the conclusions in geometry and arithmetic brought out in his mind the contrast between the true and false ways of seeking the truth. He thereupon attempted to apply mathematical reasoning to all sciences. "Comparing the mysteries of nature with the laws of mathematics, he dared to hope that the secrets of both could be unlocked with the same key." Thus he built up a system of philosophy called Cartesianism.

Great as was Descartes' celebrity as a metaphysician, it may be fairly questioned whether his claim to be remembered by posterity as a mathematician is not greater. His philosophy has long since been superseded by other systems, but the analytical geometry of Descartes will remain a valuable possession forever. At the age of twenty-one, Descartes enlisted in the army of Prince Maurice of Orange. His years of soldiering were years of leisure, in which he had time to pursue his studies. At that time mathematics was his favorite science. But in 1625 he ceased to devote himself to pure mathematics. Sir William Hamilton [2] is in error when he states that

[1] See G. Eneström in *Bibliotheca mathematica*, 3. S., Vol. 12, 1911–12, pp. 135–148.

[2] Sir William Hamilton, the metaphysician, made a famous attack upon the study of mathematics as a training of the mind, which appeared in the *Edinburgh Review* of 1836. It was shown by A. T. Bledsoe in the *Southern Review* for July, 1877, that Hamilton misrepresented the sentiments held by Descartes and other scientists. See also J. S. Mill's *Examination of Sir William Hamilton's Philosophy;*

Descartes considered mathematical studies absolutely pernicious as a means of internal culture. In a letter to Mersenne, Descartes says: "M. Desargues puts me under obligations on account of the pains that it has pleased him to have in me, in that he shows that he is sorry that I do not wish to study more in geometry, but I have resolved to quit only abstract geometry, that is to say, the consideration of questions which *serve only to exercise the mind*, and this, in order to study another kind of geometry, which has for its object the explanation of the phenomena of nature. . . . You know that all my physics is nothing else than geometry." The years between 1629 and 1649 were passed by him in Holland in the study, principally, of physics and metaphysics. His residence in Holland was during the most brilliant days of the Dutch state. In 1637 he published his *Discours de la Mêthode*, containing among others an essay of 106 pages on geometry. His *Gêomêtrie* is not easy reading. An edition appeared subsequently with notes by his friend *De Beaune*, which were intended to remove the difficulties. The *Gêomêtrie* of Descartes is of epoch-making importance; nevertheless we cannot accept Michel Chasles' statement that this work is *proles sine matre creata*—a child brought into being without a mother. In part, Descartes' ideas are found in Apollonius; the application of algebra to geometry is found in Vieta, Ghetaldi, Oughtred, and even among the Arabs. Fermat, Descartes' contemporary, advanced ideas on analytical geometry akin to his own in a treatise entitled *Ad locos planos et solidos isagoge*, which, however, was not published until 1679 in Fermat's *Varia opera*. In Descartes' *Gêomêtrie* there is no systematic development of the method of analytics. The method must be constructed from isolated statements occurring in different parts of the treatise. In the 32 geometric drawings illustrating the text the axes of coördinates are in no case explicitly set forth. The treatise consists of three "books." The first deals with "problems which can be constructed by the aid of the circle and straight line only." The second book is "on the nature of curved lines." The third book treats of the "construction of problems solid and more than solid." In the first book it is made clear, that if a problem has a finite number of solutions, the final equation obtained will have only one unknown, that if the final equation has two or more unknowns, the problem "is not wholly determined." [1] If the final equation has two unknowns "then since there is always an infinity of different points which satisfy the demand, it is therefore required to recognize and trace the line on which all of them must be located" (p. 9). To accomplish this Descartes

C. J. Keyser, *Mathematics*, 1907, pp. 20–44; F. Cajori in *Popular Science Monthly*, 1912, pp. 360–372.

[1] Descartes' *Gêomêtrie*, ed. 1886, p. 4. We are here guided by G. Eneström in *Bibliotheca mathematica*, 3. S., Vol. 11, pp. 240–243; Vol. 12, pp. 273, 274; Vol. 14, p. 357, and by H. Wieleitner in Vol. 14, pp. 241–243, 329, 330.

selects a straight line which he sometimes calls a "diameter" (p. 31) and associates each of its points with a point sought in such a way that the latter can be constructed when the former point is assumed as known. Thus, on p. 18 he says, "Je choises une ligne droite comme *AB*, pour rapporter à ses divers points tous ceux de cette ligne courbe *EC*." Here Descartes follows Apollonius who related the points of a conic to the points of a diameter, by distances (ordinates) which make a constant angle with the diameter and are determined in length by the position of the point on the diameter. This constant angle is with Descartes usually a right angle. The new feature introduced by Descartes was the use of *an equation with more than one unknown*, so that (in case of two unknowns) for any value of one unknown (abscissa), the length of the other (ordinate) could be computed. He uses the letters x and y for the abscissa and ordinate. He makes it plain that the x and y may be represented by other distances than the ones selected by him (p. 19), that, for instance, the angle formed by x and y need not be a right angle. It is noteworthy that Descartes and Fermat, and their successors down to the middle of the eighteenth century, used oblique coördinates more frequently than did later analysts. It is also noteworthy that Descartes does not formally introduce a second axis, our y-axis. Such formal introduction is found in G. Cramer's *Introduction à l'analyse des lignes courbes algêbriques*, 1750; earlier publications by de Gua, L. Euler, W. Murdoch and others contain only occasional references to a y-axis. The words "abscissa," "ordinate" were not used by Descartes. In the strictly technical sense of analytics as one of the coördinates of a point, the word "ordinate" was used by Leibniz in 1694, but in a less restricted sense such expressions as "ordinatim applicatæ" occur much earlier in F. Commandinus and others. The technical use of "abscissa" is observed in the eighteenth century by C. Wolf and others. In the more general sense of a "distance" it was used earlier by B. Cavalieri in his Indivisibles, by *Stefano degli Angeli* (1623–1697), a professor of mathematics in Rome, and by others. Leibniz introduced the word "coördinatæ" in 1692. To guard against certain current historical errors we quote the following from P. Tannery: "One frequently attributes wrongly to Descartes the introduction of the convention of reckoning coördinates positively and negatively, in the sense in which we start them from the origin. The truth is that in this respect the *Géomêtrie* of 1637 contains only certain remarks touching the interpretation of real or false (positive or negative) roots of equations.

". . . If then we examine with care the rules given by Descartes in his *Géomêtrie*, as well as his application of them, we notice that he adopts as a principle that an equation of a geometric locus is not valid except for the angle of the coördinates (quadrant) in which it was established, and all his contemporaries do likewise. The extension of an equation to other angles (quadrants) was freely made in particular cases for the interpretation of the negative roots of equations; but while it served particular conventions (for example for reckoning distances as positive and negative), it was in reality quite long in completely establishing itself, and one cannot attribute the honor for it to any particular geometer."

Descartes' geometry was called "analytical geometry," partly because, unlike the synthetic geometry of the ancients, it is actually *analytical*, in the sense that the word is used in logic; and partly because the practice had then already arisen, of designating by the term *analysis* the calculus with general quantities.

The first important example solved by Descartes in his geometry is the "problem of Pappus"; viz. "Given several straight lines in a plane, to find the locus of a point such that the perpendiculars, or more generally, straight lines at given angles, drawn from the point to the given lines, shall satisfy the condition that the product of certain of them shall be in a given ratio to the product of the rest." Of this celebrated problem, the Greeks solved only the special case when the number of given lines is four, in which case the locus of the point turns out to be a conic section. By Descartes it was solved completely, and it afforded an excellent example of the use which can be made of his analytical method in the study of loci. Another solution was given later by Newton in the *Principia*. Descartes illustrates his analytical method also by the ovals, now named after him, "certaines ovales que vous verrez être très-utiles pour la théorie de la catoptrique." These curves were studied by Descartes, probably, as early as 1629; they were intended by him to serve in the construction of converging lenses, but yielded no results of practical value. In the nineteenth century they received much attention.[1]

The power of Descartes' analytical method in geometry has been vividly set forth recently by L. Boltzmann in the remark that the formula appears at times cleverer than the man who invented it. Of all the problems which he solved by his geometry, none gave him as great pleasure as his mode of constructing tangents. It was published earlier than the methods of Fermat and Roberval which were noticed on a preceding page.

Descartes' method consisted in first finding the normal. Through a given point x, y of the curve he drew a circle which had its centre at the intersection of the normal and the x-axis. Then he imposed the condition that the circle cut the curve in two coincident points x, y. In 1638 Descartes indicated in a letter that, in place of the circle, a straight line may be used. This idea is elaborated by *Florimond de Beaune* in his notes to the 1649 edition of Descartes' *Géométrie*. In finding the point of intersection of the normal and x-axis, Descartes used the method of *Indeterminate Coefficients*, of which he bears the honor of invention. Indeterminate coefficients were employed by

[1] See G. Loria *Ebene Curven* (F. Schütte), I, 1910, p. 174.

him also in solving biquadratic equations. Descartes' method of tangents is profound, but operose, and inferior to Fermat's method. In the third book of his *Géomêtrie* he points out that if a cubic equation (with rational coefficients) has a rational root, then it can be factored and the cubic can be solved geometrically by the use of ruler and compasses only. He derives the cubic $z^3=3z-q$ as the equation upon which the trisection of an angle depends. He effects a trisection by the aid of a parabola and circle, but does not consider the reducibility of the equation. Hence he left the question of the "insolvability" of the problem untouched. Not till the nineteenth century were conclusive proofs advanced of the impossibility of trisecting any angle and of duplicating a cube, culminating at last in the clear and simple proofs given by F. Klein in 1895 in his *Ausgewählte Fragen der Elementargeometrie*, translated into English in 1897 by W. W. Beman and D. E. Smith. Descartes proved that every geometric problem giving rise to a cubic equation can be reduced either to the duplication of a cube or to the trisection of an angle. This fact had been previously recognized by Vieta.

The essays of Descartes on dioptrics and geometry were sharply criticised by Fermat, who wrote objections to the former, and sent his own treatise on "maxima and minima" to show that there were omissions in the geometry. Descartes thereupon made an attack on Fermat's method of tangents. Descartes was in the wrong in this attack, yet he continued the controversy with obstinacy. In a letter of 1638, addressed to Mersenne and to be transmitted to Fermat, Descartes gives $x^3+y^3=axy$, now known as the "folium of Descartes," as representing a curve to which Fermat's method of tangents would not apply.[1] The curve is accompanied by a figure which shows that Descartes did not then know the shape of the curve. At that time the fundamental agreement about algebraic signs of coördinates had not yet been hit upon; only finite values of variables were used. Hence the infinite branches of the curve remained unnoticed; some investigators thought there were four leaves instead of only one. C. Huygens in 1692 gave the correct shape and the asymptote of the curve.

Parabolas of higher order, $y^n=p^{n-1}x$, are mentioned by Descartes in a letter of July 13, 1638, in which the centre of mass and the volume obtained by revolution are considered. Cognate considerations are due to G. P. Roberval, P. Fermat and B. Cavalieri. Apparently, the shapes of these curves were not studied, and it remained for C. Maclaurin (1748) and G. F. A. l'Hospital (1770) to remark that they have wholly different shapes, according to whether n is a positive or a negative integer.

Descartes had a controversy with G. P. Roberval on the cycloid. This curve has been called the "Helen of geometers," on account

[1] *Oeuvres de Descartes* (Tannery et Adam), 1897, I, 490; II, 316. See also G. Loria, *Ebene Curven* (F. Schütte), I, 1910, p. 54.

of its beautiful properties and the controversies which their discovery occasioned. Its quadrature by Roberval was generally considered a brilliant achievement, but Descartes commented on it by saying that any one moderately well versed in geometry might have done this. He then sent a short demonstration of his own. On Roberval's intimating that he had been assisted by a knowledge of the solution, Descartes constructed the tangent to the curve, and challenged Roberval and Fermat to do the same. Fermat accomplished it, but Roberval never succeeded in solving this problem, which had cost the genius of Descartes but a moderate degree of attention.

The application of algebra to the doctrine of curved lines reacted favorably upon algebra. As an abstract science, Descartes improved it by the introduction of the modern exponential notation. In his *Géométrie*, 1637, he writes "*aa* ou a^2 pour multiplier *a* par soimême; et a^3 pour le multiplier encore une fois par *a*, et ainsi à l'infini." Thus, while F. Vieta represented A^3 by "A cubus" and Stevin x^3 by a figure 3 within a small circle, Descartes wrote a^3. In his *Géométrie* he does not use negative and fractional exponents, nor literal exponents. His notation was the outgrowth and an improvement of notations employed by writers before him. Nicolas Chuquet's manuscript work, *Le Triparty en la science des nombres*,[1] 1484, gives $12x^3$ and $10x^5$, and their product $120x^8$, by the symbols 12^3, 10^5, 120^8, respectively. Chuquet goes even further and writes $12x^0$ and $7x^{-1}$ thus 12^0, 7^{1m}; he represents the product of $8x^3$ and $7x^{-1}$ by 56^2. J. Bürgi, Reymer and J. Kepler use Roman numerals for the exponential symbol. J. Bürgi writes $16x^2$ thus $\overset{\text{II}}{16}$. Thomas Harriot simply repeats the letters; he writes in his *Artis analyticæ praxis* (1631), $a^4-1024a^2+6254a$, thus: $aaaa-1024aa+6254a$.

Descartes' exponential notation spread rapidly; about 1660 or 1670 the positive integral exponent had won an undisputed place in algebraic notation. In 1656 J. Wallis speaks of negative and fractional "indices," in his *Arithmetica infinitorum*, but he does not actually write a^{-1} for $\frac{1}{a}$, or $a^{2/3}$ for $\sqrt[3]{a^2}$. It was I. Newton who, in his famous letter to H. Oldenburg, dated June 13, 1676, and containing his announcement of the binomial theorem, first uses negative and fractional exponents.

With Descartes a letter represented always only a positive number. It was Johann Hudde who in 1659 first let a letter stand for negative as well as positive values.

Descartes also established some theorems on the theory of equations. Celebrated is his "rule of signs" for determining the number

[1] Chuquet's "Le Triparty," *Bullettino Boncompagni*, Vol. 13, 1880, p. 740.

of positive and negative roots. He gives the rule after pointing out the roots 2, 3, 4, -5 and the corresponding binomial factors of the equation $x^4-4x^3-19x^2+106x-120=0$. His exact words are as follows:

"On connoît aussi de ceci combien il peut y avoir de vraies racines et combien de fausses en chaque équation: à savoir il y en peut avoir autant de vraies que les signes + et − s'y trouvent de fois être changés, et autant de fausses qu'il s'y trouve de fois deux signes + ou deux signes − qui s'entre-suivent. Comme en la dernière, à cause qu'après $+x^4$ il y a $-4x^3$, qui est un changement du signe + en −, et après $-19x^2$ il y a $+106x$, et après $+106x$ il y a -120, qui sont encore deux autres changements, ou connoît qu'il y a trois vraies racines; et une fausse, à cause que les deux signes − de $4x^3$ et $19x^2$ s'entre-suivent."

This statement lacks completeness. For this reason he has been frequently criticized. J. Wallis claimed that Descartes failed to notice that the rule breaks down in case of imaginary roots, but Descartes does not say that the equation *always has*, but that it *may have*, so many roots. Did Descartes receive any suggestion of his rule from earlier writers? He might have received a hint from H. Cardan, whose remarks on this subject have been summarized by G. Eneström [1] as follows: If in an equation of the second, third or fourth degree, (1) the last term is negative, then one variation of sign signifies one and only one positive root, (2) the last term is positive, then two variations indicate either several positive roots or none. Cardan does not consider equations having more than two variations. G. W. Leibniz was the first to erroneously attribute the rule of signs to T. Harriot. Descartes was charged by J. Wallis with availing himself, without acknowledgment, of Harriot's theory of equations, particularly his mode of generating equations; but there seems to be no good ground for the charge.

In mechanics, Descartes can hardly be said to have advanced beyond Galileo. The latter had overthrown the ideas of Aristotle on this subject, and Descartes simply "threw himself upon the enemy" that had already been "put to the rout." His statement of the first and second laws of motion was an improvement in form, but his third law is false in substance. The motions of bodies in their direct impact was imperfectly understood by Galileo, erroneously given by Descartes, and first correctly stated by C. Wren, J. Wallis, and C. Huygens.

One of the most devoted pupils of Descartes was the learned *Princess Elizabeth*, daughter of Frederick V. She applied the new analytical geometry to the solution of the "Apollonian problem." His second royal follower was *Queen Christina*, the daughter of Gustavus Adolphus. She urged upon Descartes to come to the Swedish court. After much hesitation he accepted the invitation in 1649.

[1] *Bibliotheca mathematica*, 3rd S., Vol. 7, 1906–7, p. 293.

He died at Stockholm one year later. His life had been one long warfare against the prejudices of men.

It is most remarkable that the mathematics and philosophy of Descartes should at first have been appreciated less by his countrymen than by foreigners. The indiscreet temper of Descartes alienated the great contemporary French mathematicians, Roberval, Fermat, Pascal. They continued in investigations of their own, and on some points strongly opposed Descartes. The universities of France were under strict ecclesiastical control and did nothing to introduce his mathematics and philosophy. It was in the youthful universities of Holland that the effect of Cartesian teachings was most immediate and strongest.

The only prominent Frenchman who immediately followed in the footsteps of the great master was **Florimond de Beaune** (1601–1652). He was one of the first to point out that the properties of a curve can be deduced from the properties of its tangent. This mode of inquiry has been called the *inverse method of tangents*. He contributed to the theory of equations by considering for the first time the upper and lower limits of the roots of numerical equations.

In the Netherlands a large number of distinguished mathematicians were at once struck with admiration for the Cartesian geometry. Foremost among these are *van Schooten*, John de Witt, van Heuraet, Sluze, and Hudde. **Franciscus van Schooten** (died 1660), professor of mathematics at Leyden, brought out an edition of Descartes' geometry, together with the notes thereon by De Beaune. His chief work is his *Exercitationes Mathematicæ*, 1657, in which he applies the analytical geometry to the solution of many interesting and difficult problems. The noble-hearted **Johann de Witt** (1625–1672), grand-pensioner of Holland, celebrated as a statesman and for his tragical end, was an ardent geometrician. He conceived a new and ingenious way of generating conics, which is essentially the same as that by projective pencils of rays in modern synthetic geometry. He treated the subject not synthetically, but with aid of the Cartesian analysis. **René François de Sluse** (1622–1685) and **Johann Hudde** (1633–1704) made some improvements on Descartes' and Fermat's methods of drawing tangents, and on the theory of maxima and minima. With Hudde, we find the first use of three variables in analytical geometry. He is the author of an ingenious rule for finding equal roots. We illustrate it by the equation $x^3-x^2-8x+12=0$. Taking an arithmetical progression 3, 2, 1, 0, of which the highest term is equal to the degree of the equation, and multiplying each term of the equation respectively by the corresponding term of the progression, we get $3x^3-2x^2-8x=0$, or $3x^2-2x-8=0$. This last equation is by one degree lower than the original one. Find the G.C.D. of the two equations. This is $x-2$; hence 2 is one of the two equal roots. Had there been no common divisor, then the original equation would not have possessed equal roots. Hudde gave a demonstration for this rule.[1]

Heinrich van Heuraet must be mentioned as one of the earliest geometers who occupied themselves with success in the rectification of curves. He observed in a general way that the two problems of quadrature and of rectification are really identical, and that the one can be reduced to the other. Thus he carried the rectification of the parabola back to the quadrature of the hyperbola. The earliest absolute rectification was that of the logarithmic spiral, by E. Torricelli in 1640. The curve which John Wallis named the "semi-cubical parabola," $y^3=ax^2$, was the next curve to be rectified absolutely. This appears to have been accomplished independently by P. Fermat in France, Van Heuraet in Holland and by **William Neil** (1637–1670) in England. According to J. Wallis the priority belongs to Neil. Soon after, the cycloid was rectified by C. Wren and Fermat.

A mathematician of no mean ability was *Gregory St. Vincent* (1584–1667), a Belgian, who studied under C. Clavius in Rome and was two years professor at Prague, where, during war time, his manuscript volume on geometry and statics was lost in a fire. Other papers of his were saved but carried about for ten years before they came again into his possession, at his home in Ghent. They became the groundwork of his great book, the *Opus geometricum quadraturæ circuli et sectionum coni*, Antwerp, 1647. It consists of 1225 folio pages, divided into ten books. St. Vincent proposes four methods for squaring the circle, but does not actually carry them out. The work was attacked by R. Descartes, M. Mersenne and G. P. Roberval, and defended by the Jesuit Alfons Anton de Sarasa and others. Though erroneous on the possibility of squaring the circle, the *Opus* contains solid achievements, which were the more remarkable, because at that time only four of the seven books of the conics of Apollonius of Perga were known in the Occident. St. Vincent deals with conics, surfaces and solids from a new point of view, employing infinitesimals in a way perhaps less objectionable than in B. Cavalieri's book. St. Vincent was probably the first to use the word *exhaurire* in a geometrical sense. From this word arose the name of "method of exhaustion," as applied to the method of Euclid and Archimedes. St. Vincent used a method of transformation of one conic into another, called *per subtendas* (by chords), which contains germs of analytic geometry. He created another special method which he called *Ductus plani in planum* and used in the study of solids.[2] Unlike Archimedes who kept on dividing distances, only until a certain degree of smallness was reached, St. Vincent permitted the subdivisions to continue

[1] Heinrich Suter, *Geschichte der Mathematischen Wissenschaften* Zürich, 2. Theil, 1875, p. 25.

[2] See M. Marie, *Histoire des sciences math.*, Vol. 3, 1884, pp. 186–193; Karl Bopp, *Kegelschnitte des Gregorius a St. Vincento* in *Abhandl. z. Gesch. d. math. Wissensch.*, XX Heft, 1907, pp. 83–314.

ad infinitum and obtained a geometric series that was *infinite*. However, infinite series had been obtained before him by *Alvarus Thomas*, a native of Lisbon, in a work, *Liber de triplici motu*, 1509,[1] and by others. But St. Vincent was the first to apply geometric series to the "Achilles" and to look upon the paradox as a question in the summation of an infinite series. Moreover, St. Vincent was the first to state the exact time and place of overtaking the tortoise. He spoke of the limit as an obstacle against further advance, similar to a rigid wall. Apparently he was not troubled by the fact that in his theory, the variable does not *reach* its limit. His exposition of the "Achilles" was favorably received by G. W. Leibniz and by writers over a century afterward. The fullest account and discussion of Zeno's arguments on motion that was published before the nineteenth century was given by the noted French skeptical philosopher, *Pierre Bayle*, in an article "Zenon d'Elée" in his *Dictionnaire historique et critique*, 1696.[2]

The prince of philosophers in Holland, and one of the greatest scientists of the seventeenth century, was **Christian Huygens** (1629–1695), a native of The Hague. Eminent as a physicist and astronomer, as well as mathematician, he was a worthy predecessor of Sir Isaac Newton. He studied at Leyden under *Frans Van Schooten*. The perusal of some of his earliest theorems led R. Descartes to predict his future greatness. In 1651 Huygens wrote a treatise in which he pointed out the fallacies of Gregory St. Vincent on the subject of quadratures. He himself gave a remarkably close and convenient approximation to the length of a circular arc. In 1660 and 1663 he went to Paris and to London. In 1666 he was appointed by Louis XIV member of the French Academy of Sciences. He was induced to remain in Paris from that time until 1681, when he returned to his native city, partly for consideration of his health and partly on account of the revocation of the Edict of Nantes.

The majority of his profound discoveries were made with aid of the ancient geometry, though at times he used the geometry of R. Descartes or of B. Cavalieri and P. Fermat. Thus, like his illustrious friend, Sir Isaac Newton, he always showed partiality for the Greek geometry. Newton and Huygens were kindred minds, and had the greatest admiration for each other. Newton always speaks of him as the "Summus Hugenius."

To the two curves (cubical parabola and cycloid) previously rectified he added a third,—the cissoid. A French physician, Claudius Perrault, proposed the question, to determine the path in a fixed plane of a heavy point attached to one end of a taut string whose other end moves along a straight line in that plane. Huygens and G. W. Leibniz studied this problem in 1693, generalized it, and thus worked out the

[1] H. Wieleitner, in *Bibliotheca mathematica*, 3. F., Bd. 1914, 14, p. 152.
[2] See F. Cajori in *Am. Math. Monthly*, Vol. 22, 1915, pp. 109–112.

geometry of the "tractrix." [1] Huygens solved the problem of the catenary, determined the surface of the parabolic and hyperbolic conoid, and discovered the properties of the logarithmic curve and the solids generated by it. Huygens' *De horologio oscillatorio* (Paris, 1673) is a work that ranks second only to the *Principia* of Newton and constitutes historically a necessary introduction to it. The book opens with a description of pendulum clocks, of which Huygens is the inventor. Then follows a treatment of accelerated motion of bodies falling free, or sliding on inclined planes, or on given curves,—culminating in the brilliant discovery that the cycloid is the tautochronous curve. To the theory of curves he added the important theory of "evolutes." After explaining that the tangent of the evolute is normal to the involute, he applied the theory to the cycloid, and showed by simple reasoning that the evolute of this curve is an equal cycloid. Then comes the complete general discussion of the centre of oscillation. This subject had been proposed for investigation by M. Mersenne and discussed by R. Descartes and G. P. Roberval. In Huygens' assumption that the common centre of gravity of a group of bodies, oscillating about a horizontal axis, rises to its original height, but no higher, is expressed for the first time one of the most beautiful principles of dynamics, afterwards called the principle of the conservation of *vis viva*. The thirteen theorems at the close of the work relate to the theory of centrifugal force in circular motion. This theory aided Newton in discovering the law of gravitation.[2]

Huygens wrote the first formal treatise on probability. He proposed the wave-theory of light and with great skill applied geometry to its development. This theory was long neglected, but was revived and elaborated by Thomas Young and A. J. Fresnel a century later. Huygens and his brother improved the telescope by devising a better way of grinding and polishing lenses. With more efficient instruments he determined the nature of Saturn's appendage and solved other astronomical questions. Huygens' *Opuscula posthuma* appeared in 1703.

The theory of combinations, the primitive notions of which go back to ancient Greece, received the attention of *William Buckley* of King's College, Cambridge (died 1550), and especially of Blaise Pascal who treats of it in his *Arithmetical Triangle.* Before Pascal, this Triangle had been constructed by N. Tartaglia and M. Stifel. Fermat applied combinations to the study of probability. The earliest mathematical work of Leibniz was his *De arte combinatoria.* The subject was treated by John Wallis in his *Algebra.*

John Wallis (1616–1703) was one of the most original mathematicians of his day. He was educated for the Church at Cambridge and en-

[1] G. Loria, *Ebene Curven* (F. Schütte) II, 1911, p. 188.

[2] E. Dühring, *Kritische Geschichte der Allgemeinen Principien der Mechanik,* Leipzig, 1887, p. 135.

tered Holy Orders. But his genius was employed chiefly in the study of mathematics. In 1649 he was appointed Savilian professor of geometry at Oxford. He was one of the original members of the Royal Society, which was founded in 1663. He ranks as one of the world's greatest decipherers of cryptic writing.[1] Wallis thoroughly grasped the mathematical methods both of B. Cavalieri and R. Descartes. His *Conic Sections* is the earliest work in which these curves are no longer considered as sections of a cone, but as curves of the second degree, and are treated analytically by the Cartesian method of co-ordinates. In this work Wallis speaks of Descartes in the highest terms, but in his *Algebra* (1685, Latin edition 1693), he, without good reason, accuses Descartes of plagiarizing from T. Harriot. It is interesting to observe that, in his *Algebra*, Wallis discusses the possibility of a fourth dimension. Whereas nature, says Wallis, "doth not admit of more than three (local) dimensions . . . it may justly seem very improper to talk of a solid . . . drawn into a fourth, fifth, sixth, or further dimension. . . . Nor can our fansie imagine how there should be a fourth local dimension beyond these three." [2] The first to busy himself with the number of dimensions of space was Ptolemy. Wallis felt the need of a method of representing imaginaries graphically, but he failed to discover a general and consistent representation.[3] He published Nasir-Eddin's proof of the parallel postulate and, abandoning the idea of equidistance that had been employed without success by F. Commandino, C. Clavius, P. A. Cataldi and G. A. Borelli, gave a proof of his own based on the axiom that, to every figure there exists a similar figure of arbitrary magnitude.[4] We have already mentioned (page 165) Wallis's solution of the prize questions on the cycloid, which were proposed by Pascal.

The *Arithmetica infinitorum*, published in 1655, is his greatest work. By the application of analysis to the Method of Indivisibles, he greatly increased the power of this instrument for effecting quadratures. He created the arithmetical conception of a limit by considering the successive values of a fraction, formed in the study of certain ratios; these fractional values steadily approach a limiting value, so that the difference becomes less than any assignable one and vanishes when the process is carried to infinity. He advanced beyond J. Kepler by making more extended use of the "law of continuity" and placing

[1] D. E. Smith in *Bull. Am. Math. Soc.*, Vol. 24, 1917, p. 82.

[2] G. Eneström in *Bibliotheca mathematica*, 3. S., Vol. 12, 1911–12, p. 88.

[3] See Wallis' *Algebra*, 1685, pp. 264–273; see also Eneström in *Bibliotheca mathematica*, 3. S., Vol. 7, pp. 263–269.

[4] R. Bonola, *op. cit.*, pp. 12–17. See also F. Engel u. P. Stäckel, *Theorie der Parallellinien von Euclid bis auf Gauss*, Leipzig, 1895, pp. 21–36. This treatise gives translations into German of Saccheri, also the essays of Lambert and Taurinus, and letters of Gauss.

full reliance in it. By this law he was led to regard the denominators of fractions as powers with negative exponents. Thus, the descending geometrical progression x^3, x^2, x^1, x^0, if continued, gives x^{-1}, x^{-2}, x^{-3}, etc.; which is the same thing as $\frac{1}{x}$, $\frac{1}{x^2}$, $\frac{1}{x^3}$. The exponents of this geometric series are in continued arithmetical progression, 3, 2, 1, 0, −1, −2, −3. However, Wallis does not actually use here the notation x^{-1}, x^{-2}, etc.; he merely speaks of negative exponents. He also used fractional exponents, which, like the negative, had been invented long before, but had failed to be generally introduced. The symbol ∞ for infinity is due to him. Wallis introduces the name,* "hypergeometric series" for a sequence different from a, ab, ab^2, . . . ; he did not look upon this new series as a power-series nor as a function of x.

B. Cavalieri and the French geometers had ascertained the formula for squaring the parabola of any degree, $y=x^m$, m being a positive integer. By the summation of the powers of the terms of infinite arithmetical series, it was found that the area (from 0 to x) under the curve $y=x^m$ is to the area of the parallelogram having the same base and altitude as 1 is to $m+1$.

Aided by the law of continuity, Wallis arrived at the result that this formula holds true not only when m is positive and integral, but also when it is fractional or negative. Thus, in the parabola $y=\sqrt{px}$, $m=\frac{1}{2}$; hence the area of the parabolic segment is to that of the circumscribed rectangle as $1:1\frac{1}{2}$, or as $2:3$. Again, suppose that in $y=x^m$, $m=-\frac{1}{2}$; then the curve is a kind of hyperbola referred to its asymptotes, and the hyperbolic space between the curve and its asymptotes is to the corresponding parallelogram as $1:\frac{1}{2}$. If $m=-1$, as in the common equilateral hyperbola $y=x^{-1}$ or $xy=1$, then this ratio is $1:-1+1$, or $1:0$, showing that its asymptotic space is infinite. But in the case when m is less than minus one. Wallis was unable to interpret correctly his results. For example, if $m=-3$, then the ratio becomes $1:-2$, or as unity to a negative number. What is the meaning of this? Wallis reasoned thus: If the denominator is only zero, then the area is already infinite; but if it is less than zero, then the area must be more than infinite. It was pointed out later by P. Varignon, that this space, supposed to exceed infinity, is really finite, but taken negatively; that is, measured in a contrary direction.[1] The method of Wallis was easily extended to cases such as $y=ax^{\frac{m}{n}}+bx^{\frac{p}{q}}$ by performing the quadrature for each term separately, and then adding the results.

The manner in which Wallis studied the quadrature of the circle and arrived at his expression for the value of π is extraordinary. He found that the areas comprised between the axes, the ordinate cor-

[1] J. F. Montucla, *Histoire des mathématiques*, Paris, Tome 2, An VII, p. 350.

responding to x, and the curves represented by the equations $y=(1-x^2)^0$, $y=(1-x^2)^1$, $y=(1-x^2)^2$, $y=(1-x^2)^3$, etc., are expressed in functions of the circumscribed rectangles having x and y for their sides, by the quantities forming the series

$$\begin{array}{l} x, \\ x-\tfrac{1}{3}x^3, \\ x-\tfrac{2}{3}x^3+\tfrac{1}{5}x^5, \\ x-\tfrac{3}{3}x^3+\tfrac{3}{5}x^5-\tfrac{1}{7}x^7, \text{ etc.} \end{array}$$

When $x=1$, these values become respectively 1, $\tfrac{2}{3}$, $\tfrac{8}{15}$, $\tfrac{48}{105}$, etc. Now since the ordinate of the circle is $y=(1-x^2)^{\frac{1}{2}}$, the exponent of which is $\tfrac{1}{2}$ or the mean value between 0 and 1, the question of this quadrature reduced itself to this: If 0, 1, 2, 3, etc., operated upon by a certain law, give 1, $\tfrac{2}{3}$, $\tfrac{8}{15}$, $\tfrac{48}{105}$, what will $\tfrac{1}{2}$ give, when operated upon by the same law? He attempted to solve this by *interpolation*, a method first brought into prominence by him, and arrived by a highly complicated and difficult analysis at the following very remarkable expression:

$$\frac{\pi}{2}=\frac{2.2.4.4.6.6.8.8\ldots}{1\cdot3\cdot3\cdot5\cdot5\cdot7\cdot7\cdot9\cdots}.$$

He did not succeed in making the interpolation itself, because he did not employ literal or general exponents, and could not conceive a series with more than one term and less than two, which it seemed to him the interpolated series must have. The consideration of this difficulty led I. Newton to the discovery of the Binomial Theorem. This is the best place to speak of that discovery. Newton virtually assumed that the same conditions which underlie the general expressions for the areas given above must also hold for the expression to be interpolated. In the first place, he observed that in each expression the first term is x, that x increases in odd powers, that the signs alternate + and −, and that the second terms $\tfrac{0}{3}x^3$, $\tfrac{1}{3}x^3$, $\tfrac{2}{3}x^3$, $\tfrac{3}{3}x^3$, are in arithmetical progression. Hence the first two terms of the interpolated series must be $x-\dfrac{\frac{1}{2}x^3}{3}$. He next considered that the denominators 1, 3, 5, 7, etc., are in arithmetical progression, and that the coefficients in the numerators in each expression are the digits of some power of the number 11; namely, for the first expression, 11^0 or 1; for the second, 11^1 or 1, 1; for the third, 11^2 or 1, 2, 1; for the fourth, 11^3 or 1, 3, 3, 1; etc. He then discovered that, having given the second digit (call it m), the remaining digits can be found by continual multiplication of the terms of the series $\dfrac{m-0}{1}\cdot\dfrac{m-1}{2}\cdot\dfrac{m-2}{3}\cdot\dfrac{m-3}{4}$. etc. Thus, if $m=4$, then $4\cdot\dfrac{m-1}{2}$ gives 6; $6\cdot\dfrac{m-2}{3}$ gives 4;

$4 \cdot \frac{m-3}{4}$ gives 1. Applying this rule to the required series, since the second term is $\frac{\frac{1}{2}x^3}{3}$, we have $m=\frac{1}{2}$, and then get for the succeeding coefficients in the numerators respectively $-\frac{1}{8}$, $-\frac{1}{16}$, $-\frac{5}{128}$, etc.; hence the required area for the circular segment is $x-\frac{\frac{1}{2}x^3}{3}-\frac{\frac{1}{8}x^5}{5}-\frac{\frac{1}{16}x^7}{7}-$ etc. Thus he found the interpolated expression to be an infinite series, instead of one having more than one term and less than two, as Wallis believed it must be. This interpolation suggested to Newton a mode of expanding $(1-x^2)^{\frac{1}{2}}$, or, more generally, $(1-x^2)^m$, into a series. He observed that he had only to omit from the expression just found the denominators 1, 3, 5, 7, etc., and to lower each power of x by unity, and he had the desired expression. In a letter to H. Oldenburg (June 13, 1676), Newton states the theorem as follows: The extraction of roots is much shortened by the theorem

$$(P+PQ)^{\frac{m}{n}}=P^{\frac{m}{n}}+\frac{m}{n}AQ+\frac{m-n}{2n}BQ+\frac{m-2n}{3n}CQ+\text{etc.},$$

where A means the first term, $P^{\frac{m}{n}}$, B the second term, C the third term, etc. He verified it by actual multiplication, but gave no regular proof of it. He gave it for any exponent whatever, but made no distinction between the case when the exponent is positive and integral, and the others.

It should here be mentioned that very rude beginnings of the binomial theorem are found very early. The Hindus and Arabs used the expansions of $(a+b)^2$ and $(a+b)^3$ for extracting roots; Vieta knew the expansion of $(a+b)^4$; but these were the results of simple multiplication without the discovery of any law. The binomial coefficients for positive whole exponents were known to some Arabic and European mathematicians. B. Pascal derived the coefficients from the method of what is called the "arithmetical triangle." Lucas de Burgo, M. Stifel, S. Stevin, H. Briggs, and others, all possessed something from which one would think the binomial theorem could have been gotten with a little attention, "if we did not know that such simple relations were difficult to discover."

Though Wallis had obtained an entirely new expression for π, he was not satisfied with it; for instead of a finite number of terms yielding an absolute value, it contained an infinite number, approaching nearer and nearer to that value. He therefore induced his friend, **Lord Brouncker**, the first president of the Royal Society, to investigate this subject. Of course Lord Brouncker did not find what they were after, but he obtained the following beautiful equality:—

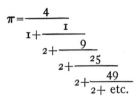

$$\pi=\cfrac{4}{1+\cfrac{1}{2+\cfrac{9}{2+\cfrac{25}{2+\cfrac{49}{2+\text{ etc.}}}}}}$$

Continued fractions, both ascending and descending, appear to have been known already to the Greeks and Hindus, though not in our present notation. Brouncker's expression gave birth to the theory of continued fractions.

Wallis' method of quadratures was diligently studied by his disciples. Lord Brouncker obtained the first infinite series for the area of the equilateral hyperbola $xy=1$ between one of its asymptotes and the ordinates for $x=1$ and $x=2$; viz. the area $\frac{1}{1.2}+\frac{1}{3.4}+\frac{1}{5.6}+\ldots$ The *Logarithmotechnia* (London, 1668) of *Nicolaus Mercator* is often said to contain the series $\log(1+a)=a-\frac{aa}{2}+\frac{a^3}{3}-\ldots$ In reality it contains the numerical values of the first few terms of that series, taking $a=.1$, also $a=.21$. He adhered to the mode of exposition which favored the concrete special case to the general formula. Wallis was the first to state Mercator's logarithmic series in general symbols. Mercator deduced his results from the grand property of the hyperbola deduced by Gregory St. Vincent in Book VII of his *Opus geometricum*, Antwerp, 1647: If parallels to one asymptote are drawn between the hyperbola and the other asymptote, so that the successive areas of the mixtilinear quadrilaterals thus formed are equal, then the lengths of the parallels form a geometric progression. Apparently the first writer to state this theorem in the language of logarithms was the Belgian Jesuit *Alfons Anton de Sarasa*, who defended Gregory St. Vincent against attacks made by Mersenne. Mercator showed how the construction of logarithmic tables could be reduced to the quadrature of hyperbolic spaces. Following up some suggestions of Wallis, *William Neil* succeeded in rectifying the cubical parabola, and *C. Wren* in rectifying any cycloidal arc. Gregory St. Vincent, in Part X of his *Opus* describes the construction of certain quartic curves, often called virtual parabolas of St. Vincent, one of which has a shape much like a lemniscate and in Cartesian co-ordinates is $d^2(y^2-x^2)=y^4$. Curves of this type are mentioned in the correspondence of C. Huygens with R. de Sluse, and with G. W. Leibniz.

A prominent English mathematician and contemporary of Wallis was **Isaac Barrow** (1630–1677). He was professor of mathematics in London, and then in Cambridge, but in 1669 he resigned his chair to his illustrious pupil, Isaac Newton, and renounced the study of mathematics for that of divinity. As a mathematician, he is most celebrated for his method of tangents. He simplified the method of P. Fermat by introducing two infinitesimals instead of one, and approximated to the course of reasoning afterwards followed by Newton in his doctrine on Ultimate Ratios. The following books are Barrow's: *Lectiones geometricæ* (1670), *Lectiones mathematicæ* (1683–1685).

He considered the infinitesimal right triangle ABB' having for its sides the difference between two successive ordinates, the distance between them, and the portion of the curve intercepted by them. This triangle is similar to BPT, formed by the ordinate, the tangent, and the sub-tangent. Hence, if we know the ratio of $B'A$ to BA, then we know the ratio of the ordinate and the sub-tangent, and the tangent can be constructed at once. For any curve, say $y^2=px$, the ratio of $B'A$ to BA is determined from its equation as follows: If x receives an infinitesimal increment $PP'=e$, then y receives an increment $B'A=a$, and the equation for the ordinate $B'P'$ becomes $y^2+2ay+a^2=px+pe$. Since $y^2=px$, we get $2ay+a^2=pe$; neglecting higher powers of the infinitesimals, we have $2ay=pe$, which gives

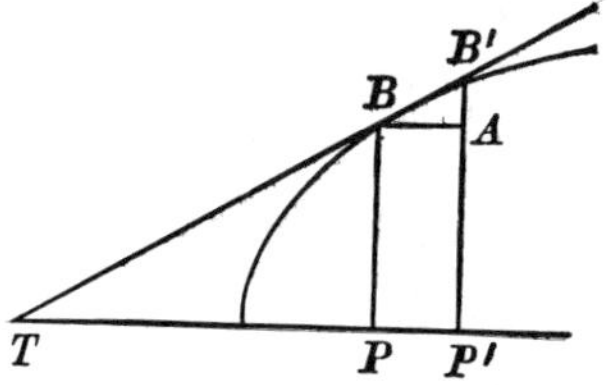

$$a:e=p:2y=p:2\sqrt{px}.$$

But a: e=the ordinate: the sub-tangent; hence

$$p:2\sqrt{px}=\sqrt{px}:\text{sub-tangent},$$

giving $2x$ for the value of the sub-tangent. This method differs from that of the differential calculus chiefly in notation. In fact, a recent investigator asserts, "Isaac Barrow was the first inventor of the infinitesimal calculus." [1]

Of the integrations that were performed before the Integral Calculus was invented, one of the most difficult grew out of a practical problem of navigation in connection with *Gerardus Mercator's* map. In 1599 *Edward Wright* published a table of latitudes giving numbers expressing the length of an arc of the nautical meridian. The table was computed by the continued addition of the secants of 1″, 2″, 3″, etc. In modern symbols this amounts to $r\int_0^\theta \sec\theta\, d\theta=r\log\tan(90^\circ-\theta)/2$.

It was Henry Bond who noticed by inspection about 1645 that Wright's table was a table of logarithmic tangents. Actual demonstrations of this, thereby really establishing the above definite integral, were given by James Gregory in 1668, Isaac Barrow in 1670, John

[1] J. M. Child, *The Geometrical Lectures of Isaac Barrow*, Chicago and London, 1916, preface.

Wallis in 1685, and Edmund Halley in 1698.[1] James Gregory and Barrow gave also the integral $\int_0^\theta \tan\theta\, d\theta = \log\sec\theta$; B. Cavalieri in 1647 established the integral of $\int_0^a x^n\, dx$. Similar results were obtained by E. Torricelli, Gregory St. Vincent, P. Fermat, G. P. Roberval and B. Pascal.[2]

Newton to Euler

It has been seen that in France prodigious scientific progress was made during the beginning and middle of the seventeenth century. The toleration which marked the reign of Henry IV and Louis XIII was accompanied by intense intellectual activity. Extraordinary confidence came to be placed in the power of the human mind. The bold intellectual conquests of R. Descartes, P. Fermat, and B. Pascal enriched mathematics with imperishable treasures. During the early part of the reign of Louis XIV we behold the sunset splendor of this glorious period. Then followed a night of mental effeminacy. This lack of great scientific thinkers during the reign of Louis XIV may be due to the simple fact that no great minds were born; but, according to Buckle, it was due to the paternalism, to the spirit of dependence and subordination, and to the lack of toleration, which marked the policy of Louis XIV.

In the absence of great French thinkers, Louis XIV surrounded himself by eminent foreigners. O. Römer from Denmark, C. Huygens from Holland, Dominic Cassini from Italy, were the mathematicians and astronomers adorning his court. They were in possession of a brilliant reputation before going to Paris. Simply because they performed scientific work in Paris, that work belongs no more to France than the discoveries of R. Descartes belong to Holland, or those of J. Lagrange to Germany, or those of L. Euler and J. V. Poncelet to Russia. We must look to other countries than France for the great scientific men of the latter part of the seventeenth century.

About the time when Louis XIV assumed the direction of the French government Charles II became king of England. At this time England was extending her commerce and navigation, and advancing considerably in material prosperity. A strong intellectual movement took place, which was unwittingly supported by the king. The age of poetry was soon followed by an age of science and philosophy. In two successive centuries England produced Shakespeare and I. Newton!

[1] See F. Cajori in *Bibliotheca mathematica*, 3. S., Vol. 14, 1915, pp. 312–319.

[2] H. G. Zeuthen, *Geschichte der Math.* (deutsch v. R. Meyer), Leipzig, 1903, pp. 256 ff.

Germany still continued in a state of national degradation. The Thirty Years' War had dismembered the empire and brutalized the people. Yet this darkest period of Germany's history produced G. W. Leibniz, one of the greatest geniuses of modern times.

There are certain focal points in history toward which the lines of past progress converge, and from which radiate the advances of the future. Such was the age of Newton and Leibniz in the history of mathematics. During fifty years preceding this era several of the brightest and acutest mathematicians bent the force of their genius in a direction which finally led to the discovery of the infinitesimal calculus by Newton and Leibniz. B. Cavalieri, G. P. Roberval, P. Fermat, R. Descartes, J. Wallis, and others had each contributed to the new geometry. So great was the advance made, and so near was their approach toward the invention of the infinitesimal analysis, that both J. Lagrange and P. S. Laplace pronounced their countryman, P. Fermat, to be the first inventor of it. The differential calculus, therefore, was not so much an individual discovery as the grand result of a succession of discoveries by different minds. Indeed, no great discovery ever flashed upon the mind at once, and though those of Newton will influence mankind to the end of the world, yet it must be admitted that Pope's lines are only a "poetic fancy":—

> "Nature and Nature's laws lay hid in night;
> God said, 'Let Newton be,' and all was light."

Isaac Newton (1642–1727) was born at Woolsthorpe, in Lincolnshire, the same year in which Galileo died. At his birth he was so small and weak that his life was despaired of. His mother sent him at an early age to a village school, and in his twelfth year to the public school at Grantham. At first he seems to have been very inattentive to his studies and very low in the school; but when, one day, the little Isaac received a severe kick upon his stomach from a boy who was above him, he labored hard till he ranked higher in school than his antagonist. From that time he continued to rise until he was the head boy.[1] At Grantham, Isaac showed a decided taste for mechanical inventions. He constructed a water-clock, a wind-mill, a carriage moved by the person who sat in it, and other toys. When he had attained his fifteenth year his mother took him home to assist her in the management of the farm, but his great dislike for farmwork and his irresistible passion for study, induced her to send him back to Grantham, where he remained till his eighteenth year, when he entered Trinity College, Cambridge (1660). Cambridge was the real birthplace of Newton's genius. Some idea of his strong intuitive powers may be drawn from the fact that he regarded the theorems of ancient geometry as self-evident truths, and that, without any preliminary study, he made himself master of Descartes' *Geometry*. He

[1] D. Brewster, *The Memoirs of Newton*, Edinburgh, Vol. I, 1855, p. 8.

afterwards regarded this neglect of elementary geometry a mistake in his mathematical studies, and he expressed to Dr. H. Pemberton his regret that "he had applied himself to the works of Descartes and other algebraic writers before he had considered the *Elements* of Euclid with that attention which so excellent a writer deserves." Besides R. Descartes' *Geometry*, he studied W. Oughtred's *Clavis*, J. Kepler's *Optics*, the works of F. Vieta, van Schooten's *Miscellanies*, I. Barrow's *Lectures*, and the works of J. Wallis. He was particularly delighted with Wallis' *Arithmetic of Infinites*, a treatise fraught with rich and varied suggestions. Newton had the good fortune of having for a teacher and fast friend the celebrated Dr. Barrow, who had been elected professor of Greek in 1660, and was made Lucasian professor of mathematics in 1663. The mathematics of Barrow and of Wallis were the starting-points from which Newton, with a higher power than his masters', moved onward into wider fields. Wallis had effected the quadrature of curves whose ordinates are expressed by any integral and positive power of $(1-x^2)$. We have seen how Wallis attempted but failed to interpolate between the areas thus calculated, the areas of other curves, such as that of the circle; how Newton attacked the problem, effected the interpolation, and discovered the Binomial Theorem, which afforded a much easier and direct access to the quadrature of curves than did the method of interpolation; for even though the binomial expression for the ordinate be raised to a fractional or negative power, the binomial could at once be expanded into a series, and the quadrature of each separate term of that series could be effected by the method of Wallis. Newton introduced the system of literal indices.

Newton's study of quadratures soon led him to another and most profound invention. He himself says that in 1665 and 1666 he conceived the method of fluxions and applied them to the quadrature of curves. Newton did not communicate the invention to any of his friends till 1669, when he placed in the hands of Barrow a tract, entitled *De Analysi per Æquationes Numero Terminorum Infinitas*, which was sent by Barrow to John Collins, who greatly admired it. In this treatise the principle of fluxions, though distinctly pointed out, is only partially developed and explained. Supposing the abscissa to increase uniformly in proportion to the time, he looked upon the area of a curve as a nascent quantity increasing by continued fluxion in the proportion of the length of the ordinate. The expression which was obtained for the fluxion he expanded into a finite or infinite series of monomial terms, to which Wallis' rule was applicable. Barrow urged Newton to publish this treatise; "but the modesty of the author, of which the excess, if not culpable, was certainly in the present instance very unfortunate, prevented his compliance." [1] Had this tract

[1] John Playfair, "Progress of the Mathematical and Physical Sciences" in *Encyclopædia Britannica*, 7th Edition.

been published then, instead of forty-two years later, there probably would have been no occasion for that long and deplorable controversy between Newton and Leibniz.

For a long time Newton's method remained unknown, except to his friends and their correspondents. In a letter to Collins, dated December 10th, 1672, Newton states the fact of his invention with one example, and then says: "This is one particular, or rather corollary, of a general method, which extends itself, without any troublesome calculation, not only to the drawing of tangents to any curve lines, whether geometrical or mechanical, or anyhow respecting right lines or other curves, but also to the resolving other abstruser kinds of problems about the crookedness, areas, lengths, centres of gravity of curves, etc.; nor is it (as Hudde's method of Maximis and Minimis) limited to equations which are free from surd quantities. This method I have interwoven with that other of working in equations, by reducing them to infinite series."

These last words relate to a treatise he composed in the year 1671, entitled *Method of Fluxions*, in which he aimed to represent his method as an independent calculus and as a complete system. This tract was intended as an introduction to an edition of Kinckhuysen's *Algebra*, which he had undertaken to publish. "But the fear of being involved in disputes about this new discovery, or perhaps the wish to render it more complete, or to have the sole advantage of employing it in his physical researches, induced him to abandon this design." [1]

Excepting two papers on optics, all of his works appear to have been published only after the most pressing solicitations of his friends and against his own wishes. His researches on light were severely criticised, and he wrote in 1675: "I was so persecuted with discussions arising out of my theory of light that I blamed my own imprudence for parting with so substantial a blessing as my quiet to run after a shadow."

The *Method of Fluxions*, translated by J. Colson from Newton's Latin, was first published in 1736, or sixty-five years after it was written. In it he explains first the expansion into series of fractional and irrational quantities,—a subject which, in his first years of study, received the most careful attention. He then proceeds to the solution of the two following mechanical problems, which constitute the pillars, so to speak, of the abstract calculus:—

"I. The length of the space described being continually (*i. e.* at all times) given; to find the velocity of the motion at any time proposed.

"II. The velocity of the motion being continually given; to find the length of the space described at any time proposed."

Preparatory to the solution, Newton says: "Thus, in the equation $y=x^2$, if y represents the length of the space at any time described,

[1] D. Brewster, *op. cit.*, Vol. 2, 1855, p. 15.

which (time) another space x, by increasing with an uniform celerity $\dot{x}$, measures and exhibits as described: then $2x\dot{x}$ will represent the celerity by which the space y, at the same moment of time, proceeds to be described; and contrarywise."

"But whereas we need not consider the time here, any farther than it is expounded and measured by an equable local motion; and besides, whereas only quantities of the same kind can be compared together, and also their velocities of increase and decrease; therefore, in what follows I shall have no regard to time formally considered, but I shall suppose some one of the quantities proposed, being of the same kind, to be increased by an equable fluxion, to which the rest may be referred, as it were to time; and, therefore, by way of analogy, it may not improperly receive the name of time." In this statement of Newton there is contained his answer to the objection which has been raised against his method, that it introduces into analysis the foreign idea of motion. A quantity thus increasing by uniform fluxion, is what we now call an independent variable.

Newton continues: "Now those quantities which I consider as gradually and indefinitely increasing, I shall hereafter call *fluents*, or *flowing quantities*, and shall represent them by the final letters of the alphabet, v, x, y, and z; . . . and the velocities by which every fluent is increased by its generating motion (which I may call *fluxions*, or simply velocities, or celerities), I shall represent by the same letters pointed, thus, $\dot{v}$, $\dot{x}$, $\dot{y}$, $\dot{z}$. That is, for the celerity of the quantity v I shall put $\dot{v}$, and so for the celerities of the other quantities x, y, and z, I shall put $\dot{x}$, $\dot{y}$, and $\dot{z}$, respectively." It must here be observed that Newton does not take the fluxions themselves infinitely small. The "moments of fluxions," a term introduced further on, are infinitely small quantities. These "moments," as defined and used in the *Method of Fluxions*, are substantially the differentials of Leibniz. De Morgan points out that no small amount of confusion has arisen from the use of the word *fluxion* and the notation $\dot{x}$ by all the English writers previous to 1704, excepting Newton and George Cheyne, in the sense of an infinitely small increment.[1] Strange to say, even in the *Commercium epistolicum* the words *moment* and *fluxion* appear to be used as synonymous.

After showing by examples how to solve the first problem, Newton proceeds to the demonstration of his solution:—

"The moments of flowing quantities (that is, their indefinitely small parts, by the accession of which, in infinitely small portions of time, they are continually increased) are as the velocities of their flowing or increasing.

"Wherefore, if the moment of any one (as x) be represented by the product of its celerity $\dot{x}$ into an infinitely small quantity o (*i. e.* by

[1] A. De Morgan, "On the Early History of Infinitesimals," in *Philosophical Magazine*, November, 1852.

$\dot{x}o$), the moments of the others, v, y, z, will be represented by $\dot{v}o$, $\dot{y}o$, $\dot{z}o$; because $\dot{v}o$, $\dot{x}o$, $\dot{y}o$, and $\dot{z}o$ are to each other as $\dot{v}$, $\dot{x}$, $\dot{y}$, and $\dot{z}$.

"Now since the moments, as $\dot{x}o$ and $\dot{y}o$, are the indefinitely little accessions of the flowing quantities x and y, by which those quantities are increased through the several indefinitely little intervals of time, it follows that those quantities, x and y, after any indefinitely small interval of time, become $x+\dot{x}o$ and $y+\dot{y}o$, and therefore the equation, which at all times indifferently expresses the relation of the flowing quantities, will as well express the relation between $x+\dot{x}o$ and $y+\dot{y}o$, as between x and y; so that $x+\dot{x}o$ and $y+\dot{y}o$ may be substituted in the same equation for those quantities, instead of x and y. Thus let any equation $x^3-ax^2+axy-y^3=0$ be given, and substitute $x+\dot{x}o$ for x, and $y+\dot{y}o$ for y, and there will arise

$$\left.\begin{array}{l} x^3+3x^2\dot{x}o+3x\dot{x}o\dot{x}o+\dot{x}^3o^3 \\ -ax^2-2ax\dot{x}o-a\dot{x}o\dot{x}o \\ +axy+ay\dot{x}o\ \ +a\dot{x}o\dot{y}o \\ \qquad +ax\dot{y}o \\ -y^3\ -3y^2\dot{y}o-3y\dot{y}o\dot{y}o-\dot{y}^3o^3 \end{array}\right\} =0.$$

"Now, by supposition, $x^3-ax^2+axy-y^3=0$, which therefore, being expunged and the remaining terms being divided by o, there will remain

$$3x^2\dot{x}-2ax\dot{x}+ay\dot{x}+ax\dot{y}-3y^2\dot{y}+3x\dot{x}\dot{x}o-a\dot{x}\dot{x}o+a\dot{x}\dot{y}o-3y\dot{y}\dot{y}o$$
$$+\dot{x}^3oo-\dot{y}^3oo=0.$$

But whereas zero is supposed to be infinitely little, that it may represent the moments of quantities, the terms that are multiplied by it will be nothing in respect of the rest (*termini in eam ducti pro nihilo possunt haberi cum aliis collati*); therefore I reject them, and there remains

$$3x^2\dot{x}-2ax\dot{x}+ay\dot{x}+ax\dot{y}-3y^2\dot{y}=0,$$

as above in Example I." Newton here uses infinitesimals.

Much greater than in the first problem were the difficulties encountered in the solution of the second problem, involving, as it does, inverse operations which have been taxing the skill of the best analysts since his time. Newton gives first a special solution to the second problem in which he resorts to a rule for which he has given no proof.

In the general solution of his second problem, Newton assumed homogeneity with respect to the fluxions and then considered three cases: (1) when the equation contains two fluxions of quantities and but one of the fluents; (2) when the equation involves both the fluents as well as both the fluxions; (3) when the equation contains the fluents and the fluxions of three or more quantities. The first case is the easiest since it requires simply the integration of $\frac{dy}{dx}=f(x)$, to which his "special solution" is applicable. The second case demanded nothing less than the general solution of a differential equation of the first order. Those who know what efforts were afterwards needed for the complete exploration of this field in analysis, will not depreciate Newton's work even though he resorted to solutions in form of infinite series. Newton's third case comes now under the solution of partial differential equations. He took the equation $2\dot{x}-\dot{z}+x\dot{y}=0$ and succeeded in finding a particular integral of it.

The rest of the treatise is devoted to the determination of maxima and minima, the radius of curvature of curves, and other geometrical applications of his fluxionary calculus. All this was done previous to the year 1672.

It must be observed that in the *Method of Fluxions* (as well as in his *De Analysi* and all earlier papers) the method employed by Newton is strictly infinitesimal, and in substance like that of Leibniz. Thus, the original conception of the calculus in England, as well as on the Continent, was based on infinitesimals. The fundamental principles of the fluxionary calculus were first given to the world in the *Principia;* but its peculiar notation did not appear until published in the second volume of Wallis' *Algebra* in 1693. The exposition given in the *Algebra* was a contribution of Newton; it rests on infinitesimals. In the first edition of the *Principia* (1687) the description of fluxions is likewise founded on infinitesimals, but in the second (1713) the foundation is somewhat altered. In Book II, Lemma II, of the first edition we read: "Cave tamen intellexeris particulas finitas. *Momenta quam primum finitæ sunt magnitudinis, desinunt esse momenta. Finiri enim repugnat aliquatenus perpetuo eorum incremento vel decremento.* Intelligenda sunt principia jamjam nascentia finitarum magnitudinum." In the second edition the two sentences which we print in italics are replaced by the following: "Particulæ finitæ non sunt momenta sed quantitates ipsæ ex momentis genitæ." Through the difficulty of the phrases in both extracts, this much distinctly appears, that in the first, moments are infinitely small quantities. What else they are in the second is not clear.[1] In the *Quadrature of Curves* of 1704, the infinitely small quantity is completely abandoned. It has been shown that in the *Method of Fluxions* Newton rejected terms involving the quantity o, because they are infinitely small compared with other terms. This reasoning is unsatisfactory; for as long as o is a quantity, though ever so small, this rejection cannot be made without affecting the result. Newton seems to have felt this, for in the *Quadrature of Curves* he remarked that "in mathematics the minutest errors are not to be neglected" (errores quam minimi in rebus mathematicis non sunt contemnendi).

The early distinction between the system of Newton and Leibniz

[1] A. De Morgan, *loc. cit.*, 1852.

lies in this, that Newton, holding to the conception of velocity or fluxion, used the infinitely small increment as a means of determining it, while with Leibniz the relation of the infinitely small increments is itself the object of determination. The difference between the two rests mainly upon a difference in the mode of generating quantities.

We give Newton's statement of the method of fluxions or rates, as given in the introduction to his *Quadrature of Curves*. "I consider mathematical quantities in this place not as consisting of very small parts, but as described by a continued motion. Lines are described, and thereby generated, not by the apposition of parts, but by the continued motion of points; superficies by the motion of lines; solids by the motion of superficies; angles by the rotation of the sides; portions of time by continual flux: and so on in other quantities. These geneses really take place in the nature of things, and are daily seen in the motion of bodies. . . .

"Fluxions are, as near as we please (*quam proxime*), as the increments of fluents generated in times, equal and as small as possible, and to speak accurately, they are in the prime ratio of nascent increments; yet they can be expressed by any lines whatever, which are proportional to them."

Newton exemplifies this last assertion by the problem of tangency: Let AB be the abscissa, BC the ordinate, VCH the tangent, Ec the increment of the ordinate, which produced meets VH at T, and Cc the increment of the curve. The right line Cc being produced to K, there are formed three small triangles, the rectilinear CEc, the mixtilinear CEc, and the rectilinear CET. Of these, the first is evidently the smallest, and the last the greatest. Now suppose the ordinate bc to move into the place BC, so that the point c exactly coincides with the point C; CK, and therefore the curve Cc, is coincident with the tangent CH, Ec is absolutely equal to ET, and the mixtilinear evanescent triangle CEc is, in the last form, similar to the triangle CET; and its evanescent sides CE, Ec, Cc, will be proportional to CE, ET, and CT, the sides of the triangle CET.

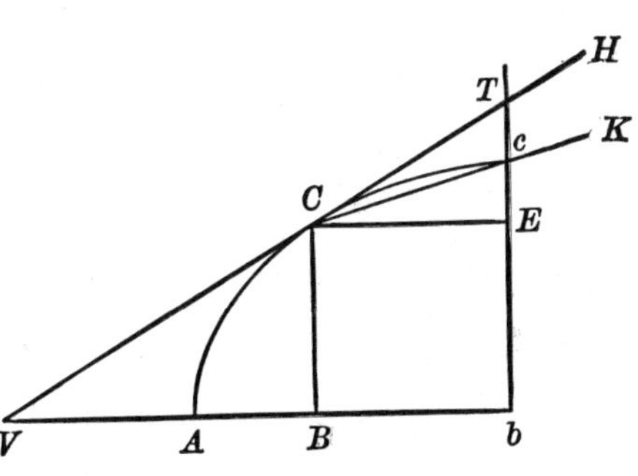

Hence it follows that the fluxions of the lines AB, BC, AC, being in the last ratio of their evanescent increments, are proportional to the sides of the triangle CET, or, which is all one, of the triangle VBC similar thereunto. As long as the points C and c are distant from each other by an interval, however small, the line CK will stand apart by a small angle from the tangent CH. But when CK coincides with CH, and the lines CE, Ec, cC reach their ultimate ratios, then the points C and c accurately coincide and are one and the same. Newton then adds that "in mathematics the minutest errors are not to be neglected." This is plainly a rejection of the postulates of Leibniz. The doctrine of infinitely small quantities is here renounced in a manner which would lead one to suppose that Newton had never held it himself. Thus it appears that Newton's doctrine was different in different periods. Though, in the above reasoning, the Charybdis of infinitesimals is safely avoided, the dangers of a Scylla stare us in the face. We are required to believe that a point may be considered a triangle, or that a triangle can be inscribed in a point; nay, that three dissimilar triangles become similar and equal when they have reached their ultimate form in one and the same point.

In the introduction to the *Quadrature of Curves* the fluxion of x^n is determined as follows:—

"In the same time that x, by flowing, becomes $x+o$, the power x^n becomes $(x+o)^n$, *i. e.* by the method of infinite series

$$x^n+nox^{n-1}+\frac{n^2-n}{2}o^2x^{n-2}+\text{etc.},$$

and the increments

$$o \text{ and } nox^{n-1}+\frac{n^2-n}{2}o^2x^{n-2}+\text{etc.},$$

are to one another as

$$1 \text{ to } nx^{n-1}+\frac{n^2-n}{2}ox^{n-2}+\text{etc.}$$

"Let now the increments vanish, and their last proportion will be 1 to nx^{n-1}: hence the fluxion of the quantity x is to the fluxion of the quantity x^n as $1 : nx^{n-1}$.

"The fluxion of lines, straight or curved, in all cases whatever, as also the fluxions of superficies, angles, and other quantities, can be obtained in the same manner by the method of prime and ultimate ratios. But to establish in this way the analysis of infinite quantities, and to investigate prime and ultimate ratios of finite quantities, nascent or evanescent, is in harmony with the geometry of the ancients; and I have endeavored to show that, in the method of fluxions, it is not necessary to introduce into geometry infinitely small quantities." This mode of differentiating does not remove all the difficulties connected with the subject. When o becomes nothing, then we get the ratio $\frac{o}{o}=nx^{n-1}$, which needs further elucidation. Indeed, the method of Newton, as delivered by himself, is encumbered with difficulties and objections. Later we shall state Bishop Berkeley's objection to this reasoning. Even among the ablest admirers of Newton, there have been obstinate disputes respecting his explanation of his method of "prime and ultimate ratios."

The so-called "method of limits" is frequently attributed to Newton, but the pure method of limits was never adopted by him as his method of constructing the calculus. All he did was to establish in his *Principia* certain principles which are applicable to that method, but which he used for a different purpose. The first lemma of the first book has been made the foundation of the method of limits:—

"Quantities and the ratios of quantities, which in any finite time converge continually to equality, and before the end of that time approach nearer the one to the other than by any given difference, become ultimately equal."

In this, as well as in the lemmas following this, there are obscurities and difficulties. Newton appears to teach that a variable quantity and its limit will ultimately coincide and be equal.

The full title of Newton's *Principia* is *Philosophiæ Naturalis Principia Mathematica.* It was printed in 1687 under the direction, and at the expense, of Edmund Halley. A second edition was brought out in 1713 with many alterations and improvements, and accompanied by a preface from Roger Cotes. It was sold out in a few months, but a pirated edition published in Amsterdam supplied the demand. The third and last edition which appeared in England during Newton's lifetime was published in 1726 by Henry Pemberton. The *Principia* consists of three books, of which the first two, constituting the great bulk of the work, treat of the mathematical principles of natural philosophy, namely, the laws and conditions of motions and forces. In the third book is drawn up the constitution of the universe as deduced from the foregoing principles. The great principle underlying this memorable work is that of universal gravitation. The first book was completed on April 28, 1686. After the remarkably short period of three months, the second book was finished. The third book is the result of the next nine or ten months' labors. It is only a sketch of a much more extended elaboration of the subject which he had planned, but which was never brought to completion.

The law of gravitation is enunciated in the first book. Its discovery envelops the name of Newton in a halo of perpetual glory. The current version of the discovery is as follows: it was conjectured by Robert Hooke (1635–1703), C. Huygens, E. Halley, C. Wren, I. Newton, and others, that, if J. Kepler's third law was true (its absolute accuracy was doubted at that time), then the attraction between the earth and other members of the solar system varied inversely as the square of the distance. But the proof of the truth or falsity of the guess was wanting. In 1666 Newton reasoned, in substance, that if g represent the acceleration of gravity on the surface of the earth, r

be the earth's radius, R the distance of the moon from the earth, T the time of lunar revolution, and a a degree at the equator, then, if the law is true,

$$g\frac{r^2}{R^2}=4\pi^2\frac{R}{T^2}, \text{ or } g=\frac{4\pi}{T^2}\left(\frac{R}{r}\right)^3 \cdot 180a.$$

The data at Newton's command gave $R=60.4r$, $T=2{,}360{,}628$ seconds, but a only 60 instead of $69\frac{1}{2}$ English miles. This wrong value of a rendered the calculated value of g smaller than its true value, as known from actual measurement. It looked as though the law of inverse squares were not the true law, and Newton laid the calculation aside. In 1684 he casually ascertained at a meeting of the Royal Society that Jean Picard had measured an arc of the meridian, and obtained a more accurate value for the earth's radius. Taking the corrected value for a, he found a figure for g which corresponded to the known value. Thus the law of inverse squares was verified. In a scholium in the *Principia*, Newton acknowledged his indebtedness to Huygens for the laws on centrifugal force employed in his calculation.

The perusal by the astronomer Adams of a great mass of unpublished letters and manuscripts of Newton forming the Portsmouth collection (which remained private property until 1872, when its owner placed it in the hands of the University of Cambridge) seems to indicate that the difficulties encountered by Newton in the above calculation were of a different nature. According to Adams, Newton's numerical verification was fairly complete in 1666, but Newton had not been able to determine what the attraction of a spherical shell upon an external point would be. His letters to E. Halley show that he did not suppose the earth to attract as though all its mass were concentrated into a point at the centre. He could not have asserted, therefore, that the assumed law of gravity was verified by the figures, though for long distances he might have claimed that it yielded close approximations. When Halley visited Newton in 1684, he requested Newton to determine what the orbit of a planet would be if the law of attraction were that of inverse squares. Newton had solved a similar problem for R. Hooke in 1679, and replied at once that it was an ellipse. After Halley's visit, Newton, with Picard's new value for the earth's radius, reviewed his early calculation, and was able to show that if the distances between the bodies in the solar system were so great that the bodies might be considered as points, then their motions were in accordance with the assumed law of gravitation. In 1685 he completed his discovery by showing that a sphere whose density at any point depends only on the distance from the centre attracts an external point as though its whole mass were concentrated at the centre.

Newton's unpublished manuscripts in the Portsmouth collection show that he had worked out, by means of fluxions and fluents, his lunar calculations to a higher degree of approximation than that given in the *Principia*, but that he was unable to interpret his results geometrically. The papers in that collection throw light upon the mode by which Newton arrived at some of the results in the *Principia*, as, for instance, the famous solution in Book II, Prop. 35, Scholium, of the problem of the solid of revolution which moves through a resisting medium with the least resistance. The solution is unproved in the *Principia*, but is demonstrated by Newton in the draft of a letter to David Gregory of Oxford, found in the Portsmouth collection.[1]

It is chiefly upon the *Principia* that the fame of Newton rests. David Brewster calls it "the brightest page in the records of human reason." Let us listen, for a moment, to the comments of P. S. Laplace, the foremost among those followers of Newton who grappled with the subtle problems of the motions of planets under the influence of gravitation: "Newton has well established the existence of the principle which he had the merit of discovering, but the development of its consequences and advantages has been the work of the successors of this great mathematician. The imperfection of the infinitesimal calculus, when first discovered, did not allow him completely to resolve the difficult problems which the theory of the universe offers; and he was oftentimes forced to give mere hints, which were always uncertain till confirmed by rigorous analysis. Notwithstanding these unavoidable defects, the importance and the generality of his discoveries respecting the system of the universe, and the most interesting points of natural philosophy, the great number of profound and original views, which have been the origin of the most brilliant discoveries of the mathematicians of the last century, which were all presented with much elegance, will insure to the *Principia* a lasting pre-eminence over all other productions of the human mind."

Newton's *Arithmetica universalis*, consisting of algebraical lectures delivered by him during the first nine years he was professor at Cambridge, were published in 1707, or more than thirty years after they were written. This work was published by William Whiston (1667–1752). We are not accurately informed how Whiston came in possession of it, but according to some authorities its publication was a breach of confidence on his part. He succeeded Newton in the Lucasian professorship at Cambridge.

The *Arithmetica universalis* contains new and important results on the theory of equations. Newton states Descartes' rule of signs in accurate form and gives formulæ expressing the sum of the powers of the roots up to the sixth power and by an "and so on" makes it evident that they can be extended to any higher power. Newton's formulæ take the implicit form, while similar formulæ given earlier

[1] O. Bolza, in *Bibliotheca mathematica*, 3. S., Vol. 13, 1913, pp. 146–149. For a bibliography of this "problem of Newton" on the surface of least resistance, see *L'Intermédiaire des mathématiciens*, Vol. 23, 1916, pp. 81–84.

by Albert Girard take the explicit form, as do also the general formulæ derived later by E. Waring. Newton uses his formulæ for fixing an upper limit of real roots; the sum of any even power of all the roots must exceed the same even power of any one of the roots. He established also another limit: A number is an upper limit, if, when substituted for x, it gives to $f(x)$ and to all its derivatives the same sign. In 1748 Colin Maclaurin proved that an upper limit is obtained by adding unity to the absolute value of the largest negative coefficient of the equation. Newton showed that in equations with real coefficients, imaginary roots always occur in pairs. His inventive genius is grandly displayed in his rule for determining the inferior limit of the number of imaginary roots, and the superior limits for the number of positive and negative roots. Though less expeditious than Descartes', Newton's rule always gives as close, and generally closer, limits to the number of positive and negative roots. Newton did not prove his rule.

Some light was thrown upon it by George Campbell and Colin Maclaurin, in the *Philosophical Transactions*, of the years 1728 and 1729. But no complete demonstration was found for a century and a half, until, at last, Sylvester established a remarkable general theorem which includes Newton's rule as a special case. Not without interest is Newton's suggestion that the conchoid be admitted as a curve to be used in geometric constructions, along with the straight line and circle, since the conchoid can be used for the duplication of a cube and trisection of an angle—to one or the other of which every problem involving curves of the third or fourth degree can be reduced.

The treatise on *Method of Fluxions* contains Newton's method of approximating to the roots of numerical equations. Substantially the same explanation is given in his *De analysi per æquationes numero terminorum infinitas*. He explains it by working one example, namely the now famous cubic [1] $y^3-2y-5=0$. The earliest printed account appeared in Wallis' *Algebra*, 1685, chapter 94. Newton assumes that an approximate value is already known, which differs from the true value by less than one-tenth of that value. He takes $y=2$ and substitutes $y=2+p$ in the equation, which becomes $p^3+6p^2+10p-1=0$. Neglecting the higher powers of p, he gets $10p-1=0$. Taking $p=.1+q$, he gets $q^3+6.3q^2+11.23q+.061=0$. From $11.23q+.061=0$ he gets $q=-.0054+r$, and by the same process, $r=-.00004853$. Finally $y=2+.1-.0054-.00004853=2.09455147$. Newton arranges his work in a paradigm. He seems quite aware that his method may fail. If there is doubt, he says, whether $p=.1$ is sufficiently close to the truth, find p from $6p^2+10p-1=0$. He does not show that even this latter method will always answer. By the same mode of pro-

[1] For quotations from Newton, see F. Cajori, "Historical Note on the Newton-Raphson Method of Approximation," *Amer. Math. Monthly*, Vol. 18, 1911, pp. 29–33.

cedure, Newton finds, by a rapidly converging series, the value of y in terms of a and x, in the equation $y^3+axy+aay-x^3-2a^3=0$.

In 1690, Joseph Raphson (1648–1715), a fellow of the Royal Society of London, published a tract, *Analysis æquationum universalis*. His method closely resembles that of Newton. The only difference is this, that Newton derives each successive step, p, q, r, of approach to the root, from a *new* equation, while Raphson finds it each time by substitution in the original equation. In Newton's cubic, Raphson would not find the second correction by the use of $x^3+6x^2+10x-1=0$, but would substitute $2.1+q$ in the original equation, finding $q=-.0054$. He would then substitute $2.0946+r$ in the original equation, finding $r=-.00004853$, and so on. Raphson does not mention Newton; he evidently considered the difference sufficient for his method to be classed independently. To be emphasized is the fact that the process which in modern texts goes by the name of "Newton's method of approximation," is really not Newton's method, but Raphson's modification of it. The *form* now so familiar, $a-\frac{f(a)}{f'(a)}$ was not used by Newton, but was used by Raphson. To be sure, Raphson does not use this notation; he writes $f(a)$ and $f'(a)$ out in full as polynomials. It is doubtful, whether this method should be named after Newton alone. Though not identical with Vieta's process, it resembles Vieta's. The chief difference lies in the divisor used. The divisor $f'(a)$ is much simpler, and easier to compute than Vieta's divisor. Raphson's version of the process represents what J. Lagrange recognized as an advance on the scheme of Newton. The method is "plus simple que celle de Newton."[1] Perhaps the name "Newton-Raphson method" would be a designation more nearly representing the facts of history. We may add that the solution of numerical equations was considered geometrically by Thomas Baker in 1684 and Edmund Halley in 1687, but in 1694 Halley "had a very great desire of doing the same in numbers." The only difference between Halley's and Newton's own method is that Halley solves a quadratic equation at each step, Newton a linear equation. Halley modified also certain algebraic expressions yielding approximate cube and fifth roots, given in 1692 by the Frenchman, *Thomas Fantet de Lagny* (1660–1734). In 1705 and 1706 Lagny outlines a method of differences; such a method, less systematically developed, had been previously explained in England by John Collins. By this method, if $a, b, c, \ldots$ are in arithmetical progression, then a root may be found approximately from the first, second, and higher differences of $f(a)$, $f(b)$, $f(c), \ldots$

Newton's *Method of Fluxions* contains also "Newton's parallelogram," which enabled him, in an equation, $f(x, y)=0$, to find a series

[1] Lagrange, *Résolution des equat. num.*, 1798, Note V, p. 138.

in powers of x equal to the variable y. The great utility of this rule lay in its determining the *form* of the series; for, as soon as the law was known by which the exponents in the series vary, then the expansion could be effected by the method of indeterminate coefficients. The rule is still used in determining the infinite branches to curves, or their figure at multiple points. Newton gave no proof for it, nor any clue as to how he discovered it. The proof was supplied half a century later, by A. G. Kästner and G. Cramer, independently.[1]

In 1704 was published, as an appendix to the *Opticks*, the *Enumeratio linearum tertii ordinis*, which contains theorems on the theory of curves. Newton divides cubics into seventy-two species, arranged in larger groups, for which his commentators have supplied the names "genera" and "classes," recognizing fourteen of the former and seven (or four) of the latter. He overlooked six species demanded by his principles of classification, and afterwards added by J. Stirling, William Murdoch (1754–1839), and G. Cramer. He enunciates the remarkable theorem that the five species which he names "divergent parabolas" give by their projection every cubic curve whatever. As a rule, the tract contains no proofs. It has been the subject of frequent conjecture how Newton deduced his results. Recently we have gotten at the facts, since much of the analysis used by Newton and a few additional theorems have been discovered among the Portsmouth papers. An account of the four holograph manuscripts on this subject has been published by W. W. Rouse Ball, in the *Transactions of the London Mathematical Society* (vol. xx, pp. 104–143). It is interesting to observe how Newton begins his research on the classification of cubic curves by the algebraic method, but, finding it laborious, attacks the problem geometrically, and afterwards returns again to analysis.

Space does not permit us to do more than merely mention Newton's prolonged researches in other departments of science. He conducted a long series of experiments in optics and is the author of the corpuscular theory of light. The last of a number of papers on optics, which he contributed to the Royal Society, 1687, elaborates the theory of "fits." He explained the decomposition of light and the theory of the rainbow. By him were invented the reflecting telescope and the sextant (afterwards re-invented by Thomas Godfrey of Philadelphia [2] and by John Hadley). He deduced a theoretical expression for the velocity of sound in air, engaged in experiments on chemistry, elasticity, magnetism, and the law of cooling, and entered upon geological speculations.

During the two years following the close of 1692, Newton suffered

[1] S. Günther, *Vermischte Untersuchungen zur Geschichte d. math. Wiss.*, Leipzig 1876, pp. 136–187.

[2] F. Cajori, *Teaching and History of Mathematics in the U. S.*, Washington, 1890, p. 42.

from insomnia and nervous irritability. Some thought that he labored under temporary mental aberration. Though he recovered his tranquillity and strength of mind, the time of great discoveries was over; he would study out questions propounded to him, but no longer did he by his own accord enter upon new fields of research. The most noted investigation after his sickness was the testing of his lunar theory by the observations of Flamsteed, the astronomer royal. In 1695 he was appointed warden, and in 1699 master of the mint, which office he held until his death. His body was interred in Westminster Abbey, where in 1731 a magnificent monument was erected, bearing an inscription ending with, "Sibi gratulentur mortales tale tantumque exstitisse humani generis decus." It is not true that the Binomial Theorem is also engraved on it.

We pass to Leibniz, the second and independent inventor of the calculus. **Gottfried Wilhelm Leibniz** (1646–1716) was born in Leipzig. No period in the history of any civilized nation could have been less favorable for literary and scientific pursuits than the middle of the seventeenth century in Germany. Yet circumstances seem to have happily combined to bestow on the youthful genius an education hardly otherwise obtainable during this darkest period of German history. He was brought early in contact with the best of the culture then existing. In his fifteenth year he entered the University of Leipzig. Though law was his principal study, he applied himself with great diligence to every branch of knowledge. Instruction in German universities was then very low. The higher mathematics was not taught at all. We are told that a certain John Kuhn lectured on Euclid's *Elements*, but that his lectures were so obscure that none except Leibniz could understand them. Later on, Leibniz attended, for a half-year, at Jena, the lectures of Erhard Weigel, a philosopher and mathematician of local reputation. In 1666 Leibniz published a treatise, *De Arte Combinatoria*, in which he does not pass beyond the rudiments of mathematics, but which contains remarkable plans for a theory of mathematical logic, a symbolic method with formal rules obviating the necessity of thinking. Vaguely such plans had been previously suggested by R. Descartes and Pierre Hérigone. In manuscripts which Leibniz left unpublished he enunciated the principal properties of what is now called logical multiplication, addition, negation, identity, class-induction and the null-class.[1] Other theses written by him at this time were metaphysical and juristical in character. A fortunate circumstance led Leibniz abroad. In 1672 he was sent by Baron Boineburg on a political mission to Paris. He there formed the acquaintance of the most distinguished men of the age. Among these was C. Huygens, who presented a copy of his work on the oscillation of the pendulum to Leibniz, and first led the gifted young German to the study of higher mathematics. In 1673 Leibniz

[1] See Philip E. B. Jourdain in *Quarterly Jour. of Math.*, Vol. 41, 1910, p. 329.

went to London, and remained there from January till March. He there became incidentally acquainted with the mathematician John Pell, to whom he explained a method he had found on the summation of series of numbers by their differences. Pell told him that a similar formula had been published by Gabriel Mouton (1618–1694) as early as 1670, and then called his attention to N. Mercator's work on the rectification of the parabola. While in London, Leibniz exhibited to the Royal Society his arithmetical machine, which was similar to B. Pascal's, but more efficient and perfect. After his return to Paris, he had the leisure to study mathematics more systematically. With indomitable energy he set about removing his ignorance of higher mathematics. C. Huygens was his principal master. He studied the geometric works of R. Descartes, Honorarius Fabri, Gregory St. Vincent, and B. Pascal. A careful study of infinite series led him to the discovery of the following expression for the ratio of the circumference to the diameter of the circle, previously discovered by James Gregory:—

$$\frac{\pi}{4}=1-\frac{1}{3}+\frac{1}{5}-\frac{1}{7}+\frac{1}{9}-\text{etc.}$$

This elegant series was found in the same way as N. Mercator's on the hyperbola. C. Huygens was highly pleased with it and urged him on to new investigations. In 1673 Leibniz derived the series

$$\text{arc tan } x=x-\tfrac{1}{3}x^3+\tfrac{1}{5}x^5-\ \ldots,$$

from which most of the practical methods of computing π have been obtained. This series had been previously discovered by James Gregory, and was used by *Abraham Sharp* (1651–1742) under instructions from E. Halley for calculating π to 72 places. In 1706 *John Machin* (1680–1751), professor of astronomy at Gresham College in London, obtained 100 places by using an expression that is obtained from the relation

$$\frac{\pi}{4}=4 \text{ arc tan } \tfrac{1}{5}-\text{arc tan } \tfrac{1}{239},$$

by substituting Gregory's infinite series for

$$\text{arc tan } \tfrac{1}{5} \text{ and arc tan } \tfrac{1}{239}.$$

Machin's formula was used in 1874 by *William Shanks* (1812–1882) for computing π to 707 places.

Leibniz entered into a detailed study of the quadrature of curves and thereby became intimately acquainted with the higher mathematics. Among the papers of Leibniz is still found a manuscript on quadratures, written before he left Paris in 1676, but which was never printed by him. The more important parts of it were embodied in articles published later in the *Acta eruditorum*.

In the study of Cartesian geometry the attention of Leibniz was drawn early to the direct and inverse problems of tangents. The direct problem had been solved by Descartes for the simplest curves only; while the inverse had completely transcended the power of his analysis. Leibniz investigated both problems for any curve; he constructed what he called the *triangulum characteristicum*—an infinitely small triangle between the infinitely small part of the curve coinciding with the tangent, and the differences of the ordinates and abscissas. A curve is here considered to be a polygon. The *triangulum characteristicum* is similar to the triangle formed by the tangent, the ordinate of the point of contact, and the sub-tangent, as well as to that between the ordinate, normal, and sub-normal. It was employed by I. Barrow in England, but Leibniz states that he obtained it from Pascal. From it Leibniz observed the connection existing between the direct and inverse problems of tangents. He saw also that the latter could be carried back to the quadrature of curves. All these results are contained in a manuscript of Leibniz, written in 1673. One mode used by him in effecting quadratures was as follows: The rectangle formed by a sub-normal p and an element a (*i. e.* infinitely small part of the abscissa) is equal to the rectangle formed by the ordinate y and the element l of that ordinate; or in symbols, $pa=yl$. But the summation of these rectangles from zero on gives a right triangle equal to half the square of the ordinate. Thus, using Cavalieri's notation, he gets

$$\text{omn. } pa=\text{omn. } yl=\frac{y^2}{2} \text{ (\textit{omn.} meaning \textit{omnia}, all).}$$

But $y=\text{omn. } l$; hence

$$\text{omn. } \overline{\overline{\text{omn. } l}\,\frac{l}{a}}=\frac{\overline{\text{omn. } l}^2}{2a}.$$

This equation is especially interesting, since it is here that Leibniz first introduces a new notation. He says: "It will be useful to write $\int$ for *omn.*, as $\int l$ for *omn. l*, that is, the sum of the l's"; he then writes the equation thus:—

$$\frac{\int \bar{l}^2}{2a}=\int\overline{\int \bar{l}\,\frac{l}{a}}.$$

From this he deduced the simplest integrals, such as

$$\int x=\frac{x^2}{2}, \quad \int(x+y)=\int x+\int y.$$

Since the symbol of summation $\int$ raises the dimensions, he concluded that the opposite calculus, or that of differences d, would lower them. Thus, if $\int l = ya$, then $l = \frac{ya}{d}$. The symbol d was at first placed by Leibniz in the denominator, because the lowering of the power of a term was brought about in ordinary calculation by division. The manuscript giving the above is dated October 29th, 1675.[1] This, then, was the memorable day on which the notation of the new calculus came to be,—a notation which contributed enormously to the rapid growth and perfect development of the calculus.

Leibniz proceeded to apply his new calculus to the solution of certain problems then grouped together under the name of the Inverse Problems of Tangents. He found the cubical parabola to be the solution to the following: To find the curve in which the subnormal is reciprocally proportional to the ordinate. The correctness of his solution was tested by him by applying to the result the method of tangents of Baron René François de Sluse (1622–1685) and reasoning backwards to the original supposition. In the solution of the third problem he changes his notation from $\frac{x}{d}$ to the now usual notation dx. It is worthy of remark that in these investigations, Leibniz nowhere explains the significance of dx and dy, except at one place in a marginal note: "Idem est dx et $\frac{x}{d}$, id est, differentia inter duas x proximas." Nor does he use the term *differential*, but always *difference*. Not till ten years later, in the *Acta eruditorum*, did he give further explanations of these symbols. What he aimed at principally was to determine the change an expression undergoes when the symbol $\int$ or d is placed before it. It may be a consolation to students wrestling with the elements of the differential calculus to know that it required Leibniz considerable thought and attention [2] to determine whether $dx\,dy$ is the same as $d(xy)$, and $\frac{dx}{dy}$ the same as $d\frac{x}{y}$. After considering these questions at the close of one of his manuscripts, he concluded that the expressions were not the same, though he could not give the true value for each. Ten days later, in a manuscript dated November 21, 1675, he found the equation $yd\overline{x} = d\overline{xy} - xd\overline{y}$, giving an expression for $d(xy)$, which he observed to be true for all curves. He succeeded also in eliminating dx from a differential equation, so that it contained only dy, and thereby led to the solution of the problem under consideration. "Behold, a most elegant way

[1] C. J. Gerhardt, *Entdeckung der höheren Analysis*. Halle, 1855, p. 125.

[2] C. I. Gerhardt, *Entdeckung der Differenzialrechnung durch Leibniz*, Halle, 1848, pp. 25, 41.

by which the problems of the inverse method of tangents are solved, or at least are reduced to quadratures!" Thus he saw clearly that the inverse problems of tangents could be solved by quadratures, or, in other words, by the integral calculus. In course of a half-year he discovered that the direct problem of tangents, too, yielded to the power of his new calculus, and that thereby a more general solution than that of R. Descartes could be obtained. He succeeded in solving all the special problems of this kind, which had been left unsolved by Descartes. Of these we mention only the celebrated problem proposed to Descartes by F. de Beaune, viz. to find the curve whose ordinate is to its sub-tangent as a given line is to that part of the ordinate which lies between the curve and a line drawn from the vertex of the curve at a given inclination to the axis.

Such was, in brief, the progress in the evolution of the new calculus made by Leibniz during his stay in Paris. Before his departure, in October, 1676, he found himself in possession of the most elementary rules and formulæ of the infinitesimal calculus.

From Paris, Leibniz returned to Hanover by way of London and Amsterdam. In London he met John Collins, who showed him a part of his scientific correspondence. Of this we shall speak later. In Amsterdam he discussed mathematics with R. F. de Sluse, and became satisfied that his own method of constructing tangents not only accomplished all that Sluse's did, but even more, since it could be extended to three variables, by which tangent planes to surfaces could be found; and especially, since neither irrationals nor fractions prevented the immediate application of his method.

In a paper of July 11, 1677, Leibniz gave correct rules for the differentiation of sums, products, quotients, powers, and roots. He had given the differentials of a few negative and fractional powers, as early as November, 1676, but had made some mistakes. For $d\sqrt{x}$ he had given the erroneous value $\frac{1}{\sqrt{x}}$, and in another place the value $-\frac{1}{2}x^{-\frac{1}{2}}$; for $d\frac{1}{x^2}$ occurs in one place the wrong value, $-\frac{2}{x^2}$, while a few lines lower is given $-\frac{2}{x^3}$, its correct value.

In 1682 was founded in Leipzig the *Acta eruditorum*, a journal sometimes known by the name of *Leipzig Acts*. It was a partial imitation of the French *Journal des Savans* (founded in 1665), and the literary and scientific review published in Germany. Leibniz was a frequent contributor. E. W. Tschirnhausen, who had studied mathematics in Paris with Leibniz, and who was familiar with the new analysis of Leibniz, published in the *Acta eruditorum* a paper on quadratures, which consists principally of subject-matter communicated by Leibniz to Tschirnhausen during a controversy which they had had on this subject. Fearing that Tschirnhausen might claim as his own and publish the notation and rules of the differential calculus, Leibniz decided, at last, to make public the fruits of his inventions. In 1684, or nine years after the new calculus first dawned upon the mind of Leibniz, and nineteen years after Newton first worked at fluxions, and three years before the publication of Newton's *Principia*, Leibniz published, in the *Acta eruditorum*, his first paper on the differential calculus. He was unwilling to give to the world all his treasures, but chose those parts of his work which were most abstruse and least perspicuous. This epoch-making paper of only six pages bears the title: "Nova methodus pro maximis et minimis, itemque tangentibus, quæ nec fractas nec irrationales quantitates moratur, et singulare pro illis calculi genus." The rules of calculation are briefly stated without proof, and the meaning of dx and dy is not made clear. Printer's errors increased the difficulty of comprehending the subject. It has been inferred from this that Leibniz himself had no definite and settled ideas on this subject. Are dy and dx finite or infinitesimal quantities? At first they appear, indeed, to have been taken as finite, when he says: "We now call any line selected at random dx, then we designate the line which is to dx as y is to the sub-tangent, by dy, which is the difference of y." Leibniz then ascertains, by his calculus, in what way a ray of light passing through two differently refracting media, can travel easiest from one point to another; and then closes his article by giving his solution, in a few words, of F. de Beaune's problem. Two years later (1686) Leibniz published in the *Acta eruditorum* a paper containing the rudiments of the integral calculus. The quantities dx and dy are there treated as infinitely small. He showed that by the use of his notation, the properties of curves could be fully expressed by equations. Thus the equation

$$y=\sqrt{2x-x^2}+\int\frac{dx}{\sqrt{2x-x^2}}$$

characterizes the cycloid.[1]

The great invention of Leibniz, now made public by his articles in the *Acta eruditorum*, made little impression upon the mass of mathematicians. In Germany no one comprehended the new calculus except Tschirnhausen, who remained indifferent to it. The author's statements were too short and succinct to make the calculus generally understood. The first to take up the study of it were two foreigners,—the Scotchman *John Craig*, and the Swiss *Jakob* (*James*) *Bernoulli*. The latter wrote Leibniz a letter in 1687, wishing to be initiated into the mysteries of the new analysis. Leibniz was then travelling abroad, so that this letter remained unanswered till 1690. James Bernoulli

[1] C. I. Gerhardt, *Geschichte der Mathematik in Deutschland*, Münich, 1877, p. 159.

succeeded, meanwhile, by close application, in uncovering the secrets of the differential calculus without assistance. He and his brother John proved to be mathematicians of exceptional power. They applied themselves to the new science with a success and to an extent which made Leibniz declare that it was as much theirs as his. Leibniz carried on an extensive correspondence with them, as well as with other mathematicians. In a letter to John Bernoulli he suggests, among other things, that the integral calculus be improved by reducing integrals back to certain fundamental irreducible forms. The integration of logarithmic expressions was then studied. The writings of Leibniz contain many innovations, and anticipations of since prominent methods. Thus he made use of variable parameters, laid the foundation of *analysis in situ*, introduced in a manuscript of 1678 the notion of determinants (previously used by the Japanese), in his effort to simplify the expression arising in the elimination of the unknown quantities from a set of linear equations. He resorted to the device of breaking up certain fractions into the sum of other fractions for the purpose of easier integration; he explicitly assumed the principle of continuity; he gave the first instance of a "singular solution," and laid the foundation to the theory of envelopes in two papers, one of which contains for the first time the terms *co-ordinate* and *axes of co-ordinates*. He wrote on osculating curves, but his paper contained the error (pointed out by John Bernoulli, but not admitted by Leibniz) that an osculating circle will necessarily cut a curve in four consecutive points. Well known is his theorem on the nth differential coefficient of the product of two functions of a variable. Of his many papers on mechanics, some are valuable, while others contain grave errors. Leibniz introduced in 1692 the use of the word *function*, but not in the modern sense Later, in 1694, Jakob Bernoulli used the word in the Leibnizian sense. In the appendix to a letter to Leibniz, dated July 5, 1698, John Bernoulli uses the word in a more nearly modern sense: "earum [applicatarum] quæcunque functiones per alias applicatas *PZ* expressæ." In 1718 John Bernoulli arrives at the definition of function as a "quantity composed in any manner of a variable and any constants." (On appelle ici fonction d'une grandeur variable, une quantité composée de quelque manière que ce soit de cette grandeur variable et de constantes.) [1]

Leibniz made important contributions to the notation of mathematics. Not only is our notation of the differential and integral calculus due to him, but he used the sign of equality in writing proportions, thus $a:b=c:d$. In Leibnizian manuscripts occurs $\sim$ for "similar" and $\simeq$ for "equal and similar" or "congruent." [2] Says

[1] See M. Cantor, *op. cit.*, Vol. III, 2 Ed., 1901, pp. 215, 216, 456, 457; *Encyclopédie des sciences mathématiques*, Tome II, Vol. I, pp. 3–5.

[2] Leibniz, *Werke* Ed. Gerhardt, 3. Folge, Bd. V, p. 153. See also J. Tropfke, *op. cit.*, Vol. II, 1903, p. 12.

P. E. B. Jourdain,[1] "Leibniz himself attributed all of his mathematical discoveries to his improvements in notation."

Before tracing the further development of the calculus we shall sketch the history of that long and bitter controversy between English and Continental mathematicians on the invention of the calculus. The question was, did Leibniz invent it independently of Newton, or was he a plagiarist?

We must begin with the early correspondence between the parties appearing in this dispute. Newton had begun using his notation of fluxions in 1665.[2] In 1669 I. Barrow sent John Collins Newton's tract, *De Analysi per equationes*, etc.

The first visit of Leibniz to London extended from the 11th of January until March, 1673. He was in the habit of committing to writing important scientific communications received from others. In 1890 C. J. Gerhardt discovered in the royal library at Hanover a sheet of manuscript with notes taken by Leibniz during this journey.[3] They are headed "Observata Philosophica in itinere Anglicano sub initium anni 1673." The sheet is divided by horizontal lines into sections. The sections given to Chymica, Mechanica, Magnetica, Botanica, Anatomica, Medica, Miscellanea, contain extensive memoranda, while those devoted to mathematics have very few notes. Under Geometrica he says only this: "Tangentes omnium figurarum. Figurarum geometricarum explicatio per motum puncti in moto lati." We suspect from this that Leibniz had read Isaac Barrow's lectures. Newton is referred to only under Optica. Evidently Leibniz did not obtain a knowledge of fluxions during this visit to London, nor is it claimed that he did by his opponents.

Various letters of I. Newton, J. Collins, and others, up to the beginning of 1676, state that Newton invented a method by which tangents could be drawn without the necessity of freeing their equations from irrational terms. Leibniz announced in 1674 to H. Oldenburg, then secretary of the Royal Society, that he possessed very general analytical methods, by which he had found theorems of great importance on the quadrature of the circle by means of series. In answer, Oldenburg stated Newton and James Gregory had also discovered methods of quadratures, which extended to the circle. Leibniz desired to have these methods communicated to him; and Newton, at the request of Oldenburg and Collins, wrote to the former the celebrated letters of June 13 and October 24, 1676. The first contained the Binomial Theorem and a variety of other matters relating to infinite series and quadratures; but nothing directly on the method of

[1] P. E. B. Jourdain, *The Nature of Mathematics*, London, p. 71.

[2] J. Edleston, *Correspondence of Sir Isaac Newton and Professor Cotes*, London, 1850, p. xxi; A. De Morgan, "Fluxions" and "Commercium Epistolicum" in the *Penny Cyclopædia*.

[3] C. J. Gerhardt, "Leibniz in London" in *Sitzungsberichte der K. Preussischen Academie d. Wissensch. zu Berlin*, Feb., 1891.

fluxions. Leibniz in reply speaks in the highest terms of what Newton had done, and requests further explanation. Newton in his second letter just mentioned explains the way in which he found the Binomial Theorem, and also communicates his method of fluxions and fluents in form of an anagram in which all the letters in the sentence communicated were placed in alphabetical order. Thus Newton says that his method of drawing tangents was

6a cc d æ 13e ff 7i 3l 9n 4o 4qrr 4s 9t 12vx.

The sentence was, "Data æquatione quotcunque fluentes quantitates involvente fluxiones invenire, et vice versa." ("Having any given equation involving never so many flowing quantities, to find the fluxions, and vice versa.") Surely this anagram afforded no hint. Leibniz wrote a reply to John Collins, in which, without any desire of concealment, he explained the principle, notation, and the use of the differential calculus.

The death of Oldenburg brought this correspondence to a close. Nothing material happened till 1684, when Leibniz published his first paper on the differential calculus in the *Acta eruditorum*, so that while Newton's claim to the priority of invention must be admitted by all, it must also be granted that Leibniz was the first to give the full benefit of the calculus to the world. Thus, while Newton's invention remained a secret, communicated only to a few friends, the calculus of Leibniz was spreading over the Continent. No rivalry or hostility existed, as yet, between the illustrious scientists. Newton expressed a very favorable opinion of Leibniz's inventions, known to him through the above correspondence with Oldenburg, in the following celebrated scholium (*Principia*, first edition, 1687, Book II, Prop. 7, scholium):—

"In letters which went between me and that most excellent geometer, G. W. Leibniz, ten years ago, when I signified that I was in the knowledge of a method of determining maxima and minima, of drawing tangents, and the like, and when I concealed it in transposed letters involving this sentence (Data æquatione, etc., above cited), that most distinguished man wrote back that he had also fallen upon a method of the same kind, and communicated his method, which hardly differed from mine, except in his forms of words and symbols."

As regards this passage, we shall see that Newton was afterwards weak enough, as De Morgan says: "First, to deny the plain and obvious meaning, and secondly, to omit it entirely from the third edition of the *Principia*." On the Continent, great progress was made in the calculus by Leibniz and his coadjutors, the brothers James and John Bernoulli, and Marquis de l'Hospital. In 1695 John Wallis informed Newton by letter that "he had heard that his notions of fluxions passed in Holland with great applause by the name of 'Leibniz's Calculus Differentialis.'" Accordingly Wallis stated in the preface to a volume of his works that the calculus differentialis was Newton's method of fluxions which had been communicated to Leibniz in the Oldenburg letters. A review of Wallis' works, in the *Acta eruditorum* for 1696, reminded the reader of Newton's own admission in the scholium above cited.

For fifteen years Leibniz had enjoyed unchallenged the honor of being the inventor of his calculus. But in 1699 *Fatio de Duillier* (1664–1753), a Swiss, who had settled in England, stated in a mathematical paper, presented to the Royal Society, his conviction that I. Newton was the first inventor; adding that, whether Leibniz, the second inventor, had borrowed anything from the other, he would leave to the judgment of those who had seen the letters and manuscripts of Newton. This was the first distinct insinuation of plagiarism. It would seem that the English mathematicians had for some time been cherishing suspicions unfavorable to Leibniz. A feeling had doubtless long prevailed that Leibniz, during his second visit to London in 1676, had or might have seen among the papers of John Collins, Newton's *Analysis per æquationes*, etc., which contained applications of the fluxionary method, but no systematic development or explanation of it. Leibniz certainly did see at least part of this tract. During the week spent in London, he took note of whatever interested him among the letters and papers of Collins. His memoranda discovered by C. I. Gerhardt in 1849 in the Hanover library fill two sheets.[1] The one bearing on our question is headed "Excerpta ex tractatu Newtoni Msc. de Analysi per æquationes numero terminorum infinitas." The notes are very brief, excepting those *De resolutione æquationum affectarum*, of which there is an almost complete copy. This part was evidently new to him. If he examined Newton's entire tract, the other parts did not particularly impress him. From it he seems to have gained nothing pertaining to the infinitesimal calculus. By the previous introduction of his own algorithm he had made greater progress than by what came to his knowledge in London. Nothing mathematical that he had received engaged his thoughts in the immediate future, for on his way back to Holland he composed a lengthy dialogue on mechanical subjects.

Fatio de Duillier's insinuations lighted up a flame of discord which a whole century was hardly sufficient to extinguish. Leibniz, who had never contested the priority of Newton's discovery, and who appeared to be quite satisfied with Newton's admission in his scholium, now appears for the first time in the controversy. He made an animated reply in the *Acta eruditorum* and complained to the Royal Society of the injustice done him.

Here the affair rested for some time. In the *Quadrature of Curves*, published 1704, for the first time, a formal exposition of the method and notation of fluxions was made public. In 1705 appeared an un-

[1] C. I. Gerhardt, "Leibniz in London," *loc. cit.*

favorable review of this in the *Acta eruditorum*, stating that Newton uses and always has used fluxions for the differences of Leibniz. This was considered by Newton's friends an imputation of plagiarism on the part of their chief, but this interpretation was always strenuously resisted by Leibniz. *John Keill* (1671–1721), professor of astronomy at Oxford, undertook with more zeal than judgment the defence of Newton. In a paper inserted in the *Philosophical Transactions* of 1708, he claimed that Newton was the first inventor of fluxions and "that the same calculus was afterward published by Leibniz, the name and the mode of notation being changed." Leibniz complained to the secretary of the Royal Society of bad treatment and requested the interference of that body to induce Keill to disavow the intention of imputing fraud. John Keill was not made to retract his accusation; on the contrary, was authorized by Newton and the Royal Society to explain and defend his statement. This he did in a long letter. Leibniz thereupon complained that the charge was now more open than before, and appealed for justice to the Royal Society and to Newton himself. The Royal Society, thus appealed to as a judge, appointed a committee which collected and reported upon a large mass of documents—mostly letters from and to Newton, Leibniz, Wallis, Collins, etc. This report, called the *Commercium epistolicum*, appeared in the year 1712 and again in 1722 and 1725, with a Recensio prefixed, and additional notes by Keill. The final conclusion in the *Commercium epistolicum* was that Newton was "the first inventor." But this was not to the point. The question was not whether Newton was the first inventor, but whether Leibniz had stolen the method. The committee had not formally ventured to assert their belief that Leibniz was a plagiarist. In the following sentence they insinuated that Leibniz did take or might have taken, his method from that of Newton: "And we find no mention of his (Leibniz's) having any other *Differential Method* than *Mouton's* before his Letter of 21st of *June*, 1677, which was a year after a Copy of Mr. *Newton's* Letter, of the 10th of *December*, 1672, had been sent to *Paris* to be communicated to him; and about four years after *Mr. Collins* began to communicate that Letter to his Correspondents; in which Letter the Method of *Fluxions* was sufficiently describ'd to any intelligent Person."

About 1850 it was shown that what H. Oldenburg sent to Leibniz was not Newton's letter of Dec. 10, 1672, but only excerpts from it which omitted Newton's method of drawing tangents and could not possibly convey an idea of fluxions. Oldenburg's letter was found among the Leibniz manuscripts in the Royal Library at Hanover, and was published by C. I. Gerhardt in 1846, 1848, 1849 and 1855,[1] and again later.

[1] See *Essays on the Life and Work of Newton* by Augustus De Morgan, edited, with notes and appendices, by Philip E. B. Jourdain, Chicago and London, 1914. Jourdain gives on p. 102 the bibliography of the publications of Newton and Leibniz.

Moreover, when J. Edleston in 1850 published the *Correspondence of Sir Isaac Newton and Professor Cotes*, it became known that the Royal Society in 1712 had not one, but two, parcels of Collins. One parcel contained letters of James Gregory, and Isaac Newton's letter of Dec. 10, 1672, in full; the other parcel, which was marked "To Leibnitz, the 14th of June, 1676 About Mr. Gregories remains," contained an abridgment of a part of the contents of the first parcel, with nothing but an allusion to Newton's method described in his letter of Dec. 10, 1672. In the *Commercium epistolicum* Newton's letter was printed in full and no mention was made of the existence of the second parcel that was marked "To Leibnitz. . . ." Thus the *Commercium epistolicum* conveyed the impression that Newton's uncurtailed letter of Dec. 10, 1672, had reached Leibniz in which fluxions "was sufficiently described to any intelligent person," while as a matter of fact the method is not described at all in the letter which Leibniz received.

Leibniz protested only in private letters against the proceeding of the Royal Society, declaring that he would not answer an argument so weak. John Bernoulli, in a letter to Leibniz, which was published later in an anonymous tract, is as decidedly unfair towards Newton as the friends of the latter had been towards Leibniz. John Keill replied, and then Newton and Leibniz appear as mutual accusers in several letters addressed to third parties. In a letter dated April 9, 1716, and sent to Antonio Schinella Conti (1677–1749), an Italian priest then residing in London, Leibniz again reminded Newton of the admission he had made in the scholium, which he was now desirous of disavowing; Leibniz also states that he always believed Newton, but that, seeing him connive at accusations which he must have known to be false, it was natural that he (Leibniz) should begin to doubt. Newton did not reply to this letter, but circulated some remarks among his friends which he published immediately after hearing of the death of Leibniz, November 14, 1716. This paper of Newton gives the following explanation pertaining to the scholium in question: "He [Leibniz] pretends that in my book of principles I allowed him the invention of the calculus differentialis, independently of my own; and that to attribute this invention to myself is contrary to my knowledge there avowed. But in the paragraph there referred unto I do not find one word to this purpose." In the third edition of the *Principia*, 1726, Newton omitted the scholium and substituted in its place another, in which the name of Leibniz does not appear.

National pride and party feeling long prevented the adoption of impartial opinions in England, but now it is generally admitted by

We recommend J. B. Biot and F. Lefort's edition of the *Commercium epistolicum*, Paris, 1856, which exhibits all the alterations made in the different reprints of this publication and reproduces also H. Oldenburg's letter to Leibniz of July 26, 1676, and other important documents bearing on the controversy.

nearly all familiar with the matter, that Leibniz really was an independent inventor. Perhaps the most telling evidence to show that Leibniz was an independent inventor is found in the study of his mathematical papers (collected and edited by C. I. Gerhardt, in seven volumes, Berlin, 1849–1863), which point out a gradual and natural evolution of the rules of the calculus in his own mind. "There was throughout the whole dispute," says De Morgan, "a confusion between the knowledge of fluxions or differentials and that of a *calculus* of fluxions or differentials; that is, a digested method with general rules."

This controversy is to be regretted on account of the long and bitter alienation which it produced between English and Continental mathematicians. It stopped almost completely all interchange of ideas on scientific subjects. The English adhered closely to Newton's methods and, until about 1820, remained, in most cases, ignorant of the brilliant mathematical discoveries that were being made on the Continent. The loss in point of scientific advantage was almost entirely on the side of Britain. The only way in which this dispute may be said, in a small measure, to have furthered the progress of mathematics, is through the challenge problems by which each side attempted to annoy its adversaries.

The recurring practice of issuing challenge problems was resumed at this time by Leibniz. They were, at first, not intended as defiances, but merely as exercises in the new calculus. Such was the problem of the isochronous curve (to find the curve along which a body falls with uniform velocity), proposed by him to the Cartesians in 1687, and solved by Jakob Bernoulli, himself, and Johann Bernoulli. Jakob Bernoulli proposed in the *Acta eruditorum* of 1690 the question to find the curve (the catenary) formed by a chain of uniform weight suspended freely from its ends. It was resolved by C. Huygens, G. W. Leibniz, Johann Bernoulli, and Jakob Bernoulli himself; the properties of the catenary were worked out methodically by David Gregory[1] of Oxford and himself. In 1696 Johann Bernoulli challenged the best mathematicians in Europe to solve the difficult problem, to find the curve (the cycloid) along which a body falls from one point to another in the shortest possible time. Leibniz solved it the day he received it. Newton, de l'Hospital, and the two Bernoullis gave solutions. Newton's appeared anonymously in the *Philosophical Transactions*, but Johann Bernoulli recognized in it his powerful mind, "tanquam," he says, "ex ungue leonem." The problem of orthogonal trajectories (a system of curves described by a known law being given, to describe a curve which shall cut them all at right angles) was proposed by Johann Bernoulli in a letter to G. W. Leibniz in 1694. Later it was long printed in the *Acta eruditorum*, but failed at first to receive much

[1] *Phil. Trans.*, London, 1697.

attention. It was again proposed in 1716 by Leibniz, to feel the pulse of the English mathematicians.

This may be considered as the first defiance problem professedly aimed at the English. Newton solved it the same evening on which it was delivered to him, although he was much fatigued by the day's work at the mint. His solution, as published, was a general plan of an investigation rather than an actual solution, and was, on that account, criticised by Johann Bernoulli as being of no value. Brook Taylor undertook the defence of it, but ended by using very reprehensible language. Johann Bernoulli was not to be outdone in incivility, and made a bitter reply. Not long afterwards Taylor sent an open defiance to Continental mathematicians of a problem on the integration of a fluxion of complicated form which was known to very few geometers in England and supposed to be beyond the power of their adversaries. The selection was injudicious, for Johann Bernoulli had long before explained the method of this and similar integrations. It served only to display the skill and augment the triumph of the followers of Leibniz. The last and most unskilful challenge was by John Keill. The problem was to find the path of a projectile in a medium which resists proportionally to the square of the velocity. Without first making sure that he himself could solve it, Keill boldly challenged Johann Bernoulli to produce a solution. The latter resolved the question in very short time, not only for a resistance proportional to the square, but to any power of the velocity. Suspecting the weakness of the adversary, he repeatedly offered to send his solution to a confidential person in London, provided Keill would do the same. Keill never made a reply, and Johann Bernoulli abused him and cruelly exulted over him.[1]

The explanations of the fundamental principles of the calculus, as given by Newton and Leibniz, lacked clearness and rigor. For that reason it met with opposition from several quarters. In 1694 *Bernhard Nieuwentijt* (1654–1718) of Holland denied the existence of differentials of higher orders and objected to the practice of neglecting infinitely small quantities. These objections Leibniz was not able to meet satisfactorily. In his reply he said the value of $\frac{dy}{dx}$ in geometry could be expressed as the ratio of finite quantities. In the interpretation of dx and dy Leibniz vacillated.[2] At one time they appear in his writings as finite lines; then they are called infinitely small quantities, and again, *quantitates inassignabiles*, which spring from *quantitates assignabiles* by the law of continuity. In this last presentation Leibniz approached nearest to Newton.

[1] John Playfair, "Progress of the Mathematical and Physical Sciences" in *Encyclopædia Britannica*, 7th Ed., continued in the 8th Ed. by Sir John Leslie.

[2] Consult G. Vivanti, *Il concetto d'Infinitesimo. Saggio storico.* Nuova edizione. Napoli, 1901.

In England the principles of fluxions were boldly attacked by Bishop **George Berkeley** (1685–1753), the eminent metaphysician, in a publication called the *Analyst* (1734). He argued with great acuteness, contending, among other things, that the fundamental idea of supposing a finite ratio to exist between terms absolutely evanescent—"the ghosts of departed quantities," as he called them—was absurd and unintelligible. Berkeley claimed that the second and third fluxions were even more mysterious than the first fluxion. His contention that no geometrical quantity can be exhausted by division is in consonance with the claim made by Zeno in his "dichotomy," and the claim that the actual infinite cannot be realized. Most modern readers recognize these contentions as untenable. Berkeley declared as axiomatic a lemma involving the shifting of the hypothesis: If x receives an increment i, where i is expressly supposed to be some quantity, then the increment of x^n, divided by i, is found to be $nx^{n-1}+n(n-1)/2\, x^{n-2}i+\ \ldots$ If now you take $i=0$, *the hypothesis is shifted* and there is a manifest sophism in retaining any result that was obtained on the supposition that i is not zero. Berkeley's lemma found no favor among English mathematicians until 1803 when Robert Woodhouse openly accepted it. The fact that correct results are obtained in the differential calculus by incorrect reasoning is explained by Berkeley on the theory of "a compensation of errors." This theory was later advanced also by Lagrange and L. N. M. Carnot. The publication of Berkeley's *Analyst* was the most spectacular mathematical event of the eighteenth century in England. Practically all British discussions of fluxional concepts of that time involve issues raised by Berkeley. Berkeley's object in writing the *Analyst* was to show that the principles of fluxions are no clearer than those of Christianity. He referred to an "infidel mathematician" (Edmund Halley), of whom the story is told [1] that, when he jested concerning theological questions, he was repulsed by Newton with the remark, "I have studied these things; you have not." A friend of Berkeley, when on a bed of sickness, refused spiritual consolation, because the great mathematician Halley had convinced him of the inconceivability of the doctrines of Christianity. This induced Berkeley to write the *Analyst.*☼

Replies to the *Analyst* were published by *James Jurin* (1684–1750) of Trinity College, Cambridge under the pseudonym of "Philalethes Cantabrigiensis" and by *John Walton* of Dublin. There followed several rejoinders. Jurin's defence of Newton's fluxions did not meet the approval of the mathematician, **Benjamin Robins** (1707–1751). In a Journal, called the *Republick of Letters* (London) and later in the *Works of the Learned*, a long and acrimonious controversy was carried on between Jurin and Robins, and later between Jurin and Henry Pemberton (1694–1771), the editor of the third edition of

[1] Mach *Mechanics*, 1907, pp. 448–449.

Newton's *Principia.* The question at issue was the precise meaning of certain passages in the writings of Newton: Did Newton hold that there are variables which reach their limits? Jurin answered "Yes"; Robins and Pemberton answered "No." The debate between Jurin and Robins is important in the history of the theory of limits. Though holding a narrow view of the concept of a limit Robins deserves credit for rejecting all infinitely small quantities and giving a logically quite coherent presentation of fluxions in a pamphlet, called *A Discourse concerning the Nature and Certainty of Sir Isaac Newton's Methods of Fluxions*, 1735. This and Maclaurin's *Fluxions*, 1742, mark the top-notch of mathematical rigor, reached during the eighteenth century in the exposition of the calculus. Both before and after the period of eight years, 1834–1842, there existed during the eighteenth century in Great Britain a mixture of Continental and British conceptions of the new calculus, a superposition of British symbols and phraseology upon the older Continental concepts. Newton's notation was poor and Leibniz's philosophy of the calculus was poor. The mixture represented the temporary survival of the least fit of both systems. The subsequent course of events was a superposition of the Leibnizian notation and phraseology upon the limit-concept as developed by Newton, Jurin, Robins, Maclaurin, D'Alembert and later writers.

In France *Michel Rolle* for a time rejected the differential calculus and had a controversy with P. Varignon on the subject. Perhaps the most powerful argument in favor of the new calculus was the consistency of the results to which it led. Famous is D'Alemhert's advice to young students: "Allez en avant, et la foi vous viendra."

Among the most vigorous promoters of the calculus on the Continent were the Bernoullis. They and Euler made Basel in Switzerland famous as the cradle of great mathematicians. The family of Bernoullis furnished in course of a century eight members who distinguished themselves in mathematics. We subjoin the following genealogical table:—

Nicolaus Bernoulli, the Father

Jakob, 1654–1705	*Nicolaus*	**Johann,** 1667–1748
	Nicolaus, 1687–1759	*Nicolaus,* 1695–1726 **Daniel,** 1700–1782 *Johann,* 1710–1790

Daniel *Johann,* 1744–1807 *Jakob,* 1759–1789

Most celebrated were the two brothers Jakob (James) and Johann (John), and Daniel, the son of John. Jakob and Johann were staunch friends of G. W. Leibniz and worked hand in hand with him. **Jakob (James) Bernoulli** (1654–1705) was born in Basel. Becoming interested in the calculus, he mastered it without aid from a teacher. From 1687 until his death he occupied the mathematical chair at the University of Basel. He was the first to give a solution to Leibniz's problem of the isochronous curve. In his solution, published in the *Acta eruditorum*, 1690, we meet for the first time with the word *integral*. Leibniz had called the integral calculus *calculus summatorius*, but in 1696 the term *calculus integralis* was agreed upon between Leibniz and Johann Bernoulli. Jakob Bernoulli gave in 1694 in the *Acta eruditorum* the formula for the radius of curvature in rectangular co-ordinates. At the same time he gave the formula also in polar co-ordinates. He was one of the first to use polar co-ordinates in a general manner and not simply for spiral shaped curves.[1] Jakob proposed the problem of the catenary, then proved the correctness of Leibniz's construction of this curve, and solved the more complicated problems, supposing the string to be (1) of variably density, (2) extensible, (3) acted upon at each point by a force directed to a fixed centre. Of these problems he published answers without explanations, while his brother Johann gave in addition their theory. He determined the shape of the "elastic curve" formed by an elastic plate or rod fixed at one end and bent by a weight applied to the other end; of the "lintearia," a flexible rectangular plate with two sides fixed horizontally at the same height, filled with a liquid; of the "velaria," a rectangular sail filled with wind. In the *Acta eruditorum* of 1694 he makes reference to the lemniscate, a curve which "formam refert jacentis notæ octonarii ∞, seu complicitæ in nodum fasciæ, sive lemnisci." That this curve is a special case of Cassini's oval remained long unnoticed and was first pointed out by Pietro Ferroni in 1782 and G. Saladini in 1806. Jakob Bernoulli studied the loxodromic and logarithmic spirals, in the last of which he took particular delight from its remarkable property of reproducing itself under a variety of conditions. Following the example of Archimedes, he willed that the curve be engraved upon his tombstone with the inscription "*eadem mutata resurgo*." In 1696 he proposed the famous problem of isoperimetrical figures, and in 1701 published his own solution. He wrote a work on *Ars Conjectandi*, published in 1713, eight years after his death. It consists of four parts. The first contains Huygens' treatise on probability, with a valuable commentary. The second part is on permutations and combinations, which he uses in a proof of the binomial theorem for the case of positive integral exponents; it contains a formula for the sum of the rth powers of the first n integers, which involves the so-called "numbers of Bernoulli." He could boast that by means of it he calculated *intra semi-quadrantem horae* the sum of the 10*th* powers of the first thousand integers. The third part contains solutions of problems on probability. The fourth part is the most important, even though left incomplete. It contains "Ber-

[1] G. Eneström in *Bibliotheca mathematica*, 3. S., Vol. 13, 1912, p. 76.

noulli's theorem": If $(r+s)^{nt}$, where the letters are integers and $t=r+s$, is expanded by the binomial theorem, then by taking n large enough the ratio of u (denoting the sum of the greatest term and the n preceding terms and the n following terms) to the sum of the remaining terms may be made as great as we please. Letting r and s be proportional to the probability of the happening and failing of an event in a single trial, then u corresponds to the probability that in nt trials the number of times the event happens will lie between $n(r-1)$ and $n(r+1)$, both inclusive. Bernoulli's theorem "will ensure him a permanent place in the history of the theory of probability." [1] Prominent contemporary workers on probability were Montmort in France and De Moivre in England. In December, 1913, the Academy of Sciences of Petrograd celebrated the bicentenary of the "law of large numbers," Jakob Bernoulli's *Ars conjectandi* having been published at Basel in 1713. Of his collected works, in three volumes, one was printed in 1713, the other two in 1744.

Johann (John) Bernoulli (1667–1748) was initiated into mathematics by his brother. He afterwards visited France, where he met Nicolas Malebranche, Giovanni Domenico Cassini, P. de Lahire, P. Varignon, and G. F. de l'Hospital. For ten years he occupied the mathematical chair at Gröningen and then succeeded his brother at Basel. He was one of the most enthusiastic teachers and most successful original investigators of his time. He was a member of almost every learned society in Europe. His controversies were almost as numerous as his discoveries. He was ardent in his friendships, but unfair, mean, and violent toward all who incurred his dislike—even his own brother and son. He had a bitter dispute with Jakob on the isoperimetrical problem. Jakob convicted him of several paralogisms. After his brother's death he attempted to substitute a disguised solution of the former for an incorrect one of his own. Johann admired the merits of G. W. Leibniz and L. Euler, but was blind to those of I. Newton. He immensely enriched the integral calculus by his labors. Among his discoveries are the exponential calculus, the line of swiftest descent, and its beautiful relation to the path described by a ray passing through strata of variable density. In 1694 he explained in a letter to l'Hospital the method of evaluating the indeterminate form $\frac{0}{0}$. He treated trigonometry by the analytical method, studied caustic curves and trajectories. Several times he was given prizes by the Academy of Science in Paris.

Of his sons, **Nicolaus** and **Daniel** were appointed professors of mathematics at the same time in the Academy of St. Petersburg. The former soon died in the prime of life; the latter returned to Basel in 1733, where he assumed the chair of experimental philosophy. His

[1] I. Todhunter, *History of Theor. of Prob.*, p. 77.

first mathematical publication was the solution of a differential equation proposed by J. F. Riccati. He wrote a work on hydrodynamics. He was the first to use a suitable notation for inverse trigonometric functions; in 1729 he used AS. to represent arcsine; L. Euler in 1736 used At for arctangent.[1] Daniel Bernoulli's investigations on probability are remarkable for their boldness and originality. He proposed the theory of *moral expectation*, which he thought would give results more in accordance with our ordinary notions than the theory of *mathematical probability*. He applies his moral expectation to the so-called "Petersburg problem": A throws a coin in the air; if head appears at the first throw he is to receive a shilling from B, if head does not appear until the second throw he is to receive 2 shillings, if head does not appear until the third throw he is to receive 4 shillings, and so on: required the expectation of A. By the mathematical theory, A's expectation is infinite, a paradoxical result. A given sum of money not being of equal importance to every man, account should be taken of *relative* values. Suppose A starts with a sum a, then the moral expectation in the Petersburg problem is finite, according to Daniel Bernoulli, when a is finite; it is 2 when $a=0$, about 3 when $a=10$, about 6 when $a=1000$.[2] The Petersburg problem was discussed by P. S. Laplace, S. D. Poisson and G. Cramer. Daniel Bernoulli's "moral expectation" has become classic, but no one ever makes use of it. He applies the theory of probability to insurance; to determine the mortality caused by small-pox at various stages of life; to determine the number of survivors at a given age from a given number of births; to determine how much inoculation lengthens the average duration of life. He showed how the differential calculus could be used in the theory of probability. He and L. Euler enjoyed the honor of having gained or shared no less than ten prizes from the Academy of Sciences in Paris. Once, while travelling with a learned stranger who asked his name, he said, "I am Daniel Bernoulli." The stranger could not believe that his companion actually was that great celebrity, and replied "I am Isaac Newton."

Johann Bernoulli (born 1710) succeeded his father in the professorship of mathematics at Basel. He captured three prizes (on the capstan, the propagation of light, and the magnet) from the Academy of Sciences at Paris. **Nicolaus Bernoulli** (born 1687) held for a time the mathematical chair at Padua which Galileo had once filled. He proved in 1742 that $\frac{\partial^2 A}{\partial t \partial u}=\frac{\partial^2 A}{\partial u \partial t}$. **Johann Bernoulli** (born 1744) at the age of nineteen was appointed astronomer royal at Berlin, and afterwards director of the mathematical department of the Academy. His brother *Jakob* took upon himself the duties of the chair of experi-

[1] G. Eneström in *Bibliotheca mathematica*, Vol. 6, pp. 319–321; Vol. 14, p. 78.

[2] I. Todhunter, *Hist. of the Theor. of Prob.*, p. 220.

mental physics at Basel, previously performed by his uncle Jakob, and later was appointed mathematical professor in the Academy at St. Petersburg.

Brief mention will now be made of some other mathematicians belonging to the period of Newton, Leibniz, and the elder Bernoullis.

Guillaume François Antoine l'Hospital (1661–1704), a pupil of Johann Bernoulli, has already been mentioned as taking part in the challenges issued by Leibniz and the Bernoullis. He helped powerfully in making the calculus of Leibniz better known to the mass of mathematicians by the publication of a treatise thereon, the *Analyse des infiniment petits*, Paris, 1696. This contains the method of finding the limiting value of a fraction whose two terms tend toward zero at the same time, due to Johann Bernoulli.

Another zealous French advocate of the calculus was **Pierre Varignon** (1654–1722). In *Mém. de Paris*, Année MDCCIV, Paris, 1722, he follows Ja. Bernoulli in the use of polar co-ordinates, ρ and ω. Letting $x=\rho$ and $y=l\omega$, the equations thus changed represent wholly different curves. For instance, the parabolas $x^m=a^{m-1}y$ become Fermatian spirals. **Joseph Saurin** (1659–1737) solved the delicate problem of how to determine the tangents at the multiple points of algebraic curves. **François Nicole** (1683–1758) in 1717 issued an elementary treatise on finite differences, in which he finds the sums of a considerable number of interesting series. He wrote also on roulettes, particularly spherical epicycloids, and their rectification. Also interested in finite differences was **Pierre Raymond de Montmort** (1678–1719). His chief writings, on the theory of probability, served to stimulate his more distinguished successor, De Moivre. Montmort gave the first general solution of the Problem of Points. **Jean Paul de Gua** (1713–1785) gave the demonstration of Descartes' rule of signs, now given in books. This skilful geometer wrote in 1740 a work on analytical geometry, the object of which was to show that most investigations on curves could be carried on with the analysis of Descartes quite as easily as with the calculus. He shows how to find the tangents, asymptotes, and various singular points of curves of all degrees, and proves by perspective that several of these points can be at infinity. **Michel Rolle** (1652–1719) is the author of a theorem named after him. That theorem is not found in his *Traité d'algèbre* of 1690, but occurs in his *Methode pour résoudre les egalitez*, Paris, 1691.[1] The name "Rolle's theorem" was first used by M. W. Drobisch (1802–1896) of Leipzig in 1834 and by Giusto Bellavitis in 1846. His *Algèbre* contains his "method of cascades." In an equation in v which he has transformed so that its signs become alternately plus and minus, he puts $v=x+z$ and arranges the result according to the descending powers of x. The coefficients of x^n, x^{n-1}, . . ., when equated to zero, are

[1] See F. Cajori on the history of Rolle's Theorem in *Bibliotheca mathematica*, 3rd S., Vol. II, 1911, pp. 300–313.

called "cascades." They are the successive derivatives of the original equation in v, each put equal to zero. Now comes a theorem which in modern version is: Between two successive real roots of $f'(v)=0$ there cannot be more than one real root of $f(v)=0$. To ascertain the root-limits of a given equation, Rolle begins with the cascade of lowest degree and ascends, solving each as he proceeds. This process is very laborious.

Of Italian mathematicians, Riccati and Fagnano must not remain unmentioned. **Jacopo Francesco, Count Riccati** (1676–1754) is best known in connection with his problem, called Riccati's equation, published in the *Acta eruditorum* in 1724. He succeeded in integrating this differential equation for some special cases. Long before this Jakob Bernoulli had made attempts to solve this equation, but without success. A geometrician of remarkable power was **Giulio Carlo, Count de Fagnano** (1682–1766). He discovered the following formula, $\pi=2i \log \frac{1-i}{1+i}$, in which he anticipated L. Euler in the use of imaginary exponents and logarithms. His studies on the rectification of the ellipse and hyperbola are the starting-points of the theory of elliptic functions. He showed, for instance, that two arcs of an ellipse can be found in an indefinite number of ways, whose difference is expressible by a right line. In the rectification of the lemniscate he reached results which connect with elliptic functions; he showed that its arc can be divided geometrically in n equal parts, if n is $2 \cdot 2^m$, $3 \cdot 2^m$, or $5 \cdot 2^m$. He gave expert advice to Pope Benedict XIV regarding the safety of the cupola of St. Peter's at Rome. In return the Pope promised to publish his mathematical productions. For some reason the promise was not fulfilled and they were not published until 1750. Fagnano's mathematical works were re-published in 1911 and 1912 by the Italian Society for the Advancement of Science.

In Germany the only noted contemporary of Leibniz is **Ehrenfried Walter Tschirnhausen** (1651–1708), who discovered the caustic of reflection, experimented on metallic reflectors and large burning-glasses, and gave a method of transforming equations named after him. He endeavored to solve equations of any degree by removing all the terms except the first and last. This procedure had been tried before him by the Frenchman *François Dulaurens* and by the Scotchman James Gregory.[1] Gregory's *Vera circuli et hyperbolæ quadratura* (Patavii, 1667) is noteworthy as containing a novel attempt, namely, to prove that the quadrature of the circle cannot be effected by the aid of algebra. His ideas were not understood in his day, not even by C. Huygens with whom he had a controversy on this subject. James Gregory's proof could not now be considered binding. Believing that the most simple methods (like those of the ancients) are the most

[1] G. Eneström in *Bibliotheca mathematica*, 3. S., Vol. 9, 1908–9, pp. 258, 259.

correct, Tschirnhausen concluded that in the researches relating to the properties of curves the calculus might as well be dispensed with.

After the death of Leibniz there was in Germany not a single mathematician of note. **Christian Wolf** (1679–1754), professor at Halle, was ambitious to figure as successor of Leibniz, but he "forced the ingenious ideas of Leibniz into a pedantic scholasticism, and had the unenviable reputation of having presented the elements of the arithmetic, algebra, and analysis developed since the time of the Renaissance in the form of Euclid,—of course only in outward form, for into the spirit of them he was quite unable to penetrate" (H. Hankel).

The contemporaries and immediate successors of Newton in Great Britain were men of no mean merit. We have reference to R. Cotes, B. Taylor, L. Maclaurin, and A. de Moivre. We are told that at the death of **Roger Cotes** (1682–1716), Newton exclaimed, "If Cotes had lived, we might have known something." It was at the request of Dr. Bentley that R. Cotes undertook the publication of the second edition of Newton's *Principia*. His mathematical papers were published after his death by Robert Smith, his successor in the Plumbian professorship at Trinity College. The title of the work, *Harmonia Mensurarum*, was suggested by the following theorem contained in it: If on each radius vector, through a fixed point O, there be taken a point R, such that the reciprocal of OR be the arithmetic mean of the reciprocals of $OR_1, OR_2, \ldots OR_n$, then the locus of R will be a straight line. In this work progress was made in the application of logarithms and the properties of the circle to the calculus of fluents. To Cotes we owe a theorem in trigonometry which depends on the forming of factors of $x^n - 1$. In the *Philosophical Transactions* of London, published 1714, he develops an important formula, reprinted in his *Harmonia Mensurarum*, which in modern notation is $i\ \phi = \log\ (\cos\phi + i.\ \sin\phi.)$ Usually this formula is attributed to L. Euler. Cotes studied the curve $\rho^2\theta = a^2$, to which he gave the name "lituus." Chief among the admirers of Newton were B. Taylor and C. Maclaurin. The quarrel between English and Continental mathematicians caused them to work quite independently of their great contemporaries across the Channel.

Brook Taylor (1685–1731) was interested in many branches of learning, and in the latter part of his life engaged mainly in religious and philosophic speculations. His principal work, *Methodus incrementorum directa et inversa*, London, 1715–1717, added a new branch to mathematics, now called "finite differences," of which he was the inventor.☆ He made many important applications of it, particularly to the study of the form of movement of vibrating strings, the reduction of which to mechanical principles was first attempted by him. This work contains also "Taylor's theorem," and, as a special case of it, what is now called "Maclaurin's Theorem." Taylor discovered his theorem at least three years before its appearance in print. He gave it in a letter to John Machin, dated July 26, 1712. Its importance was not recognized by analysts for over fifty years, until J. Lagrange pointed out its power. His proof of it does not consider the question of convergency, and is quite worthless. The first more rigorous proof was given a century later by A. L. Cauchy. Taylor gave a singular solution of a differential equation and the method of finding that solution by differentiation of the differential equation. Taylor's work contains the first correct explanation of astronomical refraction. He wrote also a work on linear perspective, a treatise which, like his other writings, suffers for want of fulness and clearness of expression. At the age of twenty-three he gave a remarkable solution of the problem of the centre of oscillation, published in 1714. His claim to priority was unjustly disputed by Johann Bernoulli. In the *Philosophical Transactions*, Vol. 30, 1717, Taylor applies "Taylor's series" to the solution of numerical equations. He assumes that a rough approximation, a, to a root of $f(x)=0$ has been found. Let $f(a)=k$, $f'(a)=k'$, $f''(a)=k''$, and $x=a+s$. He expands $0=f(a+s)$ by his theorem, discards all powers of s above the second, substitutes the values of k, k', k'', and then solves for s. By a repetition of this process, close approximations are secured. He makes the important observation that his method solves also equations involving radicals and transcendental functions. The first application of the Newton-Raphson process to the solution of transcendental equations was made by Thomas Simpson in his *Essays . . . on Mathematicks*, London, 1740.

The earliest to suggest the method of *recurring series* for finding roots was *Daniel Bernoulli* (1700–1782) who in 1728 brought the quartic to the form $1=ax+bx^2+cx^3+ex^4$, then selected arbitrarily four numbers A, B, C, D, and a fifth, E, thus, $E=aD+bC+cB+eA$, also a sixth by the same recursion formula $F=aE+bD+cC+eB$, and so on. If the last two numbers thus found are M and N, then $x=M\div N$ is an approximate root. Daniel Bernoulli gives no proof, but is aware that there is not always convergence to the root. This method was perfected by Leonhard Euler in his *Introductio in analysin infinitorum*, 1748, Vol. I, Chap. 17, and by Joseph Lagrange in Note VI of his *Résolution des équations numériques*.

Brook Taylor in 1717 expressed a root of a quadratic equation in the form of an infinite series; for the cubic François Nicole did similarly in 1738 and Clairaut in 1746. A. C. Clairaut inserted the process in his *Elements d'algèbre*. Thomas Simpson determined roots by reversion of series in 1743 and by infinite series in 1745. Marquis de Courtivron (1715–1785) also expressed the roots in the form of infinite series, while L. Euler devoted several articles to this topic.[1]

At this time the matter of convergence of the series did not receive

[1] For references see F. Cajori, in *Colorado College Publication*, General Series 51, p. 212.

proper attention, except in some rare instances. James Gregory of Edinburgh, in his *Vera circuli et hyperbolæ quadratura* (1667), first used the terms "convergent" and "divergent" series, while William Brouncker gave an argument which amounted to a proof of the convergence of his series, noted above.

Colin Maclaurin (1698–1746) was elected professor of mathematics at Aberdeen at the age of nineteen by competitive examination, and in 1725 succeeded James Gregory at the University of Edinburgh. He enjoyed the friendship of Newton, and, inspired by Newton's discoveries, he published in 1719 his *Geometria Organica*, containing a new and remarkable mode of generating conics, known by his name, and referring to the fact which became known later as "Cramer's paradox," that a curve of the nth order is not always determined by $\frac{1}{2}n(n+3)$ points, that the number may be less. A second tract, *De Linearum geometricarum proprietatibus*, 1720, is remarkable for the elegance of its demonstrations. It is based upon two theorems: the first is the theorem of Cotes; the second is Maclaurin's: If through any point O a line be drawn meeting the curve in n points, and at these points tangents be drawn, and if any other line through O cut the curve in R_1, R_2, etc., and the system of n tangents in r_1, r_2, etc., then $\Sigma\frac{1}{OR}=\Sigma\frac{1}{Or}$. This and Cotes' theorem are generalizations of theorems of Newton. Maclaurin uses these in his treatment of curves of the second and third degree, culminating in the remarkable theorem that if a quadrangle has its vertices and the two points of intersection of its opposite sides upon a curve of the third degree, then the tangents drawn at two opposite vertices cut each other on the curve. He deduced independently B. Pascal's theorem on the hexagram. Some of his geometrical results were reached independently by *William Braikenridge* (about 1700—after 1759), a clergyman in Edinburgh. The following is known as the "Braikenridge-Maclaurin theorem": If the sides of a polygon are restricted to pass through fixed points while all the vertices but one lie on fixed straight lines, the free vertex describes a conic section or a straight line. Maclaurin's more general statement (*Phil. Trans.*, 1735) is thus: If a polygon move so that each of its sides passes through a fixed point, and if all its summits except one describe curves of the degrees m, n, p, etc., respectively, then the free summit moves on a curve of the degree $2\,mnp\ldots$, which reduces to $mnp\ldots$ when the fixed points all lie on a straight line. Maclaurin was the first to write on "pedal curves," a name due to Olry Terquem (1782–1862). Maclaurin is the author of an *Algebra*. The object of his treatise on *Fluxions* was to found the doctrine of fluxions on geometric demonstrations after the manner of the ancients, and thus, by rigorous exposition, answer such attacks as Berkeley's that the doctrine rested on false reasoning. The *Fluxions* contained for the first time the correct way of distinguishing between maxima and minima, and explained their use in the theory of multiple points. "Maclaurin's theorem" was previously given by B. Taylor and James Stirling, and is but a particular case of "Taylor's theorem." Maclaurin invented the trisectrix, $x(x^2+y^2)=a(y^2-3x^2)$, which is akin to the Folium of Descartes. Appended to the treatise on *Fluxions* is the solution of a number of beautiful geometric, mechanical, and astronomical problems, in which he employs ancient methods with such consummate skill as to induce A. C. Clairaut to abandon analytic methods and to attack the problem of the figure of the earth by pure geometry. His solutions commanded the liveliest admiration of J. Lagrange. Maclaurin investigated the attraction of the ellipsoid of revolution, and showed that a homogeneous liquid mass revolving uniformly around an axis under the action of gravity must assume the form of an ellipsoid of revolution. Newton had given this theorem without proof. Notwithstanding the genius of Maclaurin, his influence on the progress of mathematics in Great Britain was unfortunate; for, by his example, he induced his countrymen to neglect analysis and to be indifferent to the wonderful progress in the higher analysis made on the Continent.

James Stirling (1692–1770), whom we have mentioned in connection with C. Maclaurin's theorem and Newton's enumeration of 72 forms of cubic curves (to which Stirling added 4 forms), was educated at Glasgow and Oxford. He was expelled from Oxford for corresponding with Jacobites. For ten years he studied in Venice. He enjoyed the friendship of Newton. His *Methodus differentialis* appeared in 1730.

It remains for us to speak of **Abraham de Moivre** (1667–1754), who was of French descent, but was compelled to leave France at the age of eighteen, on the Revocation of the Edict of Nantes. He settled in London, where he gave lessons in mathematics. He ranked high as a mathematician. Newton himself, in the later years of his life, used to reply to inquirers respecting mathematics, even respecting his *Principia:* "Go to Mr. De Moivre; he knows these things better than I do." He lived to the advanced age of eighty-seven and sank into a state of almost total lethargy. His subsistence was latterly dependent on the solution of questions on games of chance and problems on probabilities, which he was in the habit of giving at a tavern in St. Martin's Lane. Shortly before his death he declared that it was necessary for him to sleep ten or twenty minutes longer every day. The day after he had reached the total of over twenty-three hours, he slept exactly twenty-four hours and then passed away in his sleep. De Moivre enjoyed the friendship of Newton and Halley. His power as a mathematician lay in analytic rather than geometric investigation. He revolutionized higher trigonometry by the discovery of the theorem known by his name and by extending the theorems on the multiplication and division of sectors from the circle to the hyperbola. His work on the theory of probability surpasses anything done by any other mathematician except P. S. Laplace. His principal contributions are his investigations respecting the Duration of Play, his Theory of Recurring Series, and his extension of the value of Daniel Bernoulli's theorem by the aid of Stirling's theorem.[1] His chief works are the *Doctrine of Chances*, 1716, the *Miscellanea Analytica*,✡ 1730, and his papers in the *Philosophical Transactions*. Unfortunately he did not publish the proofs of his results in the doctrine of chances, and J. Lagrange more than fifty years later found a good exercise for his skill in supplying the proofs. A generalization of a problem first stated by C. Huygens has received the name of "De Moivre's Problem:" Given n dice, each having f faces, determine the chances of throwing any given number of points. It was solved by A. de Moivre, P. R. de Montmort, P. S. Laplace and others. De Moivre also generalized the Problem on the Duration of Play, so that it reads as follows: Suppose A has m counters, and B has n counters; let their chances of winning in a single game be as a to b; the loser in each game is to give a counter to his adversary: required the probability that *when or before a certain number of games has been played*, one of the players will have won all the counters of his adversary. De Moivre's solution of this problem constitutes his most substantial achievement in the theory of chances. He employed in his researches the method of ordinary finite differences, or as he called it, the method of recurrent series.

A famous theory involving the notion of inverse probability was advanced by *Thomas Bayes*. It was published in the London *Philosophical Transactions*, Vols. 53 and 54 for the years 1763 and 1764, after the death of Bayes, which occurred in 1761. These researches originated the discussion of the probabilities of causes as inferred from observed effects, a subject developed more fully by P. S. Laplace. Using modern symbols, Bayes' fundamental theorem may be stated thus:[2] If an event has happened p times and failed q times, the probability that its chance at a single trial lies between a and b is

$$\int_a^b x^p\,(1-x)^q\,dx \div \int_0^1 x^p\,(1-x)^q\,dx.$$

A memoir of John Michell "On the probable Parallax, and Magnitude of the fixed Stars" in the London *Philosophical Transactions*, Vol. 57 I, for the year 1767, contains the famous argument for the existence of design drawn from the fact of the closeness of certain stars, like the Pleiades. "We may take the six brightest of the Pleiades, and, supposing the whole number of those stars, which are equal in splendor to the faintest of these, to be about 1500, we shall

[1] I. Todhunter, *A History of the Mathematical Theory of Probability*, Cambridge and London, 1865, pp. 135–193.
[2] I. Todhunter, *op. cit.*, p. 295.

find the odds to be near 500,000 to 1, that no six stars, out of that number, scattered at random, in the whole heavens, would be within so small a distance from each other, as the Pleiades are."

Euler, Lagrange, and Laplace

In the rapid development of mathematics during the eighteenth century the leading part was taken, not by the universities, but by the academies. Particularly prominent were the academies at Berlin and Petrograd. This fact is the more singular, because at that time Germany and Russia did not produce great mathematicians. The academies received their adornment mainly from the Swiss and French. It was after the French Revolution that schools gained their ascendancy over academies.

During the period from 1730 to 1820 Switzerland had her L. Euler; France, her J. Lagrange, P. S. Laplace, A. M. Legendre, and G. Monge. The mediocrity of French mathematics which marked the time of Louis XIV was now followed by one of the very brightest periods of all history. England, on the other hand, which during the unproductive period in France had her Newton, could now boast of no great mathematician. Except young Gauss, Germany had no great name. France now waved the mathematical sceptre. Mathematical studies among the English and German people had sunk to the lowest ebb. Among them the direction of original research was ill chosen. The former adhered with excessive partiality to ancient geometrical methods; the latter produced the combinatorial school, which brought forth nothing of great value.

The labors of L. Euler, J. Lagrange, and P. S. Laplace lay in higher analysis, and this they developed to a wonderful degree. By them analysis came to be completely severed from geometry. During the preceding period the effort of mathematicians not only in England, but, to some extent, even on the continent, had been directed toward the solution of problems clothed in geometric garb, and the results of calculation were usually reduced to geometric form. A change now took place. Euler brought about an emancipation of the analytical calculus from geometry and established it as an independent science. Lagrange and Laplace scrupulously adhered to this separation. Building on the broad foundation laid for higher analysis and mechanics by Newton and Leibniz, Euler, with matchless fertility of mind, erected an elaborate structure. There are few great ideas pursued by succeeding analysts which were not suggested by L. Euler, or of which he did not share the honor of invention. With, perhaps, less exuberance of invention, but with more comprehensive genius and profounder reasoning, J. Lagrange developed the infinitesimal calculus and put analytical mechanics into the form in which we now know it. P. S. Laplace applied the calculus and mechanics to the elaboration of the theory of universal gravitation, and thus, largely extending and supplementing the labors of Newton, gave a full analytical discussion of the solar system. He also wrote an epoch-marking work on Probability. Among the analytical branches created during this period are the calculus of Variations by Euler and Lagrange, Spherical Harmonics by Legendre and Laplace, and Elliptic Integrals by Legendre.

Comparing the growth of analysis at this time with the growth during the time of K. F. Gauss, A. L. Cauchy, and recent mathematicians, we observe an important difference. During the former period we witness mainly a development with reference to *form*. Placing almost implicit confidence in results of calculation, mathematicians did not always pause to discover rigorous proofs, and were thus led to general propositions, some of which have since been found to be true in only special cases. The Combinatorial School in Germany carried this tendency to the greatest extreme; they worshipped formalism and paid no attention to the actual contents of formulæ. But in recent times there has been added to the dexterity in the formal treatment of problems, a much-needed rigor of demonstration. A good example of this increased rigor is seen in the present use of infinite series as compared to that of Euler, and of Lagrange in his earlier works.

The ostracism of geometry, brought about by the master-minds of this period, could not last permanently. Indeed, a new geometric school sprang into existence in France before the close of this period. J. Lagrange would not permit a single diagram to appear in his *Mécanique analytique*, but thirteen years before his death, G. Monge published his epoch-making *Géométrie descriptive*.

Leonhard Euler (1707–1783) was born in Basel. His father, a minister, gave him his first instruction in mathematics and then sent him to the University of Basel, where he became a favorite pupil of Johann Bernoulli. In his nineteenth year he composed a dissertation on the masting of ships, which received the second prize from the French Academy of Sciences. When Johann Bernoulli's two sons, Daniel and Nicolaus, went to Russia, they induced Catharine I, in 1727, to invite their friend L. Euler to St. Petersburg, where Daniel, in 1733, was assigned to the chair of mathematics. In 1735 the solving of an astronomical problem, proposed by the Academy, for which several eminent mathematicians had demanded some months' time, was achieved in three days by Euler with aid of improved methods of his own. But the effort threw him into a fever and deprived him of the use of his right eye. With still superior methods this same problem was solved later by K. F. Gauss in one hour! [1] The despotism of Anne I caused the gentle Euler to shrink from public affairs and to devote all his time to science. After his call to Berlin by Frederick the

[1] W. Sartorius Waltershausen, *Gauss, zum Gedächtniss*, Leipzig, 1856.

Great in 1741, the queen of Prussia, who received him kindly, wondered how so distinguished a scholar should be so timid and reticent. Euler naïvely replied, "Madam, it is because I come from a country where, when one speaks, one is hanged." It was on the recommendation of D'Alembert that Frederick the Great had invited Euler to Berlin. Frederick was no admirer of mathematicians and, in a letter to Voltaire, spoke of Euler derisively as "un gros cyclope de géomètre." In 1766 Euler with difficulty obtained permission to depart from Berlin to accept a call by Catharine II to St. Petersburg. Soon after his return to Russia he became blind, but this did not stop his wonderful literary productiveness, which continued for seventeen years, until the day of his death. He dictated to his servant his *Anleitung zur Algebra*, 1770, which, though purely elementary, is meritorious as one of the earliest attempts to put the fundamental processes on a sound basis.

The story goes that when the French philosopher Denis Diderot paid a visit to the Russian Court, he conversed very freely and gave the younger members of the Court circle a good deal of lively atheism. Thereupon Diderot was informed that a learned mathematician was in possession of an algebraical demonstration of the existence of God, and would give it to him before all the Court, if he desired to hear it. Diderot consented. Then Euler advanced toward Diderot, and said gravely, and in a tone of perfect conviction: *Monsieur*, $(a+b^n)/_n=x$, *donc Dieu existe; répondez!* Diderot, to whom algebra was Hebrew, was embarrassed and disconcerted, while peals of laughter rose on all sides. He asked permission to return to France at once, which was granted.[1]

Euler was such a prolific writer that only in the present century have plans been brought to maturity for a complete edition of his works. In 1909 the Swiss Natural Science Association voted to publish Euler's works in their original language. The task is being carried on with the financial assistance of German, French, American and other mathematical organizations and of many individual donors. The expense of publication will greatly exceed the original estimate of 400,000 francs, owing to a mass of new manuscripts recently found in Petrograd.

The following are his chief works:[2] *Introductio in analysin infinitorum*, 1748, a work that caused a revolution in analytical mathematics, a subject which had hitherto never been presented in so general and systematic manner; *Institutiones calculi differentialis*, 1755, and *Institutiones calculi integralis*, 1768–1770, which were the most complete and accurate works on the calculus of that time, and contained not only a full summary of everything then known on this subject,

[1] From De Morgan's *Budget of Paradoxes*, 2. Ed., Chicago, 1915, Vol. II, p. 4.
[2] See G. Eneström, *Verzeichniss der Schriften Leonhard Eulers*, 1. Lieferung, 1910, 2. Lieferung, 1913, Leipzig.

but also the Beta and Gamma Functions and other original investigations; *Methodus inveniendi lineas curvas maximi minimive proprietate gaudentes*, 1744, which, displaying an amount of mathematical genius seldom rivalled, contained his researches on the calculus of variations to the invention of which Euler was led by the study of the researches of Johann and Jakob Bernoulli. One of the earliest problems bearing on this subject was Newton's solid of revolution, of least resistance, reduced by him in 1686 to a differential equation. (*Principia*, Bk. II, Sec. VII, Prop. XXXIV, Scholium.) Johann Bernoulli's problem of the brachistochrone, solved by him in 1697, and by his brother Jakob in the same year, stimulated Euler. The study of isoperimetrical curves, the brachistochrone in a resisting medium and the theory of geodesics, previously treated by the elder Bernoullis and others, led to the creation of this new branch of mathematics, the Calculus of Variations. His method was essentially geometrical, which makes the solution of the simpler problems very clear. Euler's *Theoria motuum planetarum et cometarum*, 1744, *Theoria motus lunæ*, 1753, *Theoria motuum lunæ*, 1772, are his chief works on astronomy; *Ses lettres à une princesse d'Allemagne sur quelques sujets de Physique et de Philosophie*, 1770, was a work which enjoyed great popularity.

We proceed to mention the principal innovations and inventions of Euler. In his *Introductio* (1748) every "analytical expression" in x, *i. e.* every expression made up of powers, logarithms, trigonometric functions, etc., is called a "function" of x. Sometimes Euler used another definition of "function," namely, the relation between y and x expressed in the xy-plane by any curve drawn freehand, "libero manus ductu." [1] In modified form, these two rival definitions are traceable in all later history. Thus Lagrange proceeded on the idea involved in the first definition, Fourier on the idea involved in the second.

Euler treated trigonometry as a branch of analysis and consistently treated trigonometric values as ratios. The term "trigonometric function" was introduced in 1770 by *Georg Simon Klügel* (1739–1812) of Halle, the author of a mathematical dictionary.[2] Euler developed and systematized the mode of writing trigonometric formulas, taking, for instance, the sinus totus equal to 1. He simplified formulas by the simple expedient of designating the angles of a triangle by A, B, C, and the opposite sides by a, b, c, respectively. Only once before have we encountered this simple device. It was used in a pamphlet prepared by *Ri. Rawlinson* at Oxford sometime between 1655 and 1668.[3] This notation was re-introduced simultaneously with Euler by Thomas Simpson in England. We may add here that in 1734 Euler used the notation $f(x)$ to indicate "function of x," that the use of e as the

[1] F. Klein, *Elementarmathematik v. höh. Standpunkte aus* ☼ I, Leipzig, 1908, p. 438.
[2] M. Cantor, *op. cit.*, Vol. IV, 1908, p. 413.
[3] See F. Cajori in *Nature*, Vol. 94, 1915, p. 642.

symbol for the natural base of logarithms was introduced by him in 1728,[1] that in 1750 he used S to denote the half-sum of the sides of a triangle, that in 1755 he introduced Σ to signify "summation," that in 1777 he used i for $\sqrt{-1}$, a notation used later by K. F. Gauss.

We pause to remark that in Euler's time **Thomas Simpson** (1710–1761), an able and self-taught English mathematician, for many years professor at the Royal Military Academy at Woolwich, and author of several text-books, was active in perfecting trigonometry as a science. His *Trigonometry*, London, 1748, contains elegant proofs of two formulas for plane triangles, $(a+b):c=\cos\frac{1}{2}(A-B):\sin\frac{1}{2}C$ and $(a-b):c=\sin\frac{1}{2}(A-B):\cos\frac{1}{2}C$, which have been ascribed to the German astronomer *Karl Brandan Mollweide* (1774–1825), who developed them much later. The first formula was given in different notation by I. Newton in his *Universal Arithmetique;* both formulas are given by *Friedrich Wilhelm Oppel* in 1746.[2]

Euler laid down the rules for the transformation of co-ordinates in space, gave a methodic analytic treatment of plane curves and of surfaces of the second order. He was the first to discuss the equation of the second degree in three variables, and to classify the surfaces represented by it. By criteria analogous to those used in the classification of conics he obtained five species. He devised a method of solving biquadratic equations by assuming $x=\sqrt{p}+\sqrt{q}+\sqrt{r}$, with the hope that it would lead him to a general solution of algebraic equations. The method of elimination by solving a series of linear equations (invented independently by E. Bézout) and the method of elimination by symmetric functions, are due to him. Far reaching are Euler's researches on logarithms. Euler defined logarithms as exponents,[3] thus abandoning the old view of logarithms as terms of an arithmetic series in one-to-one correspondence with terms of a geometric series. This union between the exponential and logarithmic concepts had taken place somewhat earlier. The possibility of defining logarithms as exponents had been recognized by John Wallis in 1685, by Johann Bernoulli in 1694, but not till 1742 do we find a systematic exposition of logarithms, based on this idea. It is given in the introduction to Gardiner's *Tables of Logarithms*, London, 1742. This introduction is "collected wholly from the papers" of William Jones. Euler's influence caused the ready adoption of the new definition. That this view of logarithms was in every way a step in advance has been doubted by some writers. Certain it is that it involves internal difficulties of a serious nature. Euler threw a stream of light upon the subtle subject of the logarithms of negative and imaginary numbers. In 1712 and 1713 this subject had been discussed in a

[1] G. Eneström, *Bibliotheca mathematica*, Vol. 14, 1913–1914, p. 81.

[2] A. v. Braunmühl, *op. cit.*, 2. Teil, 1903, p. 93; H. Wieleitner in *Bibliotheca mathematica*, 3. S., Vol. 14, pp. 348, 349.

[3] See. L. Euler, *Introductio*, 1748, Chap. VI, § 102.

correspondence between G. W. Leibniz and Johann Bernoulli.[1] Leibniz maintained that since a positive logarithm corresponds to a number larger than unity, and a negative logarithm to a positive number less than unity, the logarithm of -1 was not really true, but imaginary; hence the ratio $-1 \div 1$, having no logarithm, is itself imaginary. Moreover, if there really existed a logarithm of -1, then half of it would be the logarithm of $\sqrt{-1}$, a conclusion which he considered absurd. The statements of Leibniz involve a double use of the term imaginary: (1) in the sense of non-existent, (2) in the sense of a number of the type $\sqrt{-1}$. Johann Bernoulli maintained that -1 has a logarithm. Since $dx : x = -dx : -x$, there results by integration $\log(x) = \log(-x)$; the logarithmic curve $y = \log x$ has therefore two branches, symmetrical to the y=axis, as has the hyperbola. The correspondence between Leibniz and Johann Bernoulli was first published in 1745. In 1714 Roger Cotes developed in the *Philosophical Transactions* an important theorem which was republished in his *Harmonia mensurarum* (1722). In modern notation it is $i\phi = \log(\cos\phi + i\sin\phi)$. In the exponential form it was discovered again by Euler in 1748. Cotes was aware of the periodicity of the trigonometric functions. Had he applied this idea to his formula, he might have anticipated Euler by many years in showing that the logarithm of a number has an infinite number of different values. A second discussion of the logarithms of negative numbers took place in a correspondence between young Euler and his revered teacher, Johann Bernoulli, in the years 1727–1731.[2] Bernoulli argued, as before, that $\log x = \log(-x)$. Euler uncovered the difficulties and inconsistencies of his own and Bernoulli's views, without, at that time, being able to advance a satisfactory theory. He showed that Johann Bernoulli's expression for the area of a circular sector becomes for a quadrant $\frac{2^2 \log(-1)}{4\sqrt{-1}}$, which is incompatible with Bernoulli's claim that $\log(-1) = 0$. Between 1731 and 1747 Euler made steady progress in the mastery of relations involving imaginaries. In a letter of Oct. 18, 1740, to Johann Bernoulli, he stated that $y = 2\cos x$ and $y = e^{x\sqrt{-1}} + e^{-x\sqrt{-1}}$, were both integrals of the differential equation $\frac{d^2y}{dx^2} + y = 0$ and were equal to each other. Euler knew the corresponding expression for $\sin x$. Both expressions are given by him in the *Miscellanea Berolinensia*, 1743, and again in his *Introductio*, 1748, Vol. I, 104. He gave the value $\sqrt{-1}^{\sqrt{-1}} = 0{,}2078795763$ as early as 1746, in a letter to Christian Goldbach (1690–1764), but makes no reference here to the in-

[1] See F. Cajori, "History of the Exponential and Logarithmic Concepts," *American Math. Monthly*, Vol. 20, 1913, pp. 39–42.

[2] See F. Cajori in *Am. Math. Monthly*, Vol. 20, 1913, pp. 44–46.

finitely many values of this imaginary expression.[1] The creative work on this topic appears to have been done in 1747. During that year and the year following Euler debated this subject with D'Alembert in a correspondence of which only a few letters of Euler are extant.[2] In a letter of April 15, 1747, Euler disproves the conclusion upheld by D'Alembert, that $log\ (-1)=0$, and states his own results indicating that now he had penetrated the subject; $log\ n$ has an infinite number of values which are all imaginary, except when n is a positive number, in which case one logarithm out of this infinite number is real. On Aug. 19, 1747, he said that he had sent an article to the Berlin Academy; this is no doubt the article published in 1862 under the title, *Sur les logarithmes des nombres négatifs et imaginaires*. The reason why Euler did not publish it at the time when it was written can only be conjectured. Our guess is that Euler became dissatisfied with the article. At any rate, he wrote a new one in 1749, *De la controverse entre Mrs. Leibnitz et Bernoulli sur les logarithmes negatifs et imaginaires*. In 1747 he based the proof that a number has an infinity of logarithms on the relation $i\varphi=log(cos\varphi+i\ sin\varphi)$; in 1749 on the assumption $log(1+\omega)=\omega$, ω being infinitely small. He developed the theory of logarithms of complex numbers a third time in a paper of 1749 on *Recherches sur les racines imaginaires des équations*. The two papers of 1749 were published in 1751 in the Berlin Memoirs. The latter primarily aims to prove that every equation has a root; it was discussed in 1799 by K. F. Gauss in his inaugural dissertation.

Euler's papers were not fully understood and did not carry conviction. D'Alembert still felt that the question was not settled, and advanced arguments of metaphysical, analytical and geometrical nature which shrouded the subject into denser haze and helped to prolong the controversy to the end of the century. In 1759 *Daviet de Foncenex* (1734–1799), a young friend of J. Lagrange, wrote on this subject. In 1768 *W. J. G. Karsten* (1732–1787), professor at Bützow, later at Halle, wrote a long treatise which contains an interesting graphic representation of imaginary logarithms.[3] The debate on Euler's results was carried on with much warmth by the Italian mathematicians.

The subject of infinite series received new life from him. To his researches on series we owe the creation of the theory of definite integrals by the development of the so-called *Eulerian integrals*. He warns his readers occasionally against the use of divergent series, but is nevertheless very careless himself. The rigid treatment to which infinite series are subjected now was then undreamed of. No clear notions existed as to what constitutes a convergent series. Neither

[1] P. H. Fuss, *Corresp. math. et phys. de quelques célebres géomètres du xviiième siècle*, I, 1843, p. 383.

[2] See F. Cajori, *Am. Math. Monthly*, Vol. 20, 1913, pp. 76–79.

[3] *Am. Math. Monthly*, Vol. 20, 1913, p. 111.

G. W. Leibniz nor Jakob and Johann Bernoulli had entertained any serious doubt of the correctness of the expression $\frac{1}{2}=1-1+1-1+\ldots$ *Guido Grandi* (1671–1742) of Pisa went so far as to conclude from this that $\frac{1}{2}=0+0+0+\ldots$ In the treatment of series Leibniz advanced a metaphysical method of proof which held sway over the minds of the elder Bernoullis, and even of Euler.[1] The tendency of that reasoning was to justify results which seem highly absurd to followers of Abel and Cauchy. The looseness of treatment can best be seen from examples. The very paper in which Euler cautions against divergent series contains the proof that

$$\ldots\frac{1}{n^2}+\frac{1}{n}+1+n+n^2+\ldots=0 \text{ as follows:}$$

$$n+n^2+\ldots=\frac{n}{1-n},\ 1+\frac{1}{n}+\frac{1}{n^2}+\ldots=\frac{n}{n-1};$$

these added give zero. Euler has no hesitation to write $1-3+5-7+\ldots=0$, and no one objected to such results excepting Nicolaus Bernoulli, the nephew of Johann and Jakob. Strange to say, Euler finally succeeded in converting Nicolaus Bernoulli to his own erroneous views. At the present time it is difficult to believe that Euler should have confidently written $\sin\phi-2\sin 2\phi+3\sin 3\phi-4\sin 4\phi+\ldots=0$, but such examples afford striking illustrations of the want of scientific basis of certain parts of analysis at that time. Euler's proof of the binomial formula for negative and fractional exponents, which was widely reproduced in elementary text-books of the nineteenth century, is faulty. A remarkable development, due to Euler, is what is called the hypergeometric series, the summation of which he observed to be dependent upon the integration of a linear differential equation of the second order, but it remained for K. F. Gauss to point out that for special values of its letters, this series represented nearly all functions then known.

Euler gave in 1779 a series for arc tan x, different from the series of James Gregory, which he applied to the formula $\pi=20 \text{ arc tan } \frac{1}{7}+8 \text{ arc tan } \frac{3}{79}$ used for computing π. The series was published in 1798. Euler reached remarkable results on the summation of the reciprocal powers of the natural numbers. In 1736 he had found the sum of the reciprocal squares to be $\pi^2/6$, and of the reciprocal fourth powers to be $\pi^4/90$. In an article of 1743 which until recently has been generally overlooked,[2] Euler finds the sums of the reciprocal even powers of the natural numbers up to and including the 26th power. Later he showed the connection of coefficients occurring in these sums with the "Bernoullian numbers" due to Jakob Bernoulli.

Euler developed the calculus of finite differences in the first chapters

[1] R. Reiff, *Geschichte der Unendlichen Reihen*, Tübingen, 1889, p. 68.

[2] P. Stäckel in *Bibliotheca mathematica*, 3. S., Vol. 8, 1907–8, pp. 37–60.

of his *Institutiones calculi differentialis*, and then deduced the differential calculus from it. He established a theorem on homogeneous functions, known by his name, and contributed largely to the theory of differential equations, a subject which had received the attention of I. Newton, G. W. Leibniz, and the Bernoullis, but was still undeveloped. A. C. Clairaut, Alexis Fontaine des Bertins (1705–1771), and L. Euler about the same time observed criteria of integrability, but Euler in addition showed how to employ them to determine integrating factors. The principles on which the criteria rested involved some degree of obscurity. Euler was the first to make a systematic study of singular solutions of differential equations of the first order. In 1736, 1756 and 1768 he considered the two paradoxes which had puzzled A. C. Clairaut: The first, that a solution may be reached by differentiation instead of integration; the second, that a singular solution is not contained in the general solution. Euler tried to establish an *a priori* rule for determining whether a solution is contained in the general solution or not. Stimulated by researches of Count de Fagnano on elliptic integrals, Euler established the celebrated addition-theorem for these integrals. He invented a new algorithm for continued fractions, which he employed in the solution of the indeterminate equation $ax+by=c$. We now know that substantially the same solution of this equation was given 1000 years earlier, by the Hindus. Euler gave 62 pairs of amicable numbers, of which 3 were false and 3 known earlier: one pair had been discovered by the Pythagoreans, another by Fermat and a third by Descartes.[1] By giving the factors of the number $2^{2^n}+1$ when $n=5$, he pointed out that this expression did not always represent primes, as was supposed by P. Fermat. He first supplied the proof to "Fermat's theorem," and to a second theorem of Fermat, which states that every prime of the form $4n+1$ is expressible as the sum of two squares in one and only one way. A third theorem, "Fermat's last theorem," that $x^n+y^n=z^n$, has no integral solution for values of n greater than 2, was proved by Euler to be correct when $n=4$ and $n=3$. Euler discovered four theorems which taken together make out the great law of quadratic reciprocity, a law independently discovered by A. M. Legendre.[2]

In 1737 Euler showed that the sum of the reciprocals of all prime numbers is $\log_e(\log_e \infty)$, thereby initiating a line of research on the distribution of primes which is usually not carried back further than to A. M. Legendre.[3]

In 1741 he wrote on partitions of numbers ("partitio numerorum"). In 1782 he published a discussion of the problem of 36 officers of six different grades and from six different regiments, who are to be placed

[1] See *Bibliotheca mathematica*, 3. S., Vol. 9, p. 263; Vol. 14, pp. 351–354.
[2] Oswald Baumgart, *Ueber das Quadratische Reciprocitätsgesetz*. Leipzig, 1885.
[3] G. Eneström in *Bibliotheca mathematica*, 3. S., Vol. 13, 1912, p. 81.

in a square in such a way that in each row and column there are six officers, all of different grades as well as of different regiments. Euler thinks that no solution is obtainable when the order of the square is of the form 2 mod. 4. Arthur Cayley in 1890 reviewed what had been written; P. A. MacMahon solved it in 1915. It is called the problem of the "Latin squares," because Euler, in his notation, used "*n* lettres latines." Euler enunciated and proved a well-known theorem, giving the relation between the number of vertices, faces, and edges of certain polyhedra, which, however, was known to R. Descartes. The powers of Euler were directed also towards the fascinating subject of the theory of probability, in which he solved some difficult problems.

Of no little importance are Euler's labors in analytical mechanics. Says Whewell: "The person who did most to give to analysis the generality and symmetry which are now its pride, was also the person who made mechanics analytical; I mean Euler."[1] He worked out the theory of the rotation of a body around a fixed point, established the general equations of motion of a free body, and the general equation of hydrodynamics. He solved an immense number and variety of mechanical problems, which arose in his mind on all occasions. Thus, on reading Virgil's lines, "The anchor drops, the rushing keel is staid," he could not help inquiring what would be the ship's motion in such a case. About the same time as Daniel Bernoulli he published the *Principle of the Conservation of Areas* and defended the principle of "least action," advanced by P. Maupertuis. He wrote also on tides and on sound.

Astronomy owes to Euler the method of the variation of arbitrary constants. By it he attacked the problem of perturbations, explaining, in case of two planets, the secular variations of eccentricities, nodes, etc. He was one of the first to take up with success the theory of the moon's motion by giving approximate solutions to the "problem of three bodies." He laid a sound basis for the calculation of tables of the moon. These researches on the moon's motion, which captured two prizes, were carried on while he was blind, with the assistance of his sons and two of his pupils. His *Mechanica sive motus scientia analytice exposita*, Vol. I, 1736, Vol. II, 1742, is, in the language of Lagrange, "the first great work in which analysis is applied to the science of movement."

Prophetic was his study of the movements of the earth's pole. He showed that if the axis around which the earth rotates is not coincident with the axis of figure, the axis of rotation will revolve about the axis of figure in a predictable period. On the assumption that the earth is perfectly rigid he showed that the period is 305 days. The earth is now known to be elastic. From observations taken in 1884–5,

[1] W. Whewell, *History of the Inductive Sciences*, 3rd Ed., Vol. I, New York, 1858, p. 363.

S. C. Chandler of Harvard found the period to be 428 days.[1] For an earth of steel the time has been computed to be 441 days.

Euler in his *Introductio in analysin* (1748) had undertaken a classification of quartic curves, as had also a mathematician of Geneva, *Gabriel Cramer* (1704–1752), in his *Introduction à l'analyse des lignes courbes algebraiques*, Geneva, 1750. Both based their classifications on the behavior of the curves at infinity, obtaining thereby eight classes which were divided into a considerable number of species. Another classification was made by E. Waring, in his *Miscellanea analytica*, 1792, which yielded 12 main divisions and 84551 species. These classifications rest upon ideas hardly in harmony with the more recent projective methods, and have been abandoned. Cramer studied the quartic $y^4-x^4+ay^2+bx^2=0$ which later received the attention of F. Moigno (1840), Charles Briot and Jean Claude Bouquet, and B. A. Nievenglowski (1895), and because of its peculiar form was called by the French "courbe du diable." Cramer gave also a classification of quintic curves.

Most of Euler's memoirs are contained in the transactions of the Academy of Sciences at St. Petersburg, and in those of the Academy at Berlin. From 1728 to 1783 a large portion of the Petropolitan transactions were filled by his writings. He had engaged to furnish the Petersburg Academy with memoirs in sufficient number to enrich its acts for twenty years—a promise more than fulfilled, for down to 1818 the volumes usually contained one or more papers of his, and numerous papers are still unpublished. His mode of working was, first to concentrate his powers upon a special problem, then to solve separately all problems growing out of the first. No one excelled him in dexterity of accommodating methods to special problems. It is easy to see that mathematicians could not long continue in Euler's habit of writing and publishing. The material would soon grow to such enormous proportions as to be unmanageable. We are not surprised to see almost the opposite in J. Lagrange, his great successor. The great Frenchman delighted in the general and abstract, rather than, like Euler, in the special and concrete. His writings are condensed and give in a nutshell what Euler narrates at great length.

Jean-le-Rond D'Alembert (1717–1783) was exposed, when an infant, by his mother in a market by the church of St. Jean-le-Rond, near the Nôtre-Dame in Paris, from which he derived his Christian name. He was brought up by the wife of a poor glazier. It is said that when he began to show signs of great talent, his mother sent for him, but received the reply, "You are only my step-mother; the glazier's wife is my mother." His father provided him with a yearly income. D'Alembert entered upon the study of law, but such was his love for mathematics, that law was soon abandoned. At the age of twenty-four his reputation as a mathematician secured for him ad-

[1] For details see *Nature*, Vol. 97, 1916, p. 530.

mission to the Academy of Sciences. In 1754 he was made permanent secretary of the French Academy. During the last years of his life he was mainly occupied with the great French encyclopædia, which was begun by Denis Diderot and himself. D'Alembert declined, in 1762, an invitation of Catharine II to undertake the education of her son. Frederick the Great pressed him to go to Berlin. He made a visit, but declined a permanent residence there. In 1743 appeared his *Traité de dynamique*, founded upon the important general principle bearing his name: The impressed forces are equivalent to the effective forces. D'Alembert's principle seems to have been recognized before him by A. Fontaine, and in some measure by Johann Bernoulli and I. Newton. D'Alembert gave it a clear mathematical form and made numerous applications of it. It enabled the laws of motion and the reasonings depending on them to be represented in the most general form, in analytical language. D'Alembert applied it in 1744 in a treatise on the equilibrium and motion of fluids, in 1746 to a treatise on the general causes of winds, which obtained a prize from the Berlin Academy. In both these treatises, as also in one of 1747, discussing the famous problem of vibrating chords, he was led to partial differential equations. He was a leader among the pioneers in the study of such equations. To the equation $\frac{\partial^2 y}{\partial t^2}=a^2\frac{\partial^2 y}{\partial x^2}$, arising in the problem of vibrating chords, he gave as the general solution,

$$y=f(x+at)+\phi(x-at),$$

and showed that there is only one arbitrary function, if y be supposed to vanish for $x=0$ and $x=l$. Daniel Bernoulli, starting with a particular integral given by Brook Taylor, showed that this differential equation is satisfied by the trigonometric series

$$y=\alpha\sin\frac{\pi x}{l}.\cos\frac{\pi t}{l}+\beta\sin\frac{2\pi x}{l}.\cos\frac{2\pi t}{l}+\ldots,$$

and claimed this expression to be the most general solution. Thus Daniel Bernoulli was the first to introduce "Fourier's series" into physics. He claimed that his solution, being compounded of an infinite number of tones and overtones of all possible intensities, was a general solution of the problem. Euler denied its generality, on the ground that, if true, the doubtful conclusion would follow that the above series represents any arbitrary function of a variable. These doubts were dispelled by J. Fourier. J. Lagrange proceeded to find the sum of the above series, but D'Alembert objected to his process, on the ground that it involved divergent series.[1]

A most beautiful result reached by D'Alembert, with aid of his principle, was the complete solution of the problem of the precession of the equinoxes, which had baffled the talents of the best minds.

[1] R. Reiff, *op. cit.*, II. Abschnitt.

He sent to the French Academy in 1747, on the same day with A. C. Clairaut, a solution of the problem of three bodies. This had become a question of universal interest to mathematicians, in which each vied to outdo all others. The problem of two bodies, requiring the determination of their motion when they attract each other with forces inversely proportional to the square of the distance between them, had been completely solved by I. Newton. The "problem of three bodies" asks for the motion of three bodies attracting each other according to the law of gravitation. Thus far, the complete solution of this has transcended the power of analysis. The general differential equations of motion were stated by P. S. Laplace, but the difficulty arises in their integration. The "solutions" given at that time are merely convenient methods of approximation in special cases when one body is the sun, disturbing the motion of the moon around the earth, or where a planet moves under the influence of the sun and another planet. The most important eighteenth century researches on the problem of three bodies are due to J. Lagrange. In 1772 a prize was awarded him by the Paris Academy for his *Essai sur le problème des trois corps*. He shows that a complete solution of the problem requires only that we know every moment the sides of the triangle formed by the three bodies, the solution of the triangle depending upon two differential equations of the second order and one differential equation of the third. He found particular solutions when the triangles remain all similar.

In the discussion of the meaning of negative quantities, of the fundamental processes of the calculus, of the logarithms of complex numbers, and of the theory of probability, D'Alembert paid some attention to the philosophy of mathematics. In the calculus he favored the theory of limits. He looked upon infinity as nothing but a limit which the finite approaches without ever reaching it. His criticisms were not always happy. When students were halted by the logical difficulties of the calculus, D'Alembert would say, "Allez en avant, et la foi vous viendra." He argued that when the probability of an event is very small, it ought to be taken 0. A coin is to be tossed 100 times and if head appear at the last trial, and not before, A shall pay B 2^{100} crowns. By the ordinary theory B should give A 1 crown at the start,✡ which should not be, argues D'Alembert, because B will certainly lose. This view was taken also by Count de Buffon. D'Alembert raised other objections to the principles of probability.

The naturalist, *Comte de Buffon* (1707–1788), wrote an *Essai d'arithmétique morale*, 1777. In the study of the Petersburg problem, he let a child toss a coin 2084 times, which produced 10057 crowns; there were 1061 games which produced 1 crown, 494 which produced 2 crowns and so on.[1] He was one of the first to emphasize the desir-

[1] For references, see I. Todhunter, *History of Theory of Probability*, p. 346.

ability of verifying the theory by actual trial. He also introduced what is called "local probability" by the consideration of problems that require the aid of geometry. Some studies along this line had been carried on earlier by *John Arbuthnot* (1658–1735) and Thomas Simpson in England. Count de Buffon derived the probability that a needle dropped upon a plane, ruled with equidistant, parallel lines, will fall across one of the lines.

The probability of the correctness of judgments determined by a majority of votes was examined mathematically by *Jean-Antoine-Nicolas Caritat de Condorcet* (1743–1794). His general conclusions are not of great importance; they are that voters must be enlightened men in order to ensure our confidence in their decisions.[1] He held that capital punishment ought to be abolished, on the ground that, however large the probability of the correctness of a single decision, there will be a large probability that in the course of many decisions some innocent person will be condemned.[1]

Alexis Claude Clairaut (1713–1765) was a youthful prodigy. He read G. F. de l'Hospital's works on the infinitesimal calculus and on conic sections at the age of ten. In 1731 was published his *Recherches sur les courbes à double courbure*, which he had ready for the press when he was sixteen. It was a work of remarkable elegance and secured his admission to the Academy of Sciences when still under legal age. In 1731 he gave a proof of the theorem enunciated by I. Newton, that every cubic is a projection of one of five divergent parabolas. Clairaut formed the acquaintance of *Pierre Louis Moreau de Maupertuis* (1698–1759), whom he accompanied on an expedition to Lapland to measure the length of a degree of the meridian. At that time the shape of the earth was a subject of serious disagreement. I. Newton and C. Huygens had concluded from theory that the earth was flattened at the poles. About 1712 *Jean-Dominique Cassini* (1625–1712) and his son *Jacques Cassini* (1677–1756) measured an arc extending from Dunkirk to Perpignan and arrived at the startling result that the earth is elongated at the poles. To decide between the conflicting opinions, measurements were renewed. Maupertius earned by his work in Lapland the title of "earth flattener" by disproving the Cassinian tenet that the earth was elongated at the poles, and showing that Newton was right. On his return, in 1743, Clairaut published a work, *Théorie de la figure de la Terre*, which was based on the results of C. Maclaurin on homogeneous ellipsoids. It contains a remarkable theorem, named after Clairaut, that the sum of the fractions expressing the ellipticity and the increase of gravity at the pole is equal to $2\frac{1}{2}$ times the fraction expressing the centrifugal force at the equator, the unit of force being represented by the force of gravity at the equator. This theorem is independent of any hypothesis with respect to the law of densities of the successive strata of the earth. It em-

[1] I. Todhunter, *History of Theory of Prob.*, Chapter 17.

bodies most of Clairaut's researches. I. Todhunter says that "in the figure of the earth no other person has accomplished so much as Clairaut, and the subject remains at present substantially as he left it, though the form is different. The splendid analysis which Laplace supplied, adorned but did not really alter the theory which started from the creative hands of Clairaut."

In 1752 he gained a prize of the St. Petersburg Academy for his paper on *Théorie de la Lune*, in which for the first time modern analysis is applied to lunar motion. This contained the explanation of the motion of the lunar apsides. This motion, left unexplained by I. Newton, seemed to him at first inexplicable by Newton's law, and he was on the point of advancing a new hypothesis regarding gravitation, when, taking the precaution to carry his calculation to a higher degree of approximation, he reached results agreeing with observation. The motion of the moon was studied about the same time by L. Euler and D'Alembert. Clairaut predicted that "Halley's Comet," then expected to return, would arrive at its nearest point to the sun on April 13, 1759, a date which turned out to be one month too late. He applied the process of differentiation to the differential equation now known by his name and detected its singular solution. The same process had been used earlier by Brook Taylor.

In their scientific labors there was between Clairaut and D'Alembert great rivalry, often far from friendly. The growing ambition of Clairaut to shine in society, where he was a great favorite, hindered his scientific work in the latter part of his life.

The astronomer Jean-Dominique Cassini, whom we mentioned above, is the inventor of a quartic curve which was published in his son's *Elêments d'astronomie*, 1749. The curve bears the name of "Cassini's oval" or "general lemniscate." It grew out of the study of a problem in astronomy.[1] Its equation is $(x^2+y^2)^2-2a^2(x^2-y^2)+a^4-c^4=0$.

Johann Heinrich Lambert (1728–1777), born at Mühlhausen in Alsace, was the son of a poor tailor. While working at his father's trade, he acquired through his own unaided efforts a knowledge of elementary mathematics. At the age of thirty he became tutor in a Swiss family and secured leisure to continue his studies. In his travels with his pupils through Europe he became acquainted with the leading mathematicians. In 1764 he settled in Berlin, where he became member of the Academy, and enjoyed the society of L. Euler and J. Lagrange. He received a small pension, and later became editor of the Berlin *Ephemeris*. His many-sided scholarship reminds one of Leibniz. It cannot be said that he was overburdened with modesty. When Frederick the Great asked him in their first interview, which science he was most proficient in, he replied curtly, "All."

[1] G. Loria, *Ebene Curven* (F. Schütte), I, 1910, p. 208.

To the emperor's further question, how he attained this mastery, he said, "Like the celebrated Pascal, by my own self."

In his *Cosmological Letters* he made some remarkable prophecies regarding the stellar system. He entered upon plans for a mathematical symbolic logic of the nature once outlined by G. W. Leibniz. In mathematics he made several discoveries which were extended and overshadowed by his great contemporaries. His first research on pure mathematics developed in an infinite series the root x of the equation $x^m+px=q$. Since each equation of the form $ax^r+bx^s=d$ can be reduced to $x^m+px=q$ in two ways, one or the other of the two resulting series was always found to be convergent, and to give a value of x. Lambert's results stimulated L. Euler, who extended the method to an equation of four terms, and particularly J. Lagrange, who found that a function of a root of $\alpha-x+\phi(x)=0$ can be expressed by the series bearing his name. In 1761 Lambert communicated to the Berlin Academy a memoir (published 1768), in which he proves rigorously that π is irrational. It is given in simplified form in Note IV of A. M. Legendre's *Géométrie*, where the proof is extended to π^2. Lambert proved that if x is rational, but not zero, then neither e^x nor tan x can be a rational number; since tan $\pi/4=1$, it follows that $\frac{\pi}{4}$ or π cannot be rational. Lambert's proofs rest on the expression for e as a continued fraction given by L. Euler [1] who in 1737 had substantially shown the irrationality of e and e^2. There were at this time so many circle squarers that in 1755 the Paris Academy found it necessary to pass a resolution that no more solutions on the quadrature of the circle should be examined by its officials. This resolution applied also to solutions of the duplication of the cube and the trisection of an angle. The conviction had been growing that the solution of the squaring of the circle was impossible, but an irrefutable proof was not discovered until over a century later. Lambert's *Freye Perspective*, 1759 and 1773, contains researches on descriptive geometry, and entitle him to the honor of being the forerunner of Monge. In his effort to simplify the calculation of cometary orbits, he was led geometrically to some remarkable theorems on conics, for instance this: "If in two ellipses having a common major axis we take two such arcs that their chords are equal, and that also the sums of the radii vectores, drawn respectively from the foci to the extremities of these arcs, are equal to each other, then the sectors formed in each ellipse by the arc and the two radii vectores are to each other as the square roots of the parameters of the ellipses." [2]

Lambert elaborated the subject of hyperbolic functions which he designated by *sinh x*, *cosh x*, etc. He was, however, not the first to

[1] R. C. Archibald in *Am. Math. Monthly*, Vol. 21, 1914, p. 253.

[2] M. Chasles, *Geschichte der Geometrie*, 1839, p. 183.

introduce them into trigonometry. That honor falls upon *Vincenzo Riccati* (1707–1775), a son of Jacopo Riccati.[1]

In 1770 Lambert published a 7-place table of natural logarithms for numbers 1–100. In 1778 one of his pupils, *Johann Karl Schulze*, published extensive tables which included the 48-place table of natural logarithms of primes and many other numbers up to 10,009, which had been computed by the Dutch artillery officer, *Wolfram*. A feat even more remarkable than Wolfram's, was the computation of the common logarithms of numbers 1–100 and of all primes from 100 to 1100, to 61 places, by *Abraham Sharp* of Yorkshire, who was some time assistant to Flamsteed at the English Royal Observatory. They were published in Sharp's *Geometry Improv'd*, 1717.

John Landen (1719–1790) was an English mathematician whose writings served as the starting-point of investigations by L. Euler, J. Lagrange, and A. M. Legendre. Landen's capital discovery, contained in a memoir of 1755, was that every arc of the hyperbola is immediately rectified by means of two arcs of an ellipse. In his "residual analysis" he attempted to obviate the metaphysical difficulties of fluxions by adopting a purely algebraic method. J. Lagrange's *Calcul des Fonctions* is based upon this idea. Landen showed how the algebraic expression for the roots of a cubic equation could be derived by application of the differential and integral calculus. Most of the time of this suggestive writer was spent in the pursuits of active life.

Of influence in the teaching of mathematics in England was *Charles Hutton* (1737–1823), for many years professor at the Royal Military Academy of Woolwich. In 1785 he published his *Mathematical Tables*, and in 1795 his *Mathematical and Philosophical Dictionary*, the best work of its kind that has appeared in the English language. His *Elements of Conic Sections*, 1789, is remarkable as being the first work in which each equation is rendered conspicuous by being printed in a separate line by itself.[2]

It is well known that the Newton-Raphson method of approximation to the roots of numerical equations, as it was handed down from the seventeenth century, labored under the defect of insecurity in the process, so that the successive corrections did not always yield results converging to the true value of the root sought. The removal of this defect is usually attributed to J. Fourier, but he was anticipated half a century by *J. Raym. Mourraille* in his *Traité de la résolution des équations en général*, Marseille et Paris, 1768. Mourraille was for fourteen years secretary of the academy of sciences in Marseille; later he became mayor of the city. Unlike I. Newton and J. Lagrange, Mourraille and J. Fourier introduced also geometrical considerations. Mourraille concluded that security is insured if the first approximation

[1] M. Cantor, *op. cit.*, Vol. IV, 1908, p. 411.
[2] M. Cantor, *op. cit.*, Vol. IV, 1908, p. 465.

a is so selected that the curve is convex toward the axis of x for the interval between a and the root. He shows that this condition is sufficient, but not necessary.[1]

In the eighteenth century proofs were given of Descartes' Rule of Signs which its discoverer had enunciated without demonstration. G. W. Leibniz had pointed out a line of proof, but did not actually give it. In 1675 *Jean Prestet* (1648–1690) published at Paris in his *Elemens des mathématiques* a proof which he afterwards acknowledged to be insufficient. In 1728 *Johann Andreas Segner* (1704–1777) published at Jena a correct proof for equations having only real roots. In 1756 he gave a general demonstration, based on the consideration that multiplying a polynomial by $(x-a)$ increases the number of variations by at least one. Other proofs were given by *Jean Paul de Gua de Malves* (1741), *Isaac Milner* (1778), *Friedrich Wilhelm Stübner*, *Abraham Gotthelf Kästner* (1745), *Edward Waring* (1782), *J. A. Grunert* (1827), *K. F. Gauss* (1828). Gauss showed that, if the number of positive roots falls short of the number of variations, it does so by an even number. E. Laguerre later extended the rule to polynomials with fractional and incommensurable exponents, and to infinite series.[2] It was established by De Gua de Malves that the absence of $2m$ successive terms indicates $2m$ imaginary roots, while the absence of $2m+1$ successive terms indicates $2m+2$ or $2m$ imaginary roots, according as the two terms between which the deficiency occurs have like or unlike signs.

Edward Waring (1734–1798) was born in Shrewsbury, studied at Magdalene College, Cambridge, was senior wrangler in 1757, and Lucasian professor of mathematics since 1760. He published *Miscellanea analytica* in 1762, *Meditationes algebraicæ* in 1770, *Proprietatis algebraicarum curvarum* in 1772, and *Meditationes analyticæ* in 1776. These works contain many new results, but are difficult of comprehension on account of his brevity and obscurity of exposition. He is said not to have lectured at Cambridge, his researches being thought unsuited for presentation in the form of lectures. He admitted that he never heard of any one in England, outside of Cambridge, who had read and understood his researches.

In his *Meditationes algebraicæ* are some new theorems on number. Foremost among these is a theorem discovered by his friend *John Wilson* (1741–1793) and universally known as "Wilson's theorem." Waring gives the theorem, known as "Waring's theorem," that every integer is either a cube or the sum of 2, 3, 4, 5, 6, 7, 8 or 9 cubes, either a fourth power or the sum of 2, 3 . . or 19 fourth powers; this has never yet been fully demonstrated. Also without proof is given the theorem that every even integer is the sum of two primes and every

[1] See F. Cajori in *Bibliotheca mathematica*, 3rd S., Vol. 11, 1911, pp. 132–137.

[2] For references to the publications of these writers, see F. Cajori in *Colorado College Publication*, General Series No. 51, 1910, pp. 186, 187.

odd integer is a prime or the sum of three primes. The part relating to even integers is generally known as "Goldbach's theorem," but was first published by Waring. Christian Goldbach communicated the theorem to L. Euler in a letter of June 30, 1742, but the letter was not published until 1843 (*Corr. math.*, P. H. Fuss).

Waring held advanced views on the convergence of series.[1] He taught that $1+\frac{1}{2^n}+\frac{1}{3^n}+\frac{1}{4^n}+\ldots$ converges when $n>1$ and diverges when $n<1$. He gave the well-known test for convergence and divergence which is often ascribed to A. L. Cauchy, in which the limit of the ratio of the $(n+1)^{th}$ to the nth term is considered. As early as 1757 he had found the necessary and sufficient relations which must exist between the coefficients of a quartic and quintic equation, for two and for four imaginary roots. These criteria were obtained by a new transformation, namely the one which yields an equation whose roots are the squares of the differences of the roots of the given equation. To solve the important problem of the separation of the roots Waring transforms a numerical equation into one whose roots are reciprocals of the differences of the roots of the given equation. The reciprocal of the largest of the roots of the transformed equation is less than the smallest difference D, between any two roots of the given equation. If M is an upper limit of the roots of the given equation, then the subtraction of D, $2D$, $3D$, etc., from M will give values which separate all the real roots. In the *Meditationes algebraicæ* of 1770, Waring gives for the first time a process for the approximation to the values of imaginary roots. If x is approximately $a+ib$, substitute $x=a+a'+(b+b')i$, expand and reject higher powers of a' and b'. Equating real numbers to each other and imaginary numbers to each other, two equations are obtained which yield values of a' and b'.

Etienne Bézout (1730–1783) was a French writer of popular mathematical school-books. In his *Théorie générale des Equations Algébriques*, 1779, he gave the method of elimination by linear equations (invented also by L. Euler). This method was first published by him in a memoir of 1764, in which he uses determinants, without, however, entering upon their theory. A beautiful theorem as to the degree of the resultant goes by his name. He and L. Euler both gave the degree as in general $m \cdot n$, the product of the orders of the intersecting loci, and both proved the theorem by reducing the problem to one of elimination from an auxiliary set of linear equations. The determinant resulting from Bézout's method is what J. J. Sylvester and later writers call the Bézoutiant. Bézout fixed the degree of the eliminant also for a large number of particular cases. "One may say that he determined the number of finite intersections of algebraic loci, not only when all the intersections are finite, but also when singular

[1] M. Cantor, *op. cit.*, Vol. IV, 1908, p. 275.

points, or singular lines, planes, etc., at infinity occasion the withdrawal to infinity of certain of the intersection points; and this at a time when the nature of such singularities had not been developed."[1]

Louis Arbogaste (1759–1803) of Alsace was professor of mathematics at Strasburg. His chief work, the *Calcul des Dérivations*, 1800, gives the method known by his name, by which the successive coefficients of a development are derived from one another when the expression is complicated. A. De Morgan has pointed out that the true nature of derivation is differentiation accompanied by integration. In this book for the first time are the symbols of operation separated from those of quantity. The notation $D_x y$ for $\frac{dy}{dx}$ is due to him.

Maria Gaetana Agnesi (1718–1799) of Milan, distinguished as a linguist, mathematician, and philosopher, filled the mathematical chair at the University of Bologna during her father's sickness. Agnesi was a somnambulist. Several times it happened to her that she went to her study, while in the somnambulist state, made a light, and solved some problem she had left incomplete when awake. In the morning she was surprised to find the solution carefully worked out on paper.[2] In 1748 she published her *Instituzioni Analitiche*, which was translated into English in 1801. The "witch of Agnesi" or "Versiera" is a cubic curve $x^2y=a^2(a-y)$ treated therein; a related curve was given earlier by P. Fermat in the form $(a^2-x^2)\ y=a^3$. The curve was discussed by Guido Grandi in his *Quadratura circuli et hyperbolæ*, Pisa, 1703 and 1710.[3] In two letters from Grandi to Leibniz, in 1713, curves resembling flowers are discussed; in 1728 Grandi published at Florence his *Flores geometrici*. He considered curves in a plane, of the type $\rho=r\sin n\omega$, and also curves on a sphere. Recent studies along this line are due to Bodo Habenicht (1895), E. W. Hyde (1875), H. Wieleitner (1906).

The leading eighteenth century historian of mathematics was **Jean Etienne Montucla** (1725–1799) who published a *Histoire des mathématiques*, in two volumes, Paris, 1758. A second edition of these two volumes appeared in 1799. A third volume, written by Montucla, was partly printed when he died; the rest of it was seen through the press by the astronomer **Joseph Jérôme le François de Lalande** (1732–1807), who prepared a fourth volume, mainly on the history of astronomy.[4]

Joseph Louis Lagrange (1736–1813), one of the greatest mathematicians of all times, was born at Turin and died at Paris. He was of French extraction. His father, who had charge of the Sardinian

[1] H. S. White in *Bull. Am. Math. Soc.*, Vol. 15, 1909, p. 331.

[2] *L'Intermédiaire des mathématiciens*, Vol. 22, 1915, p. 241.

[3] G. Loria, *Ebene Curven* (F. Schütte), I, 1910, p. 79.

[4] For details on other mathematical historians, see S. Günther's chapter in Cantor, *op. cit.*, Vol. IV, 1908, pp. 1–36.

military chest, was once wealthy, but lost all he had in speculation. Lagrange considered this loss his good fortune, for otherwise he might not have made mathematics the pursuit of his life. While at the college in Turin his genius did not at once take its true bent. Cicero and Virgil at first attracted him more than Archimedes and Newton. He soon came to admire the geometry of the ancients, but the perusal of a tract of E. Halley roused his enthusiasm for the analytical method, in the development of which he was destined to reap undying glory. He now applied himself to mathematics, and in his seventeenth year he became professor of mathematics in the royal military academy at Turin. Without assistance or guidance he entered upon a course of study which in two years placed him on a level with the greatest of his contemporaries. With aid of his pupils he established a society which subsequently developed into the Turin Academy. In the first five volumes of its transactions appear most of his earlier papers. At the age of nineteen he communicated to L. Euler a general method of dealing with "isoperimetrical problems," known now as the Calculus of Variations. This commanded Euler's lively admiration, and he courteously withheld for a time from publication some researches of his own on this subject, so that the youthful Lagrange might complete *his* investigations and claim the invention. Lagrange did quite as much as Euler towards the creation of the Calculus of Variations. As it came from Euler it lacked an analytic foundation, and this Lagrange supplied. He separated the principles of this calculus from geometric considerations by which his predecessor had derived them. Euler had assumed as fixed the limits of the integral, *i. e.* the extremities of the curve to be determined, but Lagrange removed this restriction and allowed all co-ordinates of the curve to vary at the same time. Euler introduced in 1766 the name "calculus of variations," and did much to improve this science along the lines marked out by Lagrange. Lagrange's investigations on the calculus of variations were published in 1762, 1771, 1788, 1797, 1806.

Another subject engaging the attention of Lagrange at Turin was the propagation of sound. In his papers on this subject in the *Miscellanea Taurinensia*, the young mathematician appears as the critic of I. Newton, and the arbiter between Euler and D'Alembert. By considering only the particles which are in a straight line, he reduced the problem to the same partial differential equation that represents the motions of vibrating strings.

Vibrating strings had been discussed by Brook Taylor, Johann Bernoulli and his son Daniel, by D'Alembert and L. Euler. In solving the partial differential equations, D'Alembert restricted himself to functions which can be expanded by Taylor's series, while Euler thought that no restriction was necessary, that they could be arbitrary, discontinuous. The problem was taken up with great skill by Lagrange who introduced new points of view, but decided in favor of

Euler. Later, de Condorcet and P. S. Laplace stood on the side of D'Alembert since in their judgment some restriction upon the arbitrary functions was necessary. From the modern point of view, neither D'Alembert nor Euler was wholly in the right: D'Alembert insisted upon the needless restriction to functions with a limitless number of derivatives, while Euler assumed that the differential and integral calculus could be applied to any arbitrary function.[1]

It now appears that Daniel Bernoulli's claim that his solution was a general one (a claim disputed by D'Alembert, J. Lagrange and L. Euler) was fully justified. The problem of vibrating strings stimulated the growth of the theory of expansions according to trigonometric functions of multiples of the argument. H. Burkhardt has pointed out that there was also another line of growth of this subject, namely the growth in connection with the problem of perturbations, where L. Euler started out with the development of the reciprocal distance of two planets according to the cosine of multiples of the angle between their radii vectoris.

By constant application during nine years, Lagrange, at the age of twenty-six, stood at the summit of European fame. But his intense studies had seriously weakened a constitution never robust, and though his physicians induced him to take rest and exercise, his nervous system never fully recovered its tone, and he was thenceforth subject to fits of melancholy.

In 1764 the French Academy proposed as the subject of a prize the theory of the libration of the moon. It demanded an explanation, on the principle of universal gravitation, why the moon always turns, with but slight variations, the same face to the earth. Lagrange secured the prize. This success encouraged the Academy to propose for a prize the theory of the four satellites of Jupiter,—a problem of six bodies, more difficult than the one of three bodies previously treated by A. C. Clairaut, D'Alembert, and L. Euler. Lagrange overcame the difficulties by methods of approximation. Twenty-four years afterwards this subject was carried further by P. S. Laplace. Later astronomical investigations of Lagrange are on cometary perturbations (1778 and 1783), and on Kepler's problem. His researches on the problem of three bodies has been referred to previously.

Being anxious to make the personal acquaintance of leading mathematicians, Lagrange visited Paris, where he enjoyed the stimulating delight of conversing with A. C. Clairaut, D'Alembert, de Condorcet, the Abbé Marie, and others. He had planned a visit to London, but he fell dangerously ill after a dinner in Paris, and was compelled to return to Turin. In 1766 L. Euler left Berlin for St. Petersburg, and he pointed out Lagrange as the only man capable of filling the place.

[1] For details see H. Burkhardt's *Entwicklungen nach oscillirenden Funktionen und Integration der Differentialgleichungen der mathematischen Physik.* Leipzig, 1908, p. 18. This is an exhaustive and valuable history of this topic.

D'Alembert recommended him at the same time. Frederick the Great thereupon sent a message to Turin, expressing the wish of "the greatest king of Europe" to have "the greatest mathematician" at his court. Lagrange went to Berlin, and staid there twenty years. Finding all his colleagues married, and being assured by their wives that the marital state alone is happy, he married. The union was not a happy one. His wife soon died. Frederick the Great held him in high esteem, and frequently conversed with him on the advantages of perfect regularity of life. This led Lagrange to cultivate regular habits. He worked no longer each day than experience taught him he could without breaking down. His papers were carefully thought out before he began writing, and when he wrote he did so without a single correction.

During the twenty years in Berlin he crowded the transactions of the Berlin Academy with memoirs, and wrote also the epoch-making work called the *Mécanique Analytique*. He enriched algebra by researches on the solution of equations. There are two methods of solving directly algebraic equations,—that of substitution and that of combination. The former method was developed by L. Ferrari, F. Vieta, E. W. Tchirnhausen, L. Euler, E. Bézout, and Lagrange; the latter by C. A. Vandermonde and Lagrange.[1] In the method of substitution the original forms are so transformed that the determination of the roots is made to depend upon simpler functions (resolvents). In the method of combination auxiliary quantities are substituted for certain simple combinations ("types") of the unknown roots of the equation, and auxiliary equations (resolvents) are obtained for these quantities with aid of the coefficients of the given equation. In his *Réflexions sur la résolution algébrique des équations*, published in Memoirs of the Berlin Academy for the years 1770 and 1771, Lagrange traced all known algebraic solutions of equations to the uniform principle consisting in the formation and solution of equations of lower degree whose roots are linear functions of the required roots, and of the roots of unity. He showed that the quintic cannot be reduced in this way, its resolvent being of the sixth degree. In this connection Lagrange had occasion to consider the number of values a rational function can assume when its variables are permuted in every possible way. In these studies we see the beginnings of the theory of groups. The theorem, that the order of a subgroup is a divisor of the order of the group is practically established, and is known now as "Lagrange's theorem," although its complete proof was first given about thirty years later by *Pietro Abbati* (1768–1842) of Modena in Italy. Lagrange's researches on the theory of equations were continued after he left Berlin. In the *Résolution des équations numériques* (1798) he gave among other things, a proof that every equation must have a root,—a theorem which before this usually had been considered

[1] L. Matthiessen, *op. cit.*, pp. 80–84.

self-evident. Other proofs of this were given by J. R. Argand, K. F. Gauss, and A. L. Cauchy. In a note to the above work Lagrange uses Fermat's theorem and certain suggestions of Gauss in effecting a complete algebraic solution of any binomial equation.

In the Berlin *Mémoires* for the year 1767 Lagrange contributed a paper, *Sur la résolution des équations numériques*. He explains the separation of the real roots by substituting for x the terms of the progression, 0, D, $2D$, . . ., where D must be less than the least difference between the roots. Lagrange suggested three ways of computing D: One way in 1767, another in 1795 and a third in 1798. The first depends upon the equation of the squared differences of the roots of the given equation. E. Waring before this had derived this important equation, but in 1767 Lagrange had not yet seen Waring's writings. Lagrange finds equal roots by computing the highest common factor between $f(x)$ and $f'(x)$. He proceeds to develop a new mode of approximation, that by continued fractions. P. A. Cataldi had used these fractions in extracting square roots. Lagrange enters upon greater details in his *Additions* to his paper of 1767. Unlike the older methods of approximation, Lagrange's has no cases of failure. "Cette méthode ne laisse, ce me semble, rien à désirer," yet, though theoretically perfect, it yields the root in the form of a continued fraction which is undesirable in practice.

While in Berlin Lagrange published several papers on the theory of numbers. In 1769 he gave a solution in integers of indeterminate equations of the second degree, which resembles the Hindu cyclic method; he was the first to prove, in 1771, "Wilson's theorem," enunciated by an Englishman, John Wilson, and first published by E. Waring in his *Meditationes Algebraicæ;* he investigated in 1775 under what conditions ± 2 and ± 5 (-1 and ± 3 having been discussed by L. Euler) are quadratic residues, or non-residues of odd prime numbers, q; he proved in 1770 Bachet de Méziriac's theorem that every integer is equal to the sum of four, or a less number, of squares. He proved Fermat's theorem on $x^n+y^n=z^n$, for the case $n=4$, also Fermat's theorem that, if $a^2+b^2=c^2$, then ab is not a square.

In his memoir on Pyramids, 1773, Lagrange made considerable use of determinants of the third order, and demonstrated that the square of a determinant is itself a determinant. He never, however, dealt explicitly and directly with determinants; he simply obtained accidentally identities which are now recognized as relations between determinants.

Lagrange wrote much on differential equations. Though the subject of contemplation by the greatest mathematicians (L. Euler, D'Alembert, A. C. Clairaut, J. Lagrange, P. S. Laplace), yet more than other branches of mathematics do they resist the systematic application of fixed methods and principles. The subject of singular solutions, which had been taken up by P. S. Laplace in 1771 and 1774, was investigated by Lagrange who gave the derivation of a singular solution from the general solution as well as from the differential equation itself. Lagrange brought to view the relation of singular solutions to envelopes. Nevertheless, he failed to remove all mystery surrounding this subtle subject. An inconsistency in his theorems caused about 1870 a complete reconsideration of the entire theory of singular solutions. Lagrange's treatment is given in his *Calcul des Fonctions*, Lessons 14–17. He generalized Euler's researches on total differential equations of two variables, and of the ninth order; he gave a solution of partial differential equations of the first order (*Berlin Memoirs*, 1772 and 1774), and spoke of their singular solutions, extending their solution in *Memoirs* of 1779 and 1785 to equations of any number of variables. The *Memoirs* of 1772 and 1774 were refined in certain points by a young mathematician Paul Charpit (?–1784) whose method of solution was first printed in Lacroix's *Traité du calcul*, 2. Ed., Paris, 1814, T. II, p. 548. The discussion on partial differential equations of the second order, carried on by D'Alembert, Euler, and Lagrange, has already been referred to in our account of D'Alembert.

While in Berlin, Lagrange wrote the "*Mécanique Analytique*," the greatest of his works (Paris, 1788). From the principle of virtual velocities he deduced, with aid of the calculus of variations, the whole system of mechanics so elegantly and harmoniously that it may fitly be called, in Sir William Rowan Hamilton's words, "a kind of scientific poem." It is a most consummate example of analytic generality. Geometrical figures are nowhere allowed. "On ne trouvera point de figures dans cet ouvrage" (Preface). The two divisions of mechanics —statics and dynamics—are in the first four sections of each carried out analogously, and each is prefaced by a historic sketch of principles. Lagrange formulated the principle of least action. In their original form, the equations of motion involve the co-ordinates x, y, z, of the different particles m or dm of the system. But x, y, z, are in general not independent, and Lagrange introduced in place of them any variables ξ, ψ, ϕ, whatever, determining the position of the point at the time. These "generalized co-ordinates" may be taken to be independent. The equations of motion may now assume the form

$$\frac{d}{dt}\frac{dT}{d\xi'}-\frac{dT}{d\xi}+\Xi=0;$$

or when Ξ, ψ, ϕ, . . . are the partial differential coefficients with respect to ξ, ψ, ϕ, . . . of one and the same function V, then the form

$$\frac{d}{dt}\frac{dT}{d\xi'}-\frac{dT}{d\xi}+\frac{dV}{d\xi}=0.$$

The latter is *par excellence* the Lagrangian form of the equations of motion. With Lagrange originated the remark that mechanics may be regarded as a geometry of four dimensions. To him falls the honor of the introduction of the potential into dynamics. Lagrange was anxious to have his *Mécanique Analytique* published in Paris. The work was ready for print in 1786, but not till 1788 could he find a publisher, and then only with the condition that after a few years he would purchase all the unsold copies. The work was edited by A. M. Legendre.

After the death of Frederick the Great, men of science were no longer respected in Germany, and Lagrange accepted an invitation of Louis XVI to migrate to Paris. The French queen treated him with regard, and lodging was procured for him in the Louvre. But he was seized with a long attack of melancholy which destroyed his taste for mathematics. For two years his printed copy of the *Mécanique*, fresh from the press,—the work of a quarter of a century,—lay unopened on his desk. Through A. L. Lavoisier he became interested in chemistry, which he found "as easy as algebra." The disastrous crisis of the French Revolution aroused him again to activity. About this time the young and accomplished daughter of the astronomer P. C. Lemonnier took compassion on the sad, lonely Lagrange, and insisted upon marrying him. Her devotion to him constituted the one tie to life which at the approach of death he found it hard to break.

He was made one of the commissioners to establish weights and measures having units founded on nature. Lagrange strongly favored the decimal subdivision. Such was the moderation of Lagrange's character, and such the universal respect for him, that he was retained as president of the commission on weights and measures even after it had been *purified* by the Jacobins by striking out the names of A. L. Lavoisier, P. S. Laplace, and others. Lagrange took alarm at the fate of Lavoisier, and planned to return to Berlin, but at the establishment of the *École Normale* in 1795 in Paris, he was induced to accept a professorship. Scarcely had he time to elucidate the foundations of arithmetic and algebra to young pupils, when the school was closed. His additions to the algebra of L. Euler were prepared at this time. In 1797 the *École Polytechnique* was founded, with Lagrange as one of the professors. The earliest triumph of this institution was the restoration of Lagrange to analysis. His mathematical activity burst out anew. He brought forth the *Théorie des fonctions analytiques* (1797), *Leçons sur le calcul des fonctions*, a treatise on the same lines as the preceding (1801), and the *Résolution des équations numeriques* (1798), which includes papers published much earlier; his memoir, *Nouvelle méthode pour résoudre les équations littérales par le moyen des séries*, published 1770, gives the notation ψ' for $\frac{d\psi}{dx}$, which occurs however much earlier in a part of a memoir by François Daviet de

Foncenex in the *Miscellanea Taurinensia* for 1759, believed to have been written for Foncenex by Lagrange himself.[1] In 1810 he began a thorough revision of his *Mécanique analytique*, but he died before its completion.

The *Théorie des fonctions*, the germ of which is found in a memoir of his of 1772, aimed to place the principles of the calculus upon a sound foundation by relieving the mind of the difficult conception of a limit. John Landen's residual calculus, professing a similar object, was unknown to him. In a letter to L. Euler of Nov. 24, 1759, Lagrange says that he believed he had developed the true metaphysics of the calculus; at that time he seems to have been convinced that the use of infinitesimals was rigorous. He "used both the infinitesimal method and the method of derived functions side by side during his whole life" (Jourdain). Lagrange attempted to prove Taylor's theorem (the power of which he was the first to point out) by simple algebra, and then to develop the entire calculus from that theorem. The principles of the calculus were in his day involved in philosophic difficulties of a serious nature. The infinitesimals of G. W. Leibniz had no satisfactory metaphysical basis. In the differential calculus of L. Euler they were treated as absolute zeros. In I. Newton's limiting ratio, the magnitudes of which it is the ratio cannot be found, for at the moment when they should be caught and equated, there is neither arc nor chord. The chord and arc were not taken by Newton as equal before vanishing, nor after vanishing, but *when* they vanish. "That method," said Lagrange, "has the great inconvenience of considering quantities in the state in which they cease, so to speak, to be quantities; for though we can always well conceive the ratios of two quantities, as long as they remain finite, that ratio offers to the mind no clear and precise idea, as soon as its terms become both nothing at the same time." D'Alembert's method of limits was much the same as the method of prime and ultimate ratios. When Lagrange endeavored to free the calculus of its metaphysical difficulties, by resorting to common algebra, he avoided the whirlpool of Charybdis only to suffer wreck against the rocks of Scylla. The algebra of his day, as handed down to him by L. Euler, was founded on a false view of infinity. No rigorous theory of infinite series had then been established. Lagrange proposed to define the differential coefficient of $f(x)$ with respect to x as the coefficient of h in the expansion of $f(x+h)$ by Taylor's theorem, and thus to avoid all reference to limits. But he used infinite series without ascertaining carefully that they were convergent, and his proof that $f(x+h)$ can always be expanded in a series of ascending powers of h, labors under serious defects. Though Lagrange's method of developing the calculus was at first greatly applauded, its defects were fatal, and to-day his "method of

[1] Philip E. B. Jourdain in *Proceed. 5th Intern. Congress, Cambridge, 1912*, Cambridge, 1913, Vol. II, p. 540.

derivatives," as it was called, has been generally abandoned. He introduced a notation of his own, but it was inconvenient, and was abandoned by him in the second edition of his *Mécanique*, in which he used infinitesimals. The primary object of the *Théorie des fonctions* was not attained, but its secondary results were far-reaching. It was a purely abstract mode of regarding functions, apart from geometrical or mechanical considerations. In the further development of higher analysis a function became the leading idea, and Lagrange's work may be regarded as the starting-point of the theory of functions as developed by A. L. Cauchy, G. F. B. Riemann, K. Weierstrass, and others.

The first to doubt the rigor of Lagrange's exposition of the calculus were *Abel Bürja* (1752–1816)of Berlin, the two Polish mathematicians *H. Wronski* and *J. B. Sniadecki* (1756–1830), and the Bohemian *B. Bolzano*, who were all men of limited acquaintance and influence. It remained for A. L. Cauchy really to initiate the period of greater rigor.

Instructive is C. E. Picard's characterization of the time of Lagrange: "In all this period, especially in the second half of the eighteenth century, what strikes us with admiration and is also somewhat confusing, is the extreme importance of the applications realized, while the pure theory appeared still so ill assured. One perceives it when certain questions are raised like the degree of arbitrariness in the integral of vibrating chords, which gives place to an interminable and inconclusive discussion. Lagrange appreciated these insufficiencies when he published his theory of analytic functions, where he strove to give a precise foundation to analysis. One cannot too much admire the marvellous presentiment he had of the rôle which the functions, which with him we call analytic, were to play; but we may confess that we stand astonished before the demonstration he believed to have given of the possibility of the development of a function in Taylor's series." [1]

In the treatment of infinite series Lagrange displayed in his earlier writings that laxity common to all mathematicians of his time, excepting Nicolaus Bernoulli II and D'Alembert. But his later articles mark the beginning of a period of greater rigor. Thus, in the *Calcul des fonctions* he gives his theorem on the limits of Taylor's theorem. Lagrange's mathematical researches extended to subjects which have not been mentioned here—such as probabilities, finite differences, ascending continued fractions, elliptic integrals. Everywhere his wonderful powers of generalization and abstraction are made manifest. In that respect he stood without a peer, but his great contemporary, P. S. Laplace, surpassed him in practical sagacity. Lagrange was content to leave the application of his general results to others, and some of the most important researches of Laplace (particularly those

[1] *Congress of Arts and Science*, St. Louis, 1904, Vol. I, p. 503.

on the velocity of sound and on the secular acceleration of the moon) are implicitly contained in Lagrange's works.

Lagrange was an extremely modest man, eager to avoid controversy, and even timid in conversation. He spoke in tones of doubt, and his first words generally were, "Je ne sais pas." He would never allow his portrait to be taken, and the only ones that were secured were sketched without his knowledge by persons attending the meetings of the Institute.

Pierre Simon Laplace (1749–1827) was born at Beaumont-en-Auge in Normandy. Very little is known of his early life. When at the height of his fame he was loath to speak of his boyhood, spent in poverty. His father was a small farmer. Some rich neighbors who recognized the boy's talent assisted him in securing an education. As an extern he attended the military school in Beaumont, where at an early age he became teacher of mathematics. At eighteen he went to Paris, armed with letters of recommendation to D'Alembert, who was then at the height of his fame. The letters remained unnoticed, but young Laplace, undaunted, wrote the great geometer a letter on the principles of mechanics, which brought the following enthusiastic response: "You needed no introduction; you have recommended yourself; my support is your due." D'Alembert secured him a position at the *École Militaire* of Paris as professor of mathematics. His future was now assured, and he entered upon those profound researches which brought him the title of "the Newton of France." With wonderful mastery of analysis, Laplace attacked the pending problems in the application of the law of gravitation to celestial motions. During the succeeding fifteen years appeared most of his original contributions to astronomy. His career was one of almost uninterrupted prosperity. In 1784 he succeeded E. Bézout as examiner to the royal artillery, and the following year he became member of the Academy of Sciences. He was made president of the Bureau of Longitude; he aided in the introduction of the decimal system, and taught, with J. Lagrange, mathematics in the *École Normale*. When, during the Revolution, there arose a cry for the reform of everything, even of the calendar, Laplace suggested the adoption of an era beginning with the year 1250, when, according to his calculation, the major axis of the earth's orbit had been perpendicular to the equinoctial line. The year was to begin with the vernal equinox, and the zero meridian was to be located east of Paris by 185.30 degrees of the centesimal division of the quadrant, for by this meridian the beginning of his proposed era fell at midnight. But the revolutionists rejected this scheme, and made the start of the new era coincide with the beginning of the glorious French Republic.[1]

Laplace was justly admired throughout Europe as a most sagacious

[1] Rudolf Wolf, *Geschichte der Astronomie*, Münich, 1877, p. 334.

and profound scientist, but, unhappily for his reputation, he strove not only after greatness in science, but also after political honors. The political career of this eminent scientist was stained by servility and suppleness. After the 18th of Brumaire, the day when Napoleon was made emperor, Laplace's ardor for republican principles suddenly gave way to a great devotion to the emperor. Napoleon rewarded this devotion by giving him the post of minister of the interior, but dismissed him after six months for incapacity. Said Napoleon, "Laplace ne saisissait aucune question sous son véritable point de vue; il cherchait des subtilités partout, n'avait que des idées problematiques, et portait enfin l'esprit des infiniment petits jusque dans l'administration." Desirous to retain his allegiance, Napoleon elevated him to the Senate and bestowed various other honors upon him. Nevertheless, he cheerfully gave his voice in 1814 to the dethronement of his patron and hastened to tender his services to the Bourbons, thereby earning the title of marquis. This pettiness of his character is seen in his writings. The first edition of the *Système du monde* was dedicated to the Council of Five Hundred. To the third volume of the *Mécanique Céleste* is prefixed a note that of all the truths contained in the book, the one most precious to the author was the declaration he thus made of gratitude and devotion to the peace-maker of Europe. After this outburst of affection, we are surprised to find in the editions of the *Théorie analytique des probabilités*, which appeared after the Restoration, that the original dedication to the emperor is suppressed.

Though supple and servile in politics, it must be said that in religion and science Laplace never misrepresented or concealed his own convictions however distasteful they might be to others. In mathematics and astronomy his genius shines with a lustre excelled by few. Three great works did he give to the scientific world,—the *Mécanique Céleste*, the *Exposition du système du monde*, and the *Théorie analytique des probabilités*. Besides these he contributed important memoirs to the French Academy.

We first pass in brief review his astronomical researches. In 1773 he brought out a paper in which he proved that the mean motions or mean distances of planets are invariable or merely subject to small periodic changes. This was the first and most important step in his attempt to establish the stability of the solar system.[1] To I. Newton and also to L. Euler it had seemed doubtful whether forces so numerous, so variable in position, so different in intensity, as those in the solar system, could be capable of maintaining permanently a condition of equilibrium. Newton was of the opinion that a powerful hand must intervene from time to time to repair the derangements occasioned by the mutual action of the different bodies. This paper was the beginning of a series of profound researches by J. Lagrange and

[1] D. F. J. Arago, "Eulogy on Laplace," translated by B. Powell, *Smithsonian Report*, 1874.

Laplace on the limits of variation of the various elements of planetary orbits, in which the two great mathematicians alternately surpassed and supplemented each other. Laplace's first paper really grew out of researches on the theory of Jupiter and Saturn. The behavior of these planets had been studied by L. Euler and J. Lagrange without receiving satisfactory explanation. Observation revealed the existence of a steady acceleration of the mean motions of our moon and of Jupiter and an equally strange diminution of the mean motion of Saturn. It looked as though Saturn might eventually leave the planetary system, while Jupiter would fall into the sun, and the moon upon the earth. Laplace finally succeeded in showing, in a paper of 1784–1786, that these variations (called the "great inequality") belonged to the class of ordinary periodic perturbations, depending upon the law of attraction. The cause of so influential a perturbation was found in the commensurability of the mean motion of the two planets.

In the study of the Jovian system, Laplace was enabled to determine the masses of the moons. He also discovered certain very remarkable, simple relations between the movements of those bodies, known as "Laws of Laplace." His theory of these bodies was completed in papers of 1788 and 1789. These, as well as the other papers here mentioned, were published in the *Mémoirs présentés par divers savans*. The year 1787 was made memorable by Laplace's announcement that the lunar acceleration depended upon the secular changes in the eccentricity of the earth's orbit. This removed all doubt then existing as to the stability of the solar system. The universal validity of the law of gravitation to explain all motion in the solar system seemed established. That system, as then known, was at last found to be a complete machine.

In 1796 Laplace published his *Exposition du système du monde*, a non-mathematical popular treatise on astronomy, ending with a sketch of the history of the science. In this work he enunciates for the first time his celebrated nebular hypothesis. A similar theory had been previously proposed by I. Kant in 1755, and by E. Swedenborg; but Laplace does not appear to have been aware of this.

Laplace conceived the idea of writing a work which should contain a complete analytical solution of the mechanical problem presented by the solar system, without deriving from observation any but indispensable data. The result was the *Mécanique Céleste*, which is a systematic presentation embracing all the discoveries of I. Newton, A. C. Clairaut, D'Alembert, L. Euler, J. Lagrange, and of Laplace himself, on celestial mechanics. The first and second volumes of this work were published in 1799; the third appeared in 1802, the fourth in 1805. Of the fifth volume, Books XI and XII were published in 1823; Books XIII, XIV, XV in 1824, and Book XVI in 1825. The first two volumes contain the general theory of the motions and figure of celestial bodies. The third and fourth volumes give special theories of celestial motions,—treating particularly of motions of comets, of our moon, and of other satellites. The fifth volume opens with a brief history of celestial mechanics, and then gives in appendices the results of the author's later researches. The *Mécanique Céleste* was such a master-piece, and so complete, that Laplace's immediate successors were able to add comparatively little. The general part of the work was translated into German by *Johann Karl Burkhardt* (1773–1825), and appeared in Berlin, 1800–1802. *Nathaniel Bowditch* (1773–1838) brought out an edition in English, with an extensive commentary, in Boston, 1829–1839.[*] The *Mécanique Céleste* is not easy reading. The difficulties lie, as a rule, not so much in the subject itself as in the want of verbal explanation. A complicated chain of reasoning receives often no explanation whatever.[*] J. B. Biot, who assisted Laplace in revising the work for the press, tells that he once asked Laplace some explanation of a passage in the book which had been written not long before, and that Laplace spent an hour endeavoring to recover the reasoning which had been carelessly suppressed with the remark, "Il est facile de voir." Notwithstanding the important researches in the work, which are due to Laplace himself, it naturally contains a great deal that is drawn from his predecessors. It is, in fact, the organized result of a century of patient toil. But Laplace frequently neglects properly to acknowledge the source from which he draws, and lets the reader infer that theorems and formulæ due to a predecessor are really his own.

We are told that when Laplace presented Napoleon with a copy of the *Mécanique Céleste*, the latter made the remark, "M. Laplace, they tell me you have written this large book on the system of the universe, and have never even mentioned its Creator." Laplace is said to have replied bluntly, "Je n'avais pas besoin de cette hypothèse-la." This assertion, taken literally, is impious, but may it not have been intended to convey a meaning somewhat different from its literal one? I. Newton was not able to explain by his law of gravitation all questions arising in the mechanics of the heavens. Thus, being unable to show that the solar system was stable, and suspecting in fact that it was unstable, Newton expressed the opinion that the special intervention, from time to time, of a powerful hand was necessary to preserve order. Now Laplace thought that he had proved by the law of gravitation that the solar system is stable, and in that sense may be said to have felt no necessity for reference to the Almighty.

We now proceed to researches which belong more properly to pure mathematics. Of these the most conspicuous are on the theory of probability. Laplace has done more towards advancing this subject than any one other investigator. He published a series of papers, the main results of which were collected in his *Théorie analytique des probabilités*, 1812. The third edition (1820) consists of an introduction and two books. The introduction was published separately under the title, *Essai philosophique sur les probabilités*, and is an admirable and masterly exposition without the aid of analytical formulæ of the principles and applications of the science. The first book contains the theory of generating functions, which are applied, in the second book, to the theory of probability. Laplace gives in his work on probability his method of approximation to the values of definite integrals. The solution of linear differential equations was reduced by him to definite integrals. The use of *partial* difference equations was introduced into the study of probability by him about the same time as by J. Lagrange. One of the most important parts of the work is the application of probability to the method of least squares, which is shown to give the most probable as well as the most convenient results.

Laplace's work on probability is very difficult reading, particularly the part on the method of least squares. The analytical processes are by no means clearly established or free from error. "No one was more sure of giving the result of analytical processes correctly, and no one ever took so little care to point out the various small considerations on which correctness depends" (De Morgan). Laplace's comprehensive work contains all of his own researches and much derived from other writers. He gives masterly expositions of the Problem of Points, of Jakob Bernoulli's theorem, of the problems taken from Bayes and Count de Buffon. In this work as in his *Mécanique Céleste*, Laplace is not in the habit of giving due credit to writers that preceded him. A. De Morgan [1] says of Laplace: "There is enough originating from himself to make any reader wonder that one who could so well afford to state what he had taken from others, should have set an example so dangerous to his own claims."

Of Laplace's papers on the attraction of ellipsoids, the most important is the one published in 1785, and to a great extent reprinted in the third volume of the *Mécanique Céleste*. It gives an exhaustive treatment of the general problem of attraction of any ellipsoid upon a particle situated outside or upon its surface. Spherical harmonics, or the so-called "Laplace's coefficients," constitute a powerful analytic engine in the theory of attraction, in electricity, and magnetism. The theory of spherical harmonics for two dimensions had been previously given by A. M. Legendre. Laplace failed to make due acknowledgment of this, and there existed, in consequence, between the two great men, "a feeling more than coldness." The potential function, V, is much used by Laplace, and is shown by him to satisfy the partial differential equation $\frac{\partial^2 V}{\partial x^2}+\frac{\partial^2 V}{\partial y^2}+\frac{\partial^2 V}{\partial z^2}=0$. This is known as Laplace's

[1] A. De Morgan, *An Essay on Probabilities*, London, 1838 (date of Preface) p. II of Appendix I.

equation, and was first given by him in the more complicated form which it assumes in polar co-ordinates. The notion of potential was, however, not introduced into analysis by Laplace. The honor of that achievement belongs to J. Lagrange.

Regarding Laplace's equation, P. E. Picard said in 1904: "Few equations have been the object of so many works as this celebrated equation. The conditions at the limits may be of divers forms. The simplest case is that of the calorific equilibrium of a body of which we maintain the elements of the surface at given temperatures; from the physical point of view, it may be regarded as evident that the temperature, continuous within the interior since no source of heat is there, is determined when it is given at the surface. A more general case is that where . . . the temperature may be given on one portion, while there is radiation on another portion. These questions . . . have greatly contributed to the orientation of the theory of partial differential equations. They have called attention to types of determinations of integrals, which would not have presented themselves in remaining at a point of view purely abstract." [1]

Among the minor discoveries of Laplace are his method of solving equations of the second, third, and fourth degrees, his memoir on singular solutions of differential equations, his researches in finite differences and in determinants, the establishment of the expansion theorem in determinants which had been previously given by A. T. Vandermonde for a special case, the determination of the complete integral of the linear differential equation of the second order. In the *Mécanique Céleste* he made a generalization of Lagrange's theorem on the development of functions in series known as Laplace's theorem.

Laplace's investigations in physics were quite extensive. We mention here his correction of Newton's formula on the velocity of sound in gases by taking into account the changes of elasticity due to the heat of compression and cold of rarefaction; his researches on the theory of tides; his mathematical theory of capillarity; his explanation of astronomical refraction; his formulæ for measuring heights by the barometer.

Laplace's writings stand out in bold contrast to those of J. Lagrange in their lack of elegance and symmetry. Laplace looked upon mathematics as the tool for the solution of physical problems. The true result being once reached, he devoted little time to explaining the various steps of his analysis, or in polishing his work. The last years of his life were spent mostly at Arcueil in peaceful retirement on a country-place, where he pursued his studies with his usual vigor until his death. He was a great admirer of L. Euler, and would often say, "Lisez Euler, lisez Euler, c'est notre maître à tous."

The latter part of the eighteenth century brought forth researches on the graphic representation of imaginaries, all of which remained

[1] *Congress of Arts and Science*, St. Louis, 1904, Vol. I, p. 506.

quite unnoticed at that time. During the time of R. Descartes, I. Newton and L. Euler, the negative and the imaginary came to be accepted as numbers, but the latter was still regarded as an algebraic fiction. A little over a hundred years after J. Wallis's unsuccessful efforts along the line of graphic representation of imaginaries, "a modest scientist," *Henri Dominique Truel*, pictured imaginaries upon a line that was perpendicular to the line representing real numbers. So far as known, Truel published nothing, nor are his manuscripts extant. All we know about him is a brief reference to him made by A. L. Cauchy,[1] who says that Truel had his graphic scheme as early as 1786, and about 1810 turned his manuscripts over to Augustin Normauf, a ship builder in Havre. W. J. G. Karsten's graphic scheme of 1768 was confined to imaginary *logarithms*. The earliest printed graphic representation of $\sqrt{-1}$ and $a+b\sqrt{-1}$ was given in an "Essay on the Analytic Representation of Direction, with Applications in Particular to the Determination of Plane and Spherical Polygons" presented in 1797 by *Caspar Wessel* (1745–1818) to the Royal Academy of Sciences and Letters of Denmark and published in Vol. V of its Memoirs in 1799. Wessel was born in Jonsrud, in Norway. For many years he was in the employ of the Danish Academy of Sciences as a surveyor. His paper lay buried in the Transactions of the Danish Academy for nearly a century. In 1897 a French translation was brought out by the Danish Academy.[2] Another noteworthy publication which remained unknown for many years is an *Essay* [3] published in 1806 by *Jean Robert Argand* (1768–1822) of Geneva, containing a geometric representation of $a+\sqrt{-1}b$. Some parts of his paper are less rigorous than the corresponding parts of Wessel. Argand gave some remarkable applications to trigonometry, geometry and algebra. The word "modulus," to represent the length of the vector $a+ib$, is due to Argand. The writings of Wessel and Argand being little noticed, it remained for K. F. Gauss to break down the last opposition to the imaginary. Gauss seems to have been in possession of a graphic scheme as early as 1799, but its fuller exposition was deferred until 1831.

During the French Revolution the metric system was introduced. The general idea of decimal subdivision was obtained from a work of Thomas Williams, London, 1788.[*] On April 14, 1790, *Mathurin Jacques Brisson* (1723–1806) proposed before the Paris Academy the establishment of a system resting on a natural unit of length. A scheme was elaborated which originally included the decimal subdivision of the quadrant of a circle, as is shown by the report made to

[1] Cauchy, *Exercices d'Analyse et de phys. math.*, T. IV, 1847, p. 157.

[2] See also an address on Wessel by W. W. Beman in the *Proceedings of the Am. Ass'n Adv. of Science*, Vol. 46, 1897.

[3] *Imaginary Quantities. Their Geometrical Interpretation.* Translated from the French of M. Argand by A. S. Hardy, New York, 1881.

the Academy of Sciences on March 19, 1791, by a committee consisting of J. C. Borda, J. Lagrange, P. S. Laplace, G. Monge, de Condorcet. This subdivision is found in the *François Callet* (1744–1798) logarithmic tables of 1795, and other tables published in France and Germany. Nevertheless the decimal subdivision of the quadrant did not then prevail.[1] The commission composed of Borda, Lagrange, Laplace, Monge and Condorcet decided upon the ten-millionth part of the earth's quadrant as the primitive unit of length. The length of the second's pendulum had been under consideration, but was finally rejected, because it rested upon two dissimilar elements, gravity and time. In 1799 the measurement of the earth's quadrant was completed and the meter established as the natural unit of length.

Alexandre-Théophile Vandermonde (1735–1796) studied music during his youth in Paris and advocated the theory that all art rested upon one general law, through which any one could become a composer with the aid of mathematics. He was the first to give a connected and logical exposition of the theory of determinants, and may, therefore, almost be regarded as the founder of that theory. He and J. Lagrange originated the method of combinations in solving equations.

Adrien Marie Legendre (1752–1833) was educated at the Collège Mazarin in Paris, where he began the study of mathematics under Abbé Joseph François Marie (1738–1801). His mathematical genius secured for him the position of professor of mathematics at the military school of Paris. While there he prepared an essay on the curve described by projectiles thrown into resisting media (ballistic curve), which captured a prize offered by the Royal Academy of Berlin. In 1780 he resigned his position in order to reserve more time for the study of higher mathematics. He was then made member of several public commissions. In 1795 he was elected professor at the Normal School and later was appointed to some minor government positions. Owing to his timidity and to Laplace's unfriendliness toward him, but few important public offices commensurate with his ability were tendered to him.

As an analyst, second only to P. S. Laplace and J. Lagrange, Legendre enriched mathematics by important contributions, mainly on elliptic integrals, theory of numbers, attraction of ellipsoids, and least squares. The most important of Legendre's works is his *Fonctions elliptiques*, issued in two volumes in 1825 and 1826. He took up the subject where L. Euler, John Landen, and J. Lagrange had left it, and for forty years was the only* one to cultivate this new branch of analysis, until at last C. G. J. Jacobi and N. H. Abel stepped in with admirable new discoveries.[2] Legendre imparted to the subject that

[1] For details, see R. Mehmke in *Jahresb. d. d. Math. Vereinigung*, Leipzig, 1900, pp. 138–163.

[2] M. Élie de Beaumont, "Memoir of Legendre." Translated by C. A. Alexander, *Smithsonian Report*, 1867.

connection and arrangement which belongs to an independent science. Starting with an integral depending upon the square root of a polynomial of the fourth degree in x, he showed that such integrals can be brought back to three canonical forms, designated by $F(\phi)$, $E(\phi)$, and $\Pi(\phi)$, the radical being expressed in the form $\Delta(\phi)=\sqrt{1-k^2 \sin^2\phi}$. He also undertook the prodigious task of calculating tables of arcs of the ellipse for different degrees of amplitude and eccentricity, which supply the means of integrating a large number of differentials.

An earlier publication which contained part of his researches on elliptic functions was his *Calcul integral* in three volumes (1811, 1816, 1817), in which he treats also at length of the two classes of definite integrals named by him *Eulerian*. He tabulated the values of log $\Gamma(p)$ for values of p between 1 and 2.

One of the earliest subjects of research was the attraction of spheroids, which suggested to Legendre the function P_n, named after him. His memoir was presented to the Academy of Sciences in 1783. The researches of C. Maclaurin and J. Lagrange suppose the point attracted by a spheroid to be at the surface or within the spheroid, but Legendre showed that in order to determine the attraction of a spheroid on any external point it suffices to cause the surface of another spheroid described upon the same foci to pass through that point. Other memoirs on ellipsoids appeared later.

In a paper of 1788 Legendre published criteria for distinguishing between maxima and minima in the calculus of variations, which were shown by J. Lagrange in 1797 to be insufficient; this matter was set right by C. G. J. Jacobi in 1836.

The two household gods to which Legendre sacrificed with ever-renewed pleasure in the silence of his closet were the elliptic functions and the theory of numbers. His researches on the latter subject, together with the numerous scattered fragments on the theory of numbers due to his predecessors in this line, were arranged as far as possible into a systematic whole, and published in two large quarto volumes, entitled *Théorie des nombres*, 1830. Before the publication of this work Legendre had issued at divers times preliminary articles. Its crowning pinnacle is the theorem of quadratic reciprocity, previously indistinctly given by L. Euler without proof, but for the first time clearly enunciated and partly proved by Legendre.[1]

While acting as one of the commissioners to connect Greenwich and Paris geodetically, Legendre calculated the geodetic triangles in France. This furnished the occasion of establishing formulæ and theorems on geodesics, on the treatment of the spherical triangle as if it were a plane triangle, by applying certain corrections to the angles, and on the method of least squares, published for the first time by him without demonstration in 1806.

[1] O. Baumgart, *Ueber das Quadratische Reciprocitätsgesetz*, Leipzig, 1885.

Legendre wrote an *Éléments de Géométrie*, 1794, which enjoyed great popularity, being generally adopted on the Continent and in the United States as a substitute for Euclid. This great modern rival of Euclid passed through numerous editions; some containing the elements of trigonometry and a proof of the irrationality of π and π^2. With prophetic vision Legendre remarks: "Il est même probable que le nombre π n'est pas même compris dans les irrationelles algébriques, c'est-a-dire qu'il ne peut pas être la racine 'dune équation algébrique d'un nombre fini de termes dont les coëfficients sont rationels." Much attention was given by Legendre to the subject of parallel lines. In the earlier editions of the *Éléments*, he made direct appeal to the senses for the correctness of the "parallel-axiom." He then attempted to demonstrate that "axiom," but his proofs did not satisfy even himself. In Vol. XII of the Memoirs of the Institute is a paper by Legendre, containing his last attempt at a solution of the problem. Assuming space to be infinite, he proved satisfactorily that it is impossible for the sum of the three angles of a triangle to exceed two right angles; and that if there be any triangle the sum of whose angles is two right angles, then the same must be true of all triangles. But in the next step, to show that this sum cannot be less than two right angles, his demonstration necessarily failed. If it could be granted that the sum of the three angles is always equal to two right angles, then the theory of parallels could be strictly deduced.

Another author who made contributions to elementary geometry was the Italian *Lorenzo Mascheroni* (1750–1800). He published his *Geometria del compasso* (Pavia, 1797, Palermo, 1903;[1] French editions by A. M. Carette appeared in 1798 and 1825, a German edition by J. P. Grüson in 1825). All constructions are made with a pair of compasses, but without restriction to a fixed radius. He proved that all constructions possible with ruler and compasses are possible with compasses alone. It was J. V. Poncelet who proved in 1822 that all such construction are possible with ruler alone, if we are given a fixed circle with its centre in the plane of construction; A. Adler of Vienna proved in 1890 that these constructions are possible with ruler alone whose edges are parallel, or whose edges converge in a point. Mascheroni claimed that constructions with compasses are more accurate than those with a ruler. Napoleon proposed to the French mathematicians the problem, to divide the circumference of a circle into four equal parts by the compasses only. Mascheroni does this by applying the radius three times to the circumference; he obtains the arcs A B, B C, C D; then A D is a diameter; the rest is obvious. E. W. Hobson (*Math. Gazette*, March 1, 1913) and others have shown that all Euclidean constructions can be carried out by the use of compasses alone.

[1] A list of Mascheroni's writings is given in *L'Intermédiaire des mathématiciens*, Vol. 19, 1912, p. 92.

In 1790 Mascheroni published annotations to Euler's Integral Calculus. D'Alembert had argued "le calcul en défaut" by declaring that the astroid $x^{\frac{2}{3}}+y^{\frac{2}{3}}=1$ yielded 0 as the length of the arc from $x=-1$ to $x=1$, y being taken positive. To this Mascheroni added in his annotations another paradox, by the contention that, for $x>1$, the curve is imaginary, yet has a real length of arc.[1] These paradoxes found no adequate explanation at the time, due to an inadequate fixing of the region of variability.

Joseph Fourier (1768–1830) was born at Auxerre, in central France. He became an orphan in his eighth year. Through the influence of friends he was admitted into the military school in his native place, then conducted by the Benedictines of the Convent of St. Mark. He there prosecuted his studies, particularly mathematics, with surprising success. He wished to enter the artillery, but, being of low birth (the son of a tailor), his application was answered thus: "Fourier, not being noble, could not enter the artillery, although he were a second Newton."[2] He was soon appointed to the mathematical chair in the military school. At the age of twenty-one he went to Paris to read before the Academy of Sciences a memoir on the resolution of numerical equations, which was an improvement on Newton's method of approximation. This investigation of his early youth he never lost sight of. He lectured upon it in the Polytechnic School; he developed it on the banks of the Nile; it constituted a part of a work entitled *Analyse des equationes determines* (1831), which was in press when death overtook him. This work contained "Fourier's theorem" on the number of real roots between two chosen limits. The French physician F. D. Budan had published a theorem nearly identical in principle, although different in statement, as early as 1807, but in 1807 Budan had not only not proved the theorem known by his name, but had not yet satisfied himself that it was really true. He gave a proof in 1811, which was printed in 1822. Fourier taught his theorem to his pupils in the Polytechnic School in 1796, 1797 and 1803; he first printed the theorem and its proof in 1820. His priority over Budan is firmly established.

Fourier was anticipated in two of his important results. His improvement on the Newton-Raphson method of approximation, rendering the process applicable without the possibility of failure, was given earlier by Mourraille, as was also Fourier's method of settling the question whether two roots near the border line of equality are really equal, or perhaps slightly different, or perhaps imaginary. These theorems were eclipsed by that of Sturm, published in 1829.

About this time new upper and lower limits of the real roots were discovered. In 1815 Jean Jacques Bret (1781–?) professor in Grenoble, printed three theorems, of which the following is best known: If frac-

[1] M. Cantor, *op. cit.*, Vol. IV, 1908, p. 485.

[2] D. F. J. Arago, "Joseph Fourier," *Smithsonian Report*, 1871.

tions are formed by giving each fraction a negative coefficient in an equation for its numerator, taken positively, and for its denominator the sum of the positive coefficients preceding it, if moreover unity is added to each fraction thus formed, then the largest number thus obtainable is larger than any root of the equation. In 1822 A. A. Vène, a French officer of engineers, showed: If $-P$ is largest in absolute value of the negative coefficients, and if S be the greatest coefficient among the positive terms which precede the first negative term, then will $1+P/S$ be a superior limit.

Fourier took a prominent part at his home in promoting the Revolution. Under the French Revolution the arts and sciences seemed for a time to flourish. The reformation of the weights and measures was planned with grandeur of conception. The Normal School was created in 1795, of which Fourier became at first pupil, then lecturer. His brilliant success secured him a chair in the Polytechnic School, the duties of which he afterwards quitted, along with G. Monge and C. L. Berthollet, to accompany Napoleon on his campaign to Egypt. Napoleon founded the Institute of Egypt, of which J. Fourier became secretary. In Egypt he engaged not only in scientific work, but discharged important political functions. After his return to France he held for fourteen years the prefecture of Grenoble. During this period he carried on his elaborate investigations on the propagation of heat in solid bodies, published in 1822 in his work entitled *La Théorie Analytique de la Chaleur.*✲ This work marks an epoch in the history of both pure and applied mathematics. It is the source of all modern methods in mathematical physics involving the integration of partial differential equations in problems where the boundary values are fixed ("boundary-value problems"). Problems of this type involve L. Euler's second definition of a "function" in which the relation is not necessarily capable of being expressed analytically. This concept of a function greatly influenced P. G. L. Dirichlet. The gem of Fourier's great book is "Fourier's series." By this research a long controversy was brought to a close, and the fact recognized that any arbitrary✲ function (*i. e.* any graphically given function) of a real variable can be represented by a trigonometric series. The first announcement of this great discovery was made by Fourier in 1807, before the French Academy. The trigonometric series $\sum_{n=0}^{n=\infty}(a_n \sin nx + b_n \cos nx)$ represents the function $\phi(x)$ for every value of x, if the coefficients $a_n = \frac{1}{\pi}\int_{-\pi}^{\pi}\phi(x) \sin nx dx$, and b_n be equal to a similar integral. The weak point in Fourier's analysis lies in his failure to prove generally that the trigonometric series actually converges to the value of the function. William Thomson (later Lord Kelvin) says that on May 1, 1840 (when he was only sixteen), "I took Fourier out of the University Library; and in a fortnight I had mastered it—gone right through it." Kelvin's whole career was influenced by Fourier's work on heat, of which, he said, "it is difficult to say whether their uniquely original quality, or their transcendant interest, or their perennially important instructiveness for physical science, is most to be praised."[1] Clerk Maxwell pronounced it a great mathematical poem. In 1827 Fourier succeeded P. S. Laplace as president of the council of the Polytechnic School.

About the time of Budan and Fourier, important devices were invented in Italy and England for the solution of numerical equations. The Italian scientific society in 1802 offered a gold medal for improvements in the solution of such equations; it was awarded in 1804 to Paolo Ruffini. With aid of the calculus he develops the theory of transforming one equation into another whose roots are all diminished by a certain constant.[2] Then follows the mechanism for the practical computer, and here Ruffini has a device which is simpler than Horner's scheme of 1819 and practically identical with what is now known as Horner's procedure. Horner had no knowledge of Ruffini's memoir. Nor did either Horner or Ruffini know that their method had been given by the Chinese as early as the thirteenth century. Horner's first paper was read before the Royal Society, July 1, 1819, and published in the *Philosophical Transactions* for 1819. Horner uses L. F. A. Arbogast's derivatives. The modern reader is surprised to find that Horner's exposition involves very intricate reasoning which is in marked contrast with the simple and elementary explanations found in modern texts. Perhaps this was fortunate; a simpler treatment might have prevented publication in the Philosophical Transactions. As it was, much demur was made to the insertion of the paper. "The elementary character of the subject," said T. S. Davies, "was the professed objection; his recondite mode of treating it was the professed passport for its admission." A second article of Horner on his method was refused publication in the Philosophical Transactions, and appeared in 1865 in the *Mathematician*, after the death of Horner; a third article was printed in 1830. Both Horner and Ruffini explained their methods at first by higher analysis and later by elementary algebra; both offered their methods as substitutes for the old process of root-extraction of numbers. Ruffini's paper was neglected and forgotten. Horner was fortunate in finding two influential champions of his method—John Radford Young (1799–1885) of Belfast and A. De Morgan. The Ruffini-Horner process has been used widely in England and the United States, less widely in Germany, Austria and

[1] S. P. Thompson *Life of William Thomson*, London, 1910, pp. 14, 689.

[2] See F. Cajori, "Horner's method of approximation anticipated by Ruffini," *Bull. Am. Math. Soc.* 2d S., Vol. 17, 1911, pp. 409–414.

Italy, and not at all in France. In France the Newton-Raphson method has held almost undisputed sway.[1]

Before proceeding to the origin of modern geometry we shall speak briefly of the introduction of higher analysis into Great Britain. This took place during the first quarter of the last century. The British began to deplore the very small progress that science was making in England as compared with its racing progress on the Continent. The first Englishman to urge the study of continental writers was *Robert Woodhouse* (1773–1827) of Caius College, Cambridge. In 1813 the "Analytical Society" was formed at Cambridge. This was a small club established by George Peacock, John Herschel, Charles Babbage, and a few other Cambridge students, to promote, as it was humorously expressed by Babbage, the principles of pure "*D*-ism," that is, the Leibnizian notation in the calculus against those of "dot-age," or of the Newtonian notation. This struggle ended in the introduction into Cambridge of the notation $\frac{dy}{dx}$, to the exclusion of the fluxional notation $\dot{y}$. This was a great step in advance, not on account of any great superiority of the Leibnizian over the Newtonian notation, but because the adoption of the former opened up to English students the vast storehouses of continental discoveries. Sir William Thomson, P. G. Tait, and some other modern writers find it frequently convenient to use both notations. Herschel, Peacock, and Babbage translated, in 1816, from the French, S. F. Lacroix's briefer treatise on the differential and integral calculus, and added in 1820 two volumes of examples. Lacroix's larger work, the *Traitê du calcul différentiel et integral*, first contained the term "differential coefficient" and definitions of "definite" and "indefinite" integrals. It was one of the best and most extensive works on the calculus of that time. Of the three founders of the "Analytical Society," Peacock afterwards did most work in pure mathematics. Babbage became famous for his invention of a calculating engine superior to Pascal's. It was never finished, owing to a misunderstanding with the government, and a consequent failure to secure funds. John Herschel, the eminent astronomer, displayed his mastery over higher analysis in memoirs communicated to the Royal Society on new applications of mathematical analysis, and in articles contributed to cyclopædias on light, on meteorology, and on the history of mathematics. In the Philosophical Transactions of 1813 he introduced the notation $sin^{-1}x$, $tan^{-1}x, \ldots$ for $arcsin\ x$, $arctan\ x, \ldots$ He wrote also log^2x, $cos^2x, \ldots$ for $log\ (log\ x)$, $cos\ (cos\ x), \ldots$, but in this notation he was anticipated by Heinrich Burmann (?–1817) of Mannheim, a partisan of the combinatory analysis of C. F. Hindenburg in Germany.

[1] For references and further detail, see *Colorado College Publication*, General Series 52, 1910.

George Peacock (1791–1858) was educated at Trinity College, Cambridge, became Lowndean professor there, and later, dean of Ely. His chief publications are his *Algebra*, 1830 and 1842, and his *Report on Recent Progress in Analysis*, which was the first of several valuable summaries of scientific progress printed in the volumes of the British Association. He was one of the first to study seriously the fundamental principles of algebra, and to recognize fully its purely symbolic character. He advances, though somewhat imperfectly, the "principle of the permanence of equivalent forms." It assumes that the rules applying to the symbols of arithmetical algebra apply also in symbolical algebra. About this time *Duncan Farquharson Gregory* (1813–1844), fellow of Trinity College, Cambridge, wrote a paper "on the real nature of symbolical algebra," which brought out clearly the commutative and distributive laws. These laws had been noticed years before by the inventors of symbolic methods in the calculus. It was F. Servois who introduced the names *commutative* and *distributive in* Gergonne's *Annales*, Vol. 5, 1814–15, p. 93. The term *associative* seems to be due to W. R. Hamilton. Peacock's investigations on the foundation of algebra were considerably advanced by A. De Morgan and H. Hankel.

James Ivory (1765–1842) was a Scotch mathematician who for twelve years, beginning in 1804, held the mathematical chair in the Royal Military College at Marlow (now at Sandhurst). He was essentially a self-trained mathematician, and almost the only one in Great Britain previous to the organization of the Analytical Society who was well versed in continental mathematics. Of importance is his memoir (*Phil. Trans.*, 1809) in which the problem of the attraction of a homogeneous ellipsoid upon an external point is reduced to the simpler problem of the attraction of a related ellipsoid upon a corresponding point interior to it. This is known as "Ivory's theorem." He criticised with undue severity Laplace's solution of the method of least squares, and gave three proofs of the principle without recourse to probability; but they are far from being satisfactory.

About this time began the aggressive investigation of "curves of pursuit." The Italian painter Leonardo da Vinci seems to be the first to have directed attention to such curves. They were first investigated by *Pierre Bouguer* of Paris in 1732, then by the French collector of customs, *Dubois-Aymé* (*Corrésp. sur l'école polyt.* II, 1811, p. 275) who stimulated researches carried on by *Thomas de St. Laurent*, *Ch. Sturm*, *Jean Joseph Querret* and *Tedenat* (*Ann. de Mathém.*, Vol. 13, 1822–1823).

By the researches of R. Descartes and the invention of the calculus, the analytical treatment of geometry was brought into great prominence for over a century. Notwithstanding the efforts to revive synthetic methods made by G. Desargues, B. Pascal, De Lahire, I. Newton, and C. Maclaurin, the analytical method retained almost undisputed supremacy. It was reserved for the genius of G. Monge to bring synthetic geometry in the foreground, and to open up new avenues of progress. His *Géométrie descriptive* marks the beginning of a wonderful development of modern geometry.

Of the two leading problems of descriptive geometry, the one—to represent by drawings geometrical magnitudes—was brought to a high degree of perfection before the time of Monge; the other—to solve problems on figures in space by constructions in a plane—had received considerable attention before his time. His most noteworthy predecessor in descriptive geometry was the Frenchman Amédée François Frézier (1682–1773). But it remained for Monge to create descriptive geometry as a *distinct* branch of science by imparting to it geometric generality and elegance. All problems previously treated in a special and uncertain manner were referred back to a few general principles. He introduced the line of intersection of the horizontal and the vertical plane as the axis of projection. By revolving one plane into the other around this axis or ground-line, many advantages were gained.[1]

Gaspard Monge (1746–1818) was born at Beaune. The construction of a plan of his native town brought the boy under the notice of a colonel of engineers, who procured for him an appointment in the college of engineers at Mézières. Being of low birth, he could not receive a commission in the army, but he was permitted to enter the annex of the school, where surveying and drawing were taught. Observing that all the operations connected with the construction of plans of fortification were conducted by long arithmetical processes, he substituted a geometrical method, which the commandant at first refused even to look at; so short was the time in which it could be practised that, when once examined, it was received with avidity. Monge developed these methods further and thus created his descriptive geometry. Owing to the rivalry between the French military schools of that time, he was not permitted to divulge his new methods to any one outside of this institution. In 1768 he was made professor of mathematics at Mézières. In 1780, when conversing with two of his pupils, S. F. Lacroix and S. F. Gay de Vernon in Paris, he was obliged to say, "All that I have here done by calculation, I could have done with the ruler and compasses, but I am not allowed to reveal these secrets to you." But Lacroix set himself to examine what the secret could be, discovered the processes, and published them in 1795. The method was published by Monge himself in the same year, first in the form in which the shorthand writers took down his lessons given at the Normal School, where he had been elected professor, and then again, in revised form, in the *Journal des écoles normales*. The next edition occurred in 1798–1799. After an ephemeral existence of only four months the Normal School was closed in 1795. In the same year

[1] Christian Wiener, *Lehrbuch der Darstellenden Geometric*, Leipzig, 1884, p. 26.

the Polytechnic School was opened, in the establishing of which Monge took active part. He taught there descriptive geometry until his departure from France to accompany Napoleon on the Egyptian campaign. He was the first president of the Institute of Egypt. Monge was a zealous partisan of Napoleon and was, for that reason, deprived of all his honors by Louis XVIII. This and the destruction of the Polytechnic School preyed heavily upon his mind. He did not long survive this insult.

Monge's numerous papers were by no means confined to descriptive geometry. His analytical discoveries are hardly less remarkable. He introduced into analytic geometry the methodic use of the equation of a line. He made important contributions to surfaces of the second degree (previously studied by C. Wren and L. Euler) and discovered between the theory of surfaces and the integration of partial differential equations, a hidden relation which threw new light upon both subjects. He gave the differential of curves of curvature, established a general theory of curvature, and applied it to the ellipsoid. He found that the validity of solutions was not impaired when imaginaries are involved among subsidiary quantities. Usually attributed to Monge are the centres of similitude of circles and certain theorems, which were, however, probably known to Apollonius of Perga.[1] Monge published the following books: *Statics*, 1786; *Applications de l'algèbre à la géométrie*, 1805; *Application de l'analyse à la géométrie*. The last two contain most of his miscellaneous papers.

Monge was an inspiring teacher, and he gathered around him a large circle of pupils, among which were C. Dupin, F. Servois, C. J. Brianchon, Hachette, J. B. Biot, and J. V. Poncelet. *Jean Baptiste Biot* (1774–1862), professor at the Collége de France in Paris, came in contact as a young man with Laplace, Lagrange, and Monge. In 1804 he ascended with Gay-Lussac in a balloon. They proved that the earth's magnetism is not appreciably reduced in intensity in regions above the earth's surface. Biot wrote a popular book on analytical geometry and was active in mathematical physics and geodesy. He had a controversy with Arago who championed A. J. Fresnel's wave theory of light. Biot was a man of strong individuality and great influence.

Charles Dupin (1784–1873), for many years professor of mechanics in the Conservatoire des Arts et Métiers in Paris, published in 1813 an important work on *Développements de géométrie*, in which is introduced the conception of conjugate tangents of a point of a surface, and of the indicatrix.[2] It contains also the theorem known as "Dupin's theorem." Surfaces of the second degree and descriptive geom-

[1] R. C. Archibald in *Am. Math. Monthly*, Vol. 22, 1915, pp. 6–12; Vol. 23, pp. 159–161.

[2] Gino Loria, *Die Hauptsächlisten Theorien der Geometrie* (F. Schütte), Leipzig, 1888, p. 49.

etry were successfully studied by *Jean Nicolas Pierre Hachette* (1769–1834), who became professor of descriptive geometry at the Polytechnic School after the departure of Monge for Rome and Egypt. In 1822 he published his *Traité de géométrie descriptive.*

Descriptive geometry, which arose, as we have seen, in technical schools in France, was transferred to Germany at the foundation of technical schools there. G. Schreiber (1799–1871), professor in Karlsruhe, was the first to spread Monge's geometry in Germany by the publication of a work thereon in 1828–1829.[1] In the United States descriptive geometry was introduced in 1816 at the Military Academy in West Point by Claude Crozet, once a pupil at the Polytechnic School in Paris. Crozet wrote the first English work on the subject.[2]

Lazare Nicholas Marguerite Carnot (1753–1823) was born at Nolay in Burgundy, and educated in his native province. He entered the army, but continued his mathematical studies, and wrote in 1784 a work on machines, containing the earliest proof that kinetic energy is lost in collisions of bodies. With the advent of the Revolution he threw himself into politics, and when coalesced Europe, in 1793, launched against France a million soldiers, the gigantic task of organizing fourteen armies to meet the enemy was achieved by him. He was banished in 1796 for opposing Napoleon's *coup d'état.* The refugee went to Geneva, where he issued, in 1797, a work still frequently quoted, entitled, *Réflexions sur la Métaphysique du Calcul Infinitésimal.* He declared himself as an "irreconcilable enemy of kings." After the Russian campaign he offered to fight for France, though not for the empire. On the restoration he was exiled. He died in Magdeburg. His *Géométrie de position*, 1803, and his *Essay on Transversals*, 1806, are important contributions to modern geometry. While G. Monge revelled mainly in three-dimensional geometry, Carnot confined himself to that of two. By his effort to explain the meaning of the negative sign in geometry he established a "geometry of position," which, however, is different from the "Geometrie der Lage" of to-day. He invented a class of general theorems on projective properties of figures, which have since been pushed to great extent by J. V. Poncelet, Michel Chasles, and others.

Thanks to Carnot's researches, says J. G. Darboux,[3] "the conceptions of the inventors of analytic geometry, Descartes and Fermat, retook alongside the infinitesimal calculus of Leibniz and Newton the place they had lost, yet should never have ceased to occupy. With his geometry, said Lagrange, speaking of Monge, this demon of a man will make himself immortal."

While in France the school of G. Monge was creating modern

[1] C. Wiener, *op. cit.* p. 36.

[2] F. Cajori, *Teaching and History of Mathematics in U. S.*, Washington, 1890, pp. 114, 117.

[3] *Congress of Arts and Science*, St. Louis, 1904, Vol. 1, p. 535.

geometry, efforts were made in England to revive Greek geometry by **Robert Simson** (1687–1768) and **Matthew Stewart** (1717–1785). Stewart was a pupil of Simson and C. Maclaurin, and succeeded the latter in the chair at Edinburgh. During the eighteenth century he and Maclaurin were the only prominent mathematicians in Great Britain. His genius was ill-directed by the fashion then prevalent in England to ignore higher analysis. In his *Four Tracts, Physical and Mathematical*, 1761, he applied geometry to the solution of difficult astronomical problems, which on the Continent were approached analytically with greater success. He published, in 1746, *General Theorems*, and in 1763, his *Propositiones geometricæ more veterum demonstratæ.* The former work contains sixty-nine theorems, of which only five are accompanied by demonstrations. It gives many interesting new results on the circle and the straight line. Stewart extended some theorems on transversals due to *Giovanni Ceva* (1647–1734), an Italian, who published in 1678 at Mediolani a work, *De lineis rectis se invicem secantibus*, containing the theorem now known by his name.

THE NINETEENTH AND TWENTIETH CENTURIES

Introduction

NEVER more zealously and successfully has mathematics been cultivated than during the nineteenth and the present centuries. Nor has progress, as in previous periods, been confined to one or two countries. While the French and Swiss, who during the preceding epoch carried the torch of progress, have continued to develop mathematics with great success, from other countries whole armies of enthusiastic workers have wheeled into the front rank. Germany awoke from her lethargy by bringing forward K. F. Gauss, C. G. J. Jacobi, P. G. L. Dirichlet, and hosts of more recent men; Great Britain produced her A. De Morgan, G. Boole, W. R. Hamilton, A. Cayley, J. J. Sylvester, besides champions who are still living; Russia entered the arena with her N. I. Lobachevski; Norway with N. H. Abel; Italy with L. Cremona; Hungary with her two Bolyais; the United States with Benjamin Peirce and J. Willard Gibbs.

H. S. White of Vassar College estimated the annual rate of increase in mathematical publication from 1870 to 1909, and ascertained the periods between these years when different subjects of research received the greatest emphasis.[1] Taking the *Jahrbuch über die Fortschritte der Mathematik*, published since 1871 (founded by Carl Ohrtmann (1839–1885) of the Königliche Realschule in Berlin and since 1885 under the chief editorship of Emil Lampe of the technische Hochschule in Berlin), and also the *Revue Semestrielle*, published since 1893 (under the auspices of the Mathematical Society of Amsterdam), he counted the number of titles, and in some cases also the number of pages filled by the reviews of books and articles devoted to a certain subject of research, and reached the following approximate results: (1) The total annual publication doubled during the forty years; (2) During these forty years, 30% of the publication was on applied mathematics, 25% on geometry, 20% on analysis, 18% on algebra, 7% on history and philosophy; (3) Geometry, dominated by "Plücker, his brilliant pupil Klein, Clifford, and Cayley," doubled its rate of production from 1870 to 1890, then fell off a third, to regain most of its loss after 1899; Synthetic geometry reached its maximum in 1887 and then declined during the following twenty years; the amount of analytic geometry always exceeded that of synthetic geometry, the

[1] H. S. White, "Forty Years' Fluctuations in Mathematical Research," *Science*, N. S., Vol. 42, 1915, pp. 105–113.

excess being most pronounced since 1887; (4) Analysis, "which takes its rise equally from calculus, from the algebra of imaginaries, from the intuitions and the critically refined developments of geometry, and from abstract logic: the common servant and chief ruler of the other branches of mathematics," shows a trebling in forty years, reaching its first maximum in 1890, "probably the culmination of waves set in motion by Weierstrass and Fuchs in Berlin, by Riemann in Göttingen, by Hermite in Paris, Mittag-Leffler in Stockholm, Dini and Brioschi in Italy;" before 1887 much of the growth of analysis is due to the theory of functions which reaches a maximum about 1887, with a sweep of the curve upward again after 1900, due to the theory of integral equations and the influence of Hilbert; (5) Algebra, including series and groups, experienced during the forty years a steady gain to $2\frac{1}{2}$ times its original output; the part of algebra relating to algebraic forms, invariants, etc., reached its acme before 1890 and then declined most surprisingly; (6) Differential equations increased in amount slowly but steadily from 1870, "under the combined influence of Weierstrass, Darboux and Lie," showing a slight decline in 1886, but "followed by a marked recovery and advance during the publication of lectures by Forsyth, Picard, Goursat and Painlevé;" (7) The mathematical theory of electricity and magnetism remained less than one-fourth of the whole applied mathematics, but rose after 1873 steadily toward one-fourth, by the labors of Clerk Maxwell, W. Thomson (Lord Kelvin) and P. G. Tait; (8) The constant shifting of mathematical investigation is due partly to fashion.

The progress of mathematics has been greatly accelerated by the organization of mathematical societies issuing regular periodicals. The leading societies are as follows: *London Mathematical Society* organized in 1865, *La société mathématique de France* organized in 1872, *Edinburgh Mathematical Society* organized 1883, *Circolo matematico di Palermo* organized in 1884, *American Mathematical Society* organized in 1888 under the name of *New York Mathematical Society* and changed to its present name in 1894,[1] *Deutsche Mathematiker-Vereinigung* organized in 1890, *Indian Mathematical Society* organized in 1907, *Sociedad Metematica Española* organized in 1911, *Mathematical Association of America* organized in 1915.

The number of mathematical periodicals has enormously increased during the passed century. According to Felix Müller [2] there were, up to 1700, only 17 periodicals containing mathematical articles; there were, in the eighteenth century, 210 such periodicals, in the nineteenth century 950 of them.✡

[1] Consult Thomas S. Fiske's address in *Bull. Am. Math. Soc.*, Vol. 11, 1905, p. 238. Dr. Fiske himself was a leader in the organization of the Society.

[2] *Jahresb. d. deutsch. Mathem. Vereinigung*, Vol. 12, 1903, p. 439. See also G. A. Miller in *Historical Introduction to Mathematical Literature*, New York, 1916, Chaps. I, II.

A great stimulus toward mathematical progress have been the international congresses of mathematicians. In 1889 there was held in Paris a Congrés international de bibliographie des sciences mathématiques. In 1893, during the Columbian Exposition, there was held in Chicago an International Mathematical Congress. But, by common agreement, the gathering held in 1897 at Zurich, Switzerland, is called the "first international mathematical congress." The second was held in 1900 at Paris, the third in 1904 at Heidelberg, the fourth in 1908 at Rome, the fifth in 1912 at Cambridge in England. The object of these congresses has been to promote friendly relations, to give reviews of the progress and present state of different branches of mathematics, and to discuss matters of terminology and bibliography.

One of the great co-operative enterprises intended to bring the results of modern research in digested form before the technical reader is the *Encyklopädie der Mathematischen Wissenschaften*, the publication of which was begun in 1898 under the editorship of *Wilhelm Franz Meyer* of Königsberg. Prominent as joint editor was *Heinrich Burkhardt* (1861–1914) of Zurich, later of Munich. In 1904 was begun the publication of the French revised and enlarged edition under the editorship of *Jules Molk* (1857–1914) of the University of Nancy.

As regards the productiveness of modern writers, Arthur Cayley said in 1883:[1] "It is difficult to give an idea of the vast extent of modern mathematics. This word 'extent' is not the right one: I mean extent crowded with beautiful detail,—not an extent of mere uniformity such as an objectless plain, but of a tract of beautiful country seen at first in the distance, but which will bear to be rambled through and studied in every detail of hillside and valley, stream, rock, wood, and flower." It is pleasant to the mathematician to think that in his, as in no other science, the achievements of every age remain possessions forever; new discoveries seldom disprove older tenets; seldom is anything lost or wasted.

If it be asked wherein the utility of some modern extensions of mathematics lies, it must be acknowledged that it is at present difficult to see how some of them are ever to become applicable to questions of common life or physical science. But our inability to do this should not be urged as an argument against the pursuit of such studies. In the first place, we know neither the day nor the hour when these abstract developments will find application in the mechanic arts, in physical science, or in other branches of mathematics. For example, the whole subject of graphical statics, so useful to the practical engineer, was made to rest upon von Staudt's *Geometrie der Lage;* W. R. Hamilton's "principle of varying action" has its use in astronomy; complex quantities, general integrals, and general theorems in integration offer advantages in the study of electricity and magnetism.

[1] Arthur Cayley, Inaugural Address before the British Association, 1883, *Report*, p. 25.

"The utility of such researches," said Spottiswoode in 1878,[1] "can in no case be discounted, or even imagined beforehand. Who, for instance, would have supposed that the calculus of forms or the theory of substitutions would have thrown much light upon ordinary equations; or that Abelian functions and hyperelliptic transcendents would have told us anything about the properties of curves; or that the calculus of operations would have helped us in any way towards the figure of the earth?"

As a matter of fact in the nineteenth century, as in all centuries, practical questions have been controlling factors in the growth of mathematics. Says C. E. Picard: "The influence of physical theories has been exercised not only on the general nature of the problems to be solved, but even in the details of the analytic transformations. Thus is currently designated in recent memoirs on partial differential equations under the name of Green's formula, a formula inspired by the primitive formula of the English physicist. The theory of dynamic electricity and that of magnetism, with Ampère and Gauss, have been the origin of important progress; the study of curvilinear integrals and that of the integrals of surfaces have taken thence all their developments, and formulas, such as that of Stokes which might also be called Ampère's formula, have appeared for the first time in memoirs on physics. The equations on the propagation of electricity, to which are attached the names of Ohm and Kirchhoff, while presenting a great analogy with those of heat, offer often conditions at the limits a little different; we know all that telegraphy by cables owes to the profound discussion of a Fourier's equation carried over into electricity. The equations long ago written of hydrodynamics, the equations of the theory of electricity, those of Maxwell and of Hertz in electromagnetism, have offered problems analogous to those recalled above, but under conditions still more varied." [2]

Along similar lines are the remarks of A. R. Forsyth. In 1905 he said: [3] "The last feature of the century that will be mentioned has been the increase in the number of subjects, apparently dissimilar from one another, which are now being made to use mathematics to some extent. Perhaps the most surprising is the application of mathematics to the domain of pure thought; this was effected by George Boole in his treatise 'Laws of Thought,' published in 1854; and though the developments have passed considerably beyond Boole's researches, his work is one of those classics that mark a new departure. Political economy, on the initiative of Cournot and Jevons, has begun to employ symbols and to develop the graphical methods; but there the present use seems to be one of suggestive record and expression, rather than

[1] William Spottiswoode, Inaugural Address before British Association, 1878, *Report*, p. 25.

[2] *Congress of Arts and Science*, St. Louis, 1904, Vol. I, pp. 507–508.

[3] *Report of the British Ass'n* (South Africa), 1905, London, 1906, p. 317.

of positive construction. Chemistry, in a modern spirit, is stretching out into mathematical theories; Willard Gibbs, in his memoir on the equilibrium of chemical systems, has led the way; and, though his way is a path which chemists find strewn with the thorns of analysis, his work has rendered, incidentally, a real service in co-ordinating experimental results belonging to physics and to chemistry. A new and generalized theory of statistics is being constructed; and a school has grown up which is applying it to biological phenomena. Its activity, however, has not yet met with the sympathetic goodwill of all the pure biologists; and those who remember the quality of the discussion that took place last year at Cambridge between the biometricians and some of the biologists will agree that, if the new school should languish, it will not be for want of the tonic of criticism."

The great characteristic of modern mathematics is its generalizing tendency. Nowadays little weight is given to isolated theorems, says J. J. Sylvester, "except as affording hints of an unsuspected new sphere of thought, like meteorites detached from some undiscovered planetary orb of speculation." In mathematics, as in all true sciences, no subject is considered in itself alone, but always as related to, or an outgrowth of, other things. The development of the notion of continuity plays a leading part in modern research. In geometry the principle of continuity, the idea of correspondence, and the theory of projection constitute the fundamental modern notions. Continuity asserts itself in a most striking way in relation to the circular points at infinity in a plane. In algebra the modern idea finds expression in the theory of linear transformations and invariants, and in the recognition of the value of homogeneity and symmetry.

H. F. Baker [1] said in 1913 that, with the aid of groups "a complete theory of equations which are soluble algebraically can be given. . . . But the theory of groups has other applications. . . . The group of interchanges among four quantities which leave unaltered the product of their six differences is exactly similar to the group of rotations of a regular tetrahedron whose centre is fixed, when its corners are interchanged among themselves. Then I mention the historical fact that the problem of ascertaining when that well-known differential equation called the hypergeometric equation has all its solutions expressible in finite terms as algebraic functions, was first solved in connection with a group of similar kind. For any linear differential equation it is of primary importance to consider the group of interchanges of its solutions when the independent variable, starting from an arbitrary point, makes all possible excursions, returning to its initial value. . . . There is, however, a theory of groups different from those so far referred to, in which the variables can change continuously; this alone is most extensive, as may be judged from one of its lesser applications, the familiar theory of the invariants of quantics. Moreover, perhaps

[1] *Report British Ass'n* (Birmingham), 1913, London, 1914, p. 371.

the most masterly of the analytical discussions of the theory of geometry has been carried through as a particular application of the theory of such groups."

"If the theory of groups illustrates how a unifying plan works in mathematics beneath the bewildering detail, the next matter I refer to well shows what a wealth, what a grandeur, of thought may spring from what seem slight beginnings. Our ordinary integral calculus is well-nigh powerless when the result of integration is not expressible by algebraic or logarithmic functions. The attempt to extend the possibilities of integration to the case when the function to be integrated involves the square root of a polynomial of the fourth order, led first, after many efforts, . . . to the theory of doubly-periodic functions. To-day this is much simpler than ordinary trigonometry, and, even apart from its applications, it is quite incredible that it should ever again pass from being among the treasures of civilized man. Then, at first in uncouth form, but now clothed with delicate beauty, came the theory of general algebraical integrals, of which the influence is spread far and wide; and with it all that is systematic in the theory of plane curves, and all that is associated with the conception of a Riemann surface. After this came the theory of multiply-periodic functions of any number of variables, which, though still very far indeed from being complete, has, I have always felt, a majesty of conception which is unique. Quite recently the ideas evolved in the previous history have prompted a vast general theory of the classification of algebraical surfaces according to their essential properties, which is opening endless new vistas of thought."

The nineteenth century and the beginning of the twentieth century constitute a period when the very foundations of mathematics have been re-examined and when fundamental principles have been worked out anew. Says H. F. Baker:[1] "It is a constantly recurring need of science to reconsider the exact implication of the terms employed; and as numbers and functions are inevitable in all measurement, the precise meaning of number, of continuity, of infinity, of limit, and so on, are fundamental questions. . . . These notions have many pitfalls I may cite. . . . the construction of a function which is continuous at all points of a range, yet possesses no definite differential coefficient at any point. Are we sure that human nature is the only continuous variable in the concrete world, assuming it be continuous, which can possess such a vacillating character? . . . We could take out of our life all the moments at which we can say that our age is a certain number of years, and days, and fractions of day, and still have appreciably as long to live; this would be true, however often, to whatever exactness, we named our age, provided we were quick enough in naming it. . . . These inquiries . . . have been associated also with the theory of those series which Fourier used so boldly, and

[1] H. F. Baker, *loc. cit.*, p. 369.

so wickedly, for the conduction of heat. Like all discoverers, he took much for granted. Precisely how much is the problem. This problem has led to the precision of what is meant by a function of real variables, to the question of the uniform convergence of an infinite series, as you may see in early papers of Stokes, to new formulation of the conditions of integration and of the properties of multiple integrals, and so on. And it remains still incompletely solved.

"Another case in which the suggestions of physics have caused grave disquiet to the mathematicians is the problem of the variation of a definite integral. No one is likely to underrate the grandeur of the aim of those who would deduce the whole physical history of the world from the single principle of least action. Everyone must be interested in the theorem that a potential function, with a given value at the boundary of a volume, is such as to render a certain integral, representing, say, the energy, a minimum. But in that proportion one desires to be sure that the logical processes employed are free from objection. And, alas! to deal only with one of the earliest problems of the subject, though the finally sufficient conditions for a minimum of a simple integral seemed settled long ago, and could be applied, for example, to Newton's celebrated problem of the solid of least resistance, it has since been shown to be a general fact that such a problem cannot have any definite solution at all. And, although the principle of Thomson and Dirichlet, which relates to the potential problem referred to, was expounded by Gauss, and accepted by Riemann, and remains to-day in our standard treatise on Natural Philosophy, there can be no doubt that, in the form in which it was originally stated, it proves just nothing. Thus a new investigation has been necessary into the foundations of the principle. There is another problem, closely connected with this subject, to which I would allude: the stability of the solar system. For those who can make pronouncements in regard to this I have a feeling of envy; for their methods, as yet, I have a quite other feeling. The interest of this problem alone is sufficient to justify the craving of the Pure Mathematician for powerful methods and unexceptionable rigour."

There are others who view this struggle for absolute rigor from a different angle. Horace Lamb in 1904 spoke as follows: [1] "a traveller who refuses to pass over a bridge until he has personally tested the soundness of every part of it is not likely to go very far; something must be risked, even in Mathematics. It is notorious that even in this realm of 'exact' thought, discovery has often been in advance of strict logic, as in the theory of imaginaries, for example, and in the whole province of analysis of which Fourier's theorem is a type."

Says Maxime Bôcher: [2] "There is what may perhaps be called the

[1] Address before Section A, British Ass'n, in Cambridge, 1904.

[2] Maxime Bôcher in *Congress of Arts and Science*, St. Louis, 1904, Vol. I, p. 472.

method of optimism, which leads us either willfully or instinctively to shut our eyes to the possibility of evil. Thus the optimist who treats a problem in algebra or analytic geometry will say, if he stops to reflect on what he is doing: 'I know that I have no right to divide by zero; but there are so many other values which the expression by which I am dividing might have that I will assume that the Evil One has not thrown a zero in my denominator this time.' This method . . . has been of great service in the rapid development of many branches of mathematics."

Definitions of Mathematics

One of the phases of the quest for rigor has been the re-defining of mathematics. "Mathematics, the science of quantity" is an old idea which goes back to Aristotle. A modified form of this old definition is due to *Auguste Comte* (1798–1857), the French philosopher and mathematician, the founder of positivism. Since the most striking measurements are not direct, but are indirect, as the determination of distances and sizes of the planets, or of the atoms, he defined mathematics "the science of indirect measurement." These definitions have been abandoned for the reason that several modern branches of mathematics, such as the theory of groups, analysis situs, projective geometry, theory of numbers and the algebra of logic, have no relation to quantity and measurement. "For one thing," says C. J. Keyser,[1] "the notion of the *continuum*—the 'Grand Continuum' as Sylvester called it—that central supporting pillar of modern Analysis, has been constructed by K. Weierstrass, R. Dedekind, Georg Cantor and others, without any reference whatever to quantity, so that number and magnitude are not only independent, they are essentially disparate." Or, if we prefer to go back a few centuries and refer to a single theorem, we may quote G. Desargues as saying that if the vertices of two triangles lie in three lines meeting in a point, then their sides meet in three points lying on a line. This beautiful theorem has nothing to do with measurement.

In 1870 Benjamin Peirce wrote in his *Linear Associative Algebra* that "mathematics is the science which draws necessary conclusions." This definition has been regarded as including too much and also as in need of elucidation as to what constitutes a "necessary" conclusion. Reasoning which seemed absolutely conclusive to one generation no longer satisfies the next. According to present standards no reasoning which claims to be exact can make any use of intuition, but must proceed from definitely and completely stated premises according to certain principles of formal logic.[2] Mathematical logicians from George Boole to C. S. Peirce, E. Schröder, and G. Peano have prepared the field so well that of late years Peano and his followers, and

[1] C. J. Keyser, *The Human Worth of Rigorous Thinking*, New York, 1916, p. 277.
[2] Maxime Bôcher, *loc. cit.*, Vol. I, p. 459.

independently G. Frege, "have been able to make a rather short list of logical conceptions and principles upon which it would seem that all exact reasoning depends." But the validity of logical principles must stand the test of use, and on this point we may never be sure. Frege and Bertrand Russell independently built up a theory of arithmetic, each starting with apparently self-evident logical principles. Then Russell discovers that his principles, applied to a very general kind of logical *class*, lead to an absurdity. There is evident need of reconstruction somewhere. After all, are we merely making successive *approximations* to absolute rigor?

A. B. Kempe's definition is as follows:[1] "Mathematics is the science by which we investigate those characteristics of any subject-matter of thought which are due to the conception that it consists of a number of differing and non-differing individuals and pluralities." Ten years later Maxime Bôcher modified Kempe's definition thus:[2] "If we have a certain class of objects and a certain class of relations, and if the only questions which we investigate are whether ordered groups of those objects do or do not satisfy the relations, the results of the investigation are called mathematics." Bôcher remarks that if we restrict ourselves to exact or deductive mathematics, then Kempe's definition becomes coextensive with B. Peirce's.

Bertrand Russell, in his *Principles of Mathematics*, Cambridge, 1903, regards pure mathematics as consisting exclusively of deductions "by logical principles from logical principles." Another definition given by Russell sounds paradoxical, but really expresses the extreme generality and extreme subtleness of certain parts of modern mathematics: "Mathematics is the subject in which we never know what we are talking about nor whether what we are saying is true."[3] Other definitions along similar lines are due to E. Papperitz (1892), G. Itelson (1904), and L. Couturat (1908).

Synthetic Geometry

The conflict between synthetic and analytic methods in geometry which arose near the close of the eighteenth century and the beginning of the nineteenth has now come to an end. Neither side has come out victorious. The greatest strength is found to lie, not in the suppression of either, but in the friendly rivalry between the two, and in the stimulating influence of the one upon the other. Lagrange prided himself that in his *Mecanique Analytique* he had succeeded in avoiding all figures; but since his time mechanics has received much help from geometry.

Modern synthetic geometry was created by several investigators about the same time. It seemed to be the outgrowth of a desire for

[1] *Proceed. London Math. Soc.*, Vol. 26, 1894, p. 15.
[2] M. Bôcher, *op. cit.*, p. 466.
[3] B. Russell in *International Monthly*, Vol. 4, 1901, p. 84.

general methods which should serve as threads of Ariadne to guide the student through the labyrinth of theorems, corollaries, porisms, and problems. Modern synthetic geometry was first cultivated by G. Monge, L. N. M. Carnot, and J. V. Poncelet in France; it then bore rich fruits at the hands of A. F. Möbius and Jakob Steiner in Germany and Switzerland, and was finally developed to still higher perfection by M. Chasles in France, von Staudt in Germany, and L. Cremona in Italy.

Jean Victor Poncelet (1788–1867), a native of Metz, took part in the Russian campaign, was abandoned as dead on the bloody field of Krasnoi, and taken prisoner to Saratoff. Deprived there of all books, and reduced to the remembrance of what he had learned at the Lyceum at Metz and the Polytechnic School, where he had studied with predilection the works of G. Monge, L. N. M. Carnot, and C. J. Brianchon, he began to study mathematics from its elements. He entered upon original researches which afterwards made him illustrious. While in prison he did for mathematics what Bunyan did for literature,—produced a much-read work, which has remained of great value down to the present time. He returned to France in 1814, and in 1822 published the work in question, entitled, *Traité des Propriétés projectives des figures*. In it he investigated the properties of figures which remain unaltered by projection of the figures. The projection is not effected here by parallel rays of prescribed direction, as with G. Monge, but by central projection. Thus perspective projection, used before him by G. Desargues, B. Pascal, I. Newton, and J. H. Lambert, was elevated by him into a fruitful geometric method. Poncelet formulated the so-called *principle of continuity*, which asserts that properties of a figure which hold when the figure varies according to definite laws will hold also when the figure assumes some limiting position.

"Poncelet," says J. G. Darboux,[1] "could not content himself with the insufficient resources furnished by the method of projections; to attain imaginaries he created that famous principle of continuity which gave birth to such long discussions between him and A. L. Cauchy. Suitably enunciated, this principle is excellent and can render great service. Poncelet was wrong in refusing to present it as a simple consequence of analysis; and Cauchy, on the other hand, was not willing to recognize that his own objections, applicable without doubt to certain transcendent figures, were without force in the applications made by the author of the *Traité des propriétés projectives*." J. D. Gergonne characterized the principle as a valuable instrument for the discovery of new truths, which nevertheless did not make stringent proofs superfluous.[2] By this principle of geometric

[1] *Congress of Arts and Science*, St. Louis, 1904, Vol. I, p. 539.

[2] E. Kötter, *Die Entwickelung der synthetischen Geometrie von Monge bis auf Staudt*, Leipzig, 1901, p. 123.

continuity Poncelet was led to the consideration of points and lines which vanish at infinity or become imaginary. The inclusion of such ideal points and lines was a gift which pure geometry received from analysis, where imaginary quantities behave much in the same way as real ones. Poncelet elaborated some ideas of De Lahire, F. Servois, and J. D. Gergonne into a regular method—the method of "reciprocal polars." To him we owe the Principle of Duality as a consequence of reciprocal polars. As an independent principle it is due to Gergonne. Darboux says that the significance of the principle of duality which was "a little vague at first, was sufficiently cleared up by the discussions which took place on this subject between J. D. Gergonne, J. V. Poncelet and J. Plücker." It had the advantage of making correspond to a proposition another proposition of wholly different aspect. "This was a fact essentially new. To put it in evidence, Gergonne invented the system, which since has had so much success, of memoirs printed in double columns with correlative propositions in juxtaposition" (Darboux).

Joseph Diaz Gergonne (1771–1859) was an officer of artillery, then professor of mathematics at the lyceum in Nîmes and later professor at Montpellier. He solved the Apollonian Problem and claimed superiority of analytic methods over the synthetic. Thereupon Poncelet published a purely geometric solution. Gergonne and Poncelet carried on an intense controversy on the priority of discovering the principle of duality. No doubt, Poncelet entered this field earlier, while Gergonne had a deeper grasp of the principle. Some geometers, particularly C. J. Brianchon, entertained doubts on the general validity of the principle. The controversy led to one new result, namely, Gergonne's considerations of the *class* of a curve or surface, as well as its *order*.[1] Poncelet wrote much on applied mechanics. In 1838 the Faculty of Sciences was enlarged by his election to the chair of mechanics.

J. G. Darboux says that, "presented in opposition to analytic geometry, the methods of Poncelet were not favorably received by the French analysts. But such were their importance and their novelty, that without delay they aroused, from divers sides, the most profound researches." Many of these appeared in the *Annales de mathématiques*, published by J. D. Gergonne at Nîmes from 1810 to 1831. During over fifteen years this was the only journal in the world devoted exclusively to mathematical researches. Gergonne "collaborated, often against their will, with the authors of the memoirs sent him, rewrote them, and sometimes made them say more or less than they would have wished. . . . Gergonne, having become rector of the Academy of Montpellier, was forced to suspend in 1831 the publication of his journal. But the success it had obtained, the taste for research it had contributed to develop, had commenced to bear

[1] E. Kötter, *op. cit.*, pp. 160–164.

their fruit. L. A. J. Quételet had established in Belgium the *Correspondance mathématique et physique.* A. L. Crelle, from 1826, brought out at Berlin the first sheets of his celebrated journal, where he published the memoirs of N. H. Abel, of C. G. J. Jacobi, of J. Steiner" (Darboux).

Contemporaneous with J. V. Poncelet was the German geometer, **Augustus Ferdinand Möbius** (1790–1868), a native of Schulpforta in Prussia. He studied at Göttingen under K. F. Gauss, also at Leipzig and Halle. In Leipzig he became, in 1815, privat-docent, the next year extraordinary professor of astronomy, and in 1844 ordinary professor. This position he held till his death. The most important of his researches are on geometry. They appeared in *Crelle's Journal,* and in his celebrated work entitled *Der Barycentrische Calcul,* Leipzig, 1827, "a work truly original, remarkable for the profundity of its conceptions, the elegance and the rigor of its exposition" (Darboux). As the name indicates, this calculus is based upon properties of the centre of gravity.[1] Thus, that the point S is the centre of gravity of weights a, b, c, d placed at the points A, B, C, D respectively, is expressed by the equation

$$(a+b+c+d)S=aA+bB+cC+dD.$$

His calculus is the beginning of a quadruple algebra, and contains the germs of Grassmann's marvellous system. In designating segments of lines we find throughout this work for the first time consistency in the distinction of positive and negative by the order of letters AB, BA. Similarly for triangles and tetrahedra. The remark that it is always possible to give three points A, B, C such weights α, β, γ that any fourth point M in their plane will become a centre of mass, led Möbius to a new system of co-ordinates in which the position of a point was indicated by an equation, and that of a line by co-ordinates. By this algorithm he found by algebra many geometric theorems expressing mainly invariantal properties,—for example, the theorems on the anharmonic relation. Möbius wrote also on statics and astronomy. He generalized spherical trigonometry by letting the sides or angles of triangles exceed 180°.

Not only Möbius but also H. G. Grassmann discarded the usual co-ordinate systems, and used algebraic analysis. Later in the nineteenth century and at the opening of the twentieth century, these ideas were made use of, notably by Cyparissos Stéphanos (1857–1917) of the National University of Athens, H. Wiener, C. Segre, G. Peano, F. Aschieri, E. Study, C. Burali-Forti and Hermann Grassmann (1857–1922), a son of H. G. Grassmann. Their researches, covering the fields of binary and ternary linear transformations, were brought together by the younger Grassmann into a treatise, *Projektive Geometrie der Ebene unter Benutzung der Punktrechnung dargestellt,* 1909.

[1] J. W. Gibbs, "Multiple Algebra," *Proceedings Am. Ass'n for the Advanc. of Science,* 1886.

Jakob Steiner (1796–1863), "the greatest geometrician since the time of Euclid," was born in Utzendorf in the Canton of Bern. He did not learn to write till he was fourteen. At eighteen he became a pupil of Pestalozzi. Later he studied at Heidelberg and Berlin. When A. L. Crelle started, in 1826, the celebrated mathematical journal bearing his name, Steiner and Abel became leading contributors. Through the influence of C. G. J. Jacobi and others, the chair of geometry was founded for him at Berlin in 1834. This position he occupied until his death, which occurred after years of bad health.

In 1832 Steiner published his *Systematische Entwickelung der Abhängigkeit geometrischer Gestalten von einander*, "in which is uncovered the organism by which the most diverse phenomena (*Erscheinungen*) in the world of space are united to each other." Here for the first time, is the principle of duality introduced at the outset. This book and von Staudt's lay the foundation on which synthetic geometry in its later form rested. The researches of French mathematicians, culminating in the remarkable creations of G. Monge, J. V. Poncelet and J. D. Gergonne, suggested a unification of geometric processes. This work of "uncovering the organism by which the most different forms in the world of space are connected with each other," this exposing of "a small number of very simple fundamental relations in which the scheme reveals itself, by which the whole body of theorems can be logically and easily developed" was the task which Steiner assumed. Says H. Hankel: [1] "In the beautiful theorem that a conic section can be generated by the intersection of two projective pencils (and the dually correlated theorem referring to projective ranges), J. Steiner recognized the fundamental principle out of which the innumerable properties of these remarkable curves follow, as it were, automatically with playful ease." Not only did he fairly complete the theory of curves and surfaces of the second degree, but he made great advances in the theory of those of higher degrees.

In the *Systematische Entwickelungen* (1832) Steiner directed attention to the complete figure obtained by joining in every possible way six points on a conic and showed that in this hexagrammum mysticum the 60 "Pascal lines" pass three by three through 20 points ("Steiner points") which lie four by four upon 15 straight lines ("Plücker lines"). J. Plücker had sharply criticized Steiner for an error that had crept into an earlier statement (1828) of the last theorem. Now, Steiner gave the correct statement, but without acknowledgment to Plücker. Further properties of the hexagrammum mysticum are due to T. P. Kirkman, A. Cayley and G. Salmon. The Pascal lines of three hexagons concur in a new point ("Kirkman point"). There are 60 Kirkman points. Corresponding to three Pascal lines which concur in a Steiner point, there are three Kirkman points which lie upon a straight

[1] H. Hankel, *Elemente der Projectivischen Geometrie*, 1875, p. 26.

line ("Cayley line"). There are 20 Cayley lines which pass four by four through 15 "Salmon points." Other new properties of the mystic hexagon were obtained in 1877 by G. Veronese and L. Cremona.[1]

In Steiner's hands synthetic geometry made prodigious progress. New discoveries followed each other so rapidly that he often did not take time to record their demonstrations. In an article in *Crelle's Journal* on *Allgemeine Eigenschaften Algebraischer Curven* he gives without proof theorems which were declared by L. O. Hesse to be "like Fermat's theorems, riddles to the present and future generations." Analytical proofs of some of them have been given since by others, but L. Cremona finally proved them all by a synthetic method. Steiner discovered synthetically the two prominent properties of a surface of the third order; viz. that it contains twenty-seven straight lines and a pentahedron which has the double points for its vertices and the lines of the Hessian of the given surface for its edges. This subject will be discussed more fully later. Steiner made investigations by synthetic methods on maxima and minima, and arrived at the solution of problems which at that time surpassed the analytic power of the calculus of variations. It will appear later that his reasoning on this topic is not always free from criticism.

Steiner generalized Malfatti's problem.[2] *Giovanni Francesco Malfatti* (1731–1807) of the university of Ferrara, in 1803, proposed the problem, to cut three cylindrical holes out of a three-sided prism in such a way that the cylinders and the prism have the same altitude and that the volume of the cylinders be a maximum. This problem was reduced to another, now generally known as Malfatti's problem: to inscribe three circles in a triangle so that each circle will be tangent to two sides of the triangle and to the other two circles. Malfatti gave an analytical solution, but Steiner gave without proof a construction, remarked that there were thirty-two solutions, generalized the problem by replacing the three lines by three circles, and solved the analogous problem for three dimensions. This general problem was solved analytically by C. H. Schellbach (1809–1892) and A. Cayley and by R. F. A. Clebsch with the aid of the addition theorem of elliptic functions.[3] A simple proof of Steiner's construction was given by A. S. Hart of Trinity College, Dublin, in 1856.

Of interest is Steiner's paper, *Ueber die geometrischen Constructionen, ausgeführt mittels der geraden Linie und eines festen Kreises* (1833), in which he shows that all quadratic constructions can be effected with the aid of only a ruler, provided that a fixed circle is drawn once for all. It was generally known that all linear constructions could be effected by the ruler, without other aids of any kind. The case of

[1] G. Salmon, *Conic Sections*, 6th Ed., 1879, Notes, p. 382.

[2] Karl Fink, *A Brief History of Mathematics*, transl. by W. W. Beman and D. E. Smith, Chicago, 1900, p. 256.

[3] A. Wittstein, *zur Geschichte des Malfatti'schen Problems*, Nördlingen, 1878.

cubic constructions, calling for the determination of three unknown elements (points) was worked out in 1868 by *Ludwig Hermann Kortum* (1836–1904) of Bonn, and Stephen Smith of Oxford in two researches which received the Steiner prize of the Berlin Academy; it was shown that if a conic (not a circle) is given to start with, then all such constructions can be done with a ruler and compasses. Franz London (1863–1917) of Breslau demonstrated in 1895 that these cubic constructions can be effected with a ruler only, as soon as a fixed cubic curve is once drawn.[1]

F. Bützberger[2] has recently pointed out that in an unpublished manuscript, Steiner disclosed a knowledge of the principle of inversion as early as 1824. In 1847 Liouville called it the transformation by reciprocal radii. After Steiner this transformation was found independently by J. Bellavitis in 1836, J. W. Stubbs and J. R. Ingram in 1842 and 1843, and by William Thomson (Lord Kelvin) in 1845.

Steiner's researches are confined to synthetic geometry. He hated analysis as thoroughly as J. Lagrange disliked geometry. Steiner's *Gesammelte Werke* were published in Berlin in 1881 and 1882.

Michel Chasles (1793–1880) was born at Epernon, entered the Polytechnic School of Paris in 1812, engaged afterwards in business, which he later gave up that he might devote all his time to scientific pursuits. In 1841 he became professor of geodesy and mechanics at the École polytechnique; later, "Professeur de Géométrie supérieure à la Faculté des Sciences de Paris." He was a voluminous writer on geometrical subjects. In 1837 he published his admirable *Aperçu historique sur l'origine et le développement des méthodes en géométrie*, containing a history of geometry and, as an appendix, a treatise "sur deux principes généraux de la Science." The *Aperçu historique* is still a standard historical work; the appendix contains the general theory of Homography (Collineation) and of duality (Reciprocity). The name *duality* is due to J. D. Gergonne. Chasles introduced the term *anharmonic ratio*, corresponding to the German *Doppelverhältniss* and to Clifford's *cross-ratio*. Chasles and J. Steiner elaborated independently the modern synthetic or projective geometry. Numerous original memoirs of Chasles were published later in the *Journal de l'École Polytechnique*. He gave a reduction of cubics, different from Newton's in this, that the five curves from which all others can be projected are symmetrical with respect to a centre. In 1864 he began the publication, in the *Comptes rendus*, of articles in which he solves by his "method of characteristics" and the "principle of correspondence" an immense number of problems. He determined, for instance, the number of intersections of two curves in a plane. The method of characteristics contains the basis of enumerative geometry.

As regards Chasles' use of imaginaries, J. G. Darboux says: "Here,

[1] *Jahresb. d. d. Math. Vereinigung*, Vol. 4, p. 163.
[2] *Bull. Am. Math. Soc.*, Vol. 20, 1914, p. 414.

his method was really new. . . . But Chasles introduced imaginaries only by their symmetric functions, and consequently would not have been able to define the cross-ratio of four elements when these ceased to be real in whole or in part. If Chasles had been able to establish the notion of the cross-ratio of imaginary elements, a formula he gives in the *Géométrie supérieure* (p. 118 of the new edition) would have immediately furnished him that beautiful definition of angle as logarithm of a cross-ratio which enabled E. Laguerre, our regretted confrère, to give the complete solution, sought so long, of the problem of the transformation of relations which contain at the same time angles and segments in homography and correlation." The application of the principle of correspondence was extended by A. Cayley, A. Brill, H. G. Zeuthen, H. A. Schwarz, G. H. Halphen, and others. The full value of these principles of Chasles was not brought out until the appearance, in 1879, of the *Kalkül der Abzählenden Geometrie* by Hermann Schubert (1848–1911) of Hamburg. This work contains a masterly discussion of the problem of enumerative geometry, viz. to determine the number of points, lines, curves, etc., of a system which fulfil certain conditions. Schubert extended his enumerative geometry to *n*-dimensional space.[1] The fundamental principle of enumerative geometry is the law of the "preservation of the number," which, as stated by Schubert, was found by E. Study and by G. Kohn in 1903 to be not always valid. The particular problem examined by Study and later also by F. Severi, considers the number of projectivities of a line which transform into itself a given group of four points. If the cross-ratio of the group is not a cube root of -1, the number of projectivities is 4, otherwise there are more. A recent book on this subject is H. G. Zeuthen's *Abzählende Methoden der Geometrie*, 1914.

To Chasles we owe the introduction into projective geometry of non-projective properties of figures by means of the infinitely distant imaginary sphero-circle.[2] Remarkable is his complete solution, in 1846, by synthetic geometry, of the difficult question of the attraction of an ellipsoid on an external point. This celebrated problem was treated alternately by synthetic and by analytic methods. Colin Maclaurin's results, obtained synthetically, had created a sensation. Nevertheless, both A. M. Legendre and S. D. Poisson expressed the opinion that the resources of the synthetic method were easily exhausted. Poisson solved it analytically in 1835. Then Chasles surprised every one by his synthetic investigations, based on the consideration of confocal surfaces. Poinsot reported on the memoir and remarked on the analytic and synthetic methods: "It is certain that one cannot afford to neglect either."

[1] Gino Loria, *Die Hauptsächlichsten Theorien der Geometrie*, 1888, p. 124.

[2] F. Klein, *Vergleichende Betrachtungen über neuere geometrische Forschungen*, Erlangen, 1872, p. 12.

The labors of Chasles and Steiner raised synthetic geometry to an honored and respected position by the side of analysis.

Karl Georg Christian von Staudt (1798–1867) was born in Rothenburg on the Tauber, and at his death, was professor in Erlangen. His great works are the *Geometrie der Lage*, Nürnberg, 1847, and his *Beiträge zur Geometrie der Lage*, 1856–1860. The author cut loose from algebraic formulæ and from metrical relations, particularly the anharmonic ratio of J. Steiner and M. Chasles, and then created a geometry of position, which is a complete science in itself, independent of all measurements. He shows that projective properties of figures have no dependence whatever on measurements, and can be established without any mention of them. In his theory of "throws" or "Würfe," he even gives a geometrical definition of a number in its relation to geometry as determining the position of a point. *Gustav Kohn* of the University of Vienna about 1894 introduced the throw as a fundamental concept underlying the projective properties of a geometric configuration, such that, according to a principle of duality of this geometry, throws of figures appear in pairs of reciprocal throws; figures of reciprocal throws form a complete analogy to figures of equal throws. Referring to Von Staudt's numerical co-ordinates, defined without introducing distance as a fundamental idea, A. N. Whitehead said in 1906: "The establishment of this result is one of the triumphs of modern mathematical thought."

The *Beiträge* contains the first complete and general theory of imaginary points, lines, and planes in projective geometry. Representation of an imaginary point is sought in the combination of an involution with a determinate direction, both on the real line through the point. While purely projective, von Staudt's method is intimately related to the problem of representing by actual points and lines the imaginaries of analytical geometry. Says Kötter:[1] Staudt was the first who succeeded "in subjecting the imaginary elements to the fundamental theorem of projective geometry, thus returning to analytical geometry the present which, in the hands of geometricians, had led to the most beautiful results." Von Staudt's geometry of position was for a long time disregarded, mainly, no doubt, because his book is extremely condensed. An impulse to the study of this subject was given by Culmann, who rests his graphical statics upon the work of von Staudt. An interpreter of von Staudt was at last found in Theodor Reye of Strassburg, who wrote a *Geometrie der Lage* in 1868.

The graphic representation of the imaginaries of analytical geometry was systematically undertaken by *C. F. Maximilien Marie* (1819–1891), who worked, however, on entirely different lines from those of von Staudt. Another independent attempt was made in 1893 by F. H. Loud of Colorado College.

[1] E. Kötter, *op. cit.*, p. 123.

Synthetic geometry was studied with much success by **Luigi Cremona** (1830–1903), who was born in Pavia and became in 1860 professor of higher geometry in Bologna, in 1866 professor of geometry and graphical statics in Milan, in 1873 professor of higher mathematics and director of the engineering school at Rome. He was influenced by the writings of M. Chasles, later he recognized von Staudt as the true founder of pure geometry. A memoir of 1866 on cubic surfaces secured half of the Steiner prize from Berlin, the other half being awarded to Rudolf Sturm, then of Bromberg. Cremona used the method of enumeration with great effect. He wrote on plane curves, on surfaces, on birational transformations of plane and solid space. The birational transformations, the simplest class of which is now called the "Cremona transformation," proved of importance, not only in geometry, but in the analytical theory of algebraic functions and integrals. It was developed more fully by M. Nöther and others. H. S. White comments on this subject as follows:[1] "Beyond the linear or projective transformations of the plane there were known the quadric inversions of Ludwig Immanuel Magnus (1790–1861) of Berlin, changing lines into conics through three fundamental points and those exceptional points into singular lines, to be discarded. Cremona described at once the highest generalization of these transformations, one-to-one for all points of the plane except a finite set of fundamental points. He found that it must be mediated by a net of rational curves; any two intersecting in one variable point, and in fixed points, ordinary or multiple, which are the fundamental points and which are themselves tranformed into singular rational curves of the same orders as the indices of multiplicity of the points. When the fundamental points are enumerated by classes according to their several indices, the set of class numbers for the inverse transformation is found to be the same as for the direct, but usually related to different indices. Tables of such rational nets of low orders were made out by L. Cremona and A. Cayley, and a wide new vista seemed opening (such indeed it was and is) when simultaneously three investigators announced that the most general Cremona transformation is equivalent to a succession of quadric transformations of Magnus's type. This seemed a climax, and a set-back to certain expectations." Cremona's theory of the transformation of curves and of the correspondence of points on curves was extended by him to three dimensions. There he showed how a great variety of particular transformations can be constructed, "but anything like a general theory is still in the future." Ruled surfaces, surfaces of the second order, space-curves of the third order, and the general theory of surfaces received much attention at his hands. He was interested in map-drawing, which had engaged the attention of R. Hooke, G. Mercator, J. Lagrange, K. F. Gauss and others. For

[1] *Bull. Am. Math. Soc.*, Vol. 24, 1918, p. 242.

a one-one correspondence the surface must be unicursal, and this is sufficient. L. Cremona is associated with A. Cayley, R. F. A. Clebsch, M. Nöther and others in the development of this theory.[1] Cremona's writings were translated into German by *Maximilian Curtze* (1837–1903), professor at the gymnasium in Thorn. The *Opera matematiche di Luigi Cremona* were brought out at Milan in 1914 and 1915.

One of the pupils of Cremona was **Giovanni Battista Guccia** (1855–1914). He was born in Palermo and studied at Rome under Cremona. In 1889 he became extraordinary professor at the University of Palermo, in 1894 ordinary professor. He gave much attention to the study of curves and surfaces. He is best known as the founder in 1884 of the *Circolo matematico di Palermo*, and director of its *Rendiconti*. The society has become international and has been a powerful stimulus for mathematical research in Italy.

Karl Culmann (1821–1881), professor at the Polytechnicum in Zurich, published an epoch-making work on *Die graphische Statik*, Zurich, 1864, which has rendered graphical statics a great rival of analytical statics. Before Culmann, *Barthélémy-Edouard Cousinéry* (1790–1851) a civil engineer at Paris, had turned his attention to the graphical calculus, but he made use of perspective, and not of modern geometry.[2] Culmann is the first to undertake to present the graphical calculus as a symmetrical whole, holding the same relation to the new geometry that analytical mechanics does to higher analysis. He makes use of the polar theory of reciprocal figures as expressing the relation between the force and the funicular polygons. He deduces this relation without leaving the plane of the two figures. But if the polygons be regarded as projections of lines in space, these lines may be treated as reciprocal elements of a "Nullsystem." This was done by *Clerk Maxwell* in 1864, and elaborated further by *L. Cremona*. The graphical calculus has been applied by *O. Mohr* of Dresden to the elastic line for continuous spans. *Henry T. Eddy* (1844–1921), then of the Rose Polytechnic Institute, now of the University of Minnesota, gives graphical solutions of problems on the maximum stresses in bridges under concentrated loads, with aid of what he calls "reaction polygons." A standard work, *La Statique graphique*, 1874, was issued by Maurice Levy of Paris.

Descriptive geometry [reduced to a science by G. Monge in France, and elaborated further by his successors, *J. N. P. Hachette*, *C. Dupin*, *Théodore Olivier* (1793–1853) of Paris, *Jules de la Gournerie* of Paris] was soon studied also in other countries. The French directed their attention mainly to the theory of surfaces and their curvature; the Germans and Swiss, through Guido Schreiber (1799–1871) of Karls-

[1] *Proceedings of the Roy. Soc. of London*, Vol. 75, London, 1905, pp. 277–279.

[2] A. Jay du Bois, *Graphical Statics*, New York, 1875, p. xxxii; M. d'Ocagne, *Traité de Nomographie*, Paris, 1899, p. 5.

ruhe, Karl Pohlke (1810–1877) of Berlin,[1] Josef Schlesinger (1831–1901) of Vienna, and particularly W. Fiedler, interwove projective and descriptive geometry.

Wilhelm Fiedler (1832–1912), the son of a shoe-maker in Chemnitz, Saxony, taught mathematics and mechanics in a technical school of Chemnitz, 1853 to 1864, and studied meanwhile the works of M. Chasles, G. Lamé, B. de St.-Venant, J. V. Poncelet, J. Steiner, J. Plücker, von Staudt, G. Salmon, A. Cayley, J. J. Sylvester. He was self-taught. On the recommendation of A. F. Möbius he was awarded in 1859 the degree of doctor of philosophy by the University of Leipsic for a dissertation on central projection. At this time Fiedler made arrangements with Salmon for a German elaborated edition of Salmon's *Conic Sections;* it appeared in 1860. In the same way were brought out by Fiedler Salmon's *Higher Algebra* in 1863, Salmon's *Geometry of Three Dimensions* in 1862, Salmon's *Higher Plane Curves* in 1873. In 1864 Fiedler became professor at the technical high school in Prag, and in 1867 at the Polytechnic School in Zurich, where he was active until his retirement in 1907. The emphasis of Fiedler's activity was placed upon descriptive geometry. His *Darstellende Geometrie*, 1871, was brought, in the third edition, in organic connection with v. Staudt's geometry of position. Especially after the death of Culmann in 1881, Fiedler was criticised on pedagogic grounds for excessive emphasis upon geometric construction. A harmonizing effort was the text on descriptive geometry by *Christian Wiener* (1826–1896) of the Polytechnic School in Karlsruhe. Of interest is Fiedler's recognition in 1870 of homogeneous co-ordinates as cross-ratios, invariant in all linear transformations; this idea had been advanced in 1827 by A. F. Möbius, but had remained unnoticed.[2] Fiedler's *Zyklographie*, 1882, contained constructions of problems on circles and spheres.

The interweaving of projective and descriptive geometry was carried on in Italy by G. Bellavitis. The theory of shades and shadows was first investigated by the French writers quoted above, and in Germany treated most exhaustively by Ludwig Burmester of Munich.

Elementary Geometry of the Triangle and Circle

It is truly astonishing that during the nineteenth century new theorems should have been found relating to such simple figures as the triangle and circle, which had been subjected to such close examination by the Greeks and the long line of geometers which followed. It was *L. Euler* who proved in 1765 that the orthocenter, circumcenter

[1] F. J. Obenrauch, *Geschichte der darstellenden und projectiven Geometrie*, Brünn, 1897, pp. 350, 352.

[2] A. Voss, "Wilhelm Fiedler," *Jahresb. d. d. Math. Vereinigung*, Vol. 22, 1913, p. 107.

and centroid of a triangle are collinear, lying on the "Euler line." H. C. Gossard of the University of Oklahoma showed in 1916 that the three Euler lines of the triangles formed by the Euler line and the sides, taken by twos, of a given triangle, form a triangle triply perspective with the given triangle and having the same Euler line. Conspicuous among the new developments is the "nine-point circle," the discovery of which has been erroneously ascribed to Euler. Among the several independent discoverers is the Englishman, *Benjamin Bevan* (?–1838) who proposed in Leybourn's *Mathematical Repository*, I, 18, 1804, a theorem for proof which practically gives us the nine-point circle. The proof was supplied to the *Repository*, I, Part 1, p. 143, by *John Butterworth*, who also proposed a problem, solved by himself and *John Whitley*, from the general tenor of which it appears that they knew the circle in question to pass through all nine points. These nine points are explicitly mentioned by C. J. Brianchon and J. V. Poncelet in Gergonne's *Annales* of 1821. In 1822, *Karl Wilhelm Feuerbach* (1800–1834), professor at the gymnasium in Erlangen, published a pamphlet in which he arrives at the nine-point circle, and proves the theorem known by his name, that this circle touches the incircle and the three excircles. The Germans call it "Feuerbach's Circle." The last independent discoverer of this remarkable circle, so far as known, is *T. S. Davies*, in an article of 1827 in the *Philosophical Magazine*, II, 29–31. Feuerbach's theorem was extended by *Andrew Searle Hart* (1811–1890), fellow of Trinity College, Dublin, who showed that the circles which touch three given circles can be distributed into sets of four all touched by the same circle.

In 1816 *August Leopold Crelle* published in Berlin a paper dealing with certain properties of plane triangles. He showed how to determine a point Ω inside a triangle, so that the angles (taken in the same order) formed by the lines joining it to the vertices are equal. In the adjoining figure the three marked angles are equal. If the construction is made so that angle $\Omega'AC = \Omega'CB = \Omega'BA$, then a second point Ω' is obtained. The study of these new angles and new points led Crelle to exclaim:

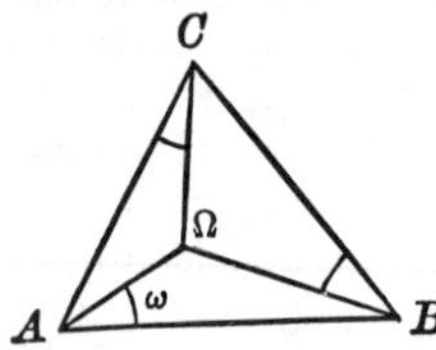

"It is indeed wonderful that so simple a figure as the triangle is so inexhaustible in properties. How many as yet unknown properties of other figures may there not be!" Investigations were made also by *Karl Friedrich Andreas Jacobi* (1795–1855) of Pforta and some of his pupils, but after his death, in 1855, the whole matter was forgotten. In 1875 the subject was again brought before the mathematical public by Henri Brocard (1845–1922) whose researches were followed up by a large number of investigators in France, England and Germany. Unfortunately, the names of geometricians which have been attached to certain remarkable points, lines and circles are not always the names of the men who first studied their properties. Thus, we speak of "Brocard points" and "Brocard angles," but historical research brought out the fact, in 1884 and 1886, that these were the points and lines which had been studied by A. L. Crelle and K. F. A. Jacobi. The "Brocard Circle" is Brocard's own creation. In the triangle ABC, let Ω and Ω′ be the first and second "Brocard point." Let A′ be the intersection of BΩ and CΩ′; B′ of

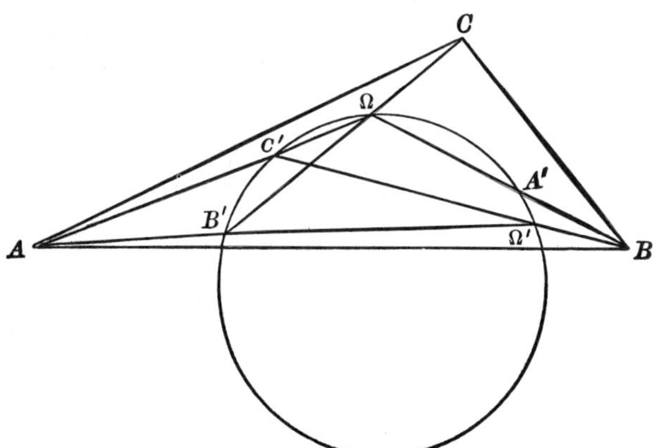

AΩ′ and CΩ; C′ of BΩ′ and AΩ. The circle passing through A′, B′, C′ is the "Brocard circle." A′B′C′ is "Brocard's first triangle." Another like triangle, A″B″C″ is called "Brocard's second triangle." The points A″, B″, C″, together with Ω, Ω′, and two other points, lie in the circumference of the "Brocard circle."

In 1873 **Émile Lemoine** (1840–1912), the editor of *l'Intermédiaire des mathématiciens*, called attention to a particular point within a plane triangle which has been variously called the "Lemoine point," "symmedian point," and "Grebe point," named after *Ernst Wilhelm Grebe* (1804–1874) of Kassel. If CD is so drawn as to make angles *a* and *b* equal, then one of the two lines AB and CD is the *anti-parallel* of the other, with reference to the angle O. Now OE, the bisector of AB, is the *median* and OF, the bisector of the anti-parallel of AB, is called the *symmedian* (abbreviated from *symétrique de la médiane*). The point of concurrence of the three symmedians in a triangle is called,

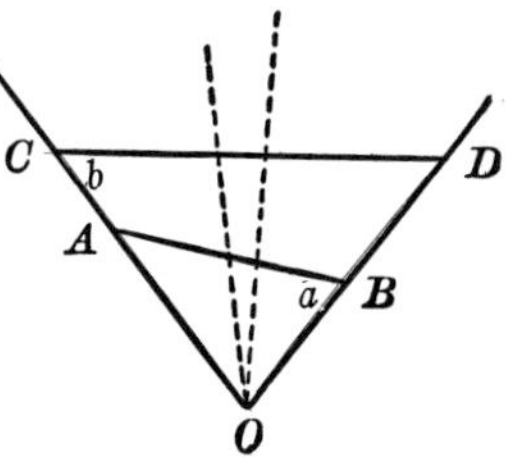

after *Robert Tucker* (1832–1905) of University College School in London, the "symmedian point." John Sturgeon Mackay (1843–1914) of Edinburgh has pointed out that some of the properties of this point, brought to light since 1873, were first discovered previously to that date. The anti-parallels of a triangle which pass through its symmedian point, meet its sides in six points which lie on a circle, called the "second Lemoine circle." The "first Lemoine circle" is a special case of a "Tucker circle" and concentric with the "Brocard circle." The "Tucker circles" may be thus defined. Let DF′=FE′=ED′; let, moreover, the following pairs of lines be anti-parallels to each other: AB and ED′, BC and FE′, CA and DF′; then the six points D, D′, E, E′, F, F′, lie on a "Tucker circle." Vary the length of the equal anti-parallels, and a family of "Tucker circles" is obtained. Allied to these are the "Taylor circles," due to H. M. Taylor of Trinity College, Cambridge. Still different types are the "Mackay circles," and the "Neuberg circles" due to Joseph Neuberg (1840–1926) of Luxemburg. A systematic treatise on this topic, *Die Brocardschen Gebilde*, was written by Albrecht Emmerich, Berlin, 1891. Of the almost innumerable mass of new theorems on the triangle and circle, a great number is given in the *Treatise on the Circle and the Sphere*, Oxford, 1916, written by J. L. Coolidge of Harvard University.

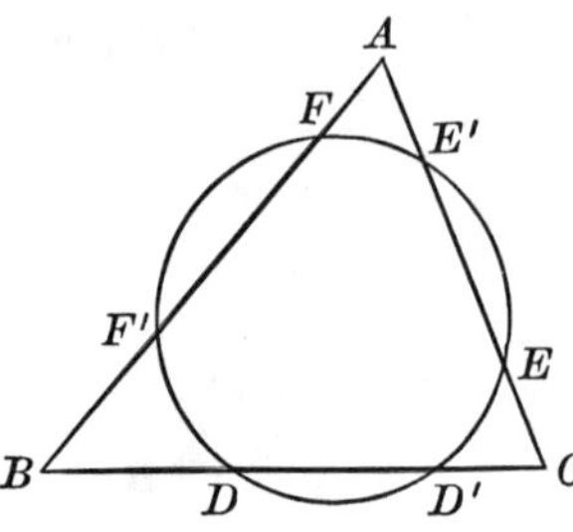

Since 1888 E. Lemoine of Paris developed a system, called geometrographics, for the purpose of numerically comparing geometric constructions with respect to their simplicity. Coolidge calls these "the best known and least undesirable tests for the simplicity of a geometrical construction"; A. Emch declares that "they are hardly of any practical value, in so far as they do not indicate how to simplify a construction or how to make it more accurate."

A new theorem upon the circumscribed tetraedron was propounded in 1897 by A. S. Bang and proved by Joh. Gehrke. The theorem is: Opposite edges of a circumscribed tetraedron subtend equal angles at the points of contact of the faces which contain them. It has been the starting-point for extended developments by Franz Meyer, J. Neuberg and H. S. White.[1]

Link-motion

The generation of rectilinear motion first arose as a practical problem in the design of steam engines. A close approximation to such motion is the "parallel motion" designed by James Watt in 1784: In a freely jointed quadrilateral ABCD, with the side AD fixed, a point M on the side BC moves in nearly a straight line. The equation of the curve traced by M, sometimes called "Watt's curve," was first derived by the French engineer, *François Marie de Prony*

[1] *Bull. Am. Math. Soc.*, Vol. 14, 1908, p. 220.

(1755–1839); the curve is of the sixth order and was studied by Darboux in 1879. A generalization of this curve is the "three-bar curve" studied by *Samuel Roberts* in 1876 and *Reinhold Müller* in 1902.[1]

A beautiful discovery in link-motion which came to attract a great deal of attention was made by A. Peaucellier, Capitaine du génie â Nice; in 1864 he proposed in the *Nouvelles annales* the problem of devising compound compasses for the generation of a straight line and also a conic. It is evident from his remarks that he himself had a solution. In 1873 he published his solution in the same journal. When Peaucellier's cell came to be appreciated, he was awarded the great mechanical prize of the Institute of France, the "Prix Montyou." The generation of exact rectilinear motion had long been believed impossible. Only recently has it been pointed out that straight-line motion had been invented before Peaucellier by another Frenchman, P. F. Sarrus of Strasbourg. He presented an article and a model to the Paris Academy of Sciences; the article, without any figure, was published in 1853 in the *Comptes Rendus* [2] of the academy, and reported on by Poncelet. The paper was entirely forgotten until attention was called to it in 1905 by G. T. Bennett of Emmanuel College, Cambridge.[3] The pieces ARSB and ATUB are each hinged by three parallel horizontal hinges, the two sets of hinges having different directions. Connected thus with B, A has a rectilinear movement, up and down. In one respect Sarrus's solution of the problem of rectilinear motion is more complete than that of Peaucellier; for, while the Peaucellier cell gives rectilinear motion only to a single point, Sarrus's apparatus gives rectilinear motion to the whole piece A. It was re-invented in 1880 by H. M. Brunel and in 1891 by Archibald Barr. Yet to this day Sarrus's device appears to remain practically unknown.

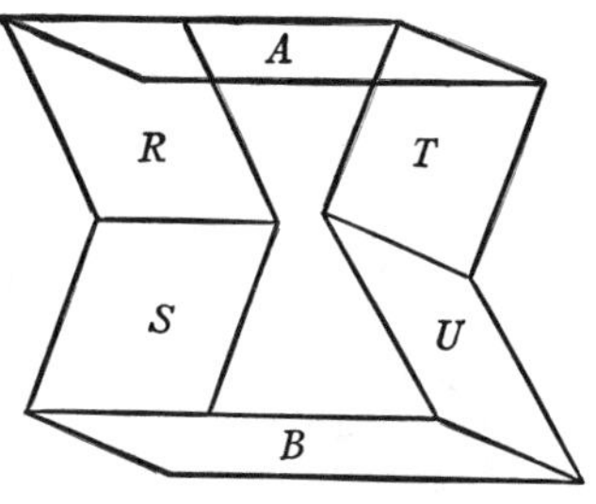

While Sarrus's device is three-dimensional, that of Peaucellier is two-dimensional. An independent solution of straight-line motion was given in 1871 by the Russian Lipkin, a pupil of P. Chebichev of the University of Petrograd. In 1874 J. J. Sylvester became interested in link-motion, and lectured on it at the Royal Institution. During the next few years several mathematicians worked on linkages. H. Hart of Woolwich reduced Peaucellier's seven links to four links. A new device by Sylvester has been called "Sylvester's linkage."

[1] G. Loria, *Ebene Curven* (F. Schütte), Vol. I, 1910, pp. 274–279.

[2] *Comptes Rendus*, Vol. 36, pp. 1036, 1125. The author's name is here spelled "Sarrut," but R. C. Archibald has pointed out that this is a misprint.

[3] See *Philosoph. Magaz.*, 6. S., Vol. 9, 1905, p. 803.

The barrister at law, Alfred Bray Kempe of London showed in 1876 that a link-motion can be found to describe any given algebraic curve; he is the author of a popular booklet, *How to draw a Straight Line*,✲ London, 1877. Other articles of note on this subject were prepared by Samuel Roberts, Arthur Cayley, W. Woolsey Johnson, V. Liguine of the University of Odessa, and G. P. X. Kœnigs of the École Polytechnique in Paris. The determination of the linkage with minimum number of pieces by which a given curve can be described is still an unsolved problem.

Parallel Lines, Non-Euclidean Geometry and Geometry of n Dimensions

During the nineteenth century very remarkable generalizations were made, which reach to the very root of two of the oldest branches of mathematics,—elementary algebra and geometry. In geometry the axioms have been searched to the bottom, and the conclusion has been reached that the space defined by Euclid's axioms is not the only possible non-contradictory space. Euclid stated (I, 27) that "if a straight line falling on two other straight lines make the alternate angels equal to one another, the two straight lines shall be parallel to one another." Being unable to prove that in every other case the two lines are not parallel, he *assumed* this to be true in what is now generally called the 5*th* "axiom," by some the 11th or the 12th "axiom."

Simpler and more obvious axioms have been advanced as substitutes. As early as 1663, John Wallis of Oxford recommended: "To any triangle another triangle, as large as you please, can be drawn, which is similar to the given triangle." G. Saccheri assumed the existence of two similar, unequal triangles. Postulates similar to Wallis' have been proposed also by J. H. Lambert, L. Carnot, P. S. Laplace, J. Delboeuf. A. C. Clairaut assumes the existence of a rectangle; W. Bolyai postulated that a circle can be passed through any three points not in the same straight line, A. M. Legendre that there existed a finite triangle whose angle-sum is two right angles, J. F. Lorenz and Legendre that through every point within an angle a line can be drawn intersecting both sides, C. L. Dodgson that in any circle the inscribed equilateral quadrangle is greater than any one of the segments which lie outside it. But probably the simplest is the assumption made by Joseph Fenn in his edition of Euclid's *Elements*, Dublin, 1769, and again sixteen years later by William Ludlam (1718–1788), vicar of Norton, and adopted by John Playfair: "Two straight lines which cut one another can not both be parallel to the same straight line." It is noteworthy[1] that this axiom is distinctly stated in Proclus's note to Euclid, I, 31.

But the most numerous efforts to remove the supposed defect in

[1] T. L. Heath, *The Thirteen Books of Euclid's Elements*, Vol. I, p. 220.

Euclid were attempts to *prove* the parallel postulate. After centuries of desperate but fruitless endeavor, the bold idea dawned upon the minds of several mathematicians that a geometry might be built up without assuming the parallel-axiom. While A. M. Legendre still endeavored to establish the axiom by rigid proof, Lobachevski brought out a publication which assumed the contradictory of that axiom, and which was the first of a series of articles destined to clear up obscurities in the fundamental concepts, and greatly to extend the field of geometry.

Nicholaus Ivanovich Lobachevski (1793–1856) was born at Makarief, in Nizhni-Novgorod, Russia, studied at Kasan, and from 1827 to 1846 was professor and rector of the University of Kasan. His views on the foundation of geometry were first set forth in a paper laid before the physico-mathematical department of the University of Kasan in February, 1826. This paper was never printed and was lost. His earliest publication was in the Kasan *Messenger* for 1829, and then in the *Gelehrte Schriften der Universität Kasan*, 1836–1838, under the title, "New Elements of Geometry, with a complete theory of Parallels." Being in the Russian language, the work remained unknown to foreigners, but even at home it attracted no notice. In 1840 he published a brief statement of his researches in Berlin, under the title, *Geometrische Untersuchungen zur Theorie der Parallellinien*. Lobachevski constructed an "imaginary geometry," as he called it, which has been described by W. K. Clifford as "quite simple, merely Euclid without the vicious assumption." A remarkable part of this geometry is this, that through a point an indefinite number of lines can be drawn in a plane, none of which cut a given line in the same plane. A similar system of geometry was deduced independently by the Bolyais in Hungary, who called it "absolute geometry."

Wolfgang Bolyai de Bolya (1775–1856) was born in Szekler-Land, Transylvania. After studying at Jena, he went to Göttingen, where he became intimate with K. F. Gauss, then nineteen years old. Gauss used to say that Bolyai was the only man who fully understood his views on the metaphysics of mathematics. Bolyai became professor at the Reformed College of Maros-Vâsârhely, where for forty-seven years he had for his pupils most of the later professors of Transylvania. The first publications of this remarkable genius were dramas and poetry. Clad in old-time planter's garb, he was truly original in his private life as well as in his mode of thinking. He was extremely modest. No monument, said he, should stand over his grave, only an apple-tree, in memory of the three apples; the two of Eve and Paris, which made hell out of earth, and that of I. Newton, which elevated the earth again into the circle of heavenly bodies.[1] His son, **Johann**

[1] F. Schmidt, "Aus dem Leben zweier ungarischer Mathematiker Johann und Wolfgang Bolyai von Bolya," *Grunert's Archiv*, 48:2, 1868. Franz Schmidt (1827–1901) was an architect in Budapest.

Bolyai (1802–1860), was educated for the army, and distinguished himself as a profound mathematician, an impassioned violin-player, and an expert fencer. He once accepted the challenge of thirteen officers on condition that after each duel he might play a piece on his violin, and he vanquished them all.

The chief mathematical work of Wolfgang Bolyai appeared in two volumes, 1832–1833, entitled *Tentamen juventutem studiosam in elementa matheseos puræ . . . introducendi.* It is followed by an appendix composed by his son Johann. Its twenty-six pages make the name of Johann Bolyai immortal. He published nothing else, but he left behind one thousand pages of manuscript.

While Lobachevski enjoys priority of publication, it may be that Bolyai developed his system somewhat earlier. Bolyai satisfied himself of the non-contradictory character of his new geometry on or before 1825; there is some doubt whether Lobachevski had reached this point in 1826. Johann Bolyai's father seems to have been the only person in Hungary who really appreciated the merits of his son's work. For thirty-five years this appendix, as also Lobachevski's researches, remained in almost entire oblivion. Finally Richard Baltzer of the University of Giessen, in 1867, called attention to the wonderful researches.

In 1866 J. Hoüel translated Lobachevski's *Geometrische Untersuchungen* into French. In 1867 appeared a French translation of Johann Bolyai's *Appendix.* In 1891 George Bruce Halsted, then of the University of Texas, rendered these treatises easily accessible to American readers by translations brought out under the titles of J. Bolyai's *The Science Absolute of Space* and N. Lobachevski's *Geometrical Researches on the Theory of Parallels* of 1840.

The Russian and Hungarian mathematicians were not the only ones to whom pangeometry suggested itself. A copy of the *Tentamen* reached K. F. Gauss, the elder Bolyai's former roommate at Göttingen, and this Nestor of German mathematicians was surprised to discover in it worked out what he himself had begun long before, only to leave it after him in his papers. As early as 1792 he had started on researches of that character. His letters show that in 1799 he was trying to prove *a priori* the reality of Euclid's system; but some time within the next thirty years he arrived at the conclusion reached by Lobachevski and Bolyai. In 1829 he wrote to F. W. Bessel, stating that his "conviction that we cannot found geometry completely *a priori* has become, if possible, still firmer," and that "if number is merely a product of our mind, space has also a *reality beyond* our mind of which we cannot fully foreordain the laws *a priori.*" The term *non-Euclidean geometry* is due to Gauss. It is surprising that the first glimpses of non-Euclidean geometry were had in the eighteenth century. *Geronnimo Saccheri* (1667–1733), a Jesuit father of Milan, in 1733 wrote

Euclides ab omni naevo vindicatus [1] (Euclid vindicated from every flaw). Starting with two equal lines AC and BD, drawn perpendicular to a line AB and on the same side of it, and joining C and D, he proves that the angles at C and D are equal. These angles must be either right, or obtuse or acute. The hypothesis of an obtuse angle is demolished by showing that it leads to results in conflict with Euclid I, 17: Any two angles of a triangle are together less than two right angles. The hypothesis of the acute angle leads to a long procession of theorems, of which the one declaring that two lines which meet in a point at infinity can be perpendicular at that point to the same straight line, is considered contrary to the nature of the straight line; hence the hypothesis of the acute angle is destroyed. Though not altogether satisfied with his proof, he declared Euclid "vindicated." Another early writer was J. H. Lambert who in 1766 wrote a paper "Zur Theorie der Parallellinien," published in the *Leipziger Magazin für reine und angewandte Mathematik*, 1786, in which: (1) The failure of the parallel-axiom in surface-spherics gives a geometry with angle-sum >2 right angles; (2) In order to make intuitive a geometry with angle-sum <2 right angles we need the aid of an "imaginary sphere" (pseudo-sphere); (3) In a space with the angle-sum differing from 2 right angles, there is an absolute measure (Bolyai's natural unit for length). Lambert arrived at no definite conclusion on the validity of the hypotheses of the obtuse and acute angles.

Among the contemporaries and pupils of K. F. Gauss, three deserve mention as writers on the theory of parallels, *Ferdinand Karl Schweikart* (1780–1859), professor of law in Marburg, *Franz Adolf Taurinus* (1794–1874), a nephew of Schweikart, and *Friedrich Ludwig Wachter* (1792–1817), a pupil of Gauss in 1809 and professor at Dantzig. Schweikart sent Gauss in 1818 a manuscript on "Astral Geometry" which he never published, in which the angle-sum of a triangle is less than two right angles and there is an absolute unit of length. He induced Taurinus to study this subject. Taurinus published in 1825 his *Theorie der Parallellinien*, in which he took the position of Saccheri and Lambert, and in 1826 his *Geometriæ prima elementa*, in an appendix of which he gives important trigonometrical formulæ for non-Euclidean geometry by using the formulæ of spherical geometry with an imaginary radius. His *Elementa* attracted no attention. In disgust he burned the remainder of his edition. Wachter's results are contained in a letter of 1816 to Gauss and in his *Demonstratio axiomatis geometrici in Euclideis undecimi*, 1817. He showed that the geometry on a sphere becomes identical with the geometry of Euclid when the radius is infinitely increased, though it is distinctly shown that the limiting surface is not a plane.[2] Elsewhere we have mentioned the

[1] See English translation by G. B. Halsted in *Am. Math. Monthly*, Vols. 1–5, 1894–1898.✡

[2] D. M. Y. Sommerville, *Elem. of Non-Euclidean Geometry*, London, 1914, p. 15.

contemporary researches on parallel lines due to A. M. Legendre in France.

The researches of K. F. Gauss, N. I. Lobachevski and J. Bolyai have been considered by F. Klein as constituting the first period in the history of non-Euclidean geometry. It is a period in which the synthetic methods of elementary geometry were in vogue. The second period embraces the researches of G. F. B. Riemann, H. Helmholtz, S. Lie and E. Beltrami, and employs the methods of differential geometry. It was in 1854 that Gauss heard from his pupil, Riemann, a marvellous dissertation✡ which considered the foundations of geometry from a new point of view. Riemann was not familiar with Lobachevski and Bolyai. He developed the notion of n-ply extended magnitude, and the measure-relations of which a manifoldness of n dimensions is capable, on the assumption that every line may be measured by every other. Riemann applied his ideas to space. He taught us to distinguish between "unboundedness" and "infinite extent." According to him we have in our mind a more general notion of space, *i. e.* a notion of non-Euclidean space; but we learn *by experience* that our physical space is, if not exactly, at least to a high degree of approximation, Euclidean space. Riemann's profound dissertation was not published until 1867, when it appeared in the *Göttingen Abhandlungen.* Before this the idea of n dimensions had suggested itself under various aspects to Ptolemy, J. Wallis, D'Alembert, J. Lagrange, J. Plücker, and H. G. Grassmann. The idea of *time* as a fourth dimension had occurred to D'Alembert and Lagrange. About the same time with Riemann's paper, others were published from the pens of *H. Helmholtz* and *E. Beltrami.* This period marks the beginning of lively discussions upon this subject. Some writers—J. Bellavitis, for example—were able to see in non-Euclidean geometry and n-dimensional space nothing but huge caricatures, or diseased outgrowths of mathematics. H. Helmholtz's article was entitled *Thatsachen, welche der Geometrie zu Grunde liegen,* 1868, and contained many of the ideas of Riemann. Helmholtz popularized the subject in lectures, and in articles for various magazines. Starting with the idea of congruence, and assuming the free mobility of a rigid body and the return unchanged to its original position after rotation about an axis, he proves that the square of the line-element is a homogeneous function of the second degree in the differentials.[1] Helmholtz's investigations were carefully examined by S. Lie who reduced the Riemann-Helmholtz problem to the following form: To determine all the continuous groups in space which, in a bounded region, have the property of displacements. There arose three types of groups,

Sommerville is the author of a *Bibliography of non-Euclidean geometry including the theory of parallels, the foundations of geometry, and space of n dimensions,* London, 1911.

[1] D. M. Y. Sommerville, *op. cit.*, p. 195.

which characterize the three geometries of Euclid, of N. I. Lobachevski and J. Bolyai, and of F. G. B. Riemann.[1]

Eugenio Beltrami (1835–1900), born at Cremona, was a pupil of F. Brioschi. He was professor at Bologna as a colleague of L. Cremona, at Pisa as an associate of E. Betti, at Pavia as a co-worker with F. Casorati, and since 1891 at Rome where he spent the last years of his career, "uno degli illustri maestri dell' analisi in Italia." Beltrami wrote in 1868 a classical paper, *Saggio di interpretazione della geometria non-euclidea* (*Giorn. di Matem.*, 6), which is analytical (and, like several other papers, should be mentioned elsewhere were we to adhere to a strict separation between synthesis and analysis). He reached the brilliant and surprising conclusion that in part the theorems of non-Euclidean geometry find their realization upon surfaces of constant negative curvature. He studied, also, surfaces of constant positive curvature, and ended with the interesting theorem that the space of constant positive curvature is contained in the space of constant negative curvature. These researches of Beltrami, H. Helmholtz, and G. F. B. Riemann culminated in the conclusion that on surfaces of constant curvature we may have three geometries,—the non-Euclidean on a surface of constant negative curvature, the spherical on a surface of constant positive curvature, and the Euclidean geometry on a surface of zero curvature. The three geometries do not contradict each other, but are members of a system,—a geometrical trinity. The ideas of hyper-space were brilliantly expounded and popularised in England by Clifford.

William Kingdon Clifford (1845–1879) was born at Exeter, educated at Trinity College, Cambridge, and from 1871 until his death professor of applied mathematics in University College, London. His premature death left incomplete several brilliant researches which he had entered upon. Among these are his paper *On Classification of Loci* and his *Theory of Graphs*. He wrote articles *On the Canonical Form and Dissection of a Riemann's Surface*, on *Biquaternions*, and an incomplete work on the *Elements of Dynamic*. He gave exact meaning in dynamics to such familiar words as "spin," "twist," "squirt," "whirl." The theory of polars of curves and surfaces was generalized by him and by Reye. His classification of loci, 1878, being a general study of curves, was an introduction to the study of n-dimensional space in a direction mainly projective. This study has been continued since chiefly by G. Veronese of Padua, C. Segre of Turin, E. Bertini, F. Aschieri, P. Del Pezzo of Naples.

Beltrami's researches on non-Euclidean geometry were followed, in 1871, by important investigations of Felix Klein, resting upon Cayley's *Sixth Memoir on Quantics*, 1859. The development of geometry in the first half of the nineteenth century had led to the separation

[1] Lie, *Theorie der Transformationsgruppen*, Bd. III, Leipzig, 1893, pp. 437–543; Bonola, *op. cit.*, p. 154.

of this science into two parts: the geometry of position or descriptive geometry which dealt with properties that are unaffected by projection, and the geometry of measurement in which the fundamental notions of distance, angle, etc., are changed by projection. Cayley's *Sixth Memoir* brought these strictly segregated parts together again by his definition of distance between two points. The question whether it is not possible so to express the metrical properties of figures that they will not vary by projection (or linear transformation) had been solved for special projections by M. Chasles, J. V. Poncelet, and E. Laguerre, but it remained for A. Cayley to give a general solution by defining the distance between two points as an arbitrary constant multiplied by the logarithm of the anharmonic ratio in which the line joining the two points is divided by the fundamental quadric. These researches, applying the principles of pure projective geometry, mark the third period in the development of non-Euclidean geometry.

Enlarging upon this notion, F. Klein showed the independence of projective geometry from the parallel-axiom, and by properly choosing the law of the measurement of distance deduced from projective geometry the spherical, Euclidean, and pseudospherical geometries, named by him respectively the elliptic, parabolic, and hyperbolic geometries. This suggestive investigation was followed up by numerous writers, particularly by G. Battaglini of Naples, E. d'Ovidio of Turin, R. de Paolis of Pisa, F. Aschieri, A. Cayley, F. Lindemann of Munich, E. Schering of Göttingen, W. Story of Clark University, H. Stahl of Tübingen, A. Voss of Munich, Homersham Cox, A. Buchheim.[1] The notion of parallelism applicable to hyperbolic space was the only extension of Euclid's notion of parallelism until Clifford discovered in elliptic space straight lines which possess most of the properties of Euclidean parallels, but differ from them in being skew. Two lines are right (or left) parallel, if they cut the same right (or left) generators of the absolute. Later F. Klein and R. S. Ball made extensive contributions to the knowledge of these lines. More recently E. Study of Bonn, J. L. Coolidge of Harvard University, W. Vogt of Heidelberg and others have been studying this subject. The methods employed have been those of analytic and synthetic geometry as well as those of differential geometry and vectorial analysis.[2] The geometry of n dimensions was studied along a line mainly metrical by a host of writers, among whom may be mentioned Simon Newcomb of the Johns Hopkins University, L. Schläfli of Bern, W. I. Stringham (1847–1909) of the University of California, W. Killing of Münster, T. Craig of the Johns Hopkins, Rudolf Lipschitz (1832–1903) of Bonn. R. S. Heath of Birmingham and W. Killing investigated the kinematics and mechanics of such a space. Regular solids in n-dimensional space were studied by Stringham, Ellery W. Davis (1857–1918) of the

[1] G. Loria, *Die hauptsächlichsten Theorien der Geometrie*, 1888, p. 102.
[2] *Bull. Am. Math. Soc.*, Vol. 17, 1911, p. 315.

University of Nebraska, Reinhold Hoppe (1816–1900) of Berlin, and others. Stringham gave pictures of projections upon our space of regular solids in four dimensions, and V. Schlegel at Hagen constructed models of such projections. These are among the most curious of a series of models published by L. Brill in Darmstadt. It has been pointed out that if a fourth dimension existed, certain motions could take place which we hold to be impossible. Thus S. Newcomb showed the possibility of turning a closed material shell inside out by simple flexure without either stretching or tearing; F. Klein pointed out that knots could not be tied; G. Veronese showed that a body could be removed from a closed room without breaking the walls; C. S. Peirce proved that a body in four-fold space either rotates about two axes at once, or cannot rotate without losing one of its dimensions.

A fourth period in the history of non-Euclidean geometry, introduced by the researches of Moritz Pasch, Giuseppe Peano, Mario Pieri, David Hilbert, Oswald Veblen, concerns itself with the logical grounding of geometry (including non-Euclidean forms) upon sets of axioms.

The geometry of hyperspace was exploited by spiritualists and mediums, of whom Henry Slade was the most notorious. He converted to spiritualism the German scientist F. Zöllner and his coterie, to whom he gave a spiritual demonstration of the existence of a fourth dimension of space. These events contributed to the severity with which the philosopher R. H. Lotze, in his *Metaphysik*, 1879, criticised the mathematical theories of hyperspace and non-Euclidean geometry.

Analytic Geometry

In the preceding chapter we endeavored to give a flashlight view of the rapid advance of synthetic geometry. In some cases we also mentioned analytical treatises. Modern synthetic and modern analytical geometry have much in common, and may be grouped together under the common name "projective geometry." Each has advantages over the other. The continual direct viewing of figures as existing in space adds exceptional charm to the study of the former, but the latter has the advantage in this, that a well-established routine in a certain degree may outrun thought itself, and thereby aid original research. While in Germany J. Steiner and von Staudt developed synthetic geometry, Plücker laid the foundation of modern analytic geometry.

Julius Plücker (1801–1868) was born at Elberfeld, in Prussia. After studying at Bonn, Berlin, and Heidelberg, he spent a short time in Paris attending lectures of G. Monge and his pupils. Between 1826 and 1836 he held positions successively at Bonn, Berlin, and Halle. He then became professor of physics at Bonn. Until 1846 his original researches were on geometry. In 1828 and in 1831 he published his

Analytisch-Geometrische Entwicklungen in two volumes. Therein he adopted the abridged notation [used before him in a more restricted way by Étienne Bobillier (1797–1832), professor of mechanics at Châlons-sur-Marne], and avoided the tedious process of algebraic elimination by a geometric consideration. In the second volume the principle of duality is formulated analytically. With him duality and homogeneity found expression already in his system of co-ordinates. The homogenous or trilinear system used by him is much the same as the co-ordinates of A. F. Möbius. In the identity of analytical operation and geometric construction Plücker looked for the source of his proofs. The *System der Analytischen Geometrie*, 1835, contains a complete classification of plane curves of the third order, based on the nature of the points at infinity. The *Theorie der Algebraischen Curven*, 1839, contains, besides an enumeration of curves of the fourth order, the analytic relations between the ordinary singularities of plane curves known as "Plücker's equations," by which he was able to explain "Poncelet's paradox." The discovery of these relations is, says A. Cayley, "the most important one beyond all comparison in the entire subject of modern geometry." The four Plücker equations have been expressed in different forms. Cayley studied higher singularities of plane curves. M. W. Haskell of the University of California, in 1914, showed from the Plücker equations that the maximum number of cusps possible for a curve of order m is the greatest integer in $m(m-2)/3$ (except when m is 4 or 6, in which case the maximum number is 3 or 9), and that there is always a self-dual curve with this maximum number of cusps.

Certain interrelations of the various geometrical researches of the first half and middle of the nineteenth century are brought out by J. G. Darboux in the following passage: [1] "While M. Chasles, J. Steiner, and, later, . . . von Staudt, were intent on constituting a rival doctrine to analysis and set in some sort altar against altar, J. D. Gergonne, E. Bobillier, C. Sturm, and above all J. Plücker, perfected the geometry of R. Descartes and constituted an analytic system in a manner adequate to the discoveries of the geometers. It is to E. Bobillier and to J. Plücker that we owe the method called *abridged notation*. Bobillier consecrated to it some pages truly new in the last volumes of the *Annales* of Gergonne. Plücker commenced to develop it in his first work, soon followed by a series of works where are established in a fully conscious manner the foundations of the modern analytic geometry. It is to him that we owe tangential co-ordinates, trilinear co-ordinates, employed with homogeneous equations, and finally the employment of canonical forms whose validity was recognized by the method, so deceptive sometimes, but so fruitful, called the *enumeration of constants*."

In Germany J. Plücker's researches met with no favor. His method

[1] *Congress of Arts and Science*, St. Louis, 1904, Vol. I, pp. 541, 542.

was declared to be unproductive as compared with the synthetic method of J. Steiner and J. V. Poncelet! His relations with C. G. J. Jacobi were not altogether friendly. Steiner once declared that he would stop writing for *Crelle's Journal* if Plücker continued to contribute to it.[1] The result was that many of Plücker's researches were published in foreign journals, and that his work came to be better known in France and England than in his native country. The charge was also brought against Plücker that, though occupying the chair of physics, he was no physicist. This induced him to relinquish mathematics, and for nearly twenty years to devote his energies to physics. Important discoveries on Fresnel's wave-surface, magnetism, spectrum-analysis were made by him. But towards the close of his life he returned to his first love,—mathematics,—and enriched it with new discoveries. By considering space as made up of lines he created a "new geometry of space." Regarding a right line as a curve involving four arbitrary parameters, one has the whole system of lines in space. By connecting them by a single relation, he got a "complex" of lines; by connecting them with a twofold relation, he got a "congruency" of lines. His first researches on this subject were laid before the Royal Society in 1865. His further investigations thereon appeared in 1868 in a posthumous work entitled *Neue Geometrie des Raumes gegründet auf die Betrachtung der geraden Linie als Raumelement*, edited by Felix Klein. Plücker's analysis lacks the elegance found in J. Lagrange, C. G. J. Jacobi, L. O. Hesse, and R. F. A. Clebsch. For many years he had not kept up with the progress of geometry, so that many investigations in his last work had already received more general treatment on the part of others. The work contained, nevertheless, much that was fresh and original. The theory of complexes of the second degree, left unfinished by Plücker, was continued by Felix Klein, who greatly extended and supplemented the ideas of his master.

Ludwig Otto Hesse (1811–1874) was born at Königsberg, and studied at the university of his native place under F. W. Bessel, C. G. J. Jacobi, F. J. Richelot, and F. Neumann. Having taken the doctor's degree in 1840, he became docent at Königsberg, and in 1845 extraordinary professor there. Among his pupils at that time were *Heinrich Durège* (1821–1893) of Prague, Carl Neumann, R. F. A. Clebsch, G. R. Kirchhoff. The Königsberg period was one of great activity for Hesse. Every new discovery increased his zeal for still greater achievement. His earliest researches were on surfaces of the second order, and were partly synthetic. He solved the problem to construct any tenth point of such a surface when nine points are given. The analogous problem for a conic had been solved by Pascal by means of the hexagram. A difficult problem confronting mathematicians of this time was that of elimination. J. Plücker had seen that the

[1] Ad. Dronke, *Julius Plücker*, Bonn, 1871.

main advantage of his special method in analytic geometry lay in the avoidance of algebraic elimination. Hesse, however, showed how by determinants to make algebraic elimination easy. In his earlier results he was anticipated by J. J. Sylvester, who published his dialytic method of elimination in 1840. These advances in algebra Hesse applied to the analytic study of curves of the third order. By linear substitutions, he reduced a form of the third degree in three variables to one of only four terms, and was led to an important determinant involving the second differential coefficient of a form of the third degree, called the "Hessian." The "Hessian" plays a leading part in the theory of invariants, a subject first studied by A. Cayley. Hesse showed that his determinant gives for every curve another curve, such that the double points of the first are points on the second, or "Hessian." Similarly for surfaces (Crelle, 1844). Many of the most important theorems on curves of the third order are due to Hesse. He determined the curve of the 14th order, which passes through the 56 points of contact of the 28 bi-tangents of a curve of the fourth order. His great memoir on this subject (Crelle, 1855) was published at the same time as was a paper by J. Steiner treating of the same subject.

Hesse's income at Königsberg had not kept pace with his growing reputation. Hardly was he able to support himself and family. In 1855 he accepted a more lucrative position at Halle, and in 1856 one at Heidelberg. Here he remained until 1868, when he accepted a position at a technic school in Munich.[1] At Heidelberg he revised and enlarged upon his previous researches, and published in 1861 his *Vorlesungen über die Analytische Geometrie des Raumes, insbesondere über Flächen 2. Ordnung*. More elementary works soon followed. While in Heidelberg he elaborated a principle, his "Uebertragungsprincip." According to this, there corresponds to every point in a plane a pair of points in a line, and the projective geometry of the plane can be carried back to the geometry of points in a line.

The researches of Plücker and Hesse were continued in England by A. Cayley, G. Salmon, and J. J. Sylvester. It may be premised here that among the early writers on analytical geometry in England was **James Booth** (1806–1878), whose chief results are embodied in his *Treatise on Some New Geometrical Methods;* and **James MacCullagh** (1809–1846), who was professor of natural philosophy at Dublin, and made some valuable discoveries on the theory of quadrics. The influence of these men on the progress of geometry was insignificant, for the interchange of scientific results between different nations was not so complete at that time as might have been desired. In further illustration of this, we mention that M. Chasles in France elaborated subjects which had previously been disposed of by J. Steiner in Germany, and Steiner published researches which had been given by

[1] Gustav Bauer, *Gedächtnissrede auf Otto Hesse*, München, 1882.

Cayley, Sylvester, and Salmon nearly five years earlier. Cayley and Salmon in 1849 determined the straight lines in a cubic surface, and studied its principal properties, while Sylvester in 1851 discovered the pentahedron of such a surface. Cayley extended Plücker's equations to curves of higher singularities. Cayley's own investigations, and those of Max Nöther (1844–1921) of Erlangen, G. H. Halphen, Jules R. M. de la Gournérie (1814–1883) of Paris, A. Brill of Tübingen, lead to the conclusion that each higher singularity of a curve is equivalent to a certain number of simple singularities,—the node, the ordinary cusp, the double tangent, and the inflection. Sylvester studied the "twisted Cartesian," a curve of the fourth order. **Georges-Henri Halphen** (1844–1889) was born at Rouen, studied at the École Polytechnique in Paris, took part in the Franco-Prussian war, then became répétiteur and examinateur at the École Polytechnique. His investigations touched mainly the geometry of algebraic curves and surfaces, differential invariants, the theory of E. Laguerre's invariants, elliptic functions and their applications. A British geometrician, Salmon, helped powerfully towards the spreading of a knowledge of the new algebraic and geometric methods by the publication of an excellent series of text-books (*Conic Sections*, *Modern Higher Algebra*, *Higher Plane Curves*, *Analytic Geometry of Three Dimensions*), which have been placed within easy reach of German readers by a free translation, with additions, made by Wilhelm Fiedler of the Polytechnicum in Zurich. Salmon's *Geometry of Three Dimensions* was brought out in the fifth and sixth editions, with much new matter, by Reginald A. P. Rogers of Trinity College, Dublin, in 1912–1915. The next great worker in the field of analytic geometry was Clebsch.

Rudolf Friedrich Alfred Clebsch (1833–1872) was born at Königsberg in Prussia, studied at the university of that place under L. O. Hesse, F. J. Richelot, F. Neumann. From 1858 to 1863 he held the chair of theoretical mechanics at the Polytechnicum in Carlsruhe. The study of Salmon's works led him into algebra and geometry. In 1863 he accepted a position at the University of Giesen, where he worked in conjunction with Paul Gordan of Erlangen. In 1868 Clebsch went to Göttingen, and remained there until his death. He worked successively at the following subjects: Mathematical physics, the calculus of variations and partial differential equations of the first order, the general theory of curves and surfaces, Abelian functions and their use in geometry, the theory of invariants, and "Flächenabbildung." [1] He proved theorems on the pentahedron enunciated by J. J. Sylvester and J. Steiner; he made systematic use of "deficiency" (*Geschlecht*) as a fundamental principle in the classification of algebraic curves. The notion of deficiency was known before him to N. H. Abel and G. F. B. Riemann. At the beginning of his career,

[1] Alfred Clebsch, *Versuch einer Darlegung und Würdigung seiner wissenschaftlichen Leistungen von einigen seiner Freunde*, Leipzig, 1873.

Clebsch had shown how elliptic functions could be advantageously applied to Malfatti's problem. The idea involved therein, viz. the use of higher transcendentals in the study of geometry, led him to his greatest discoveries. Not only did he apply Abelian functions to geometry, but conversely, he drew geometry into the service of Abelian functions.

Clebsch made liberal use of determinants. His study of curves and surfaces began with the determination of the points of contact of lines which meet a surface in four consecutive points. G. Salmon had proved that these points lie on the intersection of the surface with a derived surface of the degree $11n-24$, but his solution was given in inconvenient form. Clebsch's investigation thereon is a most beautiful piece of analysis.

The representation of one surface upon another (*Flächenabbildung*), so that they have a (1, 1) correspondence, was thoroughly studied for the first time by Clebsch. The representation of a sphere on a plane is an old problem which drew the attention of Ptolemy, Gerard Mercator, J. H. Lambert, K. F. Gauss, J. L. Lagrange. Its importance in the construction of maps is obvious. Gauss was the first to represent a surface upon another with a view of more easily arriving at its properties. J. Plücker, M. Chasles, A. Cayley, thus represented on a plane the geometry of quadric surfaces; Clebsch and L. Cremona, that of cubic surfaces. Other surfaces have been studied in the same way by recent writers, particularly *Max Nöther* of Erlangen, *Angelo Armenante* (1844–1878) of Rome, *Felix Klein*, *Georg H. L. Korndörfer*, *Ettore Caporali* (1855–1886) of Naples, *H. G. Zeuthen* of Copenhagen. A fundamental question which has as yet received only a partial answer is this: What surfaces can be represented by a (1, 1) correspondence upon a given surface? This and the analogous question for curves was studied by Clebsch. Higher correspondences between surfaces have been investigated by A. Cayley and M. Nöther. Important bearings upon geometry has Riemann's theory of birational transformations. The theory of surfaces has been studied by **Joseph Alfred Serret** (1819–1885) professor at the Sorbonne in Paris, *Jean Gaston Darboux* of Paris, *John Casey* (1820–1891) of Dublin, *William Roberts* (1817–1883) of Dublin, *Heinrich Schröter* (1829–1892) of Breslau, *Elwin Bruno Christoffel* (1829–1900), professor at Zurich, later at Strassburg. Christoffel wrote on the theory of potential, on minimal surfaces, on the so-called transformation of Christoffel, of isothermic surfaces, on the general theory of curved surfaces. His researches on surfaces were extended by *Julius Weingarten* (1836–1910) of the University of Freiburg and *Hans von Mangoldt* of Aachen, in 1882. As we shall see more fully later, surfaces of the fourth order were investigated by E. E. Kummer, and Fresnel's wave-surface, studied by W. R. Hamilton, is a particular case of Kummer's quartic surface, with sixteen double points and sixteen singular tangent planes.[1]

[1] A. Cayley, Inaugural Address, 1883.

Prominent in these geometric researches was **Jean Gaston Darboux** (1842–1917). He was born at Nimes, founded in 1870, with the collaboration of *Guillaume Jules Hoüel* (1823–1886) of Bordeaux and *Jules Tannery*, the *Bulletin des sciences mathématiques et astronomiques*, and was for half a century conspicuous as a teacher. In 1900 he became permanent secretary of the Paris Academy of Sciences, in which position he was succeeded after his death by Emil Picard. By his researches, Darboux enriched the synthetic, analytic and infinitesimal geometries, as well as rational mechanics and analysis. He wrote *Leçons sur la théorie générale des surfaces et les applications géométriques du calcul infinitésimal*, Paris, 1887–1896,☼ and *Leçons sur les systèmes orthogonaux et les coordonnees curvilignes*, Paris, 1898. He investigated triply orthogonal systems of surfaces, the deformation of surfaces and rolling of applicable surfaces, infinitesimal deformation, spherical representation of surfaces, the development of the moving axes of co-ordinates, the use of imaginary geometric elements, the use of isotropic cylinders and developables;[1] he introduced pentaspherical coördinates.

Eisenhart says: "Darboux was a strong advocate of the use of imaginary elements in the study of geometry. He believed that their use was as necessary in geometry as in analysis. He has been impressed by the success with which they have been employed in the solution of the problem of minimal surfaces. From the very beginning he made use in his papers of the isotropic line, the null sphere (the isotropic cone) and the general isotropic developable. In his first memoir on orthogonal systems of surfaces he showed that the envelope of the surfaces of such a system, when defined by a single equation, is an isotropic developable. . . . Darboux gives to Édouard Combescure (1810–?) the credit of being the first to apply the considerations of kinematics to the study of the theory of surfaces with the consequent use of moving co-ordinate axes. But to Darboux we are indebted for a realization of the power of this method, and for its systematic development and exposition. . . . Darboux's ability was based on a rare combination of geometrical fancy and analytical power. He did not sympathize with those who use only geometrical reasoning in attacking geometrical problems, nor with those who feel that there is a certain virtue in adhering strictly to analytical processes. . . . In common with Monge he was not content with discoveries, but felt that it was equally important to make disciples. Like this distinguished predecessor he developed a large group of geometers, including C. Guichard, G. Koenigs, E. Cosserat, A. Demoulin, G. Tzitzeica, and G. Demartres. Their brilliant researches are the best tribute to his teaching."

Proceeding to the fuller consideration of recent developments, we

[1] *Am. Math. Monthly*, Vol. 24, 1917, p. 354. See L. P. Eisenhart's "Darboux's Contribution to Geometry" in *Bull. Am. Math. Soc.*, Vol. 24, 1918, p. 227.

quote from H. F. Baker's address before the International Congress held at Cambridge in 1912:[1] "The general theory of Higher Plane Curves . . . would be impossible without the notion of the genus of a curve. The investigation of Abel of the number of independent integrals in terms of which his integral sums can be expressed may thus be held to be of paramount importance for the general theory. This was further emphasized by G. F. B. Riemann's consideration of the notion of birational transformation as a fundamental principle. After this two streams of thought were to be seen. First R. F. A. Clebsch remarked on the existence of invariants for surfaces, analogous to the genus of a plane curve. This number he defined by a double integral; it was to be unaltered by birational transformation of the surface. Clebsch's idea was carried on and developed by M. Noether. But also A. Brill and Noether elaborated in a geometrical form the results for plane curves which had been obtained with transcendental considerations by N. H. Abel and G. F. B. Riemann. Then the geometers of Italy took up Noether's work with very remarkable genius, and carried it to a high pitch of perfection and clearness as a geometrical theory. In connection therewith there arose the important fact, which does not occur in Noether's papers, that it is necessary to consider a surface as possessing two genera; and the names of A. Cayley and H. G. Zeuthen should be referred to at this stage. But at this time another stream was running in France. E. Picard was developing the theory of Riemann integrals—single integrals, not double integrals—upon a surface. How long and laborious was the task may be judged from the fact that the publication of Picard's book occupied ten years—and may even then have seemed to many to be an artificial and unproductive imitation of the theory of algebraic integrals for a curve. In the light of subsequent events, Picard's book appears likely to remain a permanent landmark in the history of geometry. For now the two streams, the purely geometrical in Italy, the transcendental in France, have united. The results appear to me at least to be of the greatest importance." The work of E. Picard in question is the *Théorie des fonctions algébriques de deux variables indépendantes*, which was brought out in conjunction with Georges Simart between the years 1897 and 1906.✡

H. F. Baker proceeds to the enumeration of some individual results: Guido Castelnuovo of Rome has shown that the deficiency of the characteristic series of a linear system of curves upon a surface cannot exceed the difference of the two genera of the surface. Federigo Enriques of Bologna has completed this result by showing that for an algebraic system of curves the characteristic series is complete. Upon this result, and upon E. Picard's theory of integrals of the second

[1] *Proceed. of the 5th Intern. Congress*, Vol. I, Cambridge, 1913, p. 49. For more detail, consult H. F. Baker in the *Proceed. of the London Math. Soc.*, Vol. 12, 1912.

kind Francesco Severi of Padua has constructed a proof that the number of Picard integrals of the first kind upon a surface is equal to the difference of the genera. The names of M. G. Humbert of Paris and of G. Castelnuovo also arise here. Picard's theory of integrals of the third kind has given rise in F. Severi's hands to the expression of any curve lying on a surface linearly in terms of a finite number of fundamental curves. Enriques showed that the system of curves cut upon a plane by adjoint surfaces of order $n-3$, when n is the order of the fundamental surface, if not complete, has a deficiency not exceeding the difference of the genera of the surface. Severi has given a geometrical proof that this deficiency is equal to the difference of the genera, a result previously deduced by E. Picard, with transcendental considerations, from the assumption of the number of Picard integrals of the first kind. F. Enriques and G. Castlenuovo have shown that a surface which possesses a system of curves for which what may be called the canonical number, $2\pi-2-n$, where π is the genus of the curve and n the number of intersections of two curves of the system, is negative, can be transformed birationally to a ruled surface. On the analogy of the case of plane curves, and of surfaces in three dimensions, it appears very natural to conclude that if a rational relation, connecting, say, $m+1$ variables, can be resolved by substituting for the variables rational functions of m others, then these m others can be so chosen as to be rational functions of the $m+1$ original variables. F. Enriques has recently given a case, with $m=3$, for which this is not so. To this summary of results, given by H. F. Baker, should be added that he himself has made contributions, particularly on a cubic surface and the curves which lie thereon. In reducing singularities, the Italians and French use methods of projecting from space of higher dimension which were perhaps first used in 1887 by W. K. Clifford.

A publication of wide scope on collineations and correlations is *Die Lehre von den geometrischen Verwandtschaften*, in four volumes, 1908–09, written by Rudolf Sturm (1841–1919) of the University of Breslau.

The theory of straight lines upon a cubic surface was first studied by A. Cayley and G. Salmon[1] in 1849. Cayley pointed out that there was a definite number of such lines, while Salmon found that there were exactly 27 of them. "Surely with as good reason," says J. J. Sylvester, "as had Archimedes to have the cylinder, cone and sphere engraved on his tombstone might our distinguished countrymen leave testamentary directions for the cubic eikosiheptagram to be engraved on theirs." Nor would such engraving be impossible, for in 1869 Christian Wiener made a model of a cubic surface showing 27 real lines lying upon it. J. Steiner, in 1856, studied the purely geometric

[1] These historic data are taken from A. Henderson, *The Twenty-seven Lines upon the Cubic Surface*, Cambridge, 1911, which gives bibliography and details.

theory of cubic surfaces. This was done later also by L. Cremona and R. Sturm, between whom in 1866 the "Steiner prize" was divided. An elegant notation was invented by Andrew Hart, but the notation which has met with general adoption was advanced by L. Schläfli of Bern in 1858; it is that of the double six. Schläfli's double six theorem, proved by him and by many others since, is as follows: "Given five lines a, b, c, d, e which meet the same straight line X; then may any four of the five lines be intersected by another line. Suppose that A, B, C, D, E are the other lines intersecting (b, c, d, e), (c, d, e, a), (d, e, a, b), (e, a, b, c), and (a, b, c, d) respectively. Then A, B, C, D, E will all be met by one other straight line x."

L. Schläfli first considered a division of the cubic surfaces into species, in regard to the reality of the 27 lines. His final classification was adopted by A. Cayley. In 1872 R. F. A. Clebsch constructed a model of the *diagonal surface* with 27 real lines, while F. Klein "established the fact that by the principle of continuity all forms of real surfaces of the third order can be derived from the particular surface having four conical points;" he exhibited a complete set of models of cubic surfaces at the World's Fair in Chicago in 1894. In 1869 C. F. Geiser showed that "the projection of a cubic surface from a point upon it, on a plane of projection parallel to the tangent plane at that point, is a quartic curve; and that every quartic curve can be generated in this way." "The theory of *varieties* of the third order," says A. Henderson, "that is to say, curved geometric forms of three dimensions contained in a space of four dimensions, has been the subject of a profound memoir by *Corrado Segre* (1887) of Turin. The depth and fecundity of this paper is evinced by the fact that a large proportion of the propositions upon the plane quartic and its bitangents, Pascal's theorem, the cubic surface and its 27 lines, Kummer's surface and its configuration of sixteen singular points and planes, and on the connection between these figures, are derivable from propositions relating to Segre's cubic variety, and the figure of six points or spaces from which it springs. Other investigators into the properties of this beautiful and important locus in space of four dimensions and some of its consequences are G. Castelnuovo and H. W. Richmond."

In 1869 C. Jordan first proved "that the group of the problem of the trisection of hyperelliptic functions of the first order is isomorphic with the group of the equation of the 27th degree, on which the 27 lines of the general surface of the third degree depend." In 1887 F. Klein sketched the effective reduction of the one problem to the other, while H. Maschke, H. Burkhardt, and A. Witting completed the work outlined by Klein. The Galois group of the equation of the 27 lines was investigated also by L. E. Dickson, F. Kühnen, H. Weber, E. Pascal, E. Kasner and E. H. Moore.

Surfaces of the fourth order have been studied less thoroughly than those of the third. J. Steiner worked out properties of a surface of the fourth order in 1844 when he was on a journey to Italy; that surface bears this name, and later received the attention of E. E. Kummer. In 1850 Thomas Weddle [1] remarked that the locus of the vertex of a quadric cone passing through six given points is a quartic surface and not a twisted cubic as M. Chasles had once stated. A. Cayley gave a symmetric equation of the surface in 1861. Thereupon Chasles in 1861 showed that the locus of the vertex of a cone which divides six given segments harmonically is also a quartic surface; this more general surface was identified by Cayley with the Jacobian of four quadrics, the Weddle surface corresponding to the case in which the four quadrics have six common points. Properties of the Weddle surface were studied also by H. Bateman (1905). The plane section of a Weddle surface is not an arbitrary quartic curve, but one for which an invariant vanishes. Frank Morley proved that the curve contains an infinity of configurations B^6, where it is cut by the lines on the surface.

In 1863 and 1864, E. E. Kummer entered upon an intensive study of surfaces of the fourth order. Noted is the surface named after him which has 16 nodes. The various shapes it can assume have been studied by Karl Rohn of Leipzig. It has received the attention of many mathematicians, including A. Cayley, J. G. Darboux, F. Klein, H. W. Richmond, O. Bolza, H. F. Baker, and J. I. Hutchinson.[2] It has been known for some time that Fresnel's wave surface is a case of Kummer's sixteen nodal quartic surface; also it is known that the surface of a dynamical medium possessing certain general properties is a type of Kummer's surface which can he derived from Fresnel's surface by means of a homogeneous strain. Kummer's quartic surface as a wave surface is treated by H. Bateman (1909). The general Kummer's surface appears to be the wave surface for a medium of a purely ideal character.

F. R. Sharpe and C. F. Craig of Cornell University have studied birational transformations which leave the Kummer and Weddle surfaces invariant, by the application of a theory due to F. Severi (1906).

Quintic surfaces have been investigated at intervals, since 1862, principally by L. Cremona, H. A. Schwarz, A. Clebsch, M. Nöether, R. Sturm, J. G. Darboux, E. Caporali, A. Del Re, E. Pascal, John E. Hill and A. B. Basset. No serious attempt has been made to enumerate the different forms of these surfaces.

Ruled surfaces with isotropic generators have been considered by G. Monge, J. A. Serret, S. Lie and others. L. P. Eisenhart of Princeton determines such a surface by the curve in which it is cut by a plane and the directions of the projections on the plane of the generators

[1] *Camb. & Dublin Math. Jour.*, Vol. 5, 1850, p. 69.

[2] Consult R. W. H. T. Hudson (1876–1904), *Kummer's Quartic Surface*, Cambridge, 1905.

of the surface. In this way a ruled surface of this type is determined by a set of lineal elements, in a plane, depending on one parameter.

While the classification of cubic curves was given by I. Newton and their general theory was well under way two centuries ago, the theory of quartic curves was not pursued vigorously until the time of J. Steiner and L. O. Hesse. Neglecting the classification of quartic curves due to L. Euler, G. Cramer and E. Waring, new classifications have been made, either according to their genus (Geschlecht) 3, 2, 1, 0, or according to topologic considerations studied by A. Cayley in 1865, H. G. Zeuthen (1873), Christian Crone (1877) and others. J. Steiner in 1855 and L. O. Hesse began researches on the 28 double tangents of a general quartic; 24 inflections were found, of which G. Salmon conjectured and H. G. Zeuthen proved that at most 8 are real. An enumeration (containing nearly 200 graphs) of the fundamental forms of quartic curves "when projected so as to cut the line infinity the least possible number of times" was given in 1896 by Ruth Gentry (1862–1917), then of Bryn Mawr College.

Curves of the fourth order have received attention for many years. More recently a good deal has been written on special curves of the fifth order by Frank Morley, Alfred B. Basset, Virgil Snyder, Peter Field, and others.

Gino Loria of the University of Genoa, who has written extensively on the history of geometry, and the history of curves in particular, has advanced a theory of panalgebraic curves, which are in general transcendental curves. By definition, a panalgebraic curve must satisfy a certain differential equation. A book of reference on curves was published by *Gomes Teixeira* in 1905 at Madrid under the title *Tratado de las curvas especiales notables.*✲

The infinitesimal calculus was first applied to the determination of the measure of curvature of surfaces by J. Lagrange, L. Euler, and Jean Baptiste Marie Meusnier (1754–1793) of Paris, noted for his military as well as scientific career. Meusnier's theorem, relating to curves drawn on an arbitrary surface, was extended by S. Lie and in 1908 by E. Kasner. The researches of G. Monge and F. P. C. Dupin were eclipsed by the work of K. F. Gauss, who disposed of this difficult subject in a way that opened new vistas to geometricians. His treatment is embodied in the *Disquisitiones generales circa superficies curvas* (1827) and *Untersuchungen über Gegenstände der höheren Geodäsie* of 1843 and 1846. In 1827 he established the idea of curvature as it is understood to-day. Both before and after the time of Gauss various definitions of curvature of a surface had been advanced by L. Euler, Meusnier, Monge, and Dupin, but these did not meet with general adoption. From Gauss' measure of curvature flows the theorem of *Johann August Grunert* (1797–1872), professor in Greifswald, and founder in 1841 of the *Archiv der Mathematik und Physik*, that the arithmetical mean of the radii of curvature of all normal sections through a point is the radius of a sphere which has the same measure of curvature as has the surface at that point. Gauss's deduction of the formula of curvature was simplified through the use of determinants by *Heinrich Richard Baltzer* (1818–1887) of Giessen.[1] Gauss obtained an interesting theorem that if one surface be developed (*abgewickelt*) upon another, the measure of curvature remains unaltered at each point. The question whether two surfaces having the same curvature in corresponding points can be unwound, one upon the other, was answered by F. Minding in the affirmative only when the curvature is constant. Surfaces of constant and negative curvature were called pseudo-spherical surfaces by E. Beltrami in 1868, in order, as he says, "to avoid circumlocution." The case of variable curvature is difficult, and was studied by F. Minding, *Joseph Liouville* (1809–1882) of the Polytechnic School in Paris, Ossian Bonnet (1819–1892) of Paris. Gauss's measure of curvature, expressed as a function of curvilinear co-ordinates, gave an impetus to the study of differential-invariants, or differential-parameters, which have been investigated by C. G. J. Jacobi, C. Neumann, Sir James Cockle (1819–1895) of London, G. H. Halphen, and elaborated into a general theory by E. Beltrami, S. Lie, and others. Beltrami showed also the connection between the measure of curvature and the geometric axioms.

In 1899 *Claude Guichard* of Rennes announced two theorems relating to a quadric of revolution which marked a new epoch in the theory of deformation of surfaces. Researches along this line by Guichard and Luigi Bianchi of Pisa are embodied in the second edition of Bianchi's *Lezioni di geometria differenziale*, Pisa, 1902. Another treatise on metric differential geometry was brought out in 1908 by *Reinhold v. Lilienthal* of the University of Münster. Not only does he give geometric interpretations of the first and second derivatives by means of the tangent and the circle of curvature, but he revives a notion due to *Abel Transon* (1805–1876) of Paris which gives a geometric interpretation of the third derivative in terms of the abberancy of a curve and the axis of abberancy. A still later work is the *Treatise on Differential Geometry of Curves and Surfaces* (1909) by *L. P. Eisenhart* of Princeton which possesses the interesting feature of movable axes (the so-called "moving trihedrals" used extensively by J. G. Darboux), applied to twisted curves as well as surfaces; he gives the four transformations of surfaces of constant curvature, due to N. Hatzidakis of Athens, L. Bianchi of Pisa, A. V. Bäcklund of Lund, and S. Lie. Eisenhart developed a theory of transformations of a conjugate system of curves on any surface into conjugate systems on other surfaces, and also of transformations of conjugate nets on two-dimensional spreads in space of any order.[2]

[1] August Haas, *Versuch einer Darstellung der Geschichte des Krümmungsmasses*, Tübingen, 1881.

[2] *Bull. Am. Math. Soc.*, Vol. 24, 1917, p. 68.

The metric part of differential geometry occupied the attention of mathematicians since the time of G. Monge and K. F. Gauss, and has reached a high degree of perfection. Less attention has been given until recently to projective differential geometry, particularly to the differential geometry of surfaces. G. H. Halphen started with the equation $y=f(x)$ of a curve and determined functions of y, dy/dx, d^2y/dx^2, etc., which are left invariant when x and y are subjected to a general projective transformation. His early formulation of the problem is unsymmetrical and unhomogeneous. Using a certain system of partial differential equations and the geometrical theory of semi-covariants, E. J. Wilczynski obtained homogeneous forms, such forms being deduced later also by Halphen.[1] Wilczynski treats of the projective differential geometry of curves and ruled surfaces, these surfaces being prerequisite for his theory of space curves. Wilczynski treated ruled surfaces by a system of two linear homogeneous differential equations of the second order. The method was extended to five-dimensional space by E. B. Stouffer of the University of Kansas.[2] Developable surfaces were studied by W. W. Denton of the University of Illinois. Belonging to projective differential geometry are J. G. Darboux's conjugate triple systems which are generalized notions of the orthogonal triple systems. The projective differential geometry of triple systems of surfaces, of one-parameter families of space curves and conjugate nets on a curved surface, and allied topics, were studied by **Gabriel Marcus Green** (1891–1919), of Harvard University.

Differential projective geometry of hyperspace was greatly advanced by C. Guichard who introduced two elements depending on two variables; they are the *reseau* and the *congruence*. Differential geometry of hyperspace was greatly enriched since 1906 by Corrado Segre of Turin, and by other geometers of Italy, particularly Gino Fano of Turin and Federigo Enriques of Bologna;[3] also by A. Ranum of Cornell, C. H. Sisam then of Illinois and C. L. E. Moore of the Massachusetts Institute of Technology.

The use of vector analysis in differential geometry goes back to H. G. Grassmann and W. R. Hamilton, to their successors P. G. Tait, C. Maxwell, C. Burali-Forti, R. Rothe and others. These men have introduced the terms "grad," "div," "rot." A geometric study of trajectories with the aid of analytic and chiefly contact transformations was made by Edward Kasner of Columbia University in his Princeton Colloquium lectures of 1909 on the "differential-geometric aspects of dynamics."

[1] See E. J. Wilczynski in *New Haven Colloquium, 1906*. New Haven, 1910, p. 156; also his *Projective Differential Geometry of Curves and Ruled Surfaces*, Leipzig, 1906.

[2] *Bull. Am. Math. Soc.*, Vol. 18, p. 444.

[3] See Enrico Bompiani in *Proceed. 5th Intern. Congress, Cambridge*, Cambridge, 1913, Vol. II, p. 22.

Analysis Situs

Various researches have been brought under the head of "analysis situs." The subject was first investigated by G. W. Leibniz, and was later treated by L. Euler who was interested in the problem to cross all of the seven bridges over the Pregel river at Königsberg without passing twice over any one, then by K. F. Gauss, whose theory of knots (*Verschlingungen*) has been employed more recently by Johann Benedict Listing (1808–1882) of Göttingen, Oskar Simony of Vienna, F. Dingeldey of Darmstadt, and others in their "topologic studies." P. G. Tait was led to the study of knots by Sir William Thomson's theory of vortex atoms. Through Rev. T. P. Kirkman who had studied the properties of polyhedra, Tait was led to study knots also by the polyhedral method; he gave the number of forms of knots of the first ten orders of knottiness. Higher orders were treated by Kirkman and C. N. Little. **Thomas Penyngton Kirkman** [1] (1806–1895) was born at Bolton, near Manchester. During boyhood he was forced to follow his father's business as dealer in cotton and cotton waste. Later he tore away, entered the University of Dublin, then became vicar of a parish in Lancashire. As a mathematician he was almost entirely self-taught. He wrote on pluquaternions involving more imaginaries than i, j, k, on group theory, on mathematical mnemonics producing what De Morgan called "the most curious crocket I ever saw," on the problem of the "fifteen school girls" who walk out three abreast for seven days, where it is required to arrange them daily so that no two shall walk abreast more than once. This problem was studied also by A. Cayley and Sylvester, and is related to researches of J. Steiner.

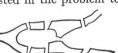

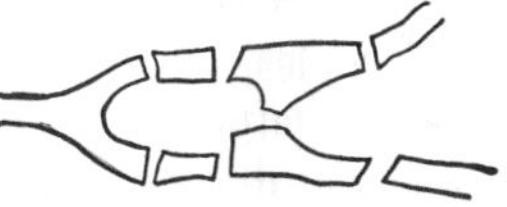

Another unique problem was the one on the coloring of maps, first mentioned by A. F. Möbius in 1840 and first seriously considered by Francis Guthrie and A. De Morgan. How many colors are necessary to draw any map so that no two countries having a line of boundary in common shall appear in the same color? Four different colors are found experimentally to be necessary and sufficient, but the proof is difficult. A. Cayley in 1878 declared that he had not succeeded in obtaining a general proof. Nor have the later demonstrations by A. B. Kempe, P. G. Tait, P. J. Heawood of the University of Durham, W. E. Story of Clark University, and J. Petersen of Copenhagen removed the difficulty.[2] Tait's proof leads to the interesting conclusion that four colors may not be sufficient for a map drawn on a multiply-connected surface like that of an anchor ring. Further studies of maps on such surfaces, and of the problem in general, are

[1] A. Macfarlane, *Ten British Mathematicians*, 1916, p. 122.

[2] W. Ahrens, *Unterhaltungen und Spiele*, 1901, p. 340.

due to O. Veblen (1912) and G. D. Birkhoff (1913). On a surface of genus zero "it is not known whether or not only four colors always suffice." A similar question considers the maximum possible number of countries, when every country touches every other along a line. Lothar Heffter wrote on this conundrum, in 1891 and again in later articles, as did also A. B. Kempe and others. In the hands of Riemann the analysis situs had for its object the determination of what remains unchanged under transformations brought about by a combination of infinitesimal distortions. In continuation of his work, Walter Dyck of Munich wrote on the analysis situs of three-dimensional spaces. Researches of this sort have important bearings in modern mathematics, particularly in connection with correspondences and differential equations.[1]

Intrinsic Co-ordinates

As a reaction against the use of the arbitrary Cartesian and polar co-ordinates there came the suggestion from the philosophers K. C. F. Krause (1781–1832), A. Peters (1803–1876) that magnitudes inherent to a curve be used, such as s, the length of arc measured from a fixed point, and φ, the angle which the tangent at the end of s makes with a fixed tangent. *William Whewell* (1794–1866) of Cambridge, the author of the *History of the Inductive Sciences*, 1837–1838, introduced in 1849 the name "intrinsic equation" and pointed out its use in studying successive evolutes and involutes. The method was used by William Walton (1813–1901) of Cambridge, J. J. Sylvester in 1868, J. Casey in 1866, and others. Instead of using s and φ, other writers have introduced the radius of curvature ρ, and have used either s and ρ, or φ and ρ. The co-ordinates (φ, ρ) were employed by L. Euler and several nineteenth century mathematicians, but altogether the co-ordinates (s, ρ) have been used most. The latter were used by L. Euler in 1741, by Sylvestre François Lacroix (1765–1843), by Thomas Hill (1818–1891, who was at one time president of Harvard College), and in recent years especially by Ernesto Cesàro of the University of Naples who published in 1896 his *Geometria intrinseca* which was translated into German in 1901 by G. Kowalewski under the title, *Vorlesungen über natürliche Geometrie*.[2] Researches along this line are due also to **Amédée Mannheim** (1831–1906) of Paris, the designer of a well-known slide rule.

The application of intrinsic or natural co-ordinates to surfaces is less common. Edward Kasner[3] said in 1904 that in the "theory of surfaces, natural co-ordinates may be introduced so as to fit into the

[1] See J. Hadamard, *Four Lectures on Mathematics delivered at Columbia University in 1911*, New York, 1915, Lecture III.

[2] Our information is drawn from E. Wölffing's article on "Natürliche Koordinaten" in *Bibliotheca mathematica*, 3. S., Vol. I, 1900, pp. 142–159.

[3] *Bull. Am. Math. Soc.*, Vol. 11, 1905, p. 303.

so-called geometry of a flexible but inextensible surface, originated by K. F. Gauss, in which the criterion of equivalence is applicability or, according to the more accurate phraseology of A. Voss, isometry. Intrinsic co-ordinates must then be invariant with respect to bending. . . . The simplest example of a complete isometric group is the group typified by the plane, consisting of all the developable surfaces. In this case the equations of the group may be obtained explicitly, in terms of eliminations, differentiations and quadratures. . . . Until the year 1866, not a single case analogous to that of the developable surfaces was discovered. Julius Weingarten, by means of his theory of evolutes, then succeeded in determining the complete group of the catenoid and of the paraboloid of revolution, and, some twenty years later, a fourth group defined in terms of minimal surfaces. During the past decade, the French geometers have concentrated their efforts in this field mainly on the arbitrary paraboloid (and to some extent on the arbitrary quadric). The difficulties even in this extremely restricted and apparently simple case are great, and are only gradually being conquered by the use of almost the whole wealth of modern analysis and the invention of new methods which undoubtedly have wider fields of application. The results obtained exhibit, for example, connections with the theories of surfaces of constant curvature, isometric surfaces, Backlund transformations, and motions with two degrees of freedom. The principal workers are J. G. Darboux, E. J. B. Goursat, L. Bianchi, A. L. Thybaut, E. Cosserat, M. G. Servant, C. Guichard, and L. Raffy."

Definition of a Curve

The theory of sets of points, originated by G. Cantor, has given rise to new views on the theory of curves and on the meaning of content. What is a curve? Camille Jordan in his *Cours d'analyse* defined it tentatively as a "continuous line." W. H. Young and Grace Chisholm Young in their *Theory of Sets of Points*, 1906,✡ p. 222, define a "Jordan curve" as "a plane set of points which can be brought into continuous (1, 1) correspondence with the points of a closed segment (a, z) of a straight line." A circle is a closed Jordan curve. Jordan asked the question, whether it was possible for a curve to fill up a space. G. Peano answered that a "continuous line" may do so and constructed in *Math. Annalen*, Vol. 36, 1890, the so-called "space-filling curve" (the "Peano curve") to fortify his assertion. His mode of construction has been modified in several ways since. The most noted of these are due to E. H. Moore [1] and D. Hilbert. In 1916 R. L. Moore of the University of Pennsylvania proved that every two points of a continuous curve, no matter how crinkly, can be joined by a simple continuous arc that lies wholly in the curve. As it does not

[1] *Trans. Am. Math. Soc.*, Vol. I, 1900, pp. 72–90.

seem desirable to depart from our empirical notions so far as to allow the term curve to be applied to a region, more restricted definitions of it become necessary. C. Jordan demanded that a curve $x=\varphi(t)$, $y=\psi(t)$ should have no double points in the interval $a \leqq t \leqq b$. Schönflies regards a curve as the frontier of a region. O. Veblen defines it in terms of order and linear continuity. W. H. Young and Grace C. Young in their *Theory of Sets of Points* define a curve as a plane set of points, dense nowhere in the plane and bearing other restrictions, yet such that it may consist of a net-work of arcs of Jordan curves.

Other curves of previously unheard of properties were created as the result of the generalization of the function concept. The continuous curve represented by $y=\sum_{n=0}^{n=\infty} b^n \cos \pi(a^n x)$, where a is an even integer >1, b a real positive number <1, was shown by Weierstrass[1] to possess no tangents at any of its points when the product ab exceeds

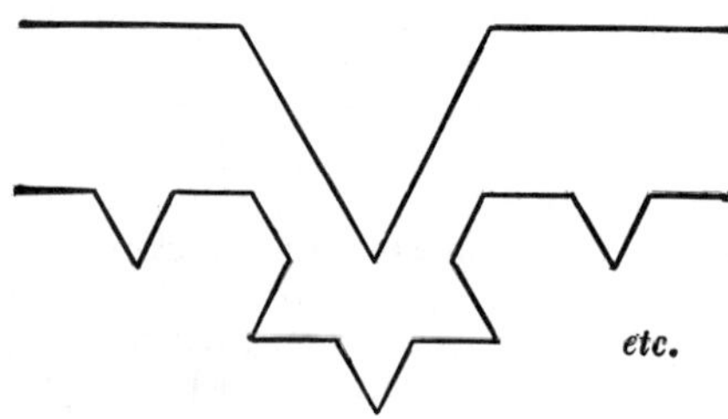

a certain limit; that is, we have here the startling phenomenon of a continuous function which has no derivative. As Christian Wiener explained in 1881, this curve has countless oscillations within every finite interval. An intuitively simpler curve was invented by Helge v. Koch of the University of Stockholm in 1904 (*Acta math.*, Vol. 30, 1906, p. 145) which is constructed by elementary geometry, is continuous, yet has no tangent at any of its points; the arc between any two of its points is infinite in length. While this curve has been represented analytically, no such representation has yet been found for the so-called H-curve of Ludwig Boltzmann (1844–1906) of Vienna, in *Math. Annalen*, Vol. 50, 1898, which is continuous, yet tangentless. The adjoining figure shows its construction. Boltzmann used it to visualize theorems in the theory of gases.

Fundamental Postulates

The foundations of mathematics, and of geometry in particular, received marked attention in Italy. In 1889 G. Peano took the novel

[1] P. du Bois-Reymond "Versuch einer Klassification der willkürlichen Funktionen reeller Argumente," *Crelle*, Vol. 74, 1874, p. 29.

view that geometric elements are *mere things*, and laid down the principle that there should be as few undefined symbols as possible. In 1897–9 his pupil *Mario Pieri* (1860–1904) of Catania used only two undefined symbols for projective geometry and but two for metric geometry. In 1894 Peano considered the independence of axioms. By 1897 the Italian mathematicians had gone so far as to make it a postulate that points are classes. These fundamental features elaborated by the Italian school were embodied✡ by David Hilbert of Göttingen, along with important novel considerations of his own, in his famous *Grundlagen der Geometrie*, 1899. A fourth enlarged edition of this appeared in 1913.✡ E. B. Wilson says in praise of Hilbert: "The archimedean axiom, the theorems of B. Pascal and G. Desargues, the analysis of segments and areas, and a host of things are treated either for the first time or in a new way, and with consummate skill. We should say that it was in the technique rather than in the philosophy of geometry that Hilbert created an epoch."[1] In Hilbert's space of 1899 are not all the points which are in our space, but only those that, starting from two given points, can be constructed with ruler and compasses. In his space, remarks Poincaré, there is no angle of 10°. So in the second edition of the *Grundlagen*, Hilbert introduced the assumption of completeness, which renders his space and ours the same. Interesting is Hilbert's treatment of non-Archimedean geometry where all his assumptions remain true save that of Archimedes, and for which he created a system of non-Archimedean numbers. This non-Archimedean geometry was first conceived by *Giuseppe Veronese* (1854–1917), professor of geometry at the University of Padua. Our common space is only a part of non-Archimedean space. Non-Archimedean theories of proportion were given in 1902 by A. Kneser of Breslau and in 1904 by P. J. Mollerup of Copenhagen. Hilbert devoted in his *Grundlagen* a chapter to Desargues' theorem. In 1902 F. R. Moulton of Chicago outlined a simple non-Desarguesian plane geometry.

In the United States, George Bruce Halsted based his *Rational Geometry*, 1904, upon Hilbert's foundations. A second, revised edition appeared in 1907. One of Hilbert's pupils, Max W. Dehn, showed that the omission of the axiom of Archimedes (Eudoxus) gives rise to a semi-Euclidean geometry in which similar triangles exist and their angle sum is two right angles, yet an infinity of parallels to any straight line may be drawn through any given point.

Systems of axioms upon which to build projective geometry were first studied more particularly by the Italian school—G. Peano, M. Pieri, Gino Fano of Turin. This subject received the attention also of Theodor Vahlen of Vienna and Friedrich Schur of Strassburg. Axioms of descriptive geometry have been considered mainly by

[1] *Bull. Am. Math. Soc.*, Vol. 11, 1904, p. 77. Our remarks on the Italian school are drawn from Wilson's article.

Italian and American mathematicians, and by D. Hilbert. The introduction of order was achieved by G. Peano by taking the *class* of points which lie *between* any two points as the fundamental idea, by G. Vailati and later by B. Russell, on the fundamental conception of a class of *relations* or class of points on a straight line, by O. Veblen (1904) on the study of the properties of one single three-term relation of order. A. N. Whitehead [1] refers to O. Veblen's method: "This method of conceiving the subject results in a notable simplification, and combines advantages from the two previous methods." While D. Hilbert has six undefined terms (point, straight line, plane, between, parallel, congruent) and twenty-one assumptions, Veblen gives only two undefined terms (point, between) and only twelve assumptions. However, the derivation of fundamental theorems is somewhat harder by Veblen's axioms. R. L. Moore showed that any plane satisfying Veblen's axioms I–VIII, XI is a number-plane and contains a system of continuous curves such that, with reference to these curves regarded as straight lines, the plane is an ordinary Euclidean plane.

In 1907, Oswald Veblen and J. W. Young gave a completely independent set of assumptions for projective geometry, in which points and undefined classes of points called lines have been taken as the undefined elements. Eight of these assumptions characterize general projective spaces; the addition of a ninth assumption yields properly projective spaces.[2]

Axioms for line geometry based upon the "line" as an undefined element and "intersection" as an undefined relation between unordered pairs of elements, were given in 1901 by M. Pieri of Catania, and in simpler form, in 1914 by E. R. Hedrick and L. Ingold of the University of Missouri.

Text-books built upon some such system of axioms and possessing great generality and scientific interest have been written by Federigo Enriques of the University of Bologna in 1898, and by O. Veblen and J. W. Young in 1910.

Geometric Models

Geometrical models for advanced students began to be manufactured about 1879 by the firm of L. Brill in Darmstadt. Many of the early models, such as Kummer's surface, twisted cubics, the tractrix of revolution, were made under the direction of F. Klein and Alexander von Brill. Since about 1890 this firm developed into that of Martin Schilling (1866–1908) of Leipzig. The catalogue of the firm for 1911 described some 400 models. Since 1905 the firm of B. G. Teubner in Leipzig has offered models designed by Hermann

[1] *The Axioms of Descriptive Geometry*, Cambridge, 1907, p. 2.
[2] *Bull. Am. Math. Soc.*, Vol. 14, 1908, p. 251.

Wiener, many of which are intended for secondary instruction. Valuable in this connection is the *Katalog mathematischer und mathematisch-physikalischer Modelle, Apparate und Instrumente* by Walter v. Dyck, professor in Munich. At the Napier tercentenary celebration in Edinburgh, in 1914, Crum Brown exhibited models of various sorts, including models of cubic and quartic surfaces, interlacing surfaces, regular solids and related forms, and thermodynamic models; D. M. Y. Sommerville displayed models of the projection, on three-dimensional space, of a four-dimensional figure; Lord Kelvin's tide-calculating machine illustrated the combination of simple harmonic motions.[1]

Algebra

The progress of algebra in recent times may be considered under three principal heads: the study of fundamental laws and the birth of new algebras, the growth of the theory of equations, and the development of what is called modern higher algebra.

The general theory of a^b, where both a and b are complex numbers, was outlined by L. Euler in 1749 in his paper, *Recherches sur les racines imaginaries des équations*, but it failed to command attention. At the beginning of the nineteenth century the theory of the general power was elaborated in Germany, England, France and the Netherlands. In the early history of logarithms of positive numbers it was found surprising that logarithms were defined independently of exponents. Now we meet a second surprise in finding that the theory of a^b is made to depend upon logarithms. Historically the logarithmic concept is the more primitive. The general theory of a^b was developed by *Martin Ohm* (1792–1872), professor in Berlin and a brother of the physicist to whom we owe "Ohm's law." Martin Ohm is the author of a much criticised series of books, *Versuch eines vollkommen consequenten Systems der Mathematik*, Nürnberg, 1822–1852. Our topic was treated in the second volume, dated 1823, second edition 1829. After having developed the Eulerian theory of logarithms Ohm takes up a^x, where $a=p+qi$, $x=\alpha+\beta i$. Assuming e^z as always single-valued and letting $v=\sqrt{p^2+q^2}$, $\log a=Lv+(\pm 2m\pi+\varphi)i$, he takes $a^x=e^{x \log a}=e^{\alpha \mathrm{L}v-\beta(\pm 2m\pi+\phi)}.\{\cos[\beta. Lv+\alpha(\pm 2m\pi+\phi)]+i. \sin[\beta Lv+\alpha(\pm 2m\pi+\phi)]\}$, where $m=0, +1, +2, \ldots$ and L. signifies the tabular logarithm. Thus the general power has an infinite number of values, but all are of the form $a+bi$. Ohm shows (1) that all of the values (infinite in number) are equal when x is an integer, (2) that there are n distinct values when x is a real, rational fraction, (3) that some of the values are equal, though the number of distinct values is infinite, when x is real but irrational, (4) that the values are all distinct when x is imaginary. He inquires next, how the formulas (A) $a^x.a^y=a^{x+y}$, (B) a^x

[1] Consult E. M. Horsburgh *Handbook of the Exhibition of Napier relics*, etc., 1914, p. 302.

$\div a^y = a^{x-y}$, (C) $a^x . b^x = (ab)^x$, (D) $a^x \div b^x = (a \div b)^x$, (E) $(a^x)^y = a^{xy}$ apply to the general exponent a^x, and finds that (A), (B) and (E) are incomplete equations, since the left members have "many, many more" values than the right members, although the right-hand values (infinite in number) are all found among the "infinite times infinite" values on the left; that (C) and (D) are complete equations for the general case. A failure to recognize that equation E is incomplete led Thomas Clausen (1801–1885) of Altona to a paradox (*Crelle's Journal*, Vol. 2, 1827, p. 286) which was stated by E. Catalan in 1869 in more condensed form, thus: $e^{2m\pi i} = e^{2n\pi i}$, where m and n are distinct integers. Raising both sides to the power $\frac{1}{2}$, there results the absurdity, $e^{-m\pi} = e^{-n\pi}$. Ohm introduced a notation to designate some particular value of a^x, but he did not introduce especially the particular value which is now called the "principal value." Otherwise his treatment of the general power is mainly that of the present time, except, of course, in the explanation of the irrational. From the general power Ohm proceeds to the general logarithm, having a complex number as its base. It is seen that the Eulerian logarithms served as a stepladder leading to the theory of the general power; the theory of the general power, in turn, led to a more general theory of logarithms having a complex base.

The *Philosophical Transactions* (London, 1829) contain two articles on general powers and logarithms—one by John Graves, the other by John Warren of Cambridge. Graves, then a young man of 23, was a class-fellow of William Rowan Hamilton in Dublin. Graves became a noted jurist. Hamilton states that reflecting on Graves's ideas on imaginaries led him to the invention of quaternions. Graves obtains $\log 1 = (2m'\pi i)/(1+2m\pi i)$. Thus Graves claimed that general logarithms involve two arbitrary integers, m and m', instead of simply one, as given by Euler. Lack of explicitness involved Graves in a discussion with A. De Morgan and G. Peacock, the outcome of which was that Graves withdrew the statement contained in the title of his paper and implying an error in the Eulerian theory, while De Morgan admitted that if Graves desired to extend the idea of a logarithm so as to use the base $e^{1+2m\pi i}$, there was no error involved in the process. Similar researches were carried on by A. J. H. Vincent at Lille, D. F. Gregory, De Morgan, W. R. Hamilton and G. M. Pagani (1796–1855), but their general logarithmic systems, involving complex numbers as bases, failed of recognition as useful mathematical inventions.[1] We pause to sketch the life of De Morgan.

Augustus De Morgan (1806–1871) was born at Madura (Madras), and educated at Trinity College, Cambridge. For the determination of the year of his birth (assumed to be in the nineteenth century) he proposed the conundrum, "I was x years of age in the year x^2." His

[1] For references and fuller details see F. Cajori in *Am. Math. Monthly*, Vol. 20, 1913, pp. 175–182.

scruples about the doctrines of the established church prevented him from proceeding to the M. A. degree, and from sitting for a fellowship. It is said of him, "he never voted at an election, and he never visited the House of Commons, or the Tower, or Westminster Abbey." In 1828 he became professor at the newly established University of London, and taught there until 1867, except for five years, from 1831–1835. He was the first president of the London Mathematical Society which was founded in 1865 De Morgan was a unique, manly character, and pre-eminent as a teacher. The value of his original work lies not so much in increasing our stock of mathematical knowledge as in putting it all upon a more logical basis. He felt keenly the lack of close reasoning in mathematics as he received it. He said once: "We know that mathematicians care no more for logic than logicians for mathematics. The two eyes of exact science are mathematics and logic: the mathematical sect puts out the logical eye, the logical sect puts out the mathematical eye; each believing that it can see better with one eye than with two." De Morgan analyzed logic mathematically, and studied the logical analysis of the laws, symbols, and operations of mathematics; he wrote a *Formal Logic* as well as a *Double Algebra*, and corresponded both with Sir William Hamilton, the metaphysician, and Sir William Rowan Hamilton, the mathematician. Few contemporaries were as profoundly read in the history of mathematics as was De Morgan. No subject was too insignificant to receive his attention. The authorship of "Cocker's Arithmetic" and the work of circle-squarers was investigated as minutely as was the history of the calculus. Numerous articles of his lie scattered in the volumes of the *Penny* and *English Cyclopædias*. In the article "Induction (Mathematics)," first printed in 1838, occurs, apparently for the first time, the *name* "mathematical induction"; it was adopted and popularized by I. Todhunter, in his *Algebra*. The term "induction" had been used by John Wallis in 1656, in his *Arithmetica infinitorum;* he used the "induction" known to natural science. In 1686 Jacob Bernoulli criticised him for using a process which was not binding logically and then advanced in place of it the proof from n to $n+1$. This is one of the several origins of the *process* of mathematical induction. From Wallis to De Morgan, the term "induction" was used occasionally in mathematics, and in a double sense, (1) to indicate incomplete inductions of the kind known in natural science, (2) for the proof from n to $n+1$. De Morgan's "mathematical induction" assigns a distinct name for the latter process. The Germans employ more commonly the name "vollständige Induktion," which became current among them after the use of it by R. Dedekind in his *Was sind und was sollen die Zahlen*, 1887. De Morgan's *Differential Calculus*, 1842, is still a standard work, and contains much that is original with the author. For the *Encyclopædia Metropolitana* he wrote on the Calculus of Functions (giving principles of symbolic reasoning) and on the theory of probability. In the Calculus of Functions he proposes the use of the slant line or "solidus" for printing fractions in the text; this proposal was adopted by G. G. Stokes in 1880.[1] Cayley wrote Stockes, "I think the 'solidus' looks very well indeed . . . ; it would give you a strong claim to be President of a Society for the prevention of Cruelty to Printers."[2]

Celebrated is De Morgan's *Budget of Paradoxes*, London, 1872, a second edition of which was edited by David Eugene Smith in 1915. De Morgan published memoirs "On the Foundation of Algebra" in the *Trans. of the Cambridge Phil. Soc.*, 1841, 1842, 1844 and 1847.

The ideas of George Peacock and De Morgan recognize the possibility of algebras which differ from ordinary algebra. Such algebras were indeed not slow in forthcoming, but, like non-Euclidean geometry, some of them were slow in finding recognition. This is true of H. G. Grassmann's, G. Bellavitis's, and B. Peirce's discoveries, but W. R. Hamilton's quaternions met with immediate appreciation in England. These algebras offer a geometrical interpretation of imaginaries.

William Rowan Hamilton (1805–1865) was born of Scotch parents in Dublin. His early education, carried on at home, was mainly in languages. At the age of thirteen he is said to have been familiar with as many languages as he had lived years. About this time he came across a copy of I. Newton's *Universal Arithmetic*. After reading that, he took up successively analytical geometry, the calculus, Newton's *Principia*, Laplace's *Mécanique Céleste*. At the age of eighteen he published a paper correcting a mistake in Laplace's work. In 1824 he entered Trinity College, Dublin, and in 1827, while he was still an undergraduate, he was appointed to the chair of astronomy. C. G. J. Jacobi met Hamilton at the meeting of the British Association at Manchester in 1842 and, addressing Section A, called Hamilton "le Lagrange de votre pays." Hamilton's early papers were on optics. In 1832 he predicted conical refraction, a discovery by aid of mathematics which ranks with the discovery of Neptune by U. J. J. Le Verrier and J. C. Adams. Then followed papers on the *Principle of Varying Action* (1827) and a general method of dynamics (1834–1835). He wrote also on the solution of equations of the fifth degree, the hodograph, fluctuating functions, the numerical solution of differential equations.

The capital discovery of Hamilton is his quaternions, in which his study of algebra culminated. In 1835 he published in the *Transactions of the Royal Irish Academy* his Theory of Algebraic Couples. He regarded algebra "as being no mere art, nor language, nor primarily

[1] G. G. Stokes, *Math. and Phys. Papers*, Vol. I, Cambridge, 1880, p. vii; see also J. Larmor, *Memoir and Scie. Corr. of G. G. Stokes*, Vol. I, 1907, p. 397.

[2] An earlier use of the solidus in designating fractions occurs in one of the very first text books published in California, viz., the *Definicion de las principales operaciones de arismética* by Henri Cambuston, 26 pages printed at Monterey in 1843. The solidus appears slightly curved.

a science of quantity, but rather as the science of order of progression." Time appeared to him as the picture of such a progression. Hence his definition of algebra as "the science of pure time." It was the subject of years' meditation for him to determine what he should regard as the product of each pair of a system of perpendicular directed lines. At last, on the 16th of October, 1843, while walking with his wife one evening, along the Royal Canal in Dublin, the discovery of quaternions flashed upon him, and he then engraved with his knife on a stone in Brougham Bridge the fundamental formula $i^2=j^2=k^2=ijk=-1$. At the general meeting of the Irish Academy, a month later, he made the first communication on quaternions. An account of the discovery was given the following year in the *Philosophical Magazine*. Hamilton displayed wonderful fertility in their development. His *Lectures on Quaternions*, delivered in Dublin, were printed in 1852.

In 1858 P. G. Tait was introduced to Hamilton and a correspondence was carried on between them which brought Hamilton back to the further development of quaternions along the lines of quaternion differentials, the linear vector function and A. Fresnel's wave surface, and led him to prepare the *Elements of Quaternions*, 1866, which he did not live to complete. Only 500 copies were printed. A new edition has been published recently by Charles Jasper Joly (1864–1906), his successor in Dunsink Observatory. Tait's own *Elementary Treatise on Quaternions* was projected in 1859, but was withheld from publication until Hamilton's work should appear; it was finally published in 1867. P. G. Tait's chief accomplishment was the development of the operator ∇, which was done in the later, greatly enlarged editions.[1] Tait submitted his quaternionic theorems to the judgment of Clerk Maxwell, and Maxwell came to recognize the power of the quaternion calculus in dealing with physical problems. "Tait brought out the real physical significance of the quantities $S\nabla\sigma$, $V\nabla\sigma$, ∇u. Maxwell's expressive names, Convergence (or Divergence) and Curl, have sunk into the very heart of electromagnetic theory."[2] In 1913 J. B. Shaw generalized the Hamiltonian ∇ for space of n dimensions, which may be either flat or curved. Related memoirs are due to G. Ricci (1892), T. Levi-Civita (1900), H. Maschke, and L. Ingold (1910). Quaternions were greatly admired in England from the start, but on the Continent they received less attention. P. G. Tait's *Elementary Treatise* helped powerfully to spread a knowledge of them in England. A. Cayley, W. K. Clifford, and Tait advanced the subject somewhat by original contributions. But there has been little progress in recent years, except that made by J. J. Sylvester in the solution of quaternion equations, nor has the application of quaternions to physics been as

[1] C. G. Knott, *Life and Scientific Work of Peter Guthrie Tait*, Cambridge, 1911, pp. 143, 148.

[2] C. G. Knott, *op. cit.*, p. 167.

extended as was predicted. The change in notation made in France by Jules Hoüel and by C. A. Laisant has been considered in England as a wrong step, but the true cause for the lack of progress is perhaps more deep-seated. There is indeed great doubt as to whether the quaternionic product can claim a necessary and fundamental place in a system of vector analysis. Physicists claim that there is a loss of naturalness in taking the square of a vector to be negative.

Widely different opinions have been expressed on the value of quaternions. While P. G. Tait was an enthusiastic champion of this science, his great friend, William Thomson (Lord Kelvin), declared that they, "though beautifully ingenious, have been an unmixed evil to those who have touched them in any way, including Clerk Maxwell."[1] A. Cayley, writing to Tait in 1874, said, "I admire the equation $d\sigma = uqd\rho q^{-1}$ extremely—it is a grand example of the pocket map." Cayley admitted the conciseness of quaternion formulas, but they had to be unfolded into Cartesian form before they could be made use of or even understood. Cayley wrote a paper "On Co-ordinates versus Quaternions" in the *Proceedings* of the Royal Society of Edinburgh, Vol. 20, to which Tait replied "On the Intrinsic Nature of the Quaternion Method."

In order to meet more adequately the wants of physicists, *J. W. Gibbs* and *A. Macfarlane* have each suggested an algebra of vectors with a new notation. Each gives a definition of his own for the product of two vectors, but in such a way that the square of a vector is positive. A third system of vector analysis has been used by *Oliver Heaviside* in his electrical researches.

What constitutes the most desirable notation in vector analysis is still a matter of dispute. Chief, among the various suggestions, are those of the American school, started by J. W. Gibbs and those of the German-Italian school. The cleavage is not altogether along lines of nationality. L. Prandl of Hanover said in 1904: "After long deliberation I have adopted the notation of Gibbs, writing $a \,.\, b$ for the inner (scalar), and $a \times$b for the outer (vector) product. If one observes the rule that in a multiple product the outer product must be taken before the inner, the inner product before the scalar, then one can write with Gibbs $a \,.\, b \times c$ and $ab \,.\, c$ without giving rise to doubt as to the meaning."[2]

In the following we give German-Italian notations first, the equivalent American notation (Gibbs') second. Inner product $\mathbf{a} \mid \mathbf{b}$, $\mathbf{a}.\,\mathbf{b}$; vector-product $\mid \mathbf{ab}$, $\mathbf{a} \times \mathbf{b}$; also $\mathbf{abc}$, $\mathbf{a}\,.\,\mathbf{b} \times \mathbf{c}$; $\mathbf{ab} \mid \mathbf{c}$, $(\mathbf{a} \times \mathbf{b}) \times \mathbf{c}$; $\mathbf{ab} \mid \mathbf{cd}$, $(\mathbf{a} \times \mathbf{b})\,.\,(\mathbf{c} \times \mathbf{d})$; $\mathbf{ab}^2$, $(\mathbf{a} \times \mathbf{b})^2$; $\mathbf{ab}\,.\,\mathbf{cd}$, $(\mathbf{a} \times \mathbf{b}) \times (\mathbf{c} \times \mathbf{d})$. R. Mehmke said in 1904: "The notation of the German-Italian school is far preferable to that of Gibbs not only in logical and methodical, but also in practical respects."

[1] S. P. Thompson, *Life of Lord Kelvin*, 1910, p. 1138.

[2] *Jahresb. d. d. Math. Vereinig.*, Vol. 13, p. 39.

In 1895 P. Molenbroek of The Hague and S. Kimura, then at Yale University, took the first steps in the organization of an International Association for Promoting the Study of Quaternions and Allied Systems of Mathematics. P. G. Tait was elected the first president, but could not accept on account of failing health. **Alexander Macfarlane** (1851–1913) of the University of Edinburgh, later of the University of Texas and of Lehigh University, served as secretary of the Association and was its president at the time of his death.

At the international congress held in Rome in 1908 a committee was appointed on the unification of vectorial notations but at the time of the congress held in Cambridge in 1912 no definite conclusions had been reached.

Vectorial notations were subjects of extended discussion in *L'Enseignement mathématique*, Vols. 11–14, 1909–1912, between C. Burali-Forti of Turin, R. Marcolongo of Naples, G. Comberiac of Bourges, H. C. F. Timerding of Strassburg, F. Klein of Göttingen, E. B. Wilson of Boston, G. Peano of Turin, C. G. Knott of Edinburgh, Alexander Macfarlane of Chatham in Canada, E. Carvallo of Paris, and E. Jahnke of Berlin. In America the relative values of notations were discussed in 1916 by E. B. Wilson and V. C. Poor.

We mention two topics outside of ordinary physics in which vector analysis has figured. The generalization of A. Einstein, known as the principle of relativity, and its interpretation by H. Minkowski, have opened new points of view. Some of the queer consequences of this theory disappear when kinematics is regarded as identical with the geometry of four-dimensional space. H. Minkowski and, following him, Max Abraham, used vector analysis in a limited degree, Minkowski usually preferring the matrix calculus of A. Cayley. A more extended use of vector analysis was made by Gilbert N. Lewis of the University of California who introduced in his extension to four dimensions some of the original features of H. G. Grassmann's system.

A "dyname" is, according to J. Plücker (and others) a system of forces applied to a rigid body. The English and French call it a "torsor." In 1899 this subject was treated by the Russian A. P. Kotjelnikoff under the name of projective theory of vectors. In 1903 E. Study of Greifswald brought out his book, *Geometrie der Dynamen*, in which a line-geometry and kinematics are elaborated, partly by the use of group theory, which are carried over to non-Euclidean spaces; Study claims for his system somewhat greater generality than is found in Hamilton's quaternions and W. K. Clifford's biquarternions.

Hermann Günther Grassmann (1809–1877) was born at Stettin, attended a gymnasium at his native place (where his father was teacher of mathematics and physics), and studied theology in Berlin for three years. His intellectual interests were very broad. He started as a theologian, wrote on physics, composed texts for the study of

German, Latin, and mathematics, edited a political paper and a missionary paper, investigated phonetic laws, wrote a dictionary to the Rig-Veda, translated the Rig-Veda in verse, harmonized folk songs in three voices, carried on successfully the regular work of a teacher and brought up nine of his eleven children—all this in addition to the great mathematical creations which we are about to describe. In 1834 he succeeded J. Steiner as teacher of mathematics in an industrial school in Berlin, but returned to Stettin in 1836 to assume the duties of teacher of mathematics, the sciences, and of religion in a school there.[1] Up to this time his knowledge of mathematics was pretty much confined to what he had learned from his father, who had written two books on "Raumlehre" and "Grössenlehre." But now he made his acquaintance with the works of S. F. Lacroix, J. L. Lagrange, and P. S. Laplace. He noticed that Laplace's results could be reached in a shorter way by some new ideas advanced in his father's books, and he proceeded to elaborate this abridged method, and to apply it in the study of tides. He was thus led to a new geometric analysis. In 1840 he had made considerable progress in its development, but a new book of Schleiermacher drew him again to theology. In 1842 he resumed mathematical research, and becoming thoroughly convinced of the importance of his new analysis, decided to devote himself to it. It now became his ambition to secure a mathematical chair at a university, but in this he never succeeded. In 1844 appeared his great classical work, the *Lineale Ausdehnungslehre*, which was full of new and strange matter, and so general, abstract, and out of fashion in its mode of exposition, that it could hardly have had less influence on European mathematics during its first twenty years, had it been published in China. K. F. Gauss, J. A. Grunert, and A. F. Möbius glanced over it, praised it, but complained of the strange terminology and its "philosophische Allgemeinheit." Eight years afterwards, C. A. Bretschneider of Gotha was said to be the only man who had read it through. An article in *Crelle's Journal*, in which Grassmann eclipsed the geometers of that time by constructing, with aid of his method, geometrically any algebraic curve, remained again unnoticed. Need we marvel if Grassmann turned his attention to other subjects,—to Schleiermacher's philosophy, to politics, to philology? Still, articles by him continued to appear in *Crelle's Journal*, and in 1862 came out a second version of his *Ausdehnungslehre*. It was intended to show better than the first the broad scope of the Ausdehnungslehre, by considering not only geometric applications, but by treating also of algebraic functions, infinite series, and the differential and integral calculus. But the second was no more appreciated than the first. At the age of fifty-three, this wonderful man, with heavy heart, gave up mathematics, and directed his energies to the study of Sanskrit, achieving in philology results which

[1] Victor Schlegel, *Hermann Grassmann*, Leipzig, 1878.

were better appreciated, and which vie in splendor with those in mathematics.

Common to the Ausdehnungslehre and to quaternions are geometric addition, the function of two vectors represented in quaternions by $S\alpha\beta$ and $V\alpha\beta$, and the linear vector functions. The quaternion is peculiar to W. R. Hamilton, while with Grassmann we find in addition to the algebra of vectors a geometrical algebra of wide application, and resembling A. F. Möbius's *Barycentrische Calcul*, in which the point is the fundamental element. Grassmann developed the idea of the "external product," the "internal product," and the "open product." The last we now call a matrix. His Ausdehnungslehre has very great extension, having no limitation to any particular number of dimensions. Only in recent years has the wonderful richness of his discoveries begun to be appreciated. A second edition of the *Ausdehnungslehre* of 1844 was printed in 1877.✡ C. S. Peirce gave a representation of Grassmann's system in the logical notation, and E. W. Hyde of the University of Cincinnati wrote the first text-book on Grassmann's calculus in the English language.

Discoveries of less value, which in part covered those of Grassmann and Hamilton, were made by *Barré de Saint-Venant* (1797–1886), who described the multiplication of vectors, and the addition of vectors and oriented areas; by *A. L. Cauchy*, whose "clefs algébriques" were units subject to combinatorial multiplication, and were applied by the author to the theory of elimination in the same way as had been done earlier by Grassmann; by **Giusto Bellavitis** (1803–1880), who published in 1835 and 1837 in the *Annali delle Scienze* his calculus of æquipollences. Bellavitis, for many years professor at Padua, was a self-taught mathematician of much power, who in his thirty-eighth year laid down a city office in his native place, Bassano, that he might give his time to science.

The first impression of H. G. Grassmann's ideas is marked in the writings of *Hermann Hankel*, who published in 1867 his *Vorlesungen über die Complexen Zahlen.* Hankel, then docent in Leipzig, had been in correspondence with Grassmann. The "alternate numbers" of Hankel are subject to his law of combinatorial multiplication. In considering the foundations of Algebra Hankel affirms the principle of the permanence of formal laws previously enunciated incompletely by G. Peacock. His *Complexe Zahlen* was at first little read, and we must turn to **Victor Schlegel** (1843–1905) of Hagen as the successful interpreter of Grassmann. Schlegel was at one time a young colleague of Grassmann at the Marienstifts-Gymnasium in Stettin. Encouraged by R. F. A. Clebsch, Schlegel wrote a *System der Raumlehre*, 1872–1875, which explained the essential conceptions and operations of the Ausdehnungslehre.

Grassmann's ideas spread slowly. In 1878 Clerk Maxwell wrote P. G. Tait: "Do you know Grassmann's Ausdehnungslehre? Spottiswoode spoke of it in Dublin as something above and beyond 4nions. I have not seen it, but Sir W. Hamilton of Edinburgh used to say that the greater the extension the smaller the intention."

Multiple algebra was powerfully advanced by B. Peirce, whose theory is not geometrical, as are those of W. R. Hamilton and H. G. Grassmann. **Benjamin Peirce** (1809–1880) was born at Salem, Mass., and graduated at Harvard College, having as undergraduate carried the study of mathematics far beyond the limits of the college course. When N. Bowditch was preparing his translation and commentary of the *Mécanique Céleste*, young Peirce helped in reading the proof-sheets. He was made professor at Harvard in 1833, a position which he retained until his death. For some years he was in charge of the *Nautical Almanac* and superintendent of the United States Coast Survey. He published a series of college text-books on mathematics, an *Analytical Mechanics*, 1855, and calculated, together with Sears C. Walker of Washington, the orbit of Neptune. Profound are his researches on *Linear Associative Algebra*. The first of several papers thereon was read at the meeting of the American Association for the Advancement of Science in 1864. Lithographed copies of a memoir were distributed among friends in 1870, but so small seemed to be the interest taken in this subject that the memoir was not printed until 1881 (*Am. Jour. Math.*, Vol. IV, No. 2). Peirce works out the multiplication tables, first of *single* algebras, then of *double* algebras, and so on up to sextuple, making in all 162 algebras, which he shows to be possible on the consideration of symbols A, B, etc., which are linear functions of a determinate number of letters or units i, j, k, l, etc., with coefficients which are ordinary analytical magnitudes, real or imaginary,—the letters i, j, etc., being such that every binary combination i^2, ij, ji, etc., is equal to a linear function of the letters, but under the restriction of satisfying the associative law.[1] *Charles S. Peirce*, a son of Benjamin Peirce, and one of the foremost writers on mathematical logic, showed that these algebras were all defective forms of quadrate algebras which he had previously discovered by logical analysis, and for which he had devised a simple notation. Of these quadrate algebras quaternions is a simple example; nonions is another. C. S. Peirce showed that of all linear associative algebras there are only three in which division is unambiguous. These are ordinary single algebra, ordinary double algebra, and quaternions, from which the imaginary scalar is excluded. He showed that his father's algebras are operational and matricular. Lectures on multiple algebra were delivered by J. J. Sylvester at the Johns Hopkins University, and published in various journals. They treat largely of the algebra of matrices.

While Benjamin Peirce's comparative anatomy of linear algebras was favorably received in England, it was criticised in Germany as

[1] A. Cayley, Address before British Association, 1883.

being vague and based on arbitrary principles of classification. German writers along this line are Eduard Study and Georg W. Scheffers. An estimate of B. Peirce's linear associative algebra was given in 1902 by H. E. Hawkes,[1] who extends Peirce's method and shows its full power. In 1898 Élie Cartan of the University of Lyon used the characteristic equation to develop several general theorems; he exhibits the *semi-simple*, or Dedekind, and the *pseudo-nul*, or nilpotent, sub-algebras; he shows that the structure of every algebra may be represented by the use of double units, the first factor being quadrate, the second non-quadrate. Extensions of B. Peirce's results were made also by Henry Taber. Olive C. Hazlett gave a classification of nilpotent alegbras.

As shown above, C. S. Peirce advanced this algebra by using the matrix theory. Papers along this line are due to F. G. Frobenius and J. B. Shaw. The latter "shows that the equation of an algebra determines its quadrate units, and certain of the *direct* units; that the other units form a nilpotent system which with the quadrates may be reduced to certain canonical forms. The algebra is thus made a sub-algebra under the algebra of the *associative units* used in these canonical forms. Frobenius proves that every algebra has a Dedekind sub-algebra, whose equation contains all factors in the equation of the algebra. This is the semi-simple algebra of Cartan. He also showed that the remaining units form a nilpotent algebra whose units may be *regularized*" (J. B. Shaw). More recently, J. B. Shaw has extended the general theorems of linear associative algebras to such algebras as have an infinite number of units.

Besides the matrix theory, the theory of continuous groups has been used in the study of linear associative algebra. This isomorphism was first pointed out by H. Poincaré (1884); the method was followed by Georg W. Scheffers who classified algebras as quaternionic and non-quaternionic and worked out complete lists of all algebras to order five. Theodor Molien, in 1893, then in Dorpat, demonstrated "that quaternionic algebras contain independent quadrates, and that quaternionic algebras can be classified according to non-quaternionic types" (J. B. Shaw). An elementary exposition of the relation between linear algebras and continuous groups was given by L. E. Dickson [2] of Chicago. This relation "enables us to translate the concepts and theorems of the one subject into the language of the other subject. It not only doubles our total knowledge, but gives us a better insight into either subject by exhibiting it from a new point of view." The theory of matrices was developed as early as 1858 by A. Cayley in an important memoir which, in the opinion of J. J. Sylvester, ushered in

[1] H. E. Hawkes in *Am. Jour. Math.*, Vol. 24, 1902, p. 87. We are using also J. B. Shaw's *Synopsis of Linear Associative Algebra*, Washington, D. C., 1907, Introduction. Shaw gives bibliography.

[2] *Bull. Am. Math. Soc.*, Vol. 22, 1915, p. 53.

the reign of Algebra the Second. W. K. Clifford, Sylvester, H. Taber, C. H. Chapman, carried the investigations much further. The originator of matrices is really W. R. Hamilton, but his theory, published in his *Lectures on Quaternions*, is less general than that of Cayley. The latter makes no reference to Hamilton.

The theory of determinants [1] was studied by *Hoëné Wronski* (1778–1853), a poor Polish enthusiast, living most the time in France, whose egotism and wearisome style tended to attract few followers, but who made some incisive criticisms bearing on the philosophy of mathematics.[2] He studied four special forms of determinants, which were extended by *Heinrich Ferdinand Scherk* (1798–1885) of Bremen and *Ferdinand Schweins* (1780–1856) of Heidelberg. In 1838 Liouville demonstrated a property of the special forms which were called "wronskians" by Thomas Muir in 1881. Determinants received the attention of *Jacques P. M. Binet* (1786–1856) of Paris, but the great master of this subject was A. L. Cauchy. In a paper (*Jour. de l'ecole Polyt.*, IX., 16) Cauchy developed several general theorems. He introduced the name *determinant*, a term used by K. F. Gauss in 1801 in the functions considered by him. In 1826 C. G. J. Jacobi began using this calculus, and he gave brilliant proof of its power. In 1841 he wrote extended memoirs on determinants in *Crelle's Journal*, which rendered the theory easily accessible. In England the study of linear transformations of quantics gave a powerful impulse. A. Cayley developed skew-determinants and Pfaffians, and introduced the use of determinant brackets, or the familiar pair of upright lines. The more general consideration of determinants whose elements are formed from the elements of given determinants was taken up by J. J. Sylvester (1851) and especially by L. Kronecker who gave an elegant theorem known by his name.[3] Orthogonal determinants received the attention of A. Cayley in 1846, in the study of n^2 elements related to each other by $\frac{1}{2}n(n+1)$ equations, also of L. Kronecker, F. Brioschi and others. Maximal values of determinants received the attention of J. J. Sylvester (1867), and especially of J. Hadamard (1893) who proved that the square of a determinant is never greater than the norm-product of the lines.

Anton Puchta (1851–1903) of Czernowitz in 1878 and M. Noether in 1880 showed that a symmetric determinant may be expressed as the product of a certain number of factors, linear in the elements. Determinants which are formed from the minors of a determinant were investigated by J. J. Sylvester in 1851, to whom we owe the

[1] Thomas Muir, *The Theory of Determinants in the Historical Order of Development* [Vol. I], 2nd Ed., London, 1906; Vol. II, Period 1841 to 1860, London, 1911. ✡ Muir was Superintendent-General of Education in Cape Colony.

[2] On Wronski, see J. Bertrand in *Journal des Savants*, 1897, and in *Revue des Deux-Mondes*, Feb., 1897. See also *L'Intermédiaire des Mathématiciens*, Vol. 23, 1916, pp. 113, 164–167, 181–183.

[3] E. Pascal, *Die Determinanten*, transl. by H. Leitzmann, 1900, p. 107.

"umbral notation," by W. Spottiswoode in 1856, and later by G. Janni, M. Reiss (1805–1869), E. d'Ovidio, H. Picquet, E. Hunyadi (1838–1889) E. Barbier, C. A. Van Velzer, E. Netto, G. Frobenius, and others. Many researches on determinants appertain to special forms. "Continuants" are due to J. J. Sylvester; "alternants," originated by A. L. Cauchy, have been developed by C. G. J. Jacobi, Nicolò Trudi (1811–1884) of Naples, H. Nägelsbach, and G. Garbieri; "axisymmetric determinants," first used by Jacobi, have been studied by V. A. Lebesgue, J. J. Sylvester, and L. O. Hesse; "circulants" are due to Eugène Charles Catalan (1814–1894) of Liège, William Spottiswoode (1825–1883) of Oxford, J. W. L. Glaisher, and R. F. Scott; for "centro-symmetric determinants" we are indebted to G. Zehfuss of Heidelberg. V. Nachreiner and S. Günther, both of Munich, pointed out relations between determinants and continued fractions; R. F. Scott uses H. Hankel's alternate numbers in his treatise. A class of determinants which have the same importance in linear integral equations as do ordinary determinants for linear equations with n unknowns was worked out by E. Fredholm (*Acta math.*, 1903) and again by D. Hilbert who reaches them as limiting expressions of ordinary determinants.

An achievement of considerable significance was the introduction in 1860 of infinite determinants by *Eduard Fürstenau* in a method of approximation to the roots of algebraic equations. Determinants of an infinite order were used by *Theodor Kötteritzsch* of Grimma in Saxony, in two papers on the solution of an infinite system of linear equations (*Zeitsch. f. Math. u. Physik*, Vol. 14, 1870). Independently, infinite determinants were introduced in 1877 by George William Hill of Washington in an astronomical paper (*Collected Works*, Vol. I, 1905, p. 243). In 1884 and 1885 H. Poincaré called attention to these determinants as developed by Hill and investigated them further. Their theory was elaborated later by Helge von Koch and Erhard Schmidt (1908).

In recent years the theory of the solution of a system of linear equations has been presented in an elegant form by means of what is known as the rank of a determinant. In particular, G. A. Miller has thus developed a necessary and sufficient condition that a given unknown in a consistent system of linear equations have only one value while some of the other unknowns may assume an infinite number of values.[1]

Text-books on determinants were written by W. Spottiswoode (1851), F. Brioschi (1854), R. Baltzer (1857), S. Günther (1875), G. J. Dostor (1877), R. F. Scott (1880), T. Muir (1882), P. H. Hanus (1886), G. W. H. Kowalewski (1909; 2. ed., 1924; 3. ed., 1942).

The symbol $n!$ for "factorial n," now universally used in algebra, is due to Christian Kramp (1760–1826) of Strassburg. who used it in

[1] *Am. Math. Monthly*, Vol. 17, 1910, p. 137

1808. The symbol $\equiv$ to express identity was first used by G. F. B. Riemann.[1]

Modern higher algebra☆ is especially occupied with the theory of linear transformations. Its development is mainly the work of A. Cayley and J. J. Sylvester.

Arthur Cayley (1821–1895), born at Richmond, in Surrey, was educated at Trinity College, Cambridge. He came out Senior Wrangler in 1842. He then devoted some years to the study and practice of law. While a student at the bar he went to Dublin and, alongside of G. Salmon, heard W. R. Hamilton's lectures on quaternions. On the foundation of the Sadlerian professorship at Cambridge, he accepted the offer of that chair, thus giving up a profession promising wealth for a very modest provision, but which would enable him to give all his time to mathematics. Cayley began his mathematical publications in the *Cambridge Mathematical Journal* while he was still an undergraduate. Some of his most brilliant discoveries were made during the time of his legal practice. There is hardly any subject in pure mathematics which the genius of Cayley has not enriched, but most important is his creation of a new branch of analysis by his theory of invariants. Germs of the principle of invariants are found in the writings of J. L. Lagrange, K. F. Gauss, and particularly of G. Boole, who showed, in 1841, that invariance is a property of discriminants generally, and who applied it to the theory of orthogonal substitution. Cayley set himself the problem to determine *a priori* what functions of the coefficients of a given equation possess this property of invariance, and found, to begin with, in 1845, that the so-called "hyper-determinants" possessed it. G. Boole made a number of additional discoveries. Then J. J. Sylvester began his papers in the *Cambridge and Dublin Mathematical Journal* on the Calculus of Forms. After this, discoveries followed in rapid succession. At that time Cayley and Sylvester were both residents of London, and they stimulated each other by frequent oral communications. It has often been difficult to determine how much really belongs to each. In 1882, when Sylvester was professor at the Johns Hopkins University, Cayley lectured there on Abelian and theta functions.

Of interest is Cayley's method of work. A. R. Forsyth describes it thus: "When Cayley had reached his most advanced generalizations he proceeded to establish them directly by some method or other, though he seldom gave the clue by which they had first been obtained: a proceeding which does not tend to make his papers easy reading. . . . His literary style is direct, simple and clear. His legal training had an influence, not merely upon his mode of arrangement but also upon his expression; the result is that his papers are severe and present a curious contrast to the luxuriant enthusiasm which pervades so many

[1] L. Kronecker, *Vorlesungen über Zahlentheorie*, 1901, p. 86.

of Sylvester's papers."[1] Curiously, Cayley took little interest in quaternions.

James Joseph Sylvester (1814–1897) was born in London. His father's name was Abraham Joseph; his eldest brother assumed in America the name of Sylvester, and he adopted this name too. About the age of 16 he was awarded a prize of $500 for solving a question in arrangements for contractors of lotteries in the United States.[2] In 1831 he entered St. John's College, Cambridge, and came out Second Wrangler in 1837, George Green being fourth. Sylvester's Jewish origin incapacitated him from taking a degree. From 1838 to 1840 he was professor of natural philosophy at what is now University College, London; in 1841 he became professor of mathematics at the University of Virginia. In a quarrel with two of his students he slightly wounded one of them with a metal pointed cane, whereupon he returned hurriedly, to England. In 1844 he served as an actuary; in 1846 he became a student at the Inner Temple and was called to the bar in 1850. In 1846 he became associated with A. Cayley; often they walked round the Courts of Lincoln's Inn, perhaps discussing the theory of invariants, and Cayley (says Sylvester) "habitually discoursing pearls and rubies." Sylvester resumed mathematical research. He, Cayley and William Rowan Hamilton entered upon discoveries in pure mathematics that are unequalled in Great Britain since the time of I. Newton. Sylvester made the friendship of G. Salmon whose books contributed greatly to bring the results of Cayley and Sylvester within easier reach of the mathematical public. From 1855 to 1870 Sylvester was professor at the Royal Military Academy at Woolwich, but showed no great efficiency as an elementary teacher. There are stories of his housekeeper pursuing him from home carrying his collar and necktie. From 1876 to 1883, he was professor at the Johns Hopkins University, where he was happy in being free to teach whatever he wished in the way he thought best. He became the first editor of the *American Journal of Mathematics* in 1878. In 1884 he was elected to succeed H. J. S. Smith in the chair of Savilian professor of geometry at Oxford, a chair once occupied by Henry Briggs, John Wallis and Edmund Halley.

Sylvester sometimes amused himself writing poetry. His *Laws of Verse* is a curious booklet. At the reading, at the Peabody Institute in Baltimore, of his Rosalind poem, consisting of about 400 lines all rhyming with "Rosalind," he first read all his explanatory footnotes, so as not to interrupt the poem; these took one hour and a-half. Then he read the poem itself to the remnant of his audience.

[1] *Proceed. London Royal Society*, Vol. 58, 1895, pp. 23, 24.

[2] H. F. Baker's Biographical Notice in *The Collected Math. Papers of J. J. Sylvester*, Vol. IV, Cambridge, 1912. We have used also P. A. MacMahon's notice in *Proceed. Royal Soc. of London*, Vol. 63, 1898, p. ix. For Sylvester's activities in Baltimore, see Fabian Franklin in *Johns Hopkins Univ. Circulars*, June, 1897; F.

Sylvester's first papers were on Fresnel's optic theory, 1837. Two years later he wrote on C. Sturm's memorable theorem. Sturm once told him that the theorem originated in the theory of the compound pendulum. Stimulated by A. Cayley he made important investigations on modern algebra. He wrote on elimination, on transformation and canonical forms, in which the expression of a cubic surface by five cubes is given, on the relation between the minor determinants of linearly equivalent quadratic functions, in which the notion of invariant factors is implicit, while in 1852 appeared the first of his papers on the principles of the calculus of forms. In a reply that he made in 1869 to Huxley who had claimed that mathematics was a science that knows nothing of observation, induction, invention and experimental verification, Sylvester narrated his personal experience: "I discovered and developed the whole theory of canonical binary forms for odd degrees, and, as far as yet made out, for even degrees too, at one evening sitting, with a decanter of port wine to sustain nature's flagging energies, in a back office in Lincoln's Inn Fields. The work was done, and well done, but at the usual cost of racking thought—a brain on fire, and feet feeling, or feelingless, as if plunged in an ice-pail. *That night we slept no more.*" His reply to Huxley is interesting reading and bears strongly on the qualities of mental activity involved in mathematical research. In 1859 he gave lectures on partitions, not published until 1897. He wrote on partitions again in Baltimore. In 1864 followed his famous proof of Newton's rule. A certain fundamental theorem in invariants which had formed the basis of an important section of A. Cayley's work, but had resisted proof for a quarter of a century was demonstrated by Sylvester in Baltimore. Noteworthy are his memoirs on Chebichev's method concerning the totality of prime numbers within certain limits, and his latent roots of matrices. His researches on invariants, theory of equations, multiple algebra, theory of numbers, linkages, probability, constitute important contributions to mathematics. His final studies, entered upon after his return to Oxford, were on reciprocants or functions of differential coefficients whose form is unaltered by certain linear transformations of the variables, and a generalization of the theory of concomitants. In 1911, G. Greenhill told reminiscently[1] how Sylvester got everybody interested in reciprocants, "now clean forgotten"; "One day after, Sylvester was noticed walking alone, addressing the sky, asking it: 'Are Reciprocants Bosh? Berry of King's says the Reciprocant is all Bosh!' There was no reply, and Sylvester himself was tiring of the subject, and so Berry escaped a castigation. But recently I had occasion from the Aeronautical point of view to work out the theory of a Vortex

Cajori, *Teaching and History of Mathematics in the United States*, Washington, 1890, pp. 261–272.

[1] *Mathematical Gazette*, Vol. 6, 1912, p. 108.

inside a Polygon, an eddy whirlwind such as Chavez had to encounter in the angle of the precipices, flying over the Simplon Pass. The analysis in some cases seemed strangely familiar, and at last I recognized the familiar Reciprocant. . . . Difference in Similarity and Similarity in Difference has been called the motto of our science." [1]

In the *American Journal of Mathematics* are memoirs on binary and ternary quantics, elaborated partly with aid of *F. Franklin,* then professor at the Johns Hopkins University. The theory of reciprocants is more general than one on differential invariants by G. H. Halphen (1878), and has been developed further by J. Hammond of Cambridge, P. A. McMahon (1854–1929) of Woolwich, A. R. Forsyth, and others. Sylvester playfully lays claim to the appellation of the Mathematical Adam, for the many names he has introduced into mathematics. Thus the terms *invariant, discriminant, Hessian, Jacobian,* are his. That not only elementary pupils, but highly trained mathematicians as well, may be attracted or repelled by the kind of symbols used, is illustrated by the experience of K. Weierstrass who related that he followed Sylvester's papers on the theory of algebraic forms very attentively until Sylvester began to employ Hebrew characters. That was more than he could stand and after that he quit him.[2]

The great theory of invariants, developed in England mainly by A. Cayley and J. J. Sylvester, came to be studied earnestly in Germany, France, and Italy. Ch. Hermite discovered evectants, and the theorem of reciprocity named after him, whereby "to every covariant of degree n in the coefficients of the quantic of order m, there corresponds a covariant of degree m in the coefficients of a quantic of order n. He discovered the skew invariant of the quintic, which was the first example of any skew invariant. He discovered the linear covariants belonging to quantics of odd order greater than 3, and he applied them to obtain the typical expression of the quantic in which the coefficients are invariants. He also invented the associated covariants of a quantic; these constitute the simplest set of algebraically complete systems as distinguished from systems that are linearly complete." [3]

In Italy, F. Brioschi of Milan and *Faà de Bruno* (1825–1888) contributed to the theory of invariants, the latter writing a text-book on binary forms (1876), which ranks by the side of G. Salmon's treatise and those of R. F. A. Clebsch and P. Gordan.

Francesco Brioschi (1824–1897) in 1852 became professor of applied mathematics at the University of Pavia and in 1862 was commissioned by the government to organize the Instituto tecnico superiore at Milan, where he filled until his death the chair of hydraulics and

[1] *Mathematical Gazette,* Vol. 6, 1912, p. 108.

[2] E. Lampe (1840–1918), in *Naturwissenschaftliche Rundschau,* Bd. 12, 1897, p. 361; quoted by R. E. Moritz, *Memorabilia Mathematica,* 1914, p. 180.

[3] *Proceed. of the Roy. Soc. of London,* Vol. 45, 1905, p. 144. Obituary notice of Hermite by A. R. Forsyth.

analysis. With Abbé *Barnaba Tortolini* (1808–1874) he founded in 1858 the *Annali di matematica pura ed applicata*. Among his pupils at Pavia were L. Cremona and E. Beltrami. V. Volterra narrates [1] how F. Brioschi in 1858, with two other young Italians, *Enrico Betti* (1823–1892), later professor at the University of Pisa, and *Felice Casorati* (1835–1890), later professor at the University of Pavia, started on a journey to enter into relations with the foremost mathematicians of France and Germany. "The scientific existence of Italy as a nation" dates from this journey. "It is to the teaching, labors, and devotion of these three, to their influence in the organization of advanced studies, to the friendly scientific relations that they instituted between Italy and foreign countries, that the existence of a school of analysts in Italy is due."

In Germany the early theory of invariants, as developed by Cayley, Sylvester, and Salmon in England, Hermite in France and F. Brioschi in Italy, did not draw attention until 1858 when **Siegfried Heinrich Aronhold** (1819–1884) of the technical high school in Berlin pointed out that Hesse's theory of ternary cubic forms of 1844 involved invariants by which that theory could be rounded out. F. G. Eisenstein and J. Steiner had also given early publication to isolated developments involving the invariantal idea. In 1863 Aronhold gave a systematic and general exposition of invariant heory (*Crelle*, 62). He and Clebsch used a notation of their own, the symbolical notation, different from Cayley's, which was used in the further developments of the theory in Germany. Great developments were started about 1868, when R. F. A. Clebsch and P. Gordan wrote on types of binary forms, L. Kronecker and E. B. Christoffel on bilinear forms, F. Klein and S. Lie on the invariant theory connected with any group of linear substitutions. **Paul Gordan** (1837–1912) was born at Erlangen and became professor there. He produced papers on finite groups, particularly on the simple group of order 168 and its associated curve $y^3z+z^3x+x^3y=0$. His best known achievement is the proof of the existence of a complete system of concomitants for any given binary form.[2] While Clebsch aimed in his researches to devise methods by which he could study the relationships between invariantal forms (Formenverwandtschaft), the chief aim of Aronhold was to examine the equivalence or the linear transformation of one form into another.[3] Investigations along this line are due to E. B. Christoffel, who showed that the number of arbitrary parameters contained in the substitution coefficients equals that of the absolute invariants of the form, K. Weierstrass who gave a general treatment of the equivalence of two

[1] *Bull. Am. Math. Soc.*, Vol. 7, 1900, p. 60.

[2] *Nature*, Vol. 90, 1913, p. 597.

[3] We are using Franz Meyer, "Bericht über den gegenwärtigen Stand der Invariantentheorie" in the *Jahresb. d. d. Math. Vereinigung*, Vol. I, 1890–91, pp. 79–292. See p. 99.

linear systems of bilinear and quadratic forms, L. Kronecker who extended the researches of Weierstrass and had a controversy with C. Jordan on certain discordant results, J. G. Darboux who in 1874 gave a general and elegant derivation of theorems due to K. Weierstrass and L. Kronecker, G. Frobenius who applied the transformation of bilinear forms to "Pfaff's problem": To determine when two given linear differential expressions of n terms can be converted one into the other by subjecting the variables to general point transformations. The study of invariance of quadratic and bilinear forms from the stand-point of group theory was pursued by H. Werner (1889), S. Lie (1885) and W. Killing (1890). Finite binary groups were examined by H. A. Schwarz (1871), and Felix Klein. Schwarz is led to the problem, to find "all spherical triangles whose symmetric repetitions on the surface of a sphere give rise to a finite number of spherical triangles differing in position," and deduces the forms belonging thereto. Without a knowledge of what Schwarz and W. R. Hamilton had done, Klein was led to a determination of the finite binary linear groups and their forms. Representing transformations as motions and adopting Riemann's interpretation of a complex variable on a spherical surface, F. Klein sets up the groups of those rotations which bring the five regular solids into coincidence with themselves, and the accompanying forms. The tetrahedron, octahedron and icosahedron lead respectively to 12, 24 and 60 rotations; the groups in question were studied by Klein. The icosahedral group led to an icosahedral equation which stands in intimate relation with the general equation of the fifth degree. Klein made the icosahedron the centre of his theory of the quintic as given in his *Vorlesungen über das Ikosaeder und die Auflösung der Gleichungen fünften Grades*,✲ Leipzig, 1884.

Finite substitution groups and their forms, as related to linear differential equations, were investigated by R. Fuchs (*Crelle*, 66, 68) in 1866 and later. If the equation has only algebraic integrals, then the group is finite, and conversely. Fuchs's researches on this topic were continued by C. Jordan, F. Klein, and F. Brioschi. Finite ternary and higher groups have been studied in connection with invariants by F. Klein who in 1887 made two such groups the basis for the solution of general equations of the sixth and seventh degrees. In 1886 F. N. Cole, under the guidance of Klein, had treated the sextic equation in the *Am. Jour. of Math.*, Vol. 8. The second group used by Klein was studied with reference to the 140 lines in space, to which it leads, by H. Maschke in 1890.

The relationship of invariantal forms, the study of which was initiated by A. Cayley and J. J. Sylvester, received since 1868 emphasis in the writings of R. F. A. Clebsch and P. Gordan. Gordan proved in *Crelle*, Vol. 69, the finiteness of the system for a single binary form. This is known as "Gordan's theorem." Even in the later simplified forms the proof of it is involved, but the theorem yields practical methods to determine the existing systems. G. Peano in 1881 generalized the theorem and applied it to the "correspondences" represented by certain double-binary forms. In 1890 D. Hilbert, by using only rational processes, demonstrated the finiteness of the system of invariants arising from a given series of any forms in n variables. A modification of this proof which has some advantages was given by W. E. Story of Clarke University. Hilbert's research bears on the number of relations called syzygies, a subject treated before this time by A. Cayley, C. Hermite, F. Brioschi, C. Stephanos of Athens, J. Hammond, E. Stroh, and P. A. MacMahon.

The symbolic notation in the theory of invariants, introduced by S. H. Aronhold and R. F. A. Clebsch, was developed further by P. Gordan, E. Stroh, and E. Study in Germany. English writers endeavored to make the expressions in the theory of forms intuitively evident by graphic representation, as when Sylvester in 1878 uses the atomic theory, an idea applied further by W. K. Clifford. The symbolic method in the theory of invariants has been used by P. A. MacMahon in the article "Algebra" in the eleventh edition of the *Encyclopædia Britannica*, and by J. H. Grace and A. Young in their *Algebra of Invariants*, Cambridge, 1903. Using a method in C. Jordan's great memoirs on invariants, these authors are led to novel results, notably to "an exact formula for the maximum order of an irreducible covariant of a system of binary forms." A complete syzygetic theory of the absolute orthogonal concomitants of binary quantics was constructed by Edwin B. Elliott of Oxford by a method that is not symbolic, while P. A. MacMahon in 1905 employs a symbolic calculus involving imaginary umbræ for similar purposes. While the theory of invariants has played an important rôle in modern algebra and analytic projective geometry, attention has been directed also to its employment in the theory of numbers. Along this line are the researches of L. E. Dickson in the *Madison Colloquium* of 1913.

The establishment of criteria by means of which the irreducibility of expressions in a given domain may be ascertained has been investigated by F. T. v. Schubert (1793), K. F. Gauss, L. Kronecker, T. Schönemann, F. G. M. Eisenstein, R. Dedekind, G. Floquet, L. Königsberger, E. Netto, O. Perron, M. Bauer, W. Dumas and H. Blumberg. The theorem of Schönemann and Eisenstein declares that if the polynomial $x^n+c_1 x^{n-1}+ \dots +c_n$ with integral coefficients is such that a prime p divides every coefficient $c_1, \dots, c_n$, but p^2 does not go into c_n, then the polynomial is irreducible in the domain of rational numbers. This theorem may be regarded as the nucleus of the work of the later authors. Floquet and Königsberger do not limit themselves to polynomials, but consider also linear homogeneous differential expressions. Blumberg gives a general theorem which practically includes all earlier results as special cases.[1]

[1] For bibliography see *Trans. Am. Math. Soc.*, Vol. 17, 1916, pp. 517–544.

Theory of Equations and Theory of Groups

A notable event was the reduction of the quintic equation to the trinomial form, effected by *George Birch Jerrard* (?–1863) in his *Mathematical Researches* (1832–1835). Jerrard graduated B. A. at Trinity College, Dublin, in 1827. It was not until 1861 that it became generally known that this reduction had been effected as early as 1786 by *Erland Samuel Bring* (1736–1798), a Swede, and brought out in a publication of the University of Lund. Both Bring and Jerrard used the method of E. W. Tschirnhausen. Bring never claimed that his transformation led to the general algebraic solution of the quintic, but Jerrard persisted in making such a claim even after N. H. Abel and others had offered proofs establishing the impossibility of a general solution. In 1836, William R. Hamilton made a report on the validity of Jerrard's method, and showed that by his process the quintic could be transformed to any one of the four trinomial forms. Hamilton defined the limits of its applicability to higher equations. J. J. Sylvester investigated this question, What is the lowest degree an equation can have in order that it may admit of being deprived of i consecutive terms by aid of equations not higher than ith degree. He carried the investigation as far as $i=8$, and was led to a series of numbers which he named "Hamilton's numbers." A transformation of equal importance to Jerrard's is that of Sylvester, who expressed the quintic as the sum of three fifth-powers. The covariants and invariants of higher equations have been studied much in recent years.

In the theory of equations J. L. Lagrange, J. R. Argand, and K. F. Gauss furnished proof to the important theorem that every algebraic equation has a real or a complex root. N. H. Abel proved rigorously that the general algebraic equation of the fifth or of higher degrees cannot be solved by radicals (*Crelle*, I, 1826). Before Abel, an Italian physician, *Paolo Ruffini* (1765–1822), had printed a proof of the insolvability. It appears in his book, *Teoria generale delle equazioni*, Bologna, 1799, and in later articles on this subject. Ruffini's proof was criticised by his countryman, G. F. Malfatti. L. N. M. Carnot, A. M. Legendre, and S. D. Poisson, in a report of 1813 on a paper of A. L. Cauchy, had occasion to refer to Ruffini's proof as "fondée sur des raisonnemens trop vagues, et n'ait pas été généralement admise." [1] N. H. Abel remarked that Ruffini's reasoning did not always seem rigorous. But Cauchy in 1821 wrote to Ruffini that he had "demontre completement l'insolubilité algebrique des équations générales d' un dégré supérieur au quatrième." [2] J. Hecker showed in 1886 that Ruffini's proof was sound in general outline, but faulty in some of the detail.[3] E. Bortolotti in 1902 stated that Ruffini's proof, as given

[1] E. Bortolotti, *Carteggio di Paolo Ruffini*, Roma, 1906, p. 32.

[2] E. Bortolotti, *Influenza dell' opera mat. di P. Ruffini*, 1902, p. 34.

[3] J. Hecker, *Ueber Ruffini's Beweis* (Dissertation), Bonn, 1886.

in 1813 in his book *Reflessioni intorno alla soluzione dell' equazioni algebraiche*, was substantially the same as that given later by *Pierre Laurent Wantzel* [1] (1814–1848), but only the second part of Wantzel's simplified proof resembles Ruffini's; the first part is modelled after Abel's. Wantzel, by the way, deserves credit for having given the first rigorous proofs (*Liouville*, Vol. 2, 1837, p. 366.) of the impossibility of the trisection of any given angle by means of ruler and compasses, and of avoiding the "irreducible case" in the algebraic solution of irreducible cubic equations. Wantzel was répétiteur at the Polytechnic School in Paris. As a student he excelled both in mathematics and languages. Saint-Venant said of him: "Ordinarily he worked evenings, not lying down until late; then he read, and took only a few hours of troubled sleep, making alternately wrong use of coffee and opium, and taking his meals at irregular hours until he was married. He put unlimited trust in his constitution, very strong by nature, which he taunted at pleasure by all sorts of abuse. He brought sadness to those who mourn his premature death."

Ruffini's researches on equations are remarkable as containing anticipations of the algebraic theory of groups.[2] Ruffini's "permutation" corresponds to our term "group." He divided groups into "simple" and "complex," and the latter into intransitive, transitive imprimitive, and transitive primitive groups. He established the important theorem for which the name "Ruffini's theorem" has been suggested,[3] that a group does not necessarily have a subgroup whose order is an arbitrary divisor of the order of the group. The collected works of Ruffini are published under the auspices of the Circolo Matematico di Palermo; the first volume appeared in 1915 with notes by Ettore Bortolotti of Bologna. A transcendental solution of the quintic involving elliptic integrals was given by Ch. Hermite (*Compt. Rend.*, 1858, 1865, 1866). After Hermite's first publication, L. Kronecker, in 1858, in a letter to Hermite, gave a second solution in which was obtained a simple resolvent of the sixth degree.

Abel's proof that higher equations cannot always be solved algebraically led to the inquiry as to what equations of a given degree can be solved by radicals. Such equations are the ones discussed by K. F. Gauss in considering the division of the circle. Abel advanced one step further by proving that an irreducible equation can always be solved in radicals, if, of two of its roots, the one can be expressed rationally in terms of the other, provided that the degree of the equation is prime; if it is not prime, then the solution depends upon that of equations of lower degree. Through geometrical considerations,

[1] E. Bortolotti, *Influenza*, etc., 1902, p. 26. Wantzel's proof is given in *Nouvelles Annales Mathématiques*, Vol. 4, 1845, pp. 57–65. See also Vol. 2, pp. 117–127. The second part of Wantzel's proof, involving substitution-theory, is reproduced in J. A. Serret's *Algèbre supérieure*.

[2] H. Burkhardt, in *Zeitschr. f. Mathematik u. Physik*, Suppl., 1892.

[3] G. A. Miller, in *Bibliotheca mathematica*, 3. F. Vol. 10, 1909–1910, p. 318.

L. O. Hesse came upon algebraically solvable equations of the ninth degree, not included in the previous groups. The subject was powerfully advanced in Paris by the youthful *Évariste Galois* (1811–1832).[1] He was born at Bourg-la-Reine, near Paris. He began to exhibit most extraordinary mathematical genius after his fifteenth year. His was a short, sad and pestered life. He was twice refused admittance to the École Polytechnique, on account of inability to meet the (to him) trivial demands of examiners who failed to recognize his genius. He entered the École Normale in 1829, then an inferior school. Proud and arrogant, and unable to see the need of the customary detailed explanations, his career in that school was not smooth. Drawn into the turmoil of the revolution of 1830, he was forced to leave the École Normale. After several months spent in prison, he was killed in a duel over a love affair. Ordinary text-books he disposed of as rapidly as one would a novel. He read J. L. Lagrange's memoirs on equations, also writings of A. M. Legendre, C. G. J. Jacobi, and N. H. Abel. As early as the seventeenth year he reached results of the highest importance. Two memoirs presented to the Academy of Sciences were lost. A brief paper on equations in the *Bulletin de Férussac*, 1830, Vol. XIII, p. 428, gives results which seem to be applications of a general theory. The night before the duel he wrote his scientific testament in the form of a letter to Auguste Chevalier, containing a statement of the mathematical results he had reached and asking that the letter be published, that "Jacobi or Gauss pass judgment, not on their correctness, but on their importance." Two memoirs found among his papers were published by J. Liouville in 1846. Further manuscripts were published by J. Tannery at Paris in 1908. As a rule Galois did not fully prove his theorems. It was only with difficulty that Liouville was able to penetrate into Galois' ideas. Several commentators worked on the task of filling out the lacunæ in Galois' exposition. Galois was the first to use the word "group" in a technical sense, in 1830. He divided groups into simple and compound, and observed that there is no simple group of any composite order less than 60. The word "group" was used by A. Cayley in 1854, by T. F. Kirkman and J. J. Sylvester in 1860.[2] Galois proved the important theorem that every invariant subgroup gives rise to a quotient group which exhibits many fundamental properties of the group. He showed that to each algebraic equation corresponds a group of substitutions which reflects the essential character of the equation. In a paper published in 1846 he established the beautiful theorem: In order that an irreducible equation of prime degree be solvable by radicals, it is necessary and sufficient that all its roots be

[1] See life by Paul Dupuy in *Annales de l'école normale supérieure*, 3. S., Vol. XIII, 1896. See also E. Picard, *Oeuvres math. d' Évariste Galois*, Paris, 1897; J. Pierpont, *Bulletin Am. Math. Soc.*, 2. S., Vol. IV, 1898, pp. 332–340.

[2] G. A. Miller in *Am. Math. Monthly*, Vol. XX, 1913, p. 18.

rational in any two of them. Galois' use of substitution groups to determine the algebraic solvability of equations, and N. H. Abel's somewhat earlier use of these groups to prove that *general* equations of degrees higher than the fourth cannot be solved by radicals, furnished strong incentives to the vigorous cultivation of group theory. It was A. L. Cauchy who entered this field next. To Galois are due also some valuable results in relation to another set of equations, presenting themselves in the theory of elliptic functions, viz., the modular equations. To Cauchy has been given the credit of being the founder of the theory of groups of finite order,[1] even though fundamental results had been previously reached by J. L. Lagrange, Pietro Abbati (1768–1842), P. Ruffini, N. H. Abel, and Galois. Cauchy's first publication was in 1815, when he proved the theorem that the number of distinct values of a non-symmetric function of degree n cannot be less than the largest prime that divides n, without becoming equal to 2. Cauchy's great researches on groups appeared in his *Exercises d'analyse et de physique mathématique*, 1844, and in articles in the Paris *Comptes Rendus*, 1845–1846. He did not use the term "group," but he uses (x y z u v w) and other devices to denote substitutions, uses the terms "cyclic substitution," "order of a substitution," "identical substitution," "transposition," "transitive," "intransitive." In 1844 he proved the fundamental theorem (stated but not proved by E. Galois) which is known as "Cauchy's theorem": Every group whose order is divisible by a given prime number p must contain at least one subgroup of order p. This theorem was later extended by L. Sylow. A. L. Cauchy was the first to enumerate the orders of the possible groups whose degrees do not exceed six, but this enumeration was incomplete. At times he fixed attention on properties of groups without immediate concern as regards applications, and thereby took the first steps toward the consideration of abstract groups. In 1846 J. Liouville made E. Galois' researches better known by publication of two manuscripts. At least as early as 1848 J. A. Serret taught group theory in Paris. In 1852, Enrico Betti of the University of Pisa published in the *Annali* of B. Tortolini the first rigorous exposition of Galois' theory of equations that made the theory intelligible to the general public. The first account of it given in a text-book on algebra is in the third edition of J. A. Serret's *Algèbre*, 1866.

In England the earliest studies in group theory are due to Arthur Cayley and William R. Hamilton. In 1854 A. Cayley published a paper in the *Philosophical Magazine* which is usually accepted as founding the theory of abstract groups, although the idea of abstract groups occurs earlier in the papers of A. L. Cauchy, and Cayley's

[1] Our account of Cauchy's researches on groups is drawn from the article of G. A. Miller in *Bibliotheca Mathematica*, Vol. X, 1909–1910, pp. 317–329, and that of Josephine E. Burns in *Am. Math. Monthly*, Vol. XX, 1913, pp. 141–148.

article is not entirely abstract. Formal definitions of abstract groups were not given until later, by L. Kronecker (1870), H. Weber (1882), and G. Frobenius (1887). The transition from substitution groups to abstract groups was gradual.[1] It may be recalled here that, before 1854, there were two sources from which the theory of groups of finite order originated. In the writings of J. Lagrange, P. Ruffini, N. H. Abel, and E. Galois it sprang from the theory of algebraic equations. A second source is the theory of numbers; the group concept is fundamental in some of L. Euler's work on power residues and in some of the early work of K. F. Gauss. It has been pointed out more recently that the group idea really underlies geometric transformations and is implied in Euclid's demonstrations.[2] Abstract groups are considered apart from any of their applications.

A. Cayley illustrates his paper of 1854 by means of the laws of combination of quaternion imaginaries, quaternions having been invented by William R. Hamilton eleven years previously. In 1859 Cayley pointed out that the quaternion units constitute a group of order 8, now known as the quaternion group, when they are multiplied together.[3] William R. Hamilton, without using the technical language of group theory, developed in 1856, in his study of a new system of roots of unity, the properties of the groups of the regular solids, as generated by two operators or elements, and he proved that these groups may be completely defined by the orders of their two generating operators and the order of their product.

E. Picard puts the matter thus: "A regular polyhedron, say an icosahedron, is on the one hand the solid that all the world knows; it is also, for the analyst, a group of finite order, corresponding to the divers ways of making the polyhedron coincide with itself. The investigation of all the types of groups of motion of finite order interests not alone the geometers, but also the crystallographers; it goes back essentially to the study of groups of ternary linear substitutions of determinant +1, and leads to the thirty-two systems of symmetry of the crystallographers for the particular complex."

In 1858 the Institute of France offered a prize for a research on group theory which, though not awarded, stimulated research. In 1859 *Émile Léonard Mathieu* (1835–1890) of the University of Nancy wrote a thesis on substitution groups, while in 1860 *Camille Jordan* (1838–1922) of the École Polytechnique in Paris contributed the first of a series of papers which culminated in his great *Traité des substitutions*, 1870. Jordan received his doctorate in 1860 in Paris; he is editor of the *Journal de mathématiques pures et appliquées*. His first paper on groups gives the fundamental theorem that the total number of

[1] G. A. Miller, in *Bibliotheca Mathematica*, 3. Ed., Vol. XX, 1909–1910, p. 326. We are making much use of Miller's historical sketch.

[2] H. Poincaré in *Monist*, Vol. 9, 1898, p. 34.

[3] G. A. Miller in *Bibliotheca Mathematica*, Vol. XI, 1910–1911, pp. 314–315.

substitutions of n letters which are commutative with every substitution of a regular group G on the same n letters constitute a group which is similar to G. To Jordan is due the fundamental concept of class of a substitution group and he proved the constancy of the factors of composition. He also proved that there is a finite number of primitive groups whose class is a given number greater than 3, and that the necessary and sufficient condition that a group be solvable is that its factors of composition are prime numbers.[1] Prominent among C. Jordan's pupils is *Edmont Maillet* (1865–194?), editor of *L'Intermédiaire des mathématiciens*, who has made extensive contributions.

In Germany L. Kronecker and R. Dedekind were the earliest to become acquainted with the Galois theory. Kronecker refers to it in an article published in 1853 in the *Berichte* of the Berlin Academy. Dedekind lectured on it in Göttingen in 1858. In 1879–1880 E. Netto gave lectures in Strassburg. His *Substitutionstheorie*, 1882, was translated into Italian in 1885 by *Giuseppe Battaglini* (1826–1894) of the University of Rome, and into English in 1892 by *F. N. Cole*, then at Ann Arbor. The book placed the subject within easier reach of the mathematical public.

In 1862–1863 **Ludwig Sylow** (1832–1918) gave lectures on substitution groups in Christiania, Norway, which were attended by Sophus Lie. Extending a theorem given nearly thirty years earlier by A. L. Cauchy, Sylow obtained the theorem known as "Sylow's theorem": Every group whose order is divisible by p^m, but not by p^{m+1}, p being a prime number, contains $1+kp$ subgroups of order p^m. About twenty years later this theorem was extended still further by *Georg Frobenius* (1849–1917) of the University of Berlin, to the effect that the number of subgroups is $kp+1$, k being an integer, even when the order of the group is divisible by a higher power of p than p^m. Sophus Lie took a very important step by the explicit application of the group concept to new domains and the creation of the theory of continuous groups. **Marius Sophus Lie** (1842–1899)[2] was born in Nordfjordeide in Norway. In 1859 he entered the University of Christiania, but not until 1868 did this slowly developing youth display marked interest in mathematics. The writings of J. V. Poncelet and J. Plücker awakened his genius. In the winter of 1869–1870 he met Felix Klein in Berlin and they published some papers of joint authorship. The summer of 1870 they were together in Paris where they were in close touch with C. Jordan and J. G. Darboux. It was then that Lie discovered his contact-transformation which changes the straight lines of ordinary space over into spheres. This led him to a general theory of transformation. At the outbreak of the Franco-Prussian war, F. Klein

[1] G. A. Miller in *Bibliotheca mathematica*, 3. S., Vol. X, 1909–1910, p. 323.

[2] F. Engel in *Bibliotheca mathematica*, 3. S., Vol. I, 1900, pp. 166–204; M. Nöther in *Math. Annalen*, Vol. 53, pp. 1–41.

left Paris; S. Lie started to travel afoot through France into Italy, but was arrested as a spy and imprisoned for a month until Darboux was able to secure his release. In 1872 he was elected professor at the University of Christiania, with all his time available for research. In 1871–1872 he entered upon the study of partial differential equations of the first order, and in 1873 he arrived at the theory of transformation groups, according to which finite continuous groups are applied to infinitesimal transformations. He considered a very general and important kind of transformations called contact-transformations, and their application in the theory of partial differential equations of the first and second orders. As his group theory and theories of integration met with no appreciation, he returned in 1876 to the study of geometry—minimal surfaces, the classification of surfaces according to the transformation group of their geodetic lines. The starting of a new journal, the *Archiv for Mathematik og Naturvidenskab*, in 1876, enabled him to publish his results promptly. G. H. Halphen's publications of 1882 on differential invariants induced Lie to direct attention to his own earlier researches and their greater generality. In 1884 Friedrich Engel was induced by F. Klein and A. Mayer to go to Christiania to assist Lie in the preparation of a treatise, the *Theorie der Transformationsgruppen*, 1888–1893. Lie accepted in 1886 a professorship at the University of Leipzig. In 1889–1890 over-work led to insomnia and depression of spirits. While he soon recovered his power for work, he ever afterwards was over-sensitive and mistrustful of his best friends. With the aid of Engel he published in 1891 a memoir on the theory of infinite continuous transformation groups. In 1898 he returned to Norway where he died the following year. Lie's lectures on *Differentialgleichungen*, given in Leipzig, were brought out in book form by his pupil, *Georg Scheffers*, in 1891. In 1895 F. Klein declared that Lie and H. Poincaré were the two most active mathematical investigators of the day. The following quotation from an article written by Lie in 1895 indicates how his whole soul was permeated by the group concept:[1] "In this century the concepts known as substitution and substitution group, transformation and transformation group, operation and operation group, invariant, differential invariant, and differential parameter, appear continually more clearly as the most important concepts of mathematics. While the curve as the representation of a function of a single variable has been the most important object of mathematical investigation for nearly two centuries from Descartes, while on the other hand, the concept of transformation first appeared in this century as an expedient in the study of curves and surfaces, there has gradually developed in the last decades a general theory of transformations whose elements are represented by the transformation

[1] *Berichte d. Koenigl. Saechs. Gesellschaft*, 1895; translated by G. A. Miller in *Am. Math. Monthly*, Vol. III, 1896, p. 296.

itself while the series of transformations, in particular the transformation groups, constitute the object."

In close association with S. Lie in the advancement of group theory and its applications was **Felix* Klein** (1849–1925). He was born at Düsseldorf in Prussia and secured his doctorate at Bonn in 1868. After studying in Paris, he became privat-docent at Göttingen in 1871, professor at Erlangen in 1872, at the Technical High School in Munich in 1875, at Leipzig in 1880 and at Göttingen in 1886. He has been active not only in the advancement of various branches of mathematics, but also in work of organization. Famous for laying out lines of research is his Erlangen paper of 1872, *Vergleichende Betrachtungen über neuere geometrische Forschungen.* He became member of the commission on the publication of the *Encyklopädie der mathematischen Wissenschaften* and editor of the fourth volume on mechanics, also editor of *Mathematische Annalen*, 1877, and in 1908 president of the International Commission on the Teaching of Mathematics. As an inspiring lecturer on mathematics he has wielded a wide influence upon German and American students. About 1912 he was forced by ill-health to discontinue his lectures at Göttingen, but in 1914 the excitement of the war roused him to activity, much as J. Lagrange was aroused at the outbreak of the French Revolution, and Klein resumed lecturing. He has constantly emphasized the importance of both schools of mathematical thought, namely, the intuitional school, and the school that rests everything on abstract logic. In his opinion, "the intuitive grasp and the logical treatment should not exclude, but should supplement each other."

S. Lie's method of treating differential invariants was further investigated by K. Zorawski in *Acta Math.*, Vol. XVI, 1892–1893. In 1902 C. N. Haskins determined the number of functionally independent invariants of any order, while A. R. Forsyth obtained the invariants for ordinary Euclidean space. Differential parameters have been investigated by J. Edmund Wright of Bryn Mawr College.[1] Lie's theory of invariants of finite continuous groups was attacked on logical grounds by E. Study of Bonn, in 1908. The validity of this criticism was partly admitted by F. Engel.

Another method of treating differential invariants, originally due to E. B. Christoffel, has been called by G. Ricci and T. Levi-Civita of Padua "covariant derivation," (*Mathematische Annalen*, Vol. 54, 1901). A third method was introduced by H. Maschke[2] who used a symbolism similar to that for algebraic invariants.

Henry W. Stager published in 1916 *A Sylow Factor Table for the first Twelve Thousand Numbers:* For every number up to 1200 the divisors of the form $p(kp+1)$ are given, where p is a prime greater than 2 and

[1] We are using J. E. Wright's *Invariants of Quadratic Differential Forms*, 1908, pp. 5–8.

[2] *Trans. Am. Math. Soc.*, Vol. 1, 1900, pp. 197–204.

k is a positive integer. These divisors aid in the determination of the number of Sylow subgroups.

Solvable (H. Weber's "metacyclic") groups have been studied by G. Frobenius who proved that every group of composite order that is not divisible by the square of a prime number must be compound, and that all these groups are solvable, for the orders of their self-conjugate subgroups and of their quotient groups cannot be divisible by the square of a prime number.[1] The study of solvable groups has been pursued also by L. Sylow, W. Burnside, R. Dedekind (who investigated what he called the Hamiltonian group), and G. A. Miller who with Frobenius developed about 1893–1896 elegant methods for proving the solvability of a given group. In 1895 O. Hölder enumerated all the insolvable groups whose order does not exceed 479. In 1898 G. A. Miller gave the numbers of all primitive solvable groups whose degree is less than 25, also the number of insolvable groups which may be represented as substitution groups whose degree is less than 12. G. A. Miller (1899) and Umberto Scarpio (1901) of Verona considered properties of commutators and commutator subgroups, and proved that the question of solvability can be decided by means of commutator subgroups.[2] Commutator groups have been studied also by W. B. Fite and Ernst Wendt. The characteristics of non-abelian groups were investigated since 1896 by G. Frobenius in Berlin, the characteristics of abelian groups having been already employed by J. Lagrange and P. Dirichlet. Characteristics of solvable groups were studied in 1901 by Frobenius. An enumeration of abstract groups was made in 1901 by R. P. Le Vavasseur of Toulouse. The list of intransitive substitution groups of degree eleven was shown by G. A. Miller and G. H. Ling in 1901 to include 1492 distinct substitution groups, which is about 500 more than the number of degree ten. H. L. Rietz proved that a primitive group of degree n and order g contains more than $g/x+1$ substitutions of degree less than n, x being the number of transitive constituents in a maximal subgroup of degree $n-1$. This result is closely related to investigations of C. Jordan, A. Bochert, and E. Maillet on the class of a primitive group.[3]

The definition of a group was simplified in 1902 by E. V. Huntington of Harvard University. He pointed out that the usual definition, as given for instance in H. Weber's *Algebra*, contains several redundancies, that only three postulates (four for finite groups) are necessary, the independence of which he established.[4] Later discussions of definitions are due to Huntington and E. H. Moore.

[1] See G. A. Miller's "Report of Recent Progress in the Theory of Groups of a Finite Order" in *Bull. Am. Math. Soc.*, Vol. 5, 1899, pp. 227–249, which we are using.

[2] G. A. Miller, "Second Report on Recent Progress in the Theory of Groups of Finite Order" in *Bull. Am. Math. Soc.*, Vol. 9, 1902, p. 108.

[3] *Loc. cit.*, p. 118.

[4] *Bull. Am. Math. Soc.*, Vol. 8, 1902, p. 296.

L. E. Dickson[1] said in 1900: "When a problem has been exhibited in group phraseology, the possibility of a solution of a certain character or the exact nature of its inherent difficulties is determined by a study of the group of the problem. . . . As the chemist analyzes a compound to determine the ultimate elements composing it, so the group-theorist decomposes the group of a given problem into a chain of simple groups. . . . Much labor has been expended in the determination of simple groups. For continuous groups of a finite number of parameters, the problem has been completely solved by W. Killing and E. J. Cartan (1894), with the result that all such simple groups, aside from five isolated ones, belong to the systems investigated by Sophus Lie, viz., the general projective group, the projective group of a linear complex, and the projective group leaving invariant a non-degenerate surface of the second order. The corresponding problem for infinite continuous groups remains to be solved. With regard to finite simple groups, the problem has been attacked in two directions. O. Hölder,[2] F. N. Cole,[3] W. Burnside,[4] G. H. Ling, and G. A. Miller have shown that the only simple groups of composite orders less than 2000 are the previously known simple groups of orders 60, 168, 360, 504, 660, 1092. On the other hand, various infinite systems of finite simple groups have been determined. The cyclic groups of prime orders and the alternating group of n letters ($n>4$) have long been recognized as simple groups. The other known systems of finite simple groups have been discovered in the study of linear groups. Four systems were found by C. Jordan, (*Traité des substitutions*) in his study of the general linear, the abelian, and the two hypoabelian groups, the field of reference being the set of residues of integers with respect to a prime modulus p. Generalizations may be made by employing the Galois field of order p^n (designated $GF\ [p^n]$), composed of the p^n Galois complexes formed with a root of a congruence of degree n irreducible modulo p. Groups of linear substitutions in a Galois field were studied by E. Betti, E. Mathieu, and C. Jordan; but the structure of such groups has been determined only in the past decade. The simplicity of the group of unary linear fractional substitutions in a Galois field was first proved by E. H. Moore (*Bulletin Am. Math. Soc.*, Dec., 1893) and shortly afterward by W. Burnside. The complete generalization of C. Jordan's four systems of simple groups and the determination of three new triply-infinite systems have been made by the writer" (*i. e.* by

[1] See L. E. Dickson in *Compte vendu du II. Congr. intern., Paris, 1900.* Paris, 1902, pp. 225, 226.

[2] O. Hölder proved in *Math. Annalen*, 1892, that there are only two simple groups of composite order less than 200, viz., those of order 60 and 168.

[3] F. N. Cole in *Am. Jour. Math.*, 1893 found that there could be only three such groups between orders 200 and 661, viz., of orders 360, 504, 660.

[4] W. Burnside showed that there was no simple group of composite order between 661 and 1092.

L. E. Dickson in 1896). Aside from the cyclic and alternating groups, the known systems of finite simple groups have been derived as quotient-groups in the series of composition of certain linear groups. Miss I. M. Schottenfels of Chicago showed that it is possible to construct two simple groups of the same order.

The determination of the smallest degree (the "class") of any of the non-identical substitutions of primitive groups which do not include the alternating group was taken up by C. Jordan and has been called "Jordan's problem." It was continued by Alfred Bochert of Breslau and E. Maillet. Bochert proved in 1892: If a substitution group of degree n does not include the alternating group and is more than simply transitive, its class exceeds $\frac{1}{4}n-1$, if it is more than doubly transitive its class exceeds $\frac{1}{3}n-1$, and if it is more than triply transitive its class is not less than $\frac{1}{2}n-1$. E. Maillet showed that when the degree of a primitive group is less than 202 its class cannot be obtained by diminishing the degree by unity unless the degree is a power of a prime number. In 1900 W. Burnside proved that every transitive permutation group in p symbols, p being prime, is either solvable or doubly transitive.

As regards linear groups, G. A. Miller wrote in 1899 as follows: "The linear groups are of extreme importance on account of their numerous direct applications. Every group of a finite order can clearly be represented in many ways as a linear substitution group since the ordinary substitution (permutation) groups are merely very special cases of the linear groups. The general question of representing such a group with the least number of variables seems to be far from a complete solution. It is closely related to that of determining all the linear groups of a finite order that can be represented with a small number of variables. Klein was the first to determine all the finite binary groups (in 1875) while the ternary ones were considered independently by C. Jordan (1880) and H. Valentiner (1889). The latter discovered the important group of order 360 which was omitted by Jordan and has recently been proved (by A. Wiman of Lund) simply isomorphic to the alternating group of degree 6. H. Maschke has considered many quaternary groups and established, in particular, a complete form system of the quaternary group of 51840 linear substitutions." *Heinrich Maschke* (1853–1908) was born in Breslau, studied in Berlin under K. Weierstrass, E. E. Kummer, and L. Kronecker, later in Göttingen under H. A. Schwarz, J. B. Listing, and F. Klein. He entered upon the study of group theory under Klein. In 1891 he came to the United States, worked a year with the Weston Electric Co., then accepted a place at the University of Chicago.

Linear groups of finite order, first treated by Felix Klein, were later used by him in the extension of the Galois theory of algebraic equations, as seen in his *Ikosaeder*. As stated above, Klein's determination of the linear groups in two variables was followed by groups in three variables, developed by C. Jordan and H. Valentiner (1889), and by groups of any number of variables, treated by C. Jordan. Special linear groups in four variables were discussed by E. Goursat (1889) and G. Bagnera of Palermo (1905). The complete determination of the groups in four variables, aside from intransitive and monomial types, was carried through by H. F. Blichfeldt of Leland Stanford University.[1] Says Blichfeldt: "There are, in the main, four distinct principles employed in the determination of the groups in 2, 3 or 4 variables: (a) the original geometrical process of Klein . . . ; (b) the processes leading to a diophantine equation, which may be approached analytically (C. Jordan . . .), or geometrically (H. Valentiner, G. Bagnera, H. H. Mitchell); (c) a process involving the relative geometrical properties of transformations which represent 'homologies' and like forms (H. Valentiner, G. Bagnera, H. H. Mitchell . . . ; (d) a process developed from the properties of the multipliers of the transformations, which are roots of unity (H. F. Blichfeldt). A new principle has been added recently by L. Bieberbach, though it had already been used by H. Valentiner in a certain form. . . . Independent of these principles stands the theory of group characteristics, of which G. Frobenius is the discoverer."

There is a marked difference between finite groups of even and of odd order.[2] As W. Burnside points out, the latter admit no self-inverse irreducible representation, except the identical one; all irreducible groups of odd order in 3, 5 or 7 symbols are soluble. G. A. Miller proved in 1901 that no group of odd order with a conjugate set of operations containing fewer than 50 members could be simple. W. Burnside proved in 1901 that transitive groups of odd order whose degree is less than 100 are soluble. H. L. Rietz in 1904 extended this last result to groups whose degrees are less than 243. W. Burnside has shown that the number of prime factors in the order of a simple group of odd order cannot be less than 7 and that 40,000 is a lower limit for the order of a group of odd order, if simple. These results suggest that, perhaps, simple groups of odd order do not exist. Recent researches on groups, mainly abstract groups, are due to L. E. Dickson, Le Vavasseur, M. Potron, L. I. Neikirk, G. Frobenius, H. Hilton, A. Wiman, J. A. de Séguier, H. W. Kuhn, A. Loewy, H. F. Blichfeldt,[3] W. A. Manning, and many others. Extensive researches on abstract groups have been carried on by G. A. Miller of the University of Illinois. In 1914 he showed, for instance, that a non-abelian group can have an abelian group of isomorphisms by proving the existence

[1] We are using H. F. Blichfeldt's *Finite Collineation Groups*, Chicago, 1917, pp. 174–177.

[2] W. Burnside, *Theory of Groups of Finite Order*, 2. Ed., Cambridge, 1911, p. 503.

[3] Consult G. A. Miller's "Third Report on Recent Progress in the Theory of Groups of Finite Order" in *Bull. Am. Math. Soc.*, Vol. 14, 1907, p. 124.

of this relation in a group of order 64. He proved the existence of a group of order p^9, p being any prime number whatever, whose group of isomorphisms has an order which is a power [1] of p. He proved also the existence of a group G of order 128 which admits of an outer isomorphism which changes each conjugate set of operations into itself. Among other results due to G. A. Miller are these: The number of independent generators of every prime power group is an invariant of the group; a necessary and sufficient condition that a solvable group is a direct product of a Sylow subgroup and another subgroup is that its group of inner isomorphisms involves the corresponding Sylow subgroup as a factor of a direct product, whenever it involves such a subgroup.[2]

A work which embodied modern researches in algebra was the *Lehrbuch der Algebra*, issued by H. Weber in 1895–1896 in two volumes, and in three volumes in the revised edition of 1898 and 1899. **Heinrich Weber** (1842–1913) was born in Heidelberg and studied at Heidelberg, Leipzig, and Königsberg. Since 1869 he was successively professor at Heidelberg, Königsberg, Berlin, Marburg, Göttingen and (since 1895) at Strassburg. He was editor of Riemann's Collected Works (1876; 2. ed. 1892). He carried on researches in algebra, theory of numbers, theory of functions, mechanics and mathematical physics. In 1911 he mourned the loss of a gifted daughter who had translated Poincaré's Valeur de la science and other French books into German.

The symmetric functions of the sums of powers of the roots of an equation, studied by I. Newton and E. Waring, were considered more recently by K. F. Gauss, A. Cayley, J. J. Sylvester, and F. Brioschi. Cayley gives rules for the "weight" and "order" of symmetric functions.

The theory of elimination was greatly advanced by J. J. Sylvester, A. Cayley, G. Salmon, C. G. J. Jacobi, L. O. Hesse, A. L. Cauchy, E. Brioschi, and P. Gordan. Sylvester gave the dialytic method (*Philosophical Magazine*, 1840), and in 1852 established a theorem relating to the expression of an eliminant as a determinant. A. Cayley made a new statement of Bézout's method of elimination and established a general theory of elimination (1852).

Contributions to the theory of equations, based on Descartes' rule of signs and especially on its application to infinite series were made by **Edmond Laguerre** (1834–1886), professor in the Collège de France in Paris. An upper limit for the number of real roots of a polynomial with real coefficients, $f(x)$, in an interval (o, a) results from the application of the rule of signs to a product $f_2(x)=f_1(x)\ f(x)$ developed in a power series which converges for $|\ x\ |<a$, but diverges for $x=a$. In particular, he proved that if z in $e^{zx}f(x)$ is taken sufficiently large,

[1] *Bull. Am. Math. Soc.*, Vol. 20, 1914, pp. 310, 311.

[2] *Ibid*, Vol. 18, 1912, p. 440.

then the exact number of positive roots is ascertainable from the variations of sign in the series. Michel Fekete and Georg Pólya, both of Budapest, use $f(x)/(1-x)^n$ for the same purpose.[1]

The theory of equations commanded the attention of **Leopold Kronecker** (1823–1891). He was born in Liegnitz near Breslau, studied at the gymnasium of his native town under Kummer, later in Berlin under C. G. J. Jacobi, J. Steiner, and P. Dirichlet, then in Breslau again under E. E. Kummer. Though for eleven years after 1844 engaged in business and the care of his estates, he did not neglect mathematics, and his fame grew apace. In 1855 he went to Berlin where he began to lecture at the University in 1861. He was a very stimulating and interesting lecturer. Kummer, K. Weierstrass, and L. Kronecker constitute the triumvirate of the second mathematical school in Berlin. This school emphasized severe rigor in demonstrations. L. Kronecker dwelt intensely upon arithmetization which repressed as far as possible all space representations and rested solely upon the concept of number, particularly the positive integer. He displayed manysided talent and extraordinary ability to penetrate new fields of thought. "But," says G. Frobenius,[2] "conspicuous as his achievements are in the different fields of number research, he does not quite reach up to A. L. Cauchy and C. G. J. Jacobi in analysis, nor to B. Riemann and Weierstrass in function-theory, nor to Dirichlet and Kummer in number-theory." Kronecker's papers on algebra, the theory of equations and elliptic functions proved to be difficult reading. A more complete and simplified exposition of his results was given by R. Dedekind and H. Weber. "Among the finest of Kronecker's achievements," says Fine,[3] "were the connections which he established among the various disciplines in which he worked: notably that between the theory of quadratic forms of negative determinant and elliptic functions, through the singular moduli which give rise to the complex multiplication of the elliptic functions, and that between the theory of numbers and algebra, by his arithmetical theory of the algebraic equation." He held to the view that the theory of fractional and irrational numbers could be built upon the integral numbers alone. "Die ganze zahl," said he, "schuf der liebe Gott, alles Uebrige ist Menschenwerk." Later he even denied the existence of irrational numbers. He once paradoxically remarked to Lindemann: "Of what use is your beautiful research on the number π? Why cogitate over such problems, when really there are no irrational numbers whatever?"

In 1890–1891 L. Kronecker developed a theory of the algebraic equation with numerical coefficients, which he did not live to publish. From notes of Kronecker's lectures, H. B. Fine of Princeton prepared

[1] *Bull. Am. Math. Soc.*, Vol. 20, 1913, p. 20.

[2] G. Frobenius, *Gedächtnissrede auf Leopold Kronecker*, Berlin, 1893, p. 1.

[3] *Bull. Am. Math. Soc.*, Vol. I, 1892, p. 175.

an address in 1913 giving Kronecker's unpublished results.[1] "All who have read Kronecker's later writings," says Fine, "are familiar with his contention that the theory of the algebraic equation in its final form must be based solely on the rational integer, algebraic numbers being excluded and only such relations and operations being admitted as can be expressed in finite terms by means of rational numbers and therefore ultimately by means of integers. These lectures of 1890–91 are chiefly concerned with the development of such a theory, and in particular with the proof of two theorems which therein take the place of the fundamental theorem of algebra as commonly stated."

Solution of Numerical Equations

Jacques Charles François Sturm (1803–1855), a native of Geneva, Switzerland, and the successor of Poisson in the chair of mechanics at the Sorbonne, published in 1829 his celebrated theorem determining the number and situation of real roots of an equation comprised between given limits. De Morgan has said that this theorem "is the complete *theoretical* solution of a difficulty upon which energies of every order have been employed since the time of Descartes." Sturm explains in that article that he enjoyed the privilege of reading Fourier's researches while they were still in manuscript and that his own discovery was the result of the close study of the principles set forth by Fourier. In 1829 Sturm published no proof. Proofs were given in 1830 by Andreas von Ettinghausen (1796–1878) of Vienna, in 1832 by Charles Choquet et Mathias Mayer in their Algèbre, and in 1835 by Sturm himself. According to J. M. C. Duhamel, Sturm's discovery was not the result of observation, but of a well-ordered line of thought as to the kind of function that would meet the requirements. According to J. J. Sylvester, the theorem "stared him (Sturm) in the face in the midst of some mechanical investigations connected with the motion of compound pendulums." Duhamel and Sylvester both state that they received their information from Sturm directly. Yet their statements do not agree. Perhaps both statements are correct, but represent different stages in the evolution of the discovery in Sturm's mind.[2]

By the theorem of Sturm one can ascertain the number of complex roots, but not their location. That limitation was removed in a brilliant research by another great Frenchman, A. L. Cauchy. He discovered in 1831 a general theorem which reveals the number of roots, whether real or complex, which lie within a given contour. This theorem makes heavier demands upon the mathematical attainments

[1] *Bull. Am. Math. Soc.*, Vol. 20, 1914, p. 339.

[2] Consult also M. Bôcher, "The published and unpublished Work of Charles Sturm on algebraic and differential Equations" in *Bull. Am. Math. Soc.*, Vol. 18, 1912, pp. 1–18.

of the reader, and for that reason has not the celebrity of Sturm's theorem. But it enlisted the lively interest of men like Sturm, J. Liouville, and F. Moigno.

A remarkable article was published in 1826 by *Germinal Dandelin* (1794–1847) in the memoirs of the Academy of Sciences of Brussels. He gave the conditions under which the Newton-Raphson method of approximation can be used with security. In this part of his research he was anticipated by both Mourraille and J. Fourier. In another part of his paper (the second supplement) he is more fortunate; there he describes a new and masterly device for approximating to the roots of an equation, which constitutes an anticipation of the famous method of C. H. Gräffe. We must add here that the fundamental idea of Gräffe's method is found even earlier, in the *Miscellanea analytica*, 1762, of Edward Waring. If a root lies between a and b, $a-b<1$, and a is on the convex side of the curve, then Dandelin puts $x=a+y$ and transforms the equation into one whose root y is small. He then multiplies $f(y)$ by $f(-y)$ and obtains, upon writing $y^2=z$, an equation of the same degree as the original one, but whose roots are the squares of the roots of the equation $f(y)=0$. He remarks that this transformation may be repeated, so as to get the fourth, eighth, and higher powers, whereby the moduli of the powers of the roots diverge sufficiently to make the transformed equation separable into as many polygons as there are roots of distinct moduli. He explains how the real and imaginary roots can be obtained. Dandelin's research had the misfortune of being buried in the ponderous tomes of a royal academy. Only accidentally did we come upon this anticipation of the method of C. H. Gräffe. Later the Academy of Sciences of Berlin offered a prize for the invention of a practical method of computing imaginary roots. The prize was awarded to **Carl Heinrich Gräffe** (1799–1873), professor of mathematics in Zurich, for his paper, published in 1837 in Zurich, entitled, *Die Auflösung der höheren numerischen Gleichungen*. This contains the famous "Gräffe method," to which reference has been made. Gräffe proceeds from the same principle as did *Moritz Abraham Stern* (1807–1894), of Göttingen in the method of recurrent series, and as did Dandelin. By the process of involution to higher and higher powers, the smaller roots are caused to vanish in comparison to the larger. The law by which the new equations are constructed is exceedingly simple. If, for example, the coefficient of the fourth term of the given equation is a_3, then the corresponding coefficient of the first transformed equation is $a_3^2-2a_2a_4+2a_1a_5-2a_6$. In the computation of the new coefficients, Gräffe uses logarithms. By this remarkable method all the roots, both real and imaginary, are found simultaneously, without the necessity of determining beforehand the number of real roots and the location of each root. The discussion of the case of equal imaginary roots, omitted by Gräffe, was taken up by the astronomer J. F. Encke in 1841. A simplified exposition of the Dandelin-Gräffe method was given by Emmanuel Carvallo in 1896; it resembles in some parts that of Dandelin, although Carvallo had not seen Dandelin's paper. For didactic purposes, an able explanation is given in Gustav Bauer's *Vorlesungen über Algebra*, 1903.

In 1860 E. Fürstenau expressed any definite real root of an algebraic equation with numerical or literal coefficients, in terms of its coefficients, through the aid of infinite determinants, a kind of determinant then used for the first time. In 1867 he extended his results to imaginary roots. The approximation is made to depend upon the fact used by Daniel Bernoulli, L. Euler, J. Fourier, M. A. Stern, G. Dandelin, and C. H. Gräffe, that high powers of the smaller roots are negligible in comparison with high powers of the greater roots. E. Fürstenau's process was elaborated by E. Schröder (1870), Siegmund Günther, (1874), and Hans Naegelsbach (1876).

Worthy of notice is "Weddle's method" of solving numerical equations, devised by **Thomas Weddle** (1817–1853) of Newcastle in England, in 1842. It is kindred to that of W. G. Horner. The successive approximations are effected by multiplications instead of additions. The method is advantageous when the degree of the equation is high and some of the terms are missing. It has received some attention in Italy and Germany. In 1851 Simon Spitzer extended it to the computation of complex roots.

The solution of equations by infinite series which was a favorite subject of research during the eighteenth century (Thomas Simpson, L. Euler, J. Lagrange, and others), received considerable attention during the nineteenth. Among the early workers were C. G. J. Jacobi (1830), W. S. B. Woolhouse (1868), O. Schlömilch (1849), but none of their devices were satisfactory to the practical computer. Later writers aimed at the simultaneous calculation of all the roots by infinite series. This was achieved for a three-term equation by R. Dietrich in 1883 and by P. Nekrasoff in 1887. For the general equation it was accomplished in 1895 by **Emory McClintock** (1840–1916), an actuary in New York, who was president of the American Mathematical Society from 1890 to 1894. He used a series derived by his Calculus of Enlargement, but which may be derived also by applying "Lagrange's series." A prominent part in McClintock's treatment is his theory of "dominant" coefficients, which theory lacks precision, inasmuch as no criterion is given to ascertain whether a coefficient is dominant or not, which is both necessary and sufficient. Preston A. Lambert of Lehigh University used Maclaurin's series in 1903; in 1908 he paid special attention to convergency conditions, pointing out that the conditions for a t-term equation can be set up when those of a $(t-1)$-term equation are known. In Italy Lambert's papers were studied in 1906 by C. Rossi and in 1907 by *Alfredo Capelli* (1855–1910) of Naples. These recent researches of American and Italian mathematicians have placed the determination of real and imaginary roots of numerical equations by the method of infinite series within reach of the practical computer. The methods themselves indicate the number of real and imaginary roots, so that one can dispense with the application of Sturm's theorem here just as easily as one can in the Dandelin-Gräffe method. Considerable attention has been given to the solution of special types—trinomial equations—by G. Dandelin (1826), K. F. Gauss (1840, 1843), J. Bellavitis (1846), Lord John M'Laren (1890). The last three used logarithms of sums and differences, which were first suggested by G. Z. Leonelli in 1802 and are often called "Gaussian logarithms." The extension of the Gaussian method to quadrinomials was undertaken by S. Gundelfinger in 1884 and 1885, Carl Faerber in 1889, and Alfred Wiener in 1886. The extension of the Gaussian method to any equation was taken up by R. Mehmke, professor in Darmstadt, who published in 1889 a logarithmic-graphic method of solving numerical equations, and in 1891 a more nearly arithmetical method of solution by logarithms. The method is essentially a mixture of the Newton-Raphson method and the regula falsi, as regards its theoretical basis. Well known is R. Mehmke's article on methods of computation in the *Encyklopädie der mathematischen Wissenschaften*, Vol. 1, p. 938.

Magic Squares and Combinatory Analysis

The latter part of the nineteenth century witnesses a revival of interest in methods of constructing magic squares. Chief among the writers on this subject are J. Horner (1871), S. M. Drach (1873), Th. Harmuth (1881), W. W. R. Ball (1893); E. Maillet (1894), E. M. Laquière (1880), E. Lucas (1882), E. McClintock (1897).[1] Magic squares of the "diabolic" type, as Lucas calls them, are designated "pandiagonal" by McClintock. These and similar forms are called "Nasik squares" by A. H. Frost. An interesting book, *Magic Squares and Cubes*, Chicago, 1908, was prepared by the American electrical engineer, W. S. Andrews. Still more recent is the *Combinatory Analysis*, Vol. I, Cambridge, 1915, Vol. II, 1916, by P. A. MacMahon, which touches the subject of magic squares. Says MacMahon: "In fact, the whole subject of Magic Squares and connected arrangements of numbers appears at first sight to occupy a position which is completely isolated from other departments of pure mathematics. The object of Chapters II and III is to establish connecting links where none previously existed. This is accomplished by selecting a certain differential operation and a certain algebraical function," I, p. VIII.

"The 'Problème des Rencontres' . . . can be discussed in the same manner. The reader will be familiar with the old question of the

[1] *Encyclopédie des sciences mathém.* T. I, Vol. 2, 1906, pp. 67–75.

letters and envelopes. A given number of letters are written to different persons and the envelopes correctly addressed but the letters are placed at random in the envelopes. The question is to find the probability that not one letter is put into the right envelope. The enumeration connected with this probability question is the first step that must be taken in the solution of the famous problem of the Latin Square," I, p. IX.

The problem of the Latin Square: "The question is to place n different letters $a, b, c, \ldots$ in each row of a square of n^2 compartments in such wise that, one letter being in each compartment, each column involves the whole of the letters. The number of arrangements is required. The question is famous because, from the time of Euler to that of Cayley inclusive, its solution was regarded as being beyond the powers of mathematical analysis. It is solved without difficulty by the method of differential operators of which we are speaking. In fact it is one of the simplest examples of the method which is shewn to be capable of solving questions of a much more recondite character."[1]

The extension of the principle of magic squares of the plane to three-dimensional space has commanded the attention of many. Most successful in this field were the Austrian Jesuit Adam Adamandus Kochansky (1686), the Frenchman Josef Sauveur (1710), the Germans Th. Hugel, (1859) and Hermann Scheffler (1882).

In Vol. II, Major MacMahon gives a remarkable group of identities discovered by S. Ramanujan✡ of Cambridge which have applications in the partitions of numbers, but have not yet been established by rigorous demonstration.

Analysis

Under this head we find it convenient to consider the subjects of the differential and integral calculus, the calculus of variations, infinite series, probability, differential equations and integral equations.

An early representative of the critical and philosophical school of mathematicians of the nineteenth century was **Bernard Bolzano** (1781–1848), professor of the philosophy of religion at Prague. In 1816 he gave a proof of the binomial formula and exhibited clear notions on the convergence of series. He held advanced views on variables, continuity and limits. He was a forerunner of G. Cantor. Noteworthy is his posthumous tract, *Paradoxien des Unendlichen* (Preface, 1850), edited by his pupil, Fr. Přihonsky. Bolzano's writings were overlooked by mathematicians until H. Hankel called attention to them. "He has everything," says Hankel, "that can place him in this respect [notions on infinite series] on the same level with Cauchy, only not the art peculiar to the French of refining their ideas and communicating them in the most appropriate and taking manner.

[1] P. A. MacMahon, *Combinatory Analysis*, Vol. I, Cambridge, 1915, p. ix.

So it came about that Bolzano remained unknown and was soon forgotten." H. A. Schwarz in 1872 looked upon Bolzano as the inventor of a line of reasoning further developed by K. Weierstrass. In 1881 O. Stolz declared that all of Bolzano's writings are remarkable "inasmuch as they start with an unbiassed and acute criticism of the contributions of the older literature." [1]

A reformer of our science who was eminently successful in reaching the ear of his contemporaries was Cauchy.

Augustin-Louis Cauchy [2] (1789–1857) was born in Paris, and received his early education from his father. J. Lagrange and P. S. Laplace, with whom the father came in frequent contact, foretold the future greatness of the young boy. At the École Centrale du Panthéon he excelled in ancient classical studies. In 1805 he entered the École Polytechnique, and two years later the École des Ponts et Chaussées. Cauchy left for Cherbourg in 1810, in the capacity of engineer. Laplace's *Mécanique Céleste* and Lagrange's *Fonctions Analytiques* were among his book companions there. Considerations of health induced him to return to Paris after three years. Yielding to the persuasions of Lagrange and Laplace, he renounced engineering in favor of pure science. We find him next holding a professorship at the École Polytechnique. On the expulsion of Charles X, and the accession to the throne of Louis Philippe in 1830, Cauchy, being exceedingly conscientious, found himself unable to take the oath demanded of him. Being, in consequence, deprived of his positions, he went into voluntary exile. At Fribourg in Switzerland, Cauchy resumed his studies, and in 1831 was induced by the king of Piedmont to accept the chair of mathematical physics, especially created for him at the University of Turin. In 1833 he obeyed the call of his exiled king, Charles X, to undertake the education of a grandson, the Duke of Bordeaux. This gave Cauchy an opportunity to visit various parts of Europe, and to learn how extensively his works were being read. Charles X bestowed upon him the title of Baron. On his return to Paris in 1838, a chair in the Collège de France was offered to him, but the oath demanded of him prevented his acceptance. He was nominated member of the Bureau of Longitude, but declared ineligible by the ruling power. During the political events of 1848 the oath was suspended, and Cauchy at last became professor at the Polytechnic School. On the establishment of the second empire, the oath was reinstated, but Cauchy and D. F. J. Arago were exempt from it. Cauchy was a man of great piety, and in two of his publications staunchly defended the Jesuits.

Cauchy was a prolific and profound mathematician. By a prompt publication of his results, and the preparation of standard text-books, he exercised a more immediate and beneficial influence upon the great

[1] Consult H. Bergman, *Das Philosophische Werk Bernard Bolzanos*, Halle, 1909.
[2] C. A. Valson, *La Vie et les travaux du Baron Cauchy*, Paris, 1868.

mass of mathematicians than any contemporary writer. He was one of the leaders in infusing rigor into analysis. His researches extended over the field of series, of imaginaries, theory of numbers, differential equations, theory of substitutions, theory of functions, determinants, mathematical astronomy, light, elasticity, etc.,—covering pretty much the whole realm of mathematics, pure and applied.

Encouraged by P. S. Laplace and S. D. Poisson, Cauchy published in 1821 his *Cours d'Analyse de l'École Royale Polytechnique*, a work of great merit. Had it been studied more diligently by writers of text-books, many a lax and loose method of analysis long prevalent in elementary text-books would have been discarded half a century earlier. With him begins the process of "arithmetization." He made the first serious attempt to give a rigorous proof of Taylor's theorem. He greatly improved the exposition of fundamental principles of the differential calculus by his mode of considering limits and his new theory on the continuity of functions. Before him, the limit concept had been emphasized in France by D'Alembert, in England by I. Newton, J. Jurin, B. Robins, and C. Maclaurin.✡ The method of Cauchy was accepted with favor by J. M. C. Duhamel, G. J. Hoüel, and others. In England special attention to the clear exposition of fundamental principles was given by A. De Morgan. Cauchy re-introduced the concept of an integral of a function as the limit of a sum, a concept originally due to G. W. Leibniz, but for a time displaced by L. Euler's integral defined as the result of reversing differentiation.

Calculus of Variations

A. L. Cauchy made some researches on the calculus of variations. This subject had long remained in its essential principles the same as when it came from the hands of J. Lagrange. More recent studies pertain to the variation of a double integral when the limits are also variable, and to variations of multiple integrals in general. Memoirs were published by K. F. Gauss in 1829, S. D. Poisson in 1831, and Michel Ostrogradski (1801–1861) of St. Petersburg in 1834, without, however, determining in a general manner the number and form of the equations which must subsist at the limits in case of a double or triple integral. In 1837 C. G. J. Jacobi✡ published a memoir, showing that the difficult integrations demanded by the discussion of the second variation, by which the existence of a maximum or minimum can be ascertained, are included in the integrations of the first variation, and thus are superfluous. This important theorem, presented with great brevity by C. G. J. Jacobi, was elucidated and extended by V. A. Lebesgue, C. E. Delaunay, Friedrich Eisenlohr (1831–1904), Simon Spitzer (1826–1887) of Vienna, L. O. Hesse and R. F. A. Clebsch. A memoir by Pierre Frédéric Sarrus (1798–1861) of the University of Strasbourg on the question of determining the limiting equations which must be combined with the indefinite equations in order to determine completely the maxima and minima of multiple integrals, was awarded a prize by the French Academy in 1845, honorable mention being made of a paper by C. E. Delaunay. P. F. Sarrus's method was simplified by A. L. Cauchy. In 1852 Gaspare Mainardi (1800–1879) of Pavia attempted to exhibit a new method of discriminating maxima and minima, and extended C. G. J. Jacobi's theorem to double integrals. Mainardi and F. Brioschi showed the value of determinants in exhibiting the terms of the second variation. In 1861 *Isaac Todhunter* (1820–1884) of St. John's College, Cambridge, published his valuable work on the *History of the Progress of the Calculus of Variations*, which contains researches of his own. In 1866 he published a most important research, developing the theory of discontinuous solutions (discussed in particular cases by A. M. Legendre), and doing for this subject what P. F. Sarrus had done for multiple integrals.

The following are the more important older authors of systematic treatises on the calculus of variations, and the dates of publication: Robert Woodhouse, Fellow of Caius College, Cambridge, 1810; Richard Abbatt in London, 1837; John Hewitt Jellett (1817–1888), once Provost of Trinity College, Dublin, 1850; Georg Wilhelm Strauch (1811–1868), of Aargau in Switzerland, 1849; François Moigno (1804–1884) of Paris, and Lorentz Leonard Lindelöf (1827–1908) of the University of Helsingfors, in 1861; Lewis Buffett Carll in 1881. Carll (1844–1918), was a blind mathematician, graduated at Columbia College in 1870 and in 1891–1892 was assistant in mathematics there.

That, of all plane curves of given length, the circle includes a maximum area, and of all closed surfaces of given area, the sphere encloses a maximum volume, are theorems considered by Archimedes and Zenodorus, but not proved rigorously for two thousand years until K. Weierstrass and H. A. Schwarz. Jakob Steiner thought he had proved the theorem for the circle. On a closed plane curve different from a circle four non-cyclic points can be selected. The quadrilateral obtained by successively joining the sides has its area increased when it is so deformed (the lunes being kept rigid) that its vertices are cyclic. Hence the total area is increased, and the circle has the maximum area. Oskar Perron of Tübingen pointed out in 1913 by an example the fallacy of this proof: Let us "prove" that 1 is the largest of all positive integers. No such integer larger than 1 can be the maximum, for the reason that its square is larger than itself. Hence, 1 must be the maximum. Steiner's "proof" does not prove that among all closed plane curves of given length there exists one whose area is a maximum.[1] K. Weierstrass gave a simple general existence theorem applicable to the extremes of continuous (stetige) functions. The maximal property of the sphere was first proved rigorously in 1884 by

[1] See W. Blaschke in *Jahresb. d. deutsch. Math. Vereinig.*, Vol. 24, 1915, p. 195.

H. A. Schwarz by the aid of results reached by K. Weierstrass in the calculus of variations. Another proof, based on geometrical theorems, was given in 1901 by Hermann Minkowski.

The subject of minimal surfaces, which had received the attention of J. Lagrange, A. M. Legendre, K. F. Gauss and G. Monge, in later time commanded the special attention of H. A. Schwarz. The blind physicist of the University of Gand, *Joseph Plateau* (1801–1883), in 1873 described a way of presenting these surfaces to the eye by means of soap bubbles made of glycerine water. Soap bubbles tend to become as thick as possible at every point of their surface, hence to make their surfaces as small as possible. More recent papers on minimal surfaces are by *Harris Hancock* of the University of Cincinnati.

Ernst Pascal of the University of Pavia expressed himself in 1897 on the calculus of variations as follows:[1] "It may be said that this development [the finding of the differential equations which the unknown functions in a problem must satisfy] closes with J. Lagrange, for the later analysts turned their attention chiefly to the other, more difficult problems of this calculus. The problem is finally disposed of, if one considers the simplicity of the formulas which arise; wholly different is this matter, if one considers the subject from the standpoint of rigor of derivation of the formulas and the extension of the domain of the problems to which these formulas are applicable. This last is what has been done for some years. It has been found necessary to prove certain theorems which underlie those formulas and which the first workers looked upon as axioms, which they are not." This new field was first entered by I. Todhunter, M. Ostrogradski, C. G. J. Jacobi, J. Bertrand, P. du Bois-Reymond, G. Erdmann, R. F. A. Clebsch, but the incisive researches which mark a turning-point in the history of the subject are due to K. Weierstrass. As an illustration of Weierstrass's method of communicatiing many of his mathematical results to others, we quote the following from O. Bolza:[2] "Unfortunately they [results on the calculus of variations] were given by Weierstrass only in his lectures [since 1872], and thus became known only very slowly to the general mathematical public. . . . Weierstrass's results and methods may at present be considered as generally known, partly through dissertations and other publications of his pupils, partly through A. Kneser's *Lehrbuch der Variationsrechnung* (Braunschweig, 1900), partly through sets of notes ('Ausarbeitungen') of which a great number are in circulation and copies of which are accessible to every one in the library of the Mathematische Verein at Berlin, and in the Mathematische Lesezimmer at Göttingen. Under these circumstances I have not hesitated to make use of Weierstrass's lectures just as if they had been published in print." Weierstrass applied modern requirements of rigor to the calculus of varia-

[1] E. Pascal, *Die Variationsrechnung*, übers. v. A. Schepp, Leipzig, 1899, p. 5

[2] O. Bolza, *Lectures on the Calculus of Variations*, Chicago, 1904, pp. ix, xi.

tions in the study of the first and second variation. Not only did he give rigorous proofs for the first three necessary conditions and for the sufficiency of these conditions for the so-called "weak" extremum, but he also extended the theory of the first and second variation to the case where the curves under consideration are given in parameter representation. He discovered the fourth necessary condition and a sufficiency proof for a so-called "strong" extremum, which gave for the first time a complete solution by means of a new method, based on the so-called "Weierstrass's construction."[1] Under the stimulus of Weierstrass, new developments, were made by A. Kneser, then of Dorpat, whose theory is based on the extension of certain theorems on geodesics to extremals in general, and by David Hilbert of Göttingen, who gave an "*a priori* existence proof for an extremum of a definite integral—a discovery of far-reaching importance, not only for the Calculus of Variations, but also for the theory of differential equations and the theory of functions" (O. Bolza). In 1909 Bolza published an enlarged German edition of his calculus of variations, including the results of Gustav v. Escherich of Vienna, the Hilbert method of proving Lagrange's rule of multipliers (multiplikator-regel), and the J. W. Lindeberg of Helsingfors treatment of the isoperimetric problem. About the same time appeared J. Hadamard's *Calcul des variations recuellies par M. Fréchet*, Paris, 1910. *Jacques Hadamard* (1865–1963), born at Versailles, was editor of the *Annales scientifiques de l'école normale supérieure.* In 1912 he was appointed professor of mathematical analysis at the École Polytechnique of Paris as successor to Camille Jordan. In the above mentioned book he regards the calculus of variations as a part of a new and broader "functional calculus," along the lines followed also by V. Volterra in his functions of lines. This functional calculus was initiated by Maurice Fréchet of the University of Poitiers in France. The authors include also researches by W. F. Osgood. Other prominent researches on the calculus of variations are due to J. G. Darboux, E. Goursat, E. Zermelo, H. A. Schwarz, H. Hahn, and to the Americans H. Hancock, G. A. Bliss, E. R. Hedrick, A. L. Underhill, Max Mason. Bliss and Mason systematically extended the Weierstrassian theory of the calculus of variations to problems in space.

In 1858 *David Bierens de Haan* (1822–1895) of Leiden published his *Tables d'Intégrales Définies*. A revision and the consideration of the underlying theory appeared in 1862–1867. It contained 8339 formulas. A critical examination of the latter, made by E. W. Sheldon in 1912, showed that it was "remarkably free from error when one imposes proper limitations upon constants and functions, not stated by Haan.

The lectures on definite integrals, delivered by P. G. L. Dirichlet in 1858, were elaborated into a standard work in 1871 by Gustav Ferdinand Meyer of Munich.

[1] This summary is taken from O. Bolza, *op. cit.*, Preface.

Convergence of Series

The history of infinite series illustrates vividly the salient feature of the new era which analysis entered upon during the first quarter of this century. I. Newton and G. W. Leibniz felt the necessity of inquiring into the convergence of infinite series, but they had no proper criteria, excepting the test advanced by Leibniz for alternating series. By L. Euler and his contemporaries the *formal* treatment of series was greatly extended, while the necessity for determining the convergence was generally lost sight of. L. Euler reached some very pretty results on infinite series, now well known, and also some very absurd results, now quite forgotten. The faults of his time found their culmination in the Combinatorial School in Germany, which has now passed into oblivion. This combinatorial school was founded by *Carl Friedrich Hindenburg* (1741–1808) of Leipzig whose pupils filled many of the German University chairs during the first decennium of the nineteenth century. The first important and strictly rigorous investigation of infinite series was made by K. F. Gauss in connection with the hypergeometric series. This series,* thus named by J. F. Pfaff, had been treated by L. Euler in 1769 and 1778 from the triple standpoint of a power-series, of the integral of a certain linear differential equation of the second order, and of a definite integral. The criterion developed by K. F. Gauss settles the question of convergence of the hypergeometric series in every case which it is intended to cover, and thus bears the stamp of generality so characteristic of Gauss's writings. Owing to the strangeness of treatment and unusual rigor, Gauss's paper excited little interest among the mathematicians of that time.

More fortunate in reaching the public was A. L. Cauchy, whose *Analyse Algébrique* of 1821 contains a rigorous treatment of series. All series whose sum does not approach a fixed limit as the number of terms increases indefinitely are called divergent. Like Gauss, he institutes comparisons with geometric series, and finds that series with positive terms are convergent or not, according as the nth root of the nth term, or the ratio of the $(n+1)$th term and the nth term, is ultimately less or greater than unity. To reach some of the cases where these expressions become ultimately unity and fail, Cauchy established two other tests. He showed that series with negative terms converge when the absolute values of the terms converge, and then deduces G. W. Leibniz's test for alternating series. The product of two convergent series was found to be not necessarily convergent. Cauchy's theorem that the product of two absolutely convergent series converges to the product of the sums of the two series was shown half a century later by F. Mertens of Graz to be still true if, of the two convergent series to be multiplied together, only one is absolutely convergent.

The most outspoken critic of the old methods in series was N. H.

Abel. His letter to his friend B. M. Holmboe (1826) contains severe criticisms. It is very interesting reading, even to modern students. In his demonstration of the binomial theorem he established the theorem that if two series and their product series are all convergent, then the product series will converge towards the product of the sums of the two given series. This remarkable result would dispose of the whole problem of multiplication of series if we had a universal practical criterion of convergency for semi-convergent series. Since we do not possess such a criterion, theorems have been recently established by A. Pringsheim of Munich and A. Voss later of Munich which remove in certain cases the necessity of applying tests of convergency to the product series by the application of tests to easier related expressions. A. Pringsheim reached the following interesting conclusions: The product of two conditionally convergent series can never converge absolutely, but a conditionally convergent series, or even a divergent series, multiplied by an absolutely convergent series, *may* yield an absolutely convergent product.

The researches of N. H. Abel and A. L. Cauchy caused a considerable stir. We are told that after a scientific meeting in which Cauchy had presented his first researches on series, P. S. Laplace hastened home and remained there in seclusion until he had examined the series in his *Mécanique Céleste*. Luckily, every one was found to be convergent! We must not conclude, however, that the new ideas at once displaced the old. On the contrary, the new views were generally accepted only after a long struggle. As late as 1844 A. De Morgan began a paper on "divergent series" in this style: "I believe it will be generally admitted that the heading of this paper describes the only subject yet remaining, of an elementary character, on which a serious schism exists among mathematicians as to the absolute correctness or incorrectness of results."

First in time in the evolution of more delicate criteria of convergence and divergence come the researches of Josef Ludwig Raabe (1801–1859) of Zurich, in *Crelle*, Vol. IX; then follow those of A. De Morgan as given in his calculus. A. De Morgan established the logarithmic criteria which were discovered in part independently by J. Bertrand. The forms of these criteria, as given by J. Bertrand and by Ossian Bonnet, are more convenient than De Morgan's. It appears from N. H. Abel's posthumous papers that he had anticipated the above-named writers in establishing logarithmic criteria. It was the opinion of Bonnet that the logarithmic criteria never fail; but P. Du Bois-Reymond and A. Pringsheim have each discovered series demonstrably convergent in which these criteria fail to determine the convergence. The criteria thus far alluded to have been called by Pringsheim *special* criteria, because they all depend upon a comparison of the nth term of the series with special functions a^n, n^x, $n(\log n)^x$, etc. Among the first to suggest *general* criteria, and to consider the subject from a still wider point of view, culminating in a regular mathematical theory, was E. E. Kummer. He established a theorem yielding a test consisting of two parts, the first part of which was afterwards found to be superfluous. The study of general criteria was continued by Ulisse Dini (1845–1918) of Pisa, P. Du Bois-Reymond, G. Kohn of Vienna, and A. Pringsheim. Du Bois-Reymond divides criteria into two classes: criteria of the *first kind* and criteria of the *second kind*, according as the general nth term, or the ratio of the $(n+1)$th term and the nth term, is made the basis of research. E. E. Kummer's is a criterion of the second kind. A criterion of the first kind, analogous to this, was invented by A. Pringsheim. From the general criteria established by Du Bois-Reymond and Pringsheim respectively, all the special criteria can be derived. The theory of Pringsheim is very complete, and offers, in addition to the criteria of the first kind and second kind, entirely new criteria of a *third kind*, and also generalized criteria of the second kind, which apply, however, only to series with never increasing terms. Those of the third kind rest mainly on the consideration of the limit of the difference either of consecutive terms or of their reciprocals. In the generalized criteria of the second kind he does not consider the ratio of two consecutive terms, but the ratio of any two terms however far apart, and deduces, among others, two criteria previously given by Gustav Kohn and W. Ermakoff respectively.

It is a strange vicissitude that divergent series, which early in the nineteenth century were supposed to have been banished once for all from rigorous mathematics, should at its close be invited to return. In 1886 T. J. Stieltjes and H. Poincaré showed the importance to analysis of the asymptotic series, at that time employed in astronomy alone. In other fields of research G. H. Halphen, E. N. Laguerre, and T. J. Stieltjes have encountered particular examples in which, a whole series being divergent, the corresponding continued fraction was convergent. In 1894 H. Padé now of Bordeaux, established the possibility of defining, in certain cases, a function by an entire divergent series. This subject was taken up also by J. Hadamard in 1892, C. E. Fabry in 1896 and M. Servant in 1899. Researches on divergent series have been carried on also by H. Poincaré, E. Borel, T. J. Stieltjes, E. Cesàro, W. B. Ford of Michigan and R. D. Carmichael of Illinois. *Thomas-Jean Stieltjes* (1856–1894) was born in Zwolle in Holland, came in 1882 under the influence of Ch. Hermite, became a French citizen, and later received a professorship at the University of Toulouse. Stieltjes was interested not only in divergent and conditionally convergent series, but also in G. F. B. Riemann's ζ function and the theory of numbers.

Difficult questions arose in the study of Fourier's series.[1] A. L.

[1] Arnold Sachse, *Versuch einer Geschichte der Darstellung willkürlicher Funktionen einer variablen durch trigonometrische Reihen*, Göttingen, 1879.

Cauchy was the first who felt the necessity of inquiring into its convergence. But his mode of proceeding was found by P. G. L. Dirichlet to be unsatisfactory. Dirichlet made the first thorough researches on this subject (*Crelle*, Vol. IV). They culminate in the result that whenever the function does not become infinite, does not have an infinite number of discontinuities, and does not possess an infinite number of maxima and minima, then Fourier's series converges toward the value of that function at all places, except points of discontinuity, and there it converges toward the mean of the two boundary values. L. Schläfli of Bern and P. Du Bois-Reymond expressed doubts as to the correctness of the mean value, which were, however, not well founded. Dirichlet's conditions are sufficient, but not necessary. *Rudolf Lipschitz* (1832–1903), of Bonn, proved that Fourier's series still represents the function when the number of discontinuities is infinite, and established a condition on which it represents a function having an infinite number of maxima and minima. Dirichlet's belief that all continuous functions can be represented by Fourier's series at all points was shared by G. F. B. Riemann and H. Hankel, but was proved to be false by Du Bois-Reymond and H. A. Schwarz. A. Hurwitz showed how to express the product of two ordinary Fourier series in the form of another Fourier series. W. W. Küstermann solved the analogous problem for double Fourier series in which a relation involving Fourier constants figures vitally. For functions of a single variable an analogous relation is due to M. A. Parseval and was proved by him under certain restrictions on the nature of convergence of the Fourier series involved. In 1893 de la Vallée Poussin gave a proof requiring merely that the function and its square be integrable. A. Hurwitz in 1903 gave further developments. More recently the subject has commanded general interest through the researches of Frigyes Riesz and Ernst Fischer (Riesz-Fischer theorem).[1]

Riemann inquired what properties a function must have, so that there may be a trigonometric series which, whenever it is convergent, converges toward the value of the function. He found necessary and sufficient conditions for this. They do not decide, however, whether such a series actually represents the function or not. Riemann rejected Cauchy's definition of a definite integral on account of its arbitrariness, gave a new definition, and then inquired when a function has an integral. His researches brought to light the fact that continuous functions need not always have a differential coefficient. But this property, which was shown by K. Weierstrass to belong to large classes of functions, was not found necessarily to exclude them from being represented by Fourier's series. Doubts on some of the conclusions about Fourier's series were thrown by the observation, made by Weierstrass, that the integral of an infinite series can be shown to be equal to the sum of the integrals of the separate terms

[1] Summary taken from *Bull. Am. Math. Soc.*, Vol. 22, 1915, p. 6.

only when the series converges *uniformly* within the region in question. The subject of uniform convergence was first investigated in 1847 by G. G. Stokes of Cambridge and in 1848 by *Philipp Ludwig v. Seidel* (1821–1896). Seidel had studied under F. W. Bessel, C. G. J. Jacobi, J. F. Encke, and P. G. L. Dirichlet. He became professor at the University of Munich in 1855. Later his lecturing and scientific activity were stopped by a disease of his eyes. Uniform convergence assumed great importance in K. Weierstrass' theory of functions. It became necessary to prove that a trigonometric series representing a continuous function converges uniformly. This was done by *Heinrich Eduard Heine* (1821–1881), of Halle. Later researches on Fourier's series were made by G. Cantor and *Paul Du Bois-Reymond* (1831–1889), professor at the technical high school in Charlottenburg.

Less stringent than that of uniform convergence is U. Dini's definition [1] of "simple uniform convergence," which is as follows: The series is said to be simply uniformly convergent in the interval (a, b) when corresponding to every arbitrarily chosen positive number σ as small as we please and to every integer m', only one or several integers m exist which are not less than m', and are such that, for all the values of x in the interval (a, b), the $| R_m(x) |$ are <5. Still another kind of convergence, the "uniform convergence by segments," sometimes called "sub-uniform convergence," was introduced in 1883 by *Cesare Arzelà* (1847–1912) of the University of Bologna. He advanced the theory of functions of real variables and generalized a theorem of U. Dini on the necessary and sufficient conditions for the continuity of the sum of a convergent series of continuous functions.

Probability and Statistics✡

As compared with the vast development of other mathematical branches, the theory of probability has made✡ little progress from the time of Laplace until the 1930's. Jakob Bernoulli's Theorem which had received the careful attention of De Moivre, J. Stirling, C. Maclaurin, and L. Euler, was considered especially by P. S. Laplace who made an inverse application of it, assuming that an event had been observed to happen m times and to fail n times in μ trials, and then deducing the initially unknown probability of its happening at each trial. The result thus obtained did not agree accurately with the results gotten by the use of Bayes' Theorem. The subject was investigated by S. D. Poisson in his *Recherches sur la probabilité*, Paris, 1837, who obtained consonant results after carrying the approximations, in the use of Bayes' Theorem, to a higher degree. An endeavor to remove the obscurities in which Bayes' Theorem seemed involved was made by Poisson, and by A. De Morgan in his *Theory of Probabil-*

[1] *Grundlagen f. e. Theorie der Functionen*, by J. Lüroth u. A. Schepp, Leipzig, 1892, p. 137.

ities,[1] through the use of illustrations with urns that were exactly alike and contained black and white balls in different numbers and different ratios, the observed event being the drawing of a white ball from any one of the urns. For the same purpose Johannes von Kries, in his *Prinzipien der Wahrscheinlichkeitsrechnung*, Freiburg i. B., 1886, used as an illustration six equal cubes, of which one had the + sign on one side, another had it on two sides, a third on three sides and finally the sixth on all six sides. All other sides were marked with a o. Nevertheless, objections to certain applications of Bayes' Theorem were raised by the Danish actuary J. Bing in the *Tidsskrift for Matematik*, 1879, by Joseph Bertrand in his *Calcul des probabilités*, Paris, 1889, by Thorwald Nicolai Thiele (1838–1910) of the observatory at Copenhagen in a work published at Copenhagen in 1889 (an English edition of which appeared under the title *Theory of Observations*, London, 1903) by George Chrystal (1851–1911) of the University of Edinburgh, and others.[2] As recently as 1908 the Danish philosophic writer Kroman has come out in defence of Bayes. Thus it appears that, as yet, no unanimity of judgment has been reached in this matter. In determining the probability of alternative causes deduced from observed events there is often need of evidence other than that which is afforded by the observed event. By inverse probability some logicians have explained induction. For example, if a man, who has never heard of the tides, were to go to the shore of the Atlantic Ocean and witness on m successive days the rise of the sea, then, says *Adolphe Quetelet* of the observatory at Brussels, he would be entitled to conclude that there was a probability equal to $\frac{m+1}{m+2}$ that the sea would rise next day. Putting $m=0$, it is seen that this view rests upon the unwarrantable assumption that the probability of a totally unknown event is $\frac{1}{2}$, or that of all theories proposed for investigation one-half are true. William Stanley Jevons (1835–1882) in his *Principles of Science* founds induction upon the theory of inverse probability, and F. Y. Edgeworth also accepts it in his *Mathematical Psychics*. Daniel Bernoulli's "moral expectation," which was elaborated also by Laplace, has received little attention from more recent French writers. Bertrand emphasizes its impracticability; Poincaré, in his *Calcul des probabilités*, Paris, 1896, disposes of it in a few words.[3]

The only noteworthy pre-1930 addition to probability is the subject of "local☼ probability," developed by several English and a few American and French mathematicians. G. L. L. Buffon's needle problem is the earliest important problem on local probability; it received the

[1] *Encyclopædia Metrop.* II, 1845.

[2] We are using Emanuel Czuber's *Entwickelung der Wahrscheinlichkeitstheorie* in the *Jahresb. d. deutsch. Mathematiker-Vereinigung*, 1899, pp. 93–105; also Arne Fisher, *The Mathematical Theory of Probabilities*, New York, 1915, pp. 54–56.

[3] E. Czuber, *op. cit.*, p. 121.

consideration of P. S. Laplace, of Emile Barbier in the years 1860 and 1882, of Morgan W. Crofton (1826–1915) of the military school at Woolwich, who in 1868 contributed a paper to the London *Philosophical Transactions*, Vol. 158, and in 1885 wrote the article "Probability" in the *Encyclopedia Brittannica*, ninth edition. The name "local probability" is due to Crofton. Through considerations of local probability he was led to the evaluation of certain definite integrals.

Noteworthy is J. J. Sylvester's four point problem: To find the probability that four points taken at random within a given boundary, shall form a re-entrant quadrilateral. Local probability was studied in England also by A. R. Clarke, H. McColl, S. Watson, J. Wolstenholme, W. S. Woolhouse; in France also by C. Jordan and E. Lemoine; in America by E. B. Seitz. Rich collections of problems on local probability have been published by Emanuel Czuber of Vienna in his *Geometrische Wahrscheinlichkeiten und Mittelwerte*, Leipzig, 1884, and by G. B. M. Zerr in the *Educational Times*, Vol. 55, 1891, pp. 137–192. The fundamental concepts of local probability have received the special attention of Ernesto Cesàro (1859–1906) of Naples.[1]

Criticisms occasionally passed upon the principles of probability and lack of confidence in theoretical results have induced several scientists to take up the experimental side, which had been emphasized by G. L. L. Buffon. Trials of this sort were made by A. De Morgan, W. S. Jevons, L. A. J. Quetelet, E. Czuber, R. Wolf, and showed a remarkably close agreement with theory. In Buffon's needle problem, the theoretical probability involves π. This and similar expressions [2] have been used for the empirical determination of π. Attempts to place the theory of probability on a purely empirical basis were made by John Stuart Mill (1806–1873), John Venn (1834–1923) and G. Chrystal. Mill's induction method was put on a sounder basis by A. A. Chuproff in a brochure, *Die Statistik als Wissenschaft*. Empirical methods have commanded the attention of another Russian, v. Bortkievicz.

In 1835 and 1836 the Paris Academy was led by S. D. Poisson's researches to discuss the topic, whether questions of morality could be treated by the theory of probability. M. H. Navier argued on the affirmative, while L. Poinsot and Ch. Dupin denied the applicability as "une sorte d'aberration de l'esprit;" they declared the theory applicable only to cases where a separation and counting of the cases or events was possible. John Stuart Mill opposed it; Joseph L. F. Bertrand (1822–1900), professor at the Collége de France in Paris and J. v. Kries are among more recent writers on this topic.[1]

[1] See *Encyclopédie des sciences math.* I, 20 (1906), p. 23.

[2] E. Czuber, *op. cit.*, pp. 88–91.

[3] Consult E. Czuber, *op. cit.*, p. 141; J. S. Mill, *System of Logic*, New York, 8th Ed., 1884, Chap. 18, pp. 379–387; J. v. Kries, *op. cit.*, pp. 253–259.

Among the various applications of probability the one relating to verdicts of juries, decisions of courts and results of elections is specially interesting. This subject was studied by Marquis de Condorcet, P. S. Laplace, and S. D. Poisson. To exhibit Laplace's method of determining the worth of candidates by combining the votes, M. W. Crofton employs the fortuitous division of a straight line. This involves, however, an *a priori* distribution of values covering evenly the whole range from 0 to 100. Experience shows that the normal law of error exhibits a more correct distribution. On this point Karl Pearson produced a most important research.[1] He took a random sample of n individuals from a population of N members and derived an expression for the average difference in character between the pth and the $(p+1)$th individual when the sample is arranged in order of magnitude of the character. H. L. Moore of Columbia University has attempted to trace Pearson's theory in the statistics relating to the efficiency of wages (*Economic Journal*, Dec., 1907).

Early statistical study was carried on under the name of "political arithmetic" by such writers as Captain John Graunt of London (1662) and J. P. Süssmilch, a Prussian clergyman (1788). Application of the theory of probability to statistics was made by Edmund Halley, Jakob Bernoulli, A. De Moivre, L. Euler, P. S. Laplace, and S. D. Poisson. The establishment of official statistical societies and statistical offices was largely due to the influence of the Belgian astronomer and statistician, **Adolphe Quetelet** (1796–1874) of the observatory at Brussels, "the founder of modern statistics." Quetelet's "average man" in whom "all processes correspond to the average results obtained for society," who "could be considered as a type of the beautiful," has given rise to much critical discussion by Harold Westergaard (1890), J. Bertillon (1896), A. de Foville in his "homo medius" of 1907, Joseph Jacobs in his "the Middle American" and "the Mean Englishman."[2] Quetelet's visit in England led to the organization, in 1833, of the statistical section of the British Association for the Advancement of Science, and in 1835 of the Statistical Society of London. Soon after, in 1839, was formed the American Statistical Society. Quetelet's best researches on the application of probability to the physical and social sciences are given in a series of letters to the duke of Saxe-Coburg and Gotha, *Lettres sur la théorie des probabilités*, Brussels, 1846. He laid emphasis on the "law of large numbers," which was advanced also by the Frenchman S. D. Poisson and discussed by the German W. Lexis (1877), the Scandinavians H. Westergaard and Carl Charlier, and the Russian *Pafnuti Liwowich Chebichev* (1821–1894) of the University of Petrograd. To Chebichev we owe also an interesting problem: A proper fraction being

[1] "Note on Francis Galton's Problem," *Biometrica*, Vol. I, pp. 390–399.

[2] See Franz Žižek's *Statistical Averages*, transl. by W. M. Persons, New York, 1913, p. 374.

chosen at random, what is the probability that it is in its lowest terms?

Of the different kinds of averages, A. De Morgan concluded that the arithmetic mean represents *a priori* the most probable values. J. W. L. Glaisher took exception to this. G. T. Fechner investigated the cases where the "median" (which has the central position in a series of items arranged according to size) may be used profitably. The "mode," an average introduced by K. Pearson in 1895, and used by G. Udny Yule, has been applied in Germany and Austria to the fixing of workingmen's insurance. The theory of averages has been studied by the aid of the calculus of probabilities by W. Lexis, F. Y. Edgeworth of Oxford, H. Westergaard, L. von Bortkewich, G. T. Fechner, J. von Kries, E. Czuber of Vienna, E. Blaschke, F. Galton, K. Pearson, G. U. Yule, and A. L. Bowley of London. A few writers take the ground that it is not only unnecessary to employ probability in founding statistical theory, but that it is inadvisable to do so. Among such writers are G. F. Knapp and A. M. Guerry.[1] The Russian actuary Jastremski in 1912 applied the Lexian dispersion theory to the testing of the influence of medical selection in life insurance. Other recent publications of note are by Lexis' pupil L. von Bortkewich and by Harold Westergaard of Copenhagen. Early theories of population were involved in much confusion. E. Halley and some eighteenth century writers proceeded on the assumption of a stationary population. L. Euler adopted the hypothesis that the yearly births progress in a geometric series. This was combatted in 1839 by L. Moser, while G. F. Knapp in 1868 represented the number of births and deaths as a continuous function of the time and of the age, respectively. He made use of graphic representation. G. Zeuner in 1869 introduced additional geometric and analytic aids. In 1874 Knapp made still further modifications, allowing for discontinuous changes, such as were studied also by W. Lexis, in his *Theorie der Bevölkerungsstatistik*, Strassburg, 1875. Formal theories of population and the determination of mortality were investigated also by K. Becker in 1867 and 1874, and by Th. Wittstein, about 1881. In 1877 W. Lexis introduced the idea of "dispersion" and "normal dispersion." Wilhelm Lexis (1837–1914) became in 1872 professor at Strassburg, in 1884 at Breslau and in 1887 at Göttingen. In 1893 he was drawn into the service for the German government.

The application of statistical method to biology was begun by *Sir Francis Galton* (1822–1911), "a born statistician." Important is his *Natural Inheritance*, 1889, in which he uses the method of percentiles, with the quartile deviation as the measure of dispersion.[2] Two other Englishmen entered this field of research, **Karl Pearson** of University College and W. F. R. Weldon. Pearson developed

[1] *Encyklopädie d. Math. Wissensch*, I D 4a, p. 822.

[2] G. Udny Yule, *Theory of Statistics*, 2. Ed., London, 1912, p. 154.

general and adequate mathematical methods for the analysis of biological statistics. To him are due the terms "mode," "standard deviation" and "coefficient of variation." Before him the "normal curve" of errors had been used exclusively to describe the distribution of chance events. This curve is symmetrical, but natural phenomena sometimes indicate an asymmetrical distribution. Accordingly Pearson, in his *Contributions to the Theory of Evolution*, 1899, developed skew frequency curves. About 1890 the Georg Mendel law of inheritance became generally known and caused some modification in the application of statistics to heredity. Such a readjustment was effected by the Danish botanist W. Johannsen.[1]

The first study of the most advantageous combinations of data of observation is due to Roger Cotes, in the appendix to his *Harmonia mensurarum*, 1722, where he assigns weights to the observations. The use of the arithmetic mean was advocated by Thomas Simpson in a paper "An attempt to show the advantage arising by taking the mean of a number of observations, in practical astronomy," [2] also by J. Lagrange in 1773 and by Daniel Bernoulli in 1778. The first printed statement of the principle of least squares was made in 1806 by A. M. Legendre, without demonstration. K. F. Gauss had used it still earlier, but did not publish it until 1809. The first deduction of the law of probability of error that appeared in print was given in 1808 by Robert Adrain in the *Analyst*, a journal published by himself in Philadelphia. Of the earlier proofs given of this law, perhaps the most satisfactory is that of P. S. Laplace. K. F. Gauss gave two proofs. The first rests upon the assumption that the arithmetic mean of the observations is the most probable value. Attempts to prove this assumption have been made by Laplace, J. F. Encke (1831), A. De Morgan (1864), G. V. Schiaparelli, E. J. Stone (1873), and A. Ferrero (1876). Valid criticisms upon some of these investigations were passed by J. W. L. Glaisher.[3] The founding of the Gaussian probability law upon the nature of the observed errors was attempted by F. W. Bessel (1838), G. H. L. Hagen (1837), J. F. Encke (1853), P. G. Tait (1867), and M. W. Crofton (1870). That the arithmetic mean, taken as the most probable value, is not under all circumstances compatible with the Gaussian probabilty law has been shown by Joseph Bertrand in his *Calcul des probabilités* (1889), and by others.[4] The development of the theory of least squares along practical lines is due mainly to K. F. Gauss, J. F. Encke, P. A. Hansen, Th. Galloway, J. Bienaymé, J. Bertrand, A. Ferrero, P. Pizzetti.[5] Simon

[1] *Quart. Pub. Am. Stat. Ass'n*, N. S., Vol. XIV, 1914, p. 45.

[2] *Miscellaneous Tracts*, London, 1757.

[3] *Lond. Astr. Soc. Mem.* 39, 1872, p. 75; for further references, see *Cyklopädie d. Math. Wiss.*, I D2, p. 772.

[4] Consult E. L. Dodd, "Probability of the Arithmetic Mean, etc.," *Annals of Mathematics*, 2. S., Vol. 14, 1913, p. 186.

[5] E. Czuber, *op. cit.*, p. 179.

Newcomb of Washington advanced a "generalized theory of the combination of observations so as to obtain the best result,"[1] when large errors arise more frequently than is allowed by the Gaussian probability law. The same subject was treated by R. Lehmann-Filhés in *Astronomische Nachrichten*, 1887.

A criterion for the rejection of doubtful observations[2] was given by Benjamin Peirce of Harvard. It was accepted by the American astronomers B. A. Gould (1824–1896), W. Chauvenet (1820–1870), and J. Winlock (1826–1875), but was criticised by the English astronomer G. B. Airy. The prevailing feeling has been that there exists no theoretical basis upon which such criterion can be rightly established.

The application of probability to epidemiology was first considered by Daniel Bernoulli and has more recently commanded the attention of the English statisticians William Farr (1807–1883), John Brownlee, Karl Pearson, and Sir Ronald Ross. Pearson studied normal and abnormal frequency curves. Such curves have been fitted to epidemics by J. Brownlee in 1906, S. M. Greenwood in 1911 and 1913, and Sir Ronald Ross in 1916.[3]

Some interest attaches to the discussion of whist from the standpoint of the theory of probability, as is contained in William Pole's *Philosophy of Whist*, New York and London, 1883. The problem is a generalization of the game of "treize" or "recontre," treated by Pierre R. de Montmort in 1708.

Differential Equations. Difference Equations

Criteria for distinguishing between singular solutions and particular solutions of differential equations of the first order were advanced by A. M. Legendre, S. D. Poisson, S. F. Lacroix, A. L. Cauchy, and G. Boole. After J. Lagrange, the c-discriminant relation commanded the attention of *Jean Marie Constant Duhamel* (1797–1872) of Paris, C. L. M. H. Navier, and others. But the entire theory of singular solutions was re-investigated about 1870 along new paths by J. G. Darboux, A. Cayley, E. C. Catalan, F. Casorati, and others. The geometric side of the subject was considered more minutely and the cases were explained in which Lagrange's method does not yield singular solutions. Even these researches were not altogether satisfactory as they did not furnish necessary and sufficient conditions for singular solutions which depend on the differential equation alone and not in any way upon the general solution. Returning to more purely analytical considerations and building on work of Ch. Briot and J. C. Bouquet of 1856, Carl Schmidt of Giessen in 1884, H. B. Fine of Princeton in 1890, and Meyer Hamburger (1838–1903) of

[1] *Am. Jour. Math.*, Vol. 8, 1886, p. 343.
[2] *Gould Astr. Jour.*, II, 1852.
[3] *Nature*, Vol. 97, 1916, p. 243.

Berlin brought the problem to final solutions. Active in this line were also John Muller Hill and A. R. Forsyth.[1]

The first scientific treatment of partial differential equations was given by J. Lagrange and P. S. Laplace. These equations were investigated in more recent time by G. Monge, J. F. Pfaff, C. G. J. Jacobi, Émile Bour (1831–1866) of Paris, A. Weiler, R. F. A. Clebsch, A. N. Korkine of St. Petersburg, G. Boole, A. Meyer, A. L. Cauchy, J. A. Serret, Sophus Lie, and others. In 1873 their reseaches, on partial differential equations of the first order, were presented in text-book form by Paul Mansion, of the University of Gand. Proceeding to the consideration of some detail, we remark that the keen researches of **Johann Friedrich Pfaff** (1765–1825) marked a decided advance. He was an intimate friend of K. F. Gauss at Göttingen. Afterwards he was with the astronomer J. E. Bode. Later he became professor at Helmstädt, then at Halle. By a peculiar method, Pfaff found the general integration of partial differential equations of the first order for any number of variables. Starting from the theory of ordinary differential equations of the first order in n variables, he gives first their general integration, and then considers the integration of the partial differential equations as a particular case of the former, assuming, however, as known, the general integration of differential equations of any order between two variables. His researches led C. G. J. Jacobi to introduce the name "Pfaffian problem." From the connection, observed by W. R. Hamilton, between a system of ordinary differential equations (in analytical mechanics) and a partial differential equation, C. G. J. Jacobi drew the conclusion that, of the series of systems whose successive integration Pfaff's method demanded, all but the first system were entirely superfluous. R. F. A. Clebsch considered Pfaff's problem from a new point of view, and reduced it to systems of simultaneous linear partial differential equations, which can be established independently of each other without any integration. Jacobi materially advanced the theory of differential equations of the first order. The problem to determine unknown functions in such a way that an integral containing these functions and their differential coefficients, in a prescribed manner, shall reach a maximum or minimum value, demands, in the first place, the vanishing of the first variation of the integral. This condition leads to differential equations, the integration of which determines the functions. To ascertain whether the value is a maximum or a minimum, the second variation must be examined. This leads to new and difficult differential equations, the integration of which, for the simpler cases, was ingeniously deduced by C. G. J. Jacobi from the integration of the differential equations of the first variation. Jacobi's solution was perfected by L. O. Hesse, while R. F. A. Clebsch extended

[1] We have used S. Rothenberg, "Geschichte . . . der singulären Lösungen" in *Abh. z. Gesch. d. Math. Wissensch.* (M. Cantor), Heft XX, 3. Leipzig, 1908.

to the general case Jacobi's results on the second variation. A. L. Cauchy gave a method of solving partial differential equations of the first order having any number of variables, which was corrected and extended by J. A. Serret, J. Bertrand, O. Bonnet in France, and Wassili Grigorjewich Imshenetski (1832–1892) of the University of Charkow in Russia. Fundamental is the proposition of Cauchy that every ordinary differential equation admits in the vicinity of any non-singular point of an integral, which is synectic within a certain circle of convergence, and is developable by Taylor's theorem. Allied to the point of view indicated by this theorem is that of G. F. B. Riemann, who regards a function of a single variable as defined by the position and nature of its singularities, and who has applied this conception to that linear differential equation of the second order, which is satisfied by the hypergeometric series. This equation was studied also by K. F. Gauss and E. E. Kummer. Its general theory, when no restriction is imposed upon the value of the variable, has been considered by J. Tannery, of Paris, who employed L. Fuchs' method of linear differential equations and found all of Kummer's twenty-four integrals of this equation. This study has been continued by *Édouard Goursat* (1858–1936), professor of mathematical analysis in the University of Paris. His activities have been in the theory of functions, pseudo- and hyper-elliptic integrals, differential equations, invariants and surfaces. *Jules Tannery* (1848–1910) became professor of mechanics at the Sorbonne in 1875, and sub-director at the École Normale in Paris in 1884. His researches have been in the field of analysis and the theory of functions.

As outlined by A. R. Forsyth [1] in 1908, the status of partial differential equations is briefly as follows: Since the posthumous publication, in 1862, of C. G. J. Jacobi's treatment of partial differential equations of the first order involving only one dependent variable, or a system of such equations, it may be said that we have a complete method of formal integration of such equations. In the formal integration of a partial differential equation of the second or higher orders new difficulties are encountered. Only in rare instances is direct integration possible. The known normal types of integrals even for such equations of only the second order are few in number. The primitive may be given by means of a single relation between the variables, or by means of a number of equations involving eliminable parameters (such as the customary forms, due to A. M. Legendre, G. Monge, or K. Weierstrass, of the primitive of minimal surfaces), or by means of a relation involving definite integrals arising in problems in physics. "With all these types of primitives," says A. R. Forsyth,[2] "it being assumed that immediate and direct integration is impossible—, a

[1] A. R. Forsyth in *Atti. del IV. Congr. Intern., Roma, 1908.* Vol. I, Roma, 1909, p. 90.

[2] A. R. Forsyth, *loc. cit.*, p. 90. We are summarizing part of this article.

primitive is obtained by the use of processes, that sometimes are fragmentary in theory, usually are tentative in practice and nearly always are indirect in the sense that they are compounded of a number of formal operations having no organic relation with the primitive. In such circumstances . . . is the primitive completely comprehensive of all the integrals belonging to the equation?" A. M. Ampère in 1815 propounded a broad definition of a general integral—one in which the only relations, which subsist among the variables and the derivatives of the dependent variable and which are free from the arbitrary elements in the integral, are constituted by the differential equation itself and by equations deduced from it by differentiation. This definition is incomplete on various grounds. E. Goursat gave in 1898 a simple instance to show that an integral satisfying all of Ampère's requirements was not general. A second definition of a general integral was given in 1889 by J. G. Darboux, based on A. L. Cauchy's existence-theorem: An integral is general when the arbitrary elements which it contains can be specialized in such a way as to provide the integral established in that theorem. This definition, according to A. R. Forsyth, calls for a more careful discussion of obvious and latent singularities.

There are three principal methods of proceeding to the construction of an integral of partial differential equations of the second order, which lead to success in special cases. One method given by P. S. Laplace in 1777 applies to linear equations with two independent variables. It can be used for equations of order higher than the second. It has been developed by J. G. Darboux and V. G. Imshenetski, 1872. A second method, originated by A. M. Ampère, while general in spirit and in form, depends upon individual skill unassisted by critical tests. Later researches along this line are due to E. Borel (1895) and E. T. Whittaker (1903). A third method is due to J. G. Darboux and includes, according to A. R. Forsyth's classification, the earlier work of Monge and G. Boole. As first given by J. G. Darboux in 1870, it applied only to the case of two independent variables, but it has been extended to equations of more than two independent variables and orders higher than the second; it is not universally effective. "Such then," says Forsyth, "are the principal methods hitherto devised for the formal integration of partial equations of the second order. They have been discussed by many mathematicians and they have been subjected to frequent modifications in details: but the substance of the processes remains unaltered."

Instances are known in ordinary linear equations when the primitives can be expressed by definite integrals or by means of asymptotic expansions, the theory of which owes much to H. Poincaré. Such instances within the region of partial equations are due to E. Borel.

G. F. B. Riemann had remarked in 1857 that functions expressed by K. F. Gauss' hypergeometric series F $(\alpha, \beta, \gamma, x)$, which satisfy a homogeneous linear differential equation of the second order with rational coefficients, might be utilized in the solution of any linear differential equation.[1] Another mode of solving such equations was due to Cauchy and was extended by C. A. A. Briot and J. C. Bouquet) and consisted in the development into power series. The fertility of the conceptions of G. F. B. Riemann and A. L. Cauchy with regard to differential equations is attested by the researches to which they have given rise on the part of **Lazarus Fuchs** (1833–1902) of Berlin. Fuchs was born in Moschin, near Posen, and became professor at the University of Berlin in 1884. In 1865 L. Fuchs combined the two methods in the study of linear differential equations: One method using power-series, as elaborated by A. L. Cauchy, C. A. A. Briot, and J. C. Bouquet; the other method using the hypergeometric series as had been done by G. F. B. Riemann. By this union Fuchs initiated a new theory of linear differential equations.[2] Cauchy's development into power-series together with the calcul des limites, afforded existence theorems which are essentially the same in nature as those relating to differential equations in general. The singular points of the linear differential equation received attention also from G. Frobenius in 1874, G. Peano in 1889, M. Bôcher in 1901. A second approach to existence theorems was by successive approximation, first used in 1864 by J. Caqué, then by L. Fuchs in 1870, and later by H. Poincaré and G. Peano. A third line, by interpolation, is originally due to A. L. Cauchy and received special attention from V. Volterra in 1887. The general theory of linear differential equations received the attention of L. Fuchs, and of a large number of workers, including C. Jordan, V. Volterra, and L. Schlesinger. Singular places where the solutions are not indeterminate were investigated by J. Tannery, L. Schlesinger, G. J. Wallenberg, and many others. Ludwig Wilhelm Thomé (1841–1910) of the University of Greifswald, discovered in 1877 what he called normal integrals. Divergent series which formally satisfy differential equations, noticed by C. A. A. Briot and J. C. Bouquet in 1856, were first seriously considered by H. Poincaré in 1885 who pointed out that such series may represent certain solutions asymptotically. Asymptotic representations have been examined by A. Kneser (1896), E. Picard (1896), J. Horn (1897), and A. Hamburger (1905). A special type of linear differential equation, the "Fuchsian type," with coefficients that are single-valued (eindeutig), and the solutions of which have no points of indeterminateness, was investigated by Fuchs, and it was found that the coefficients of such an equation are rational functions of x. Studies based on analogies of linear differential equations with algebraic equations, first undertaken by

[1] We are using L. Schlesinger, *Entwickelung d. Theorie d. linearen Differentialgleichungen seit 1865*, Leipzig and Berlin, 1909.

[2] We are using here a report by L. Schlesinger in *Jahresb. d. d. Math. Vereinigung*, Vol. 18, 1909, pp. 133–266.

N. H. Abel, J. Liouville and C. G. J. Jacobi, were pursued later by P. Appell (1880), by E. Picard who worked under the influence of S. Lie's theory of transformation groups, and by an army of workers in France, England, Germany, and the United States. The consideration of differential invariants enters here. Lamé's differential equation, considered by him in 1857, was taken up by Ch. Hermite in 1877 and soon after in still more generalized form by L. Fuchs, F. Brioschi, E. Picard, G. M. Mittag-Leffler and F. Klein.

The analogies of linear differential equations with algebraic functions, problems of inversion and uniformization, as well as questions involving group theory received the attention of the analysts of the second half of the century.

The theory of invariants associated with linear differential equations as developed by Halphen and by A. R. Forsyth is closely connected with the theory of functions and of groups. Endeavors have thus been made to determine the nature of the function defined by a differential equation from the differential equation itself, and not from any analytical expression of the function, obtained first by solving the differential equation. Instead of studying the properties of the integrals of a differential equation for all the values of the variable, investigators at first contented themselves with the study of the properties in the vicinity of a given point. The nature of the integrals at singular points and at ordinary points is entirely different. *Charles Auguste Albert Briot* (1817–1882) and *Jean Claude Bouquet* (1819–1885) both of Paris, studied the case when, near a singular point, the differential equations take the form $(x-x_0)\frac{dy}{dx}=\int(xy)$. L. Fuchs gave the development in series of the integrals for the particular case of linear equations. H. Poincaré did the same for the case when the equations are not linear, as also for partial differential equations of the first order. The developments for ordinary points were given by A. L. Cauchy and *Sophie Kovalevski* (1850–1891). Madame Kovalevski was born at Moscow, was a pupil of K. Weierstrass and became professor of Analysis at Stockholm.

Henri Poincaré (1854–1912) was born at Nancy and commenced his studies at the Lycée there. While taking high rank as a student, he did not display exceptional precocity. He attended the École Polytechnique and the École Nationale Supérieure des Mines in Paris, receiving his doctorate from the University of Paris in 1879. He became instructor in mathematical analysis at the University of Caen. In 1881 he occupied the chair of physical and experimental mechanics at the Sorbonne, later the chair of mathematical physics and, after the death of F. Tisserand, the chair of mathematical astronomy and celestial mechanics. Although he did not reach old age, he published numerous books and more than 1500 memoirs. Probably neither

A. L. Cauchy nor even L. Euler equalled him in the quantity of scientific productions. P. Painlevé said that every one of his many papers carried the mark of a lion. Poincaré wrote on mathematics, physics, astronomy, and philosophy. No other scientist of his day was able to work in such a wide range of subjects. Many consider him the greatest mathematician of his time. Each year he lectured on a different subject; these lectures were reported and published by his former students. In this manner were brought out works on capillarity, elasticity, Newtonian potential, vortices, the propagation of heat, thermodynamics, light, electric oscillations, electricity and optics, Hertzian oscillations, mathematical electricity, kinematics, equilibrium of fluid masses, celestial mechanics, general astronomy, probability. His popular works on the philosophy of science, *La science et l'hypothèse* (1902), *La valeur de la science* (1905), *Science et méthode* (1908) have been translated into German, in part also into Spanish, Hungarian, and Japanese. An English translation by George Bruce Halsted appeared in one volume in 1913.

Our numerous references to Poincaré will indicate that he wrote on nearly every branch of pure mathematics. Says F. R. Moulton:[1] "The importance of his papers can be inferred from the enormous number of references to his theorems in all modern treatises, especially on the various branches of analysis. The emphasis on analysis does not mean that he neglected geometry, analysis situs, groups, number theory, or the foundations of mathematics, for he illuminated all these subjects and others; but it is placed there because this domain includes his researches on differential equations, dating from his doctor's dissertation to very recent times, his contributions to the theory of functions, and his discovery of fuchsian and theta-fuchsian functions. His command of the powerful methods of modern analysis was positively dazzling." As to his method of work E. Borel says: "The method of Poincaré is essentially active and constructive. He approaches a question, acquaints himself with its present condition without being much concerned about its history, finds out immediately the new analytical formulas by which the question can be advanced, deduces hastily the essential results, and then passes to another question. After having finished the writing of a memoir, he is sure to pause for a while, and to think out how the exposition could be improved; but he would not, for a single instance, indulge in the idea of devoting several days to didactic work. Those days could be better utilized in exploring new regions." Poincaré tells how he came to make his first mathematical discoveries: "For a fortnight I labored to demonstrate that there could exist no function analogous to those that I have since called the fuchsian functions. I was then very ig-

[1] *Popular Astronomy*, Vol. 20, 1912. We are using also Ernest Lebon, *Henri Poincaré, Biographie*, Paris, 1909; "Jules Henri Poincaré" in *Nature*, Vol. 90, London, 1912, p. 353; George Sarton, "Henri Poincaré" in *Ciel et Terre*, 1913.

norant. Every day I seated myself at my work table and spent an hour or two there, trying a great many combinations, but I arrived at no result. One night when, contrary to my custom, I had taken black coffee and I could not sleep, ideas surged up in crowds. I felt them as they struck against one another until two of them stuck together, so to speak, to form a stable combination. By morning I had established the existence of a class of fuchsian functions, those which are derived from the hypergeometric series. I had merely to put the results in shape, which only took a few hours."[1]

Poincaré enriched the theory of integrals. The attempt to express integrals by developments that are always convergent and not limited to particular points in a plane necessitates the introduction of new transcendents, for the old functions permit the integration of only a small number of differential equations. H. Poincaré tried this plan with linear equations, which were then the best known, having been studied in the vicinity of given points by L. Fuchs, L. W. Thomé, G. Frobenius, H. A. Schwarz, F. Klein, and G. H. Halphen. Confining himself to those with rational algebraical coefficients, H. Poincaré was able to integrate them by the use of functions named by him *Fuchsians*.[2] He divided these equations into "families." If the integral of such an equation be subjected to a certain transformation, the result will be the integral of an equation belonging to the same family. The new transcendents have a great analogy to elliptic functions; while the region of the latter may be divided into parallelograms, each representing a group, the former may be divided into curvilinear polygons, so that the knowledge of the function inside of one polygon carries with it the knowledge of it inside the others. Thus H. Poincaré arrives at what he calls *Fuchsian groups*. He found, moreover, that Fuchsian functions can be expressed as the ratio of two transcendents (theta-fuchsians) in the same way that elliptic functions can be. If, instead of linear substitutions with real coefficients, as employed in the above groups, imaginary coefficients be used, then discontinuous groups are obtained, which he called *Kleinians*. The extension to non-linear equations of the method thus applied to linear equations was begun by L. Fuchs and H. Poincaré.

Much interest attaches to the determination of those linear differential equations which can be integrated by simpler functions, such as algebraic, elliptic, or Abelian. This has been studied by C. Jordan, P. Appell of Paris, and H. Poincaré.

Paul Appell (1855–1930) was born in Strassburg. After the annexation of Alsace to Germany in 1871, he emigrated to Nancy to escape German citizenship. Later he studied in Paris and in 1886

[1] H. Poincaré, *The Foundations of Science*, transl. by G. B. Halsted, The Science Press, New York and Garrison, N. Y., 1913, p. 387.

[2] Henri Poincaré, *Notice sur les Travaux Scientifiques de Henri Poincaré*, Paris, 1886, p. 9.

became professor of mechanics there. His researches are in analysis, function theory, infinitesimal geometry and rational mechanics.

Whether an ordinary differential equation has one or more solutions which satisfy certain terminal or boundary conditions, and, if so, what the character of these solutions is, has received renewed attention the last quarter century by the consideration of finer and more remote questions.[1] Existence theorems, oscillation properties, asymptotic expressions, development theorems have been studied by David Hilbert of Göttingen, Maxime Bôcher of Harvard, Max Mason of the University of Wisconsin, Mauro Picone of Turin, R. M. E. Mises of Strassburg, H. Weyl of Göttingen and especially by George D. Birkhoff of Harvard. Integral equations have been used to some extent in boundary problems of one dimension; "this method would seem, however, to be chiefly valuable in the cases of two or more dimensions where many of the simplest questions are still to be treated."

A standard text-book on *Differential Equations*, including original matter on integrating factors, singular solutions, and especially on symbolical methods, was prepared in 1859 by G. Boole.☼

A *Treatise on Linear Differential Equations* (1889) was brought out by Thomas Craig of the Johns Hopkins University. He chose the algebraic method of presentation followed by Ch. Hermite and H. Poincaré, instead of the geometric method preferred by F. Klein and H. A. Schwarz. A notable work, the *Traité d'Analyse*, 1891–1896, was published by Émile Picard of Paris, the interest of which was made to centre in the subject of differential equations. A second edition has appeared.

Simple difference equations or "finite differences" were studied by eighteenth century mathematicians. When in 1882 H. Poincaré developed the novel notion of asymptotic representation, he applied it to linear difference equations. In recent years a new type of problem has arisen in connection with them. It looks now as if the continuity of nature, which has been for so long assumed to exist, were a fiction and as if discontinuities represented the realities. "It seems almost certain that electricity is done up in pellets, to which we have given the name of electrons. That heat comes in quanta also seems probable." [2] Much of theory based on the assumption of continuity may be found to be mere approximation. Homogeneous linear difference equations, not intimately bound up with continuity, were taken up independently by investigators widely apart. In 1909 Niels Erik Nörlund of the University of Lund in Sweden, Henri Galbrun of l'École Normale in Paris and, in 1911, R. D. Carmichael of the University of Illinois entered this field of research. Carmichael used a method of successive approximation and an extension of a contour integral due to

[1] See a historical summary by Maxime Bôcher in *Proceed. of the 5th Intern. Congress, Cambridge*, 1912, Vol. I, Cambridge, 1913, p. 163.

[2] R. D. Carmichael in *Science*, N. S. Vol. 45, 1917, p. 472.

C. Guichard. G. D. Birkhoff of Harvard made important contributions showing the existence of certain intermediate solutions and of the principal solutions. The asymptotic form of these solutions is determined by him throughout the complex plane. The extension to non-homogeneous equations of results reached for homogeneous ones has been made by K. P. Williams of the University of Indiana.[1]

Integral Equations, Integro-differential Equations, General Analysis, Functional Calculus

The mathematical perplexities which led to the invention of integral equations were stated by J. Hadamard [2] in 1911 as follows: "Those problems (such as Dirichlet's) exercised the sagacity of geometricians and were the object of a great deal of important and well-known work through the whole of the nineteenth century. The very variety of ingenious methods applied showed that the question did not cease to preserve its rather mysterious character. Only in the last years of the century were we able to treat it with some clearness and understand its true nature. . . . Let us therefore inquire by what device this new view of Dirichlet's problem was obtained. Its peculiar and most remarkable feature consists in the fact that the partial differential equation is put aside and replaced by a new sort of equation, namely, the integral equation. This new method makes the matter as clear as it was formerly obscure. In many circumstances in modern analysis, contrary to the usual point of view, the operation of integration proves a much simpler one than the operation of derivation. An example of this is given by integral equations where the unknown function is written under such signs of integration and not of differentiation. The type of equation which is thus obtained is much easier to treat than the partial differential equation. The type of integral equations corresponding to the plane Dirichlet problem is

$$\phi(x)-\lambda\int_A^B \phi(y)\, K\,(x,y)\, dy=f(x), \tag{1}$$

where ϕ is the unknown function of x in the interval (A, B), f and K are known functions, and λ is a known parameter. The equations of the elliptic type in many-dimensional space give similar integral equations, containing however multiple integrals and several independent variables. Before the introduction of equations of the above type, each step in the study of elliptic partial differential equations seemed to bring with it new difficulties. . . . [But] an equation such as (1) . . . gives all the required results at once and for all the possible types of such problems. . . . Previously, in the calculation of

[1] *Trans. Am. Math. Soc.*, Vol. 14, 1913, p. 209.

[2] J. Hadamard, *Four Lectures on Mathematics delivered at Columbia University in 1911*, New York, 1915, pp. 12–15.

the resonance of a room filled with air, the shape of the resonator had to be quite simple, which requirement is not a necessary one for the case where integral equations are employed. We need only make the elementary calculation of the function K and apply to the function so calculated the general method of resolution of integral equations."

The new departure in analysis was made in 1900 by **Eric Ivar Fredholm** (1866–1927), a native of Stockholm, who in 1898 was docent at the University of Stockholm and later became connected with the imperial bureau of insurance. In a paper,[1] "Sur une nouvelle méthode pour la résolution du problème de Dirichlet," 1900, he studied an integral equation from the point of view of an immediate generalization of a system of linear equations. Integral equations bear the same relation to the integral calculus as differential equations do to the differential calculus. Before this time certain integral equations had received the attention of N. H. Abel, J. Liouville, and Eugène Rouché (1832–1910) of Paris, but were quite neglected.

Abel had in 1823 proposed a generalization of the tautochrone problem, the solution of which involved an integral equation that has since been designated as of the first kind. Liouville in 1837 showed that a particular solution of a linear differential equation of the second order could be found by solving an integral equation, now designated as of the second kind. A method of solving integral equations of the second kind was given by C. Neumann (1887). The term "integral equation" is due to P. du Bois-Reymond (*Crelle*, Vol. 103, 1888, p. 228) who exemplified the danger of making predictions by the declaration that "the treatment of such equations seems to present insuperable difficulties to the analysis of to-day." The recent theory of integral equations owes its origin to specific problems in mechanics and mathematical physics. Since 1900 these equations have been used in the study of existence theorems in the theory of potential; they were employed in 1904 by W. A. Stekloff and D. Hilbert in the consideration of boundary values and in matters relating to Fourier series, by Henri Poincaré in the study of tides and Hertzian waves. Linear integral equations present many analogies with linear algebraic equations. While E. I. Fredholm used the theory of algebraic equations merely to suggest theorems on integral equations, which were proved independently, D. Hilbert in his early work on this subject followed the process of taking limits in the results of algebraic theory. Hilbert has introduced the term "kernel" of linear integral equations of the first and second kind. The theory of integral equations has been advanced by Erhard Schmidt of Breslau and Vito Volterra of Rome. Systematic treatises on integral equations have been prepared by Maxime Bôcher of Harvard (1909), Gerhard Kowalewski of Prag (1909), Adolf Kneser of Breslau (1911), T. Lalesco of Paris (1912), H. B. Heywood, and M. Fréchet (1912).

[1] *Öfversigt af akademiens förhandlingar* 57, Stockholm, 1900.

Maxime Bôcher (1867–1918) was born in Boston and graduated at Harvard in 1888. After three years of study at Göttingen he returned to Harvard where he was successively instructor, assistant-professor and professor of mathematics. He was president of the American Mathematical Society in 1909–1910. Among his works are *Reihenentwickelungen der Potential-theorie*, 1891, enlarged in 1894, and *Leçons sur les Méthodes de Sturm*, containing the author's lectures delivered at the Sorbonne in 1913–1914.

A. Voss in 1913 stresses the value of integral equations thus:[1] "In the last ten years . . . the theory of integral equations has attained extraordinary importance, because through them problems in the theory of differential equations may be solved which previously could be disposed of only in special cases. We abstain from sketching their theory, which makes use of infinite determinants that belong to linear equations with an infinite number of unknowns, of quadratic forms with infinitely many variables, and which has succeeded in throwing new light upon the great problems of pure and applied mathematics, especially of mathematical physics."

Important advances along the line of a "general analysis" and its application to a generalization of the theory of linear integral equations have been made since 1906 by E. H. Moore of the University of Chicago.[2] From the existence of analogies in different theories he infers the existence of a general theory comprising the analogous theories as special cases. He proceeds to a "unification," resulting, first, from the recent generalization of the concept of independent variable effected by passing from the consideration of variables defined for all points in a given interval to that of variables defined for all points in any given set of points lying in the range of the variable, secondly, from the consideration of functions of an infinite as well as a finite number of variables, and, thirdly, from a still further generalization which leads him to functions of a "general variable." E. H. Moore's general theory includes as special cases the theories of E. I. Fredholm, D. Hilbert, and E. Schmidt. G. D. Birkhoff in 1911 presented the following birds'-eye view of recent movements:[3] "Since the researches of G. W. Hill, V. Volterra, and E. I. Fredholm in the direction of extended linear systems of equations, mathematics has been in the way of great development. That attitude of mind which conceives of the function as a generalized point, of the method of successive approximation as a Taylor's expansion in a function variable, of the calculus of variations as a limiting form of the ordinary algebraic problem of maxima and minima is now crystallizing into a new branch of mathematics under the leadership of S. Pincherle, J.

[1] A. Voss, *Ueber das Wesen d. Math.*, 1913, p. 63.

[2] See *Proceed. 5th Intern. Congress of Mathematicians*, Cambridge, 1913, Vol. I, p. 230.

[3] *Bull. Am. Math. Soc.*, Vol. 17, 1911, p. 415.

Hadamard, D. Hilbert, E. H. Moore, and others. For this field Professor Moore proposes the term 'General Analysis,' defined as 'the theory of systems of classes of functions, functional operations, etc., involving at least one general variable on a general range.' He has fixed attention on the most abstract aspect of this field by considering functions of an absolutely general variable. The nearest approach to a similar investigation is due to M. Fréchet (Paris thesis, 1906), who restricts himself to variables for which the notion of a limiting value is valid." Researches along the line of E. H. Moore's "General Analysis" are due to A. D. Pitcher of Adelbert College and E. W. Chittenden of the University of Illinois. In his "General Analysis" Moore defines "complete independence" of postulates which has received the further attention of E. V. Huntington, R. D. Beetle, L. L. Dines, and M. G. Gaba.

V. Volterra discusses integro-differential equations which involve not only the unknown functions under signs of integration but also the unknown functions themselves and their derivatives, and shows their use in mathematical physics. G. C. Evans of the Rice Institute extended A. L. Cauchy's existence theorem for partial differential equations to integro-differential equations of the "static type" in which the variables of differentiation are different from those of integration. Mixed linear integral equations have been discussed by W. A. Hurwitz of Cornell University.

The study of integral equations and the theory of point sets has led to the development of a body of theory called functional calculus. One part of this is the theory of the functions of a line. As early as 1887 Vito Volterra of the University of Rome developed the fundamental theory of what he called functions depending on other functions and functions of curves. Any quantity which depends for its value on the arc of a curve as a whole is called a function of the line. The relationships of functions depending on other functions are called "fonctionelles" by J. Hadamard in his *Leçons sur le calcul des variations*, 1910, and "functionals" by English writers. Functional equations and systems of functional equations have received the attention of Griffith C. Evans of the Rice Institute, Luigi Sinigallia of Pavia, Giovanni L. T. C. Giorgi of Rome, A. R. Schweitzer of Chicago, Eric H. Neville of Cambridge, and others. Neville solves the racecourse puzzle of covering a circle by a set of five circular discs. Says G. B. Mathews:[1] "We must express our regret that English mathematics is so predominantly analytical. Cannot some one, for instance, give us a truly geometrical theory of J. V. Poncelet's poristic polygons, or of von Staudt's thread-constructions for conicoids?" In the theory of functional equations, "a single equation or a system of equations expressing some property is taken as the definition of a class of functions whose characteristics, particular as well as collective,

[1] *Nature*, Vol. 97, 1916, p. 398.

are to be developed as an outcome of the equations" (E. B. Van Vleck).

An important generalization of Fourier series has been made, "and we have a great class of expansions in the so-called orthogonal and biorthogonal functions arising in the study of differential and integral equations. In the field of differential equations the most important class of these functions was first defined in a general and explicit manner (in 1907) by . . . G. D. Birkhoff of Harvard University; and their leading fundamental properties were developed by him."[1] In boundary value problems of differential equations which are not self-adjoint, biorthogonal systems of functions play the same rôle as the orthogonal systems do in the self-adjoint case. Anna J. Pell established theorems for biorthogonal systems analogous to those of F. Riesz and E. Fischer for orthogonal systems.

Theories of Irrationals and Theory of Aggregates

The new non-metrical theories of the irrational were called forth by the demands for greater rigor. The use of the word "quantity" as a geometrical magnitude without reference to number and also as a number which measures some magnitude was disconcerting, especially as there existed no safe ground for the assumption that the same rules of operation applied to both. The metrical view of number involved the entire theory of measurement which assumed greater difficulties with the advent of the non-Euclidean geometries. In attempts to construct arithmetical theories of number, irrational numbers were a source of trouble. It was not satisfactory to operate with irrational numbers as if they were rational. What are irrational numbers? Considerable attention was paid to the definition of them as limits of certain sequences of rational numbers. A. L. Cauchy in his *Cours d'Analyse*, 1821, p. 4, says "an irrational number is the limit of diverse fractions which furnish more and more approximate values of it." Probably Cauchy was satisfied of the existence of irrationals on geometric grounds. If not, his exposition was a reasoning in a circle. To make this plain, suppose we have a development of rational numbers and we desire to define *limit* and also *irrational number*. With Cauchy we may say that "when the successive values attributed to a variable approach a fixed value indefinitely so as to end by differing from it as little as is wished, this fixed value is called the limit of all the others." Since we are still confined to the field of rational numbers, this limit, if not rational, is non-existent and fictitious. If now we endeavor to define irrational number as a limit, we encounter a break-down in our logical development. It became desirable to define irrational number arithmetically without reference to limits. This was achieved independently and at almost

[1] R. D. Carmichael in *Science*, N. S., Vol. 45, 1917, p. 471.

the same time by four men, **Charles Méray** (1835–1911), K. Weierstrass, R. Dedekind, and Georg Cantor. Méray's first publication was in the *Revue des sociétés savantes: sc. math.* (2) 4, 1869, p. 284; his later publications were in 1872, 1887, 1894. Méray was born at Châlons in France and was professor at the University of Dijon. The earliest publication of K. Weierstrass's presentation was made by H. Kossak in 1872. In the same year it was published in *Crelle's Journal*, Vol. 74, p. 174, by E. Heine who had received it from Weierstrass by oral communication. R. Dedekind's publication is *Stetigkeit und irrationale Zahlen*, Braunschweig, 1872. In 1888 appeared his *Was sind und was sollen die Zahlen?*☆ **Richard Dedekind** (1831–1916) was born in Braunschweig, studied at Göttingen and in 1854 became privatdocent there. From 1858 to 1862 he was professor at the Polytechnicum in Zurich as the successor of J. L. Raabe, and from 1863 to 1894 professor at the Technical High School in Braunschweig. He worked on the theory of numbers. The substance of his *Stetigkeit und Irrationale Zahlen*, was worked out by him before he left Zurich. He worked also in function theory. Georg Cantor's first printed statement was in *Mathem. Annalen*, Vol. 5, 1872, p. 123.[1] **Georg Cantor** (1845–1918) was born at Petrograd, lived from 1856 to 1863 in South Germany, studied from 1863 to 1869 at Berlin where he came under the influence of Weierstrass. While in Berlin he once defended the remarkable thesis: In mathematics the art of properly stating a question is more important than the solving of it (In re mathematica ars proponendi quæstionem pluris facienda est quam solvendi). He became privatdocent at Halle in 1869, extraordinary professor in 1872 and ordinary professor in 1879. In his final years he suffered from ill health, taking the form of mental disturbances. When emerging from such attacks, his mind was said to be most productive of scientific results. Nearly all his papers are on the development of the theory of aggregates. It had been planned to hold on March 3, 1915, an international celebration of his seventieth birthday, but on account of the war, only a few German friends gathered at Halle to do him honor.

In the theories of Ch. Méray and G. Cantor the irrational number is obtained by an endless sequence of rational numbers $a_1, a_2, a_3, \ldots$ which have the property $|a_n - a_m| < \epsilon$, provided n and m are sufficiently great. The method of K. Weierstrass is a special case of this. R. Dedekind defined every "cut" in the system of rational numbers to be a number, the "open cuts" constituting the irrational numbers. To G. Cantor and Dedekind we owe the important theory of the linear continuum which represents the culmination of efforts which go back to the church fathers of the Middle Ages and the writings of Aristotle. By this modern continuum, "the notion of number, integral or frac-

[1] For details see *Encyclopédie des sciences mathématiques*, Tome I, Vol. I, 1904, §§ 6–8, pp. 147–155.

tional, has been placed upon a basis entirely independent of measurable magnitude, and pure analysis is regarded as a scheme which deals with number only, and has, *per se*, no concern with measurable quantity. Analysis thus placed upon an arithmetical basis is characterized by the rejection of all appeals to our special intuitions of space, time and motion, in support of the possibility of its operations" (E. W. Hobson). The arithmetization of mathematics, which was in progress during the entire nineteenth century, but mainly during the time of Ch. Méray, L. Kronecker, and K. Weierstrass, was characterized by E. W. Hobson in 1902 in the following terms:[1] "In some of the textbooks in common use in this country, the symbol ∞ is still used as if it denoted a number, and one in all respects on a par with the finite numbers. The foundations of the integral calculus are treated as if Riemann had never lived and worked. The order in which double limits are taken is treated as immaterial, and in many other respects the critical results of the last century are ignored. . . .

"The theory of exact measurement in the domain of the ideal objects of abstract geometry is not immediately derivable from intuition, but is now usually regarded as requiring for its development a previous independent investigation of the nature and relations of number. The relations of number having been developed on an independent basis, the scheme is applied by the help of the principle of congruency, or other equivalent principle, to the representation of extensive or intensive magnitude. . . . This complete separation of the notion of number, especially fractional number, from that of magnitude, involves, no doubt, a reversal of the historical and psychological orders. . . . The extreme arithmetizing school, of which, perhaps, L. Kronecker was the founder, ascribes reality, whatever that may mean, to integral numbers only, and regards fractional numbers as possessing only a derivative character, and as being introduced only for convenience of notation. The ideal of this school is that every theorem of analysis should be interpretable as giving a relation between integral numbers only. . . .

"The true ground of the difficulties of the older analysis as regards the existence of limits, and in relation to the application to measurable quantity, lies in its inadequate conception of the domain of number, in accordance with which the only numbers really defined were rational numbers. This inadequacy has now been removed by means of a purely arithmetical definition of irrational numbers, by means of which the continuum of real numbers has been set up as the domain of the independent variable in ordinary analysis. This definition has been given in the main in three forms—one by E. Heine and G. Cantor, the second by R. Dedekind, and the third by K. Weierstrass. Of these the first two are the simplest for working purposes, and are essentially equivalent to one another; the difference between

[1] *Proceed. London Math. Soc.*, Vol. 35, 1902, pp. 117–139; see p. 118.

them is that, while Dedekind defines an irrational number by means of a section of all the rational numbers, in the Heine-Cantor form of definition a selected convergent aggregate of such numbers is employed. The essential change introduced by this definition of irrational numbers is that, for the scheme of rational numbers, a new scheme of numbers is substituted, in which each number, rational or irrational, is defined and can be exhibited in an indefinitely great number of ways, by means of a convergent aggregate of rational numbers. . . . By this conception of the domain of number the root difficulty of the older analysis as to the existence of a limit is turned, each number of the continuum being really defined in such a way that it itself exhibits the limit of certain classes of convergent sequences. . . . It should be observed that the criterion for the convergence of an aggregate is of such a character that no use is made in it of infinitesimals, definite finite numbers alone being used in the tests. The old attempts to prove the existence of limits of convergent aggregates were, in default of a previous arithmetical definition of irrational number, doomed to inevitable failure. . . . In such applications of analysis—as, for example, the rectification of a curve—the length of the curve is defined by the aggregate formed by the lengths of a proper sequence of inscribed polygons. . . . In case the aggregate is not convergent, the curve is regarded as not rectifiable. . . .

"It has in fact been shown that many of the properties of functions, such as continuity, differentiability, are capable of precise definition when the domain of the variable is not a continuum, provided, however, that domain is perfect; this has appeared clearly in the course of recent investigations of the properties of non-dense perfect aggregates, and of functions of a variable whose domain is such an aggregate."

In 1912 Philip E. B. Jourdain of Fleet, near London, characterized theories of the irrational substantially as follows: [1] "Dedekind's theory had not for its object to prove the existence of irrationals: it showed the necessity, as Dedekind thought, for the mathematician to create them. In the idea of the creation of numbers, Dedekind was followed by O. Stolz; but H. Weber and M. Pasch showed how the supposition of this creation could be avoided: H. Weber defined real numbers as sections (*Schnitte*) in the series of rationals; M. Pasch (like B. Russell) as the segments which generate these sections. In K. Weierstrass' theory, irrationals were defined as classes of rationals. Hence B. Russell's objections (stated in his *Principles of Mathematics*, Cambridge, 1903, p. 282) do not hold against it, nor does Russell seem to credit Weierstrass and Cantor with the avoidance of quite the contradiction that they did avoid. The real objection to Weierstrass' theory, and one of the objections to G. Cantor's theory, is

[1] P. E. B. Jourdain, "On Isoid Relations and Theories of Irrational Number" in *International Congress of Mathematicians*, Cambridge, 1912.

that equality has to be re-defined. In the various arithmetical theories of irrational numbers there are three tendencies: (a) the number is defined as a logical entity—a class or an operation—, as with K. Weierstrass, H. Weber, M. Pasch, B. Russell, M. Pieri; (b) it is "created," or, more frankly, postulated, as with R. Dedekind, O. Stolz, G. Peano, and Ch. Méray; (c) it is defined as a sign (for what, is left indeterminate), as with E. Heine, G. Cantor, H. Thomae, A. Pringsheim. . . . In the geometrical theories, as with Paul du Bois-Reymond, a real number is a sign for a length. In B. Russell's theory it appears to be equally legitimate to define a real number in various ways."

The theory of sets (Mengenlehre, théorie des ensembles, theory of aggregates) owes its development to the endeavor to clarify the concepts of independent variable and of function. Formerly the notion of an independent variable rested on the naïve concept of the geometric continuum. Now the independent variable is restricted to some aggregate of values or points selected out of the continuum. The term function was destined to receive various definitions. J. Fourier advanced the theorem that an arbitrary function can be represented by a trigonometric series. P. G. L. Dirichlet looked upon the general functional concept as equivalent to any arbitrary table of values. When G. F. B. Riemann gave an example of a function expressed analytically which was discontinuous at each rational point, the need of a more comprehensive theory became evident. The first attempts to meet the new needs were made by Hermann Hankel and Paul du Bois-Reymond. The *Allgemeine Funktionentheorie* of du Bois-Reymond brilliantly sets forth the problems in philosophical form, but it remained for Georg Cantor to advance and develop the necessary ideas, involving a treatment of infinite aggregates. Even though the infinite had been the subject of philosophic contemplation for more than two thousand years, G. Cantor hesitated for ten years before placing his ideas before the mathematical public. The theory of aggregates sprang into being, as a science, when G. Cantor introduced the notion of "enumerable" aggregates.[1] G. Cantor began his publications in 1870; in 1883 he published his *Grundlagen einer allgemeinen Mannichfaltigkeitslehre.* In 1895 and 1897 appeared in *Mathematische Annalen* his *Beiträge zur Begründung der transfiniten Mengenlehre.*[2] These researches have played a most conspicuous rôle not only in the march of mathematics toward logical exactitude, but also in the realm of philosophy.

G. Cantor's theory of the continuum was used by P. Tannery in 1885 in the search for a profounder view of Zeno's arguments against

[1] A. Schoenflies, *Entwickelung der Mengenlehre und ihrer Anwendungen, gemeinsam mit Hans Hahn herausgegeben*, Leipzig u. Berlin, 1913, p. 2.

[2] Translated into English by Philip E. B. Jourdain and published by the Open Court Publ. Co., Chicago, 1915.

motion. *Paul Tannery* (1843–1904), a brother of Jules Tannery, attended the École Polytechnique in Paris and then entered the state corps of manufacturing engineers. He devoted his days to business and his evenings to the study of the history of science. From 1892 to 1896 he held the chair of Greek and Latin philosophy at the Collège de France, but later he failed to receive the appointment to the chair of the history of the sciences, although he was the foremost French historian of his day. He was a deep student of Greek scientists, particularly of Diophantus. Other historical periods were taken up afterwards, particularly that of R. Descartes and P. Fermat. His researches, consisting mostly of separate articles, have been republished in collected form.

In 1883 G. Cantor stated that every set and in particular the continuum can be well-ordered. In 1904 and 1907 E. Zermelo gave proofs of this theorem, but they have not been generally accepted and have given rise to much discussion. G. Peano objected to Zermelo's proof because it rests on a postulate ("Zermelo's principle") expressing a property of the continuum. In 1907 E. Zermelo formulated that postulate thus: "A set S which is divided into subsets A, B, C, . . ., each containing at least one element, but containing no elements in common, contains at least one subset S, which has just one element in common with each of the subsets A, B, C, . . ." (*Math. Ann.*, Vol. 56, p. 110). To this G. Peano objects that one may not apply an infinite number of times an arbitrary law by virtue of which one correlates to a class some member of that class. E. B. Wilson comments on this: "Here are two postulates by two different authorities; the postulates are contradictory, and each thinker is at liberty to adopt whichever appears to him the more convenient." Zermelo's postulate, before it had been formulated, was tacitly assumed in the researches of R. Dedekind, G. Cantor, F. Bernstein, A. Schönflies, J. König, and others. Zermelo's proof was rejected by H. Poincaré, E. Borel, R. Baire, H. Lebesgue.[1] A third proof that every set can be well-ordered was given in 1915 by Friedrich M. Hartogs of Munich. P. E. B. Jourdain of Fleet did not consider this proof altogether satisfactory. In 1918 he gave one of his own (*Mind*, 27, 386–388) and declared, " thus any aggregate can be well ordered, Zermelo's ' axiom ' can be proved quite generally, and Hartog's work is completed."

In 1897 Cesare Burali-Forti of Turin pointed out the following paradox: The series of ordinal numbers, which is well-ordered, must have the greatest of all ordinal numbers as its order type. Yet the type of the above series of ordinal numbers, when followed by its type, must be a greater ordinal number, for $\beta+1$ is greater than β. Therefore, a well-ordered series of ordinal numbers containing all ordinal numbers itself defines a new ordinal number not included in the original series.

Another paradox, due to Jules Antoine Richard of Châteauroux

[1] See *Bulletin Am. Math. Soc.*, Vol. 14, p. 438.

in 1906, relates to the aggregate of decimal fractions between 0 and 1 which can be defined by a finite number of words; a new decimal fraction can be defined, which is not included in the previous ones.

Bertrand Russell discovered another paradox, given in his *Principles of Mathematics*, 1903, (pp. 364–368, 101–107), which is stated by Philip E. B. Jourdain thus: "If w is the class of all those terms x such that x is not a member of x, then, if w is a member of w, it is plain that w is not a member of w; while if w is not a member of w, it is equally plain that w is a member of w." [1] These paradoxes are closely allied to the "Epimenides puzzle": Epimenides was a Cretan who said that all Cretans were liars. Hence, if his statement was true he was a liar. H. Poincaré and B. Russell attribute the paradoxes to the open and clandestine use of the word "all" (in the definition of a class, or aggregate).

Noteworthy among the attempts to place the theory of aggregates upon a foundation that will exclude the paradoxes and antinomies that had arisen, was the formulation in 1907 of seven restricting axioms by E. Zermelo in *Math. Annalen*, 65, p. 261.

Julius König (1849–1913), the Hungarian mathematician, in his *Neue Grundlagen der Logik, Arithmetik und Mengenlehre*, 1914, speaks of E. Zermelo's axiom of choice (Auswahlaxiom) as being really a logical assumption, not an axiom in the old sense, whose freedom from contradiction must be demonstrated along with the other axioms. He takes pains to steer clear of the antinomies of B. Russell and C. Burali-Forti. For a discussion of the logical and philosophical questions involved in the theory of aggregates, consult the second edition of E. Borel's *Leçons sur la théorie des fonctions*, Paris, 1914, note IV, which gives letters written by J. Hadamard, E. Borel, H. Lebesgue, R. Baire, touching the validity of Zermelo's demonstration that the linear continuum is well-ordered. A set of axioms of ordinal magnitude was given by A. B. Frizell in 1912 at the Cambridge Congress.

In the treatment of the infinite there are two schools. Georg Cantor proved that the continuum is not denumerable; J. A. Richard, contending that no mathematical entity exists that is not definable in a finite number of words, argued that the continuum is denumerable. H. Poincaré claimed that this contradiction is not real, since J. A. Richard employs a non-predicative definition.[2] H. Poincaré,[3] in discussing the logic of the infinite, states that, according to the first school, the pragmatists, the infinite flows out of the finite; there is an infinite, because there is an infinity of possible finite things. According to the second school, the Cantorians, the infinite precedes the finite; the finite is obtained by cutting off a small piece of the infinite.

[1] P. E. B. Jourdain, *Contributions*, Chicago, 1915, p. 206.
[2] *Bull. Am. Math. Soc.*, Vol. 17, 1911, p. 193.
[3] *Scientia*, Vol. 12, 1912, pp. 1–11.

For pragmatists a theorem has no meaning unless it can be verified; they reject indirect proofs of existence; hence they reply to E. Zermelo who proves that space can be converted into a well-ordered aggregate (wohlgeordnete Menge): Fine, convert it! We cannot carry out this transformation because the number of operators is infinite. For Cantorians mathematical things exist independently of man who may think about them; for them cardinal number is no mystery. On the other hand, pragmatists are not sure that any aggregate has a cardinal number, and when they say that the Mächtigkeit of the continuum is not that of the whole numbers, they mean simply that it is impossible to set up a correspondence between these two aggregates, which could not be destroyed by the creation of new points in space. If mathematicians are ordinarily agreed among themselves, it is because of confirmations which pass final judgment. In the logic of infinity there are no confirmations.

L. E. J. Brouwer of the University of Amsterdam, expressing views of G. Mannoury, said in 1912 that to the psychologist belongs the task of explaining "why we are averse to the so-called contradictory systems in which the negative as well as the positive of certain propositions are valid," that "the intuitionist recognizes only the existence of denumerable sets" and "can never feel assured of the exactness of mathematical theory by such guarantees as the proof of its being non-contradictory, the possibility of defining its concepts by a finite number of words, or the practical certainty that it will never lead to misunderstanding in human relations." [1] A. B. Frizell showed in 1914 that the field of denumerably infinite processes is not a closed domain—a concept which the intuitionist refuses to recognize, but which "need not disturb an intuitionist who cuts loose from the principium contradictionis." More recent tendencies of research in this field are described by E. H. Moore: [2] "From the linear continuum with its infinite variety of functions and corresponding singularities G. Cantor developed his *theory of classes of points* (*Punktmengenlehre*) with the notions: limit-point, derived class, closed class, perfect class, etc., and his *theory of classes in general* (allgemeine Mengenlehre) with the notions: cardinal number, ordinal number, order-type, etc. These theories of G. Cantor are permeating Modern Mathematics. Thus there is a theory of functions on point-sets, in particular, on perfect point-sets, and on more general order-types, while the arithmetic of cardinal numbers and the algebra and function theory of ordinal numbers are under development.

"Less technical generalizations or analogues of functions of the continuous real variable occur throughout the various doctrines and applications of analysis. A function of several variables is a function of a single multipartite variable; a distribution of potential or a field

[1] L. E. J. Brouwer in *Bull. Am. Math. Soc.*, Vol. 20, 1913, pp. 84, 86.

[2] *New Haven Colloquium, 1906*, New Haven, 1910, p. 2–4.

of force is a function of position on a cuve or surface or region; the value of the definite integral of the Calculus of Variations is a function of the variable function entering the definite integral; a curvilinear integral is a function of the path of integration; a functional operation is a function of the argument function or functions; etc., etc.

"A multipartite variable itself is a function of the variable index of the part. Thus a finite sequence: $x_1; \ldots; x_n$, of real numbers is a function x of the index i, viz., $x(i)=x_i$ $(i=1; \ldots; n)$. Similarly, an infinite sequence: $x_1; \ldots; x_n; \ldots$, of real numbers is a function x of the index n, viz., $x(n)=x_n$ $(n=1; 2; \ldots)$. Accordingly, n-fold algebra and the theory of sequences and of series are embraced in the theory of functions.

"As apart from the determination and extension of notions and theories in analogy with simpler notions and theories, there is the extension by direct generalization. The Cantor movement is in this direction. Finite generalization, from the case $n=1$ to the case $n=n$, occurs throughout Analysis, as, for instance, in the theory of functions of several independent variables. The theory of functions of a denumerable infinity of variables is another step in this direction.[1] We notice a more general theory dating from the year 1906. Recognizing the fundamental rôle played by the notion *limit-element* (number, point, function, curve, etc.) in the various special doctrines, M. Fréchet has given, with extensive applications, an abstract generalization of a considerable part of Cantor's theory of classes of points and of the theory of continuous functions on classes of points. Fréchet considers a general class P of elements p with the notion *limit* defined for sequences of elements. The nature of the elements p is not specified; the notion *limit* is not explicitly defined; it is postulated as defined subject to specified conditions. For particular applications explicit definitions satisfying the conditions are given. . . . The functions considered are either functions μ of variables p of specified character or functions μ on ranges P with postulated features: *e. g. limit; distance; element of condensation; connection*, of specified character." E. H. Moore's own form of general analysis of 1906 considers *functions μ of a general variable p on a general range* P, where this *general* embraces every well-defined particular case of variable and range.

Early in the development of the theory of point sets it was proposed to associate with them numbers that are analogous to those representing lengths, areas, volumes.[2] On account of the great arbitrariness of this procedure, several different definitions of such numbers have been given. The earliest were given in 1882 by H. Hankel and A. Harnack. Another definition due to G. Cantor (1884) was generalized by H. Minkowski in 1900. More precise measures were

[1] D. Hilbert, 1906, 1909.

[2] *Encyclopédie des sciences mathématiques*, Tome II, Vol. 1, 1912, p. 150.

assigned in 1893 by C. Jordan in his *Cours d'analyse*, which, for plane sets, was as follows: If the plane is divided up into squares whose sides are s and if S is the sum of the squares all interior points of which belong to the set P, if, moreover, $S+S'$ is the sum of all squares which contain points of P, then, as s approaches zero, S and $S+S'$ converge to limits I and A, called the "interior" and "exterior" areas. If $I=A$, then P is said to be "measurable." Examples of plane curves whose exterior areas were not zero were given in 1903 by W. F. Osgood and H. Lebesgue. Another definition, given by E. Borel, was generalized by H. Lebesgue so as to present fewer inconveniences than the older ones. The existence of non-measurable sets, according to Lebesgue's definition, was proved by G. Vitali and Lebesgue himself; the proofs assume E. Zermelo's axiom.

Instead of sets of points, E. Borel in 1903 began to study sets of lines or planes; G. Ascoli considered sets of curves. M. Fréchet in 1906 proposed a generalization by establishing general properties without specifying the nature of the elements, and was led to the so-called "calcul fonctionnel," to which attention has been called before.

Functional equations, in which the unknown elements are one or more functions, have received renewed attention in recent years. In the eighteenth century certain types of them were treated by D'Alembert, L. Euler, J. Lagrange, and P. S. Laplace.[1] Later came the "calculus of functions," studied chiefly by C. Babbage, J. F. W. Herschel, and A. De Morgan, which was a theory of the solution of functional equations by means of known functions or symbols. A. M. Legendre and A. L. Cauchy studied the functional equation $f(x+y)=f(x)+f(y)$, which recently has been investigated by R. Volpi, G. Hamel, and R. Schimmack. Other equations, $f(x+y)=f(x).f(y)$, $f(xy)=f(x).f(y)$, and $\varphi(y+x)+\varphi(y-x)=2\varphi(x).\varphi(y)$, the last being D'Alembert's, were treated by A. L. Cauchy, R. Schimmack, and J. Andrade. The names of Ch. Babbage, N. H. Abel, E. Schröder, J. Farkas, P. Appell, and E. B. Van Vleck are associated with this subject. Said S. Pincherele of Bologna in 1912: "The study of certain classes of functional equations has given rise to some of the most important chapters of analysis. It suffices to cite the theory of differential or partial differential equations. . . . The theory of equations of finite differences, simple or partial, . . . the calculus of variations, the theory of integral equations . . . the integrodifferential equations recently considered by V. Volterra, are as many chapters of mathematics devoted to the study of functional equations."

The theory of point sets led in 1902 to a generalization of Riemann's definition of a definite integral by **Henri Lebesgue** (1875–1941) of Paris. E. B. Van Vleck describes the need of this change:[2] "This

[1] *Encyclopédie des sciences mathématiques*, Tome II, Vol. 5, 1912, p. 46.
[2] *Bull. Am. Math. Soc.*, Vol. 23, 1916, pp. 6, 7.

(Riemann's integral) admits a finite number of discontinuities but an infinite number only under certain narrow restrictions. A totally discontinuous function—for example, one equal to zero in the rational points which are everywhere dense in the interval of integration, and equal to 1 in the rational points which are likewise everywhere dense—is not integrable à la Riemann. The restriction became a very hampering one when mathematicians began to realize that the analytic world in which theorems are deducible does not consist merely of highly civilized and continuous functions. In 1902 Lebesgue with great penetration framed a new integral which is identical with the integral of Riemann when the latter is applicable but is immeasurably more comprehensive. It will, for instance, include the totally discontinuous function above mentioned. This new integral of Lebesgue is proving itself a wonderful tool. I might compare it with a modern Krupp gun, so easily does it penetrate barriers which before were impregnable." Instructive is also the description of this movement, as given by G. A. Bliss:[1] "Volterra has pointed out, in the introductory chapter of his *Leçons sur les fonctions des lignes* (1913), the rapid development which is taking place in our notions of infinite processes, examples of which are the definite integral limit, the solution of integral equations, and the transition from functions of a finite number of variables to functions of lines. In the field of integration the classical integral of Riemann, perfected by Darboux, was such a convenient and perfect instrument that it impressed itself for a long time upon the mathematical public as being something unique and final. The advent of the integrals of T. J. Stieltjes and H. Lebesgue has shaken the complacency of mathematicians in this respect, and, with the theory of linear integral equations, has given the signal for a re-examination and extension of many of the types of processes which Volterra calls passing from the finite to the infinite. It should be noted that the Lebesgue integral is only one of the evidences of this restlessness in the particular domain of the integration theory. Other new definitions of an integral have been devised by Stieltjes, W. H. Young, J. Pierpont, E. Hellinger, J. Radon, M. Fréchet, E. H. Moore, and others. The definitions of Lebesgue, Young, and Pierpont, and those of Stieltjes and Hellinger, form two rather well defined and distinct types, while that of Radon is a generalization of the integrals of both Lebesgue and Stieltjes. The efforts of Fréchet and Moore have been directed toward definitions valid on more general ranges than sets of points of a line or higher spaces, and which include the others for special cases of these ranges. Lebesgue and H. Hahn, with the help of somewhat complicated transformations, have shown that the integrals of Stieltjes and Hel-

[1] G. A. Bliss, "Integrals of Lebesgue," *Bull. Am. Math. Soc.*, Vol. 24, 1917, pp. 1–47. See also T. H. Hildebrandt, *loc. cit.*, Vol. 24, pp. 113–144, who gives bibliography.

linger are expressible as Lebesgue integrals. . . . Van Vleck has . . . remarked that a Lebesgue integral is expressible as one of Stieltjes by a transformation much simpler than that used by Lebesgue for the opposite purpose, and the Stieltjes integral so obtained is readily expressible in terms of a Riemann integral. . . . Furthermore the Stieltjes integral seems distinctly better suited than that of Lebesgue to certain types of questions, as is well indicated by the original 'problem of moments' of Stieltjes, or by a generalization of it which F. Riesz has made. . . . The conclusion then seems to be that one should reserve judgment, for the present at least, as to the final form or forms which the integration theory is to take."

Mathematical Logic

Summarizing the history of mathematical logic, P. E. B. Jourdain says:[1] "In somewhat close connection with the work of Leibniz . . . stands the work of Johann Heinrich Lambert, who sought—not very successfully—to develop the logic of relations. Toward the middle of the nineteenth century George Boole independently worked out and published his famous calculus of logic. . . . Independently of him or anybody else, Augustus De Morgan began to work out logic as a calculus, and later on, taking as his guide the maxim that logic should not consider merely certain kinds of deduction but deduction quite generally, founded all the essential parts of the logic of relations. William Stanley Jevons criticised and popularized Boole's work; and Charles S. Peirce (1839–1914), Mrs. Christine Ladd-Franklin, Richard Dedekind, Ernst Schröder (1841–1902), Hermann G. and Robert Grassmann, Hugh MacColl, John Venn, and many others, either developed the work of G. Boole and A. De Morgan or built up systems of calculative logic in modes which were largely independent of the work of others. But it was in the work of Gottlob Frege, Guiseppe Peano, Bertrand Russell, and Alfred North Whitehead, that we find a closer approach to the *lingua characteristica* dreamed of by Leibniz." We proceed to a few details.

"Pure mathematics," says B. Russell,[2] "was discovered by Boole in a work which he called *The Laws of Thought* (1854). . . . His work was concerned with formal logic, and this is the same thing as mathematics." **George Boole** (1815–1864) became in 1849 professor in Queen's College, Cork, Ireland. He was a native of Lincoln, and a self-educated mathematician of great power. In his boyhood he studied, unaided, the classical languages.[3] While teaching school he pursued modern languages and entered upon the study of J. Lagrange

[1] *The Monist*, Vol. 26, 1916, p. 522.

[2] *International Monthly*, 1901, p. 83.

[3] See A. Macfarlane, *Ten British Mathematicians*, New York, 1916. Boole's *Laws of Thought* was republished in 1917 by the Open Court Publ. Co., under the editorship of P. E. B. Jourdain.

and P. S. Laplace. His treatises on *Differential Equations* (1859), and *Finite Differences* (1860) are works of merit.

A point of view different from that of G. Boole was taken by *Hugh MacColl* (1837–1909) who was led to his system of symbolic logic by researches on the theory of probability. While Boole used letters to represent the times during which certain propositions are true, MacColl employed the proposition as the real unit in symbolic reasoning.[1] When the variables in the Boolean algebra are interpreted as propositions, C. I. Lewis of the University of California worked out a matrix algebra for implications.

When the investigation of the principles of mathematics became the chief task of logical symbolism, the aspect of symbolic logic as a calculus was eclipsed temporarily. **Friedrich Ludwig Gottlob Frege** (1848–1925) of the University of Jena entered this field. Considering the foundations of arithmetic he inquired how far one could go by conclusions which rest merely on the laws of general logic. Ordinary language was found to be unequal to the accuracy required. So knowing nothing of the work of his predecessors, except G. W. Leibniz, he devised a symbolism and in 1879 published his *Begriffsschrift*, and in 1893 his *Grundgesetze der Arithmetik*. Says P. E. B. Jourdain: "Frege criticised the notion which mathematicians denote by the word 'aggregate' (Menge), and particularly the views of Dedekind and Schröder. Neither of these authors distinguished the subordination of a concept under a concept from the falling of an object under a concept; a distinction upon which Peano rightly laid so much stress, and which is, indeed, one of the most characteristic features of Peano's system of ideography." Ernst Schröder of Karlsruhe had published in 1877 his *Algebra der Logik* and John Venn his *Symbolic Logic* in 1881.

"Peano's first publication on mathematical logic followed the lines of Schröder's work of 1877 very closely. An excellent exposition in Peano's *Calcolo geometrico secondo l'Ausdehnungslehre di H. Grassmann*, Turin, 1888, of the geometrical calculus of A. F. Möbius, H. G. Grassmann, and others was preceded by an introduction treating of the operations of deductive logic, which are very analogous to those of ordinary algebra and of the geometrical calculus. The signs of logic were sometimes used in the later parts of the book, though this was not done systematically, as it was in many of Peano's later works" (Jourdain). In 1891 appeared under G. Peano's editorship, the first volume of the *Rivista di Matematica* which contains articles on mathematical logic and its applications, but this kind of work was carried on more fully in the *Formulaire de mathématiques* of which the first volume was published in 1895. This was projected to be a classified collection of mathematical truths, written wholly in Peano's symbols;

[1] For details see Philip E. B. Jourdain in *Quarterly Jour. of Math.*, Vol. 43, 1912, p. 219.

it was prepared by Peano and his collaborators, C. Burali-Forti, G. Viviania, R. Bettazzi, F. Giudice, F. Castellano, and G. Fano. "In the later editions of the *Formulaire*," says P. E. B. Jourdain, "Peano gave up all attempts to work out which are the primitive propositions of logic; and the logical principles or theorems which are used in the various branches of mathematics were merely collected together in as small a space as possible. In the last edition (v., 1905), logic only occupied 16 pages, while mathematical theories—a fairly complete collection—occupied 463 pages. On the other hand, in the works of Frege and B. Russell the exact enumeration of the primitive propositions of logic was always one of the most important problems." In England mathematical logic has been strongly emphasized by Bertrand Russell who in 1903 published his *Principles of Mathematics* and in 1910, in conjunction with A. N. Whitehead, brought out the first volume of the *Principia mathematica*, a remarkable work. Russell and Whitehead follow in the main Peano in matters of notation, Frege in matters of logical analysis, G. Cantor in the treatment of arithmetic, and v. Staudt, M. Pasch, G. Peano, M. Pieri, and O. Veblen in the discussion of geometry. By their theory of logical types they solve the paradoxes of C. Burali-Forti, B. Russell, J. König, J. A. Richard, and others. Certain points in the logic of relations as given in the *Principia mathematica* have been simplified by Norbert Wiener in 1914 and 1915.

In France this subject has been cultivated chiefly by **Louis Couturat** (1868–1914), who, at the time of his death in an automobile accident in Paris, held high rank in the philosophy of mathematics and of language. He wrote *La logique de Leibniz* and *Les principes des mathématiques*, Paris, 1905. Sets of postulates for the Boolean algebra of logic were given by A. N. Whitehead, which were simplified in 1904 by E. V. Huntington of Harvard and B. A. Bernstein of California. Says P. E. B. Jourdain: "Frege's symbolism, though far better for logical analysis than Peano's, for instance, is far inferior to Peano's—a symbolism in which the merits of internationality and power of expressing mathematical theorems are very satisfactorily attained—in practical convenience. B. Russell, especially in the later works, used the ideas of Frege, many of which he discovered subsequently to, but independently of Frege, and modified the symbolism of Peano as little as possible. Still, the complications thus introduced take away that simple character which seems necessary to a calculus, and which Boole and others reached by passing over certain distinctions which a subtler logic has shown us must be made." [1]

In 1886 A. B. Kempe discussed the fundamental conceptions both of symbolic logic and of geometry. Later he developed this subject further in the study of the relations between the logical theory of

[1] Philip E. B. Jourdain in *Quart. Jour. of Math.*, Vol. 41, 1910, p. 331.

classes and the geometrical theory of points. This topic received a re-statement and an extension in 1905 from the pen of *Josiah Royce* (1855–1916), professor of philosophy at Harvard University. Royce contends that "the entire system of the relationships of the exact sciences stands in a much closer connection with the simple principles of symbolic logic than has thus far been generally recognized."

There existed divergence of opinion on the value of the notation of the calculus of logic. Said A. Voss: [1] "As far as I am able to survey the practical results of mathematical logic, they run aground at every real application, on account of the extreme complexity of its formulas; by a comparatively large expenditure of effort they yield almost trivial results, which, however, can be read off with absolute certainty. Only in the discussion of purely mathematical questions, *i. e.* relations between numbers, does it, in Peano's *Formulaire* . . ., prove itself to be a real power, probably replaceable by no other mode of expression. By some even this is called into question."

Alessandro Padoa of the Royal Technical Institute of Genoa said in 1912: [2] "I do not hope to suggest to you the sympathetic and touching optimism of Leibniz, who, prophesying the triumphal success of these researches, affirmed: 'I dare say that this is the last effort of the human mind, and, when this project shall have been carried out, all that men will have to do will be to be happy, since they will have an instrument that will serve to exalt the intellect not less than the telescope serves to perfect their vision.' Although for some fifteen years I have given myself up to these studies, I have not a hope so hyperbolic; but I delight in recalling the candor of this master who, absorbed in scientific and philosophic investigations, forgot that the majority of men sought and continue to seek happiness in the feverish conquest of pleasure, money, and honors. Meanwhile we should avoid an excessive scepticism, because always and everywhere, there has been an élite—to-day less restricted than in the past—which was charmed by, and delights now in, all that raises one above the confused troubles of the passions, into the imperturbable immensity of knowledge, whose horizons become the more vast as the wings of thought become more powerful and rapid."

In 1914 an international congress of mathematical philosophers was held in Paris, with Emil Boutroux as president. Unfortunately the great war nipped this promising new movement in the bud. Pre-1920 books on the philosophy of mathematics are M. Winter's *La méthode dans la philosophie des mathématics*, Paris, 1911, Léon Brunschvicg's *Les étapes de la philosophie mathématique*, Paris, 1912, and J. B. Shaw's *Lectures on the Philosophy of Mathematics*, Chicago and London, 1918.

[1] A. Voss, *Ueber das Wesen der Mathematik*, Leipzig u. Berlin, 2. Aufl., 1913, p. 28.

[2] *Bull. Am. Math. Soc.*, Vol. 20, 1913, p. 98.

Theory of Functions

We begin our sketch of the vast progress in the theory of functions by considering investigations which center about the special class called elliptic functions. These were richly developed by N. H. Abel and C. G. J. Jacobi.

Niels Henrik Abel (1802–1829) was born at Findoé in Norway, and was prepared for the university at the cathedral school in Christiania. He exhibited no interest in mathematics until 1818, when B. Holmboe became lecturer there, and aroused Abel's interest by assigning original problems to the class. Like C. G. J. Jacobi and many other young men who became eminent mathematicians, Abel found the first exercise of his talent in the attempt to solve by algebra the general equation of the fifth degree. In 1821 he entered the University in Christiania. The works of L. Euler, J. Lagrange, and A. M. Legendre were closely studied by him. The idea of the inversion of elliptic functions dates back to this time. His extraordinary success in mathematical study led to the offer of a stipend by the government, that he might continue his studies in Germany and France. Leaving Norway in 1825, Abel visited the astronomer, H. C. Schumacher, in Hamburg, and spent six months in Berlin, where he became intimate with **August Leopold Crelle** (1780–1855), and met J. Steiner. Encouraged by Abel and J. Steiner, Crelle started his journal in 1826. Abel began to put some of his work in shape for print. His proof of the impossibility of solving the general equation of the fifth degree by radicals,—first printed in 1824 in a very concise form, and difficult of apprehension,—was elaborated in greater detail, and published in the first volume. He investigated also the question, what equations are solvable by algebra and deduced important general theorems thereon. These results were published after his death. Meanwhile E. Galois traversed this field anew. Abel first used the expression, now called the "Galois resolvent"; Galois himself attributed the idea of it to Abel. Abel showed how to solve the class of equations, now called "Abelian." He entered also upon the subject of infinite series (particularly the binomial theorem, of which he gave in *Crelle's Journal* a rigid general investigation), the study of functions, and of the integral calculus. The obscurities everywhere encountered by him owing to the prevailing loose methods of analysis he endeavored to clear up. For a short time he left Berlin for Freiberg, where he had fewer interruptions to work, and it was there that he made researches on hyperelliptic and Abelian functions. In July, 1826, Abel left Germany for Paris without having met K. F. Gauss! Abel had sent to Gauss his proof of 1824 of the impossibility of solving equations of the fifth degree, to which Gauss never paid any attention. This slight, and a haughtiness of spirit which he associated with Gauss, prevented the genial Abel from going to Göttingen. A similar feeling was entertained by him later against A. L. Cauchy. Abel remained ten months in Paris. He met there P. G. L. Dirichlet, A. M. Legendre, A. L. Cauchy, and others, but was little appreciated. He had already published several important memoirs in *Crelle's Journal*, but by the French this new periodical was as yet hardly known to exist, and Abel was too modest to speak of his own work. Pecuniary embarrassments induced him to return home after a second short stay in Berlin. At Christiania he for some time gave private lessons, and served as docent. Crelle secured at last an appointment for him at Berlin; but the news of it did not reach Norway until after the death of Abel at Froland.[1] Ch. Hermite is said to have remarked: "Abel a laissé aux mathématiciens de quoi travailler pendant cent cinquante ans."

At nearly the same time with Abel, C. G. J. Jacobi published articles on elliptic functions. A. M. Legendre's favorite subject, so long neglected, was at last to be enriched by some extraordinary discoveries. The advantage to be derived by inverting the elliptic integral of the first kind and treating it as a function of its amplitude (now called elliptic function) was recognized by Abel, and a few months later also by Jacobi. A second fruitful idea, also arrived at independently by both, is the introduction of imaginaries leading to the observation that the new functions simulated at once trigonometric and exponential functions. For it was shown that while trigonometric functions had only a real period, and exponential only an imaginary, elliptic functions had both sorts of periods. These two discoveries were the foundations upon which Abel and Jacobi, each in his own way, erected beautiful new structures. Abel developed the curious expressions representing elliptic functions by infinite series or quotients of infinite products. Great as were the achievements of Abel in elliptic functions, they were eclipsed by his researches on what are now called Abelian functions. Abel's theorem on these functions was given by him in several forms, the most general of these being that in his *Mémoire sur une propriété générale d'une classe très-étendue de fonctions transcendentes* (1826). The history of this memoir is interesting. A few months after his arrival in Paris, Abel submitted it to the French Academy. A. L. Cauchy and A. M. Legendre were appointed to examine it; but said nothing about it until after Abel's death. In a brief statement of the discoveries in question, published by Abel in *Crelle's Journal*, 1829, reference is made to that memoir. This led C. G. J. Jacobi to inquire of Legendre what had become of it. Legendre says that the manuscript was so badly written as to be illegible, and that Abel was asked to hand in a better copy, which he

[1] C. A. Bjerknes, *Niels-Henrik Abel, Tableau de sa vie et de son action scientifique*, Paris, 1885. See also *Abel (N. H.) Mémorial publié à l'occasion du centenaire de sa naissance*. Kristiania [1902]; also *N. H. Abel. Sa vie et son Oeuvre*, par Ch. Lucas de Pesloüan, Paris, 1906.☆

neglected to do. Others have attributed this failure to appreciate Abel's paper to the fact that the French academicians were then interested chiefly in applied mathematics—heat, elasticity, electricity. S. D. Poisson having in a report on C. G. J. Jacobi's *Fundamenta nova* recalled the reproach made by J. Fourier to Abel and Jacobi of not having occupied themselves preferably with the movement of heat, Jacobi wrote to Legendre: "It is true that Monsieur Fourier held the view that the principal aim of mathematics was public utility, and the explanation of natural phenomena; but a philosopher such as he should have known that the unique aim of science is the honor of the human spirit, and that from this point of view a question about numbers is as important as a question about the system of the world." In 1823 Abel published a paper [1] in which he is led, by a mechanical question including as a special case the problem of the tautochrone, to what is now called an integral equation, on the solution of which the solution of the problem depends. His problem was, to determine the curve for which the time of descent is a given function of the vertical height. In view of the recent developments in integral equations, Abel's problem is of great historical interest. Independently of Abel, researches along this line were published in 1832, 1837, and 1839 by J. Liouville, who in 1837 showed that a particular solution of a certain differential equation can be obtained by the aid of an integral equation of "the second kind," somewhat different from Abel's equation "of the first kind."

Abel's *Mémoire* of 1826 remained in A. L. Cauchy's hands. It was not published until 1841. By a singular mishap, the manuscript was lost before the proof-sheets were read.

In its form, the contents of the memoir belongs to the integral calculus. Abelian integrals depend upon an irrational function y which is connected with x by an algebraic equation $F(x, y)=0$. Abel's theorem asserts that a sum of such integrals can be expressed by a definite number p of similar integrals, where p depends merely on the properties of the equation $F(x, y)=0$. It was shown later that p is the deficiency of the curve $F(x, y)=0$. The addition theorems of elliptic integrals are deducible from Abel's theorem. The hyperelliptic integrals introduced by Abel, and proved by him to possess multiple periodicity, are special cases of Abelian integrals whenever $p=$ or >3. The reduction of Abelian to elliptic integrals has been studied mainly by C. G. J. Jacobi, Ch. Hermite, Leo Königsberger, F. Brioschi, E. Goursat, E. Picard, and O. Bolza, then of the University of Chicago. Abel's theorem was pronounced by Jacobi the greatest discovery of our century on the integral calculus. The aged Legendre, who greatly admired Abel's genius, called it "*monumentum aere perennius.*" Some cases of Abel's theorem were investigated independently by William Henry Fox Talbot (1800–1877), the English pioneer of photography,

[1] N. H. Abel, *Oeuvres complètes*, 1881, Vol. 1, p. 11. See also p. 97.

who showed that the theorem is deducible from symmetric functions of the roots of equations and partial fractions.[1]

Two editions of Abel's works have been published: the first by Berndt Michael Holmboe (1795–1850) of Christiania in 1839, and the second by L. Sylow and S. Lie in 1881. During the few years of work allotted to the young Norwegian, he penetrated new fields of research. Abel's published papers stimulated researches containing certain results previously reached by Abel himself in his then unpublished Parisian memoir. We refer to papers of Christian Jürgensen (1805–1861) of Copenhagen, Ole Jacob Broch (1818–1889) of Christiania, Ferdinand Adolf Minding (1806–1885) of Dorpat, and G. Rosenhain.

Some of the discoveries of Abel and Jacobi were anticipated by K. F. Gauss. In the *Disquisitiones Arithmeticæ* he observed that the principles which he used in the division of the circle were applicable to many other functions, besides the circular, and in particular to the transcendents dependent on the integral $\int \frac{dx}{\sqrt{1-x^4}}$. From this Jacobi[2] concluded that Gauss had thirty years earlier considered the nature and properties of elliptic functions and had discovered their double periodicity. The papers in the collected works of Gauss confirm this conclusion.

Carl Gustav Jacob Jacobi (1804–1851) was born of Jewish✡ parents at Potsdam. Like many other mathematicians he was initiated into mathematics by reading L. Euler. At the University of Berlin, where he pursued his mathematical studies independently of the lecture courses, he took the degree of Ph.D. in 1825. After giving lectures in Berlin for two years, he was elected extraordinary professor at Königsberg, and two years later to the ordinary professorship there. After the publication of his *Fundamenta Nova* in 1829 he spent some time in travel, meeting Gauss in Göttingen, and A. M. Legendre, J. Fourier, S. D. Poisson, in Paris. In 1842 he and his colleague, F. W. Bessel, attended the meetings of the British Association, where they made the acquaintance of English mathematicians. Jacobi was a great teacher. "In this respect he was the very opposite of his great contemporary Gauss, who disliked to teach, and who was anything but inspiring."

Jacobi's early researches were on Gauss' approximation to the value of definite integrals, partial differential equations, Legendre's coefficients, and cubic residues. He read Legendre's *Exercises*, which give an account of elliptic integrals. When he returned the book to the library, he was depressed in spirits and said that important books generally excited in him new ideas, but that this time he had not been led to a single original thought. Though slow at first, his ideas

[1] G. B. Mathews in *Nature*, Vol. 95, 1915, p. 219.
[2] R. Tucker, "Carl Friedrich Gauss," *Nature*, April, 1877.

flowed all the richer afterwards. Many of his discoveries in elliptic functions were made independently by Abel. Jacobi communicated his first researches to *Crelle's Journal*. In 1829, at the age of twenty-five, he published his *Fundamenta Nova Theoriæ Functionum Ellipticarum*, which contains in condensed form the main results in elliptic functions. This work at once secured for him a wide reputation. He then made a closer study of theta-functions and lectured to his pupils on a new theory of elliptic functions based on the theta-functions. He developed a theory of transformation which led him to a multitude of formulæ containing q, a transcendental function of the modulus, defined by the equation $q=e^{-\pi k'/k}$. He was also led by it to consider the two new functions H and Θ, which taken each separately with two different arguments are the four (single) theta-functions designated by the Θ_1, Θ_2, Θ_3, Θ_4.[1] In a short but very important memoir of 1832, he shows that for the hyperelliptic integral of any class the direct functions to which Abel's theorem has reference are not functions of a single variable, such as the elliptic *sn*, *cn*, *dn*, but functions of p variables.[1] Thus in the case $p=2$, which Jacobi especially considers, it is shown that Abel's theorem has reference to two functions $\lambda(u, v)$, $\lambda_1(u, v)$, each of two variables, and gives in effect an addition-theorem for the expression of the functions $\lambda(u+u', v+v')$, $\lambda_1(u+u', v+v')$ algebraically in terms of the functions $\lambda(u, v)$, $\lambda_1(u, v)$, $\lambda(u', v')$, $\lambda_1(u', v')$. By the memoirs of N. H. Abel and Jacobi it may be considered that the notion of the Abelian function of p variables was established and the addition-theorem for these functions given. Recent studies touching Abelian functions have been made by K. Weierstrass, E. Picard, Madame Kovalevski, and H. Poincaré. Jacobi's work on differential equations, determinants, dynamics, and the theory of numbers is mentioned elsewhere.

In 1842 C. G. J. Jacobi visited Italy for a few months to recuperate his health. At this time the Prussian government gave him a pension, and he moved to Berlin, where the last years of his life were spent.

Among those who greatly extended the researches on functions mentioned thus far was **Charles Hermite** (1822–1901), who was born at Dieuze in Lorraine.[2] He early manifested extraordinary talent for mathematics. Neglecting the regular courses of study, he read in Paris with greatest ardor the masterpieces of L. Euler, J. Lagrange, K. F. Gauss, and C. G. J. Jacobi. In 1842 he entered the École Polytechnique. From birth he had suffered from an infirmity of the right leg and had to use a cane. On this account he was declared ineligible to any government position given to graduates of the École. Hermite, therefore, left at the end of the first year. A letter to Jacobi displayed his mathematical genius, but the necessity of taking examinations which he held *en horreur* compelled him to descend from his lofty

[1] Arthur Cayley, Inaugural Address before the British Association, 1883.

[2] *Bull. Am. Math. Soc.*, Vol. 13, 1907, p. 182.

mathematical speculations and take up the irksome details preparatory to examinations. In 1848 he became examinateur d'admission and répétiteur d'analyse at the École Polytechnique. In that position he succeeded P. L. Wantzel. That year he married a sister of his friend, Joseph Bertrand. In 1869, at the age of forty-seven, he became professor and at length reached a position befitting his talents. At the Sorbonne he succeeded J. M. C. Duhamel as professor of higher algebra. He occupied the chair at the École Polytechnique until 1876, at the Sorbonne until 1897. For many years he had been regarded as the venerated chief among French mathematicians. Hermite had no fondness for geometry. His researches are confined to algebra and analysis. He wrote on the theory of numbers, invariants and covariants, definite integrals, theory of equations, elliptic functions and the theory of functions. Of his collected works, or *Oeuvres*, Vol. III appeared in 1912, edited by E. Picard. In the theory of functions he was the foremost French writer of his day, since A. L. Cauchy. He has given an entirely new significance to the use of definite integrals in the theory of functions: we name the developments of the properties of the gamma-function which have been thus initiated.

Elliptic functions, considered on the Jacobian rather than on the Weierstrassian basis, was a favorite study of Hermite. "To him is due the reduction of an elliptic integral to its canonical form by means of the syzygy among the concomitants of a binary quartic. His investigations on modular functions and modular equations are of the highest importance. It was Hermite who discovered pseudo-periodic functions of the second kind, and developed their properties. In a memoir that may be fairly described as classical, 'Sur quelques applications des fonctions elliptiques' in the *Comptes Rendus*, 1877–1882, he applied these functions to the integration of the unspecialized form of Lamé's differential equation; and elliptic functions generally were applied in that memoir to obtain the solution of a number of physical problems" (A. R. Forsyth).

In 1858 Hermite introduced in place of the variable q of Jacobi a new variable ω connected with it by the equation $q=e^{i\pi\omega}$, so that $\omega=ik'/k$, and was led to consider the functions $\phi(\omega)$, $\psi(\omega)$, $\chi(\omega)$.[1] Henry Smith regarded a theta-function with the argument equal to zero, as a function of ω. This he called an omega-function, while the three functions $\phi(\omega)$, $\psi(\omega)$, $\chi(\omega)$, are his modular functions. Researches on theta-functions with respect to real and imaginary arguments have been made by Ernst Meissel (1826–1895) of Kiel, J. Thomae of Jena, Alfred Enneper (1830–1885) of Göttingen. A general formula for the product of two theta-functions was given in 1854 by H. Schröter (1829–1892) of Breslau. These functions have

[1] Arthur Cayley, Inaugural Address, 1883.

been studied also by Cauchy, Königsberger of Heidelberg (born 1837), Friedrich Julius Richelot (1808–1875) of Königsberg, Johann Georg Rosenhain (1816–1887) of Königsberg, Ludwig Schläfli (1814–1895) of Bern.[1]

A. M. Legendre's method of reducing an elliptic differential to its normal form has called forth many investigations, most important of which are those of F. J. Richelot and of K. Weierstrass of Berlin.

The algebraic transformations of elliptic functions involve a relation between the old modulus and the new one which C. G. J. Jacobi expressed by a differential equation of the third order, and also by an algebraic equation, called by him "modular equation." The notion of modular equations was familiar to Abel, but the development of this subject devolved upon later investigators. These equations have become of importance in the theory of algebraic equations, and have been studied by Ludwig Adolph Sohncke (1807–1853) of Halle, E. Mathieu, L. Königsberger, E. Betti of Pisa, Ch. Hermite of Paris, P. Joubert of Angers, Francesco Brioschi of Milan, L. Schläfli, H. Schröter, C. Gudermann of Cleve, Carl Eduard Gützlaff (1805–?) of Marienwerder in Prussia.

Felix Klein of Göttingen has made an extensive study of modular functions, dealing with a type of operations lying between the two extreme types, known as the theory of substitutions and the theory of invariants and covariants. Klein's theory has been presented in book-form by his pupil, Robert Fricke. The bolder features of it were first published in his *Ikosaeder*, 1884. His researches embrace the theory of modular functions as a specific class of elliptic functions, the statement of a more general problem as based on the doctrine of groups of operations, and the further development of the subject in connection with a class of Riemann's surfaces.

The elliptic functions were expressed by N. H. Abel as quotients of doubly infinite products. He did not, however, inquire rigorously into the convergency of the products. In 1845 A. Cayley studied these products, and found for them a complete theory, based in part upon geometrical interpretation, which he made the basis of the whole theory of elliptic functions. F. G. Eisenstein discussed by purely analytical methods the general doubly infinite product, and arrived at results which have been greatly simplified in form by the theory of primary factors, due to K. Weierstrass. A certain function involving a doubly infinite product has been called by Weierstrass the sigma-function, and is the basis of his beautiful theory of elliptic functions. The first systematic presentation of Weierstrass' theory of elliptic functions was published in 1886 by G. H. Halphen in his *Théorie des fonctions elliptiques et des leurs applications*. Applications of these functions have been given also by A. G. Greenhill of London. Gener-

[1] Alfred Enneper, *Elliptische Funktionen, Theorie und Geschichte,* Halle a/S, 1876.

alizations analogous to those of Weierstrass on elliptic functions have been made by Felix Klein on hyperelliptic functions.

Standard works on elliptic functions have been published by *C. A. A. Briot* and *J. C. Bouquet* (1859), by *L. Königsberger*, *A. Cayley*, *Heinrich Durège* (1821–1893) of Prague, and others.

Jacobi's work on Abelian and theta-functions was greatly extended by **Adolph Göpel** (1812)–1847), professor in a gymnasium near Potsdam, and **Johann Georg Rosenhain** (1816–1887) of Königsberg. Göpel in his *Theoriæ transcendentium primi ordinis adumbratio levis* (*Crelle*, 35, 1847) and Rosenhain in several memoirs established each independently, on the analogy of the single theta-functions, the functions of two variables, called double theta-functions, and worked out in connection with them the theory of the Abelian functions of two variables. The theta-relations established by Göpel and Rosenhain received for thirty years no further development, notwithstanding the fact that the double theta series came to be of increasing importance in analytical, geometrical, and mechanical problems, and that Ch. Hermite and L. Königsberger had considered the subject of transformation. Finally, the investigations of C. W. Borchardt, treating of the representation of Kummer's surface by Göpel's biquadratic relation between four theta-functions of two variables, and researches of H. H. Weber, F. Prym, Adolf Krazer, and Martin Krause of Dresden led to broader views. **Carl Wilhelm Borchardt** (1817–1880) was born in Berlin, studied under P. G. L. Dirichlet and C. G. J. Jacobi in Germany, and under Ch. Hermite, M. Chasles, and J. Liouville in France. He became professor in Berlin and succeeded A. L. Crelle as editor of the *Journal für Mathematik*. Much of his time was given to the applications of determinants in mathematical research.

Friedrich Prym (1841–1915) studied at Berlin, Göttingen, and Heidelberg. He became professor at the Polytechnicum in Zurich, then at Würzburg. His interest lay in the theory of functions. Researches on double theta-functions, made by A. Cayley, were extended to quadruple theta-functions by *Thomas Craig* (1855–1900), professor at the Johns Hopkins University. He was a pupil of J. J. Sylvester. While lecturing at the University he was during 1879–1881 connected with the United States Coast and Geodetic Survey. For many years he was an editor of the *American Journal of Mathematics*.

Starting with the integrals of the most general form and considering the inverse functions corresponding to these integrals (the Abelian functions of p variables), G. F. B. Riemann defined the theta-functions of p variables as the sum of a p-tuply infinite series of exponentials, the general term depending on p variables. Riemann shows that the Abelian functions are algebraically connected with theta-functions of the proper arguments, and presents the theory in the broadest form.[1]

[1] Arthur Cayley, Inaugural Address, 1883.

He rests the theory of the multiple theta-functions upon the general principles of the theory of functions of a complex variable.

Through the researches of A. Brill of Tübingen, M. Nöther of Erlangen, and Ferdinand Lindemann of Munich, made in connection with Riemann-Roch's theorem and the theory of residuation, there has grown out of the theory of Abelian functions a theory of algebraic functions and point-groups on algebraic curves.

General Theory of Functions

The history of the general theory of functions begins with the adoption of new definitions of a function. As an inheritance from the eighteenth century, y was called a function of x, if there existed an equation between these variables which made it possible to calculate y for any given value of x lying anywhere between $-\infty$ and $+\infty$. We have seen that L. Euler sometimes used a second, more general, definition, which was adopted by J. Fourier and which was translated by P. G. L. Dirichlet into the language of analysis thus: y is called a function of x, if y possess one or more definite values for each of certain values that x is assumed to take in an interval x_0 to x_1. In functions thus defined, there need be no analytical connection between y and x, and it becomes necessary to look for possible discontinuities. This definition was still further emphasized and generalized later, after the introduction of the theory of aggregates. There a function need not be defined for each point in the continuum embracing all real and complex numbers, nor for each point in an interval, but only for the points x in some particular set of points. Thus, y is a function of x, if for each point or number in any set of points or numbers x, there corresponds a point or number in a set y.

P. G. L. Dirichlet lectured on the theory of the potential and thereby made this theory more generally known in Germany. In 1839 K. F. Gauss had made researches on the potential; in England George Green had issued his fundamental memoir as early as 1828. Dirichlet's lectures on the potential became known to G. F. B. Riemann who made it of fundamental importance for the whole of mathematics. Before considering Riemann we must take up A. L. Cauchy.

J. Fourier's declaration that any given arbitrary function can be represented by a trigonometric series led Cauchy to a new formulation of the concepts "continuous," "limiting value" and "function." In his *Cours d'Analyse*, 1821, he says: "The function $f(x)$ is continuous between two given limits, if for each value of x that lies between these limits, the numerical value of the difference $f(x+\alpha)-f(x)$ diminishes with α in such a way as to become less than every finite number" (Chap. II, § 2). With S. F. Lacroix and A. L. Cauchy there are indications of a tendency to free the functional concept from an actual representation.[1] Although in his earlier writings slow to recognize the importance of imaginary variables, Cauchy later entered deeply into the treatment of functions of complex variables, not in a geometrical form as found in C. Wessel, J. R. Argand, and K. F. Gauss, but rather in analytical form. He carried on integrations through imaginary fields. While L. Euler and P. S. Laplace had declared the order of integration in double integrals to be immaterial, A. L. Cauchy showed that this was true only when the expression to be integrated does not become indeterminate in the interval (*Mémoire sur la théorie des intégrales définies*, read 1814, printed 1825).

If between two paths of integration, in the complex plane, there lies a pole, then the difference between the respective integrals can be represented by means of a "résidu de la fonction" (1826), a concept of undoubted importance known as the calculus of residues. In 1846 he showed that if X and Y are continuous functions of x and y within a closed area, then $\int (Xdx+Ydy) = \pm \int\int \left(\frac{\partial X}{\partial y} - \frac{\partial Y}{\partial x}\right) dxdy$, where the left integral extends over the boundary and the right integral over the inner area of the complex plane; he considers integration along a closed path surrounding a "pole," and later along a closed path surrounding a line on which the function is discontinuous, as for instance $\log x$ for $x<0$ when the function changes by $2\pi i$ in crossing the x-axis. The fundamental theorem of Cauchy's theory of series was given in 1837: "A function can be expanded in an ascending power series in x, as long as the modulus of x is less than that for which the function ceases to be finite and continuous." In 1840 the proof of this theorem is made to rest on the theorem of mean value. Cauchy, J. C. F. Sturm, and J. Liouville had carried on discussions as to whether the continuity of a function was sufficient to insure its expandibility or whether that of its derivative must be demanded as well. In 1851 Cauchy concluded that the continuity of the derivative must be demanded. A function $f(z)$, which is single-valued for $z=x+iy$ was called by Cauchy "monotypique," later "monodrome," by Briot and Bouquet "monotrope," by Hermite "uniforme," by the Germans "eindeutig." Cauchy called a function "monogen" when for every z in a region it had only one derivative value, "synectique" if it is monodromic and monogenic and does not become infinite. Instead of "synectique," C. A. A. Briot and J. C. Bouquet, and later French writers say "holomorph," also "meromorph" when the function has "poles" in the region.

Some parts of Cauchy's theory of functions were elaborated by P. M. H. Laurent and Victor Alexandre Puiseux (1820–1883), both

[1] A. Brill und M. Noether, "Entwicklung der Theorie der algebraischen Functionen in älterer und neuerer Zeit," *Jahresb. d. d. Math. Vereinig.*, Vol. 3, 1892–1893, p. 162. We are making extensive use of this historical monograph.

of Paris. Laurent pointed out the advantage resulting in certain cases from a mixed expansion in ascending and descending powers of a variable, while Puiseux demonstrated the advantage that may be gained by the use of series involving fractional powers of the variable. Puiseux examined many-valued algebraic functions of a complex variable, their branch-points and moduli of periodicity.

We proceed to investigations made in Germany by G. F. B. Riemann.

Georg Friedrich Bernhard Riemann (1826–1866) was born at Breselenz in Hanover. His father wished him to study theology, and he accordingly entered upon philological and theological studies at Göttingen. He attended also some lectures on mathematics. Such was his predilection for this science that he abandoned theology. After studying for a time under K. F. Gauss and M. A. Stern, he was drawn, in 1847, to Berlin by a galaxy of mathematicians, in which shone P. G. L. Dirichlet, C. G. J. Jacobi, J. Steiner, and F. G. Eisenstein. Returning to Göttingen in 1850, he studied physics under W. Weber, and obtained the doctorate the following year. The thesis presented on that occasion, *Grundlagen für eine allgemeine Theorie der Funktionen einer veränderlichen complexen Grösse*, excited the admiration of K. F. Gauss to a very unusual degree, as did also Riemann's trial lecture, *Ueber die Hypothesen welche der Geometrie zu Grunde liegen.* Influenced by Gauss and W. Weber, physical views were the mainspring of his purely mathematical investigations. Riemann's Habilitationsschrift (1854, published 1867) was on the Representation of a Function by means of a Trigonometric Series, in which he advanced materially beyond the position of Dirichlet. A. L. Cauchy had set up criteria for the existence of a definite integral defined as the limit of a sum, and had stated that such a limit always exists when the function is continuous. Riemann made a startling extension by pointing out that the existence of such a limit is not confined to cases of continuity. Riemann's new criterion placed the definite integral upon a foundation wholly independent of the differential calculus and the existence of a derivative. It led to the consideration of areas and lengths of arcs which may transcend all geometric figures within the reach of our intuitions. Half a century later the concept of a definite integral was still further extended by H. Lebesgue of Paris and others. Our hearts are drawn to Riemann, an extraordinarily gifted but shy genius, when we read of the timidity and nervousness displayed when he began to lecture at Göttingen, and of his jubilation over the unexpectedly large audience of eight students at his first lecture on differential equations.

Later he lectured on Abelian functions to a class of three only, E. C. J. Schering, Bjerknes, and Dedekind. K. F. Gauss died in 1855, and was succeeded by P. G. L. Dirichlet. On the death of the latter, in 1859, Riemann was made ordinary professor. In 1860 he visited

Paris, where he made the acquaintance of French mathematicians. The delicate state of his health induced him to go to Italy three times. He died on his last trip at Selasca, and was buried at Biganzolo.

Like all of Riemann's researches, those on functions were profound and far-reaching. A decidedly modern tendency was his mode of investigating functions. In the words of E. B. Van Vleck:[1] "He [Riemann] presents a strange antithesis to his contemporary countryman, Weierstrass. Riemann bases the function theory upon a property rather than upon an algorism—to wit, the possession of a differential coefficient by the function in the complex plane. Thus at a stroke it is freed from dependence upon a particular process like the power series of Taylor. His celebrated memoir upon the *P*-function is a characteristic development of a whole Schar (family) of functions from their mutual relations."

G. F. B. Riemann laid the foundation for a general theory of functions of a complex variable. The theory of potential, which up to that time had been used only in mathematical physics, was applied by him in pure mathematics. He accordingly based his theory of functions on the partial differential equation, $\frac{\partial^2 u}{\partial x^2}+\frac{\partial^2 u}{\partial y^2}=\Delta u=0$, which must hold for the analytical function $w=u+iv$ of $z=x+iy$. It had been proved by P. G. L. Dirichlet that (for a plane) there is always one, and only one, function of x and y, which satisfies $\Delta u=0$, and which, together with its differential quotients of the first two orders, is for all values of x and y within a given area one-valued and continuous, and which has for points on the boundary of the area arbitrarily given values.[2] Riemann called this "*Dirichlet's principle,*" but the same theorem was stated by Green and proved analytically by Sir William Thomson. It follows then that w is uniquely determined for all points within a closed surface, if u is arbitrarily given for all points on the curve, whilst v is given for one point within the curve. In order to treat the more complicated case where w has n values for one value of z, and to observe the conditions about continuity, Riemann invented the celebrated surfaces, known as "Riemann's surfaces," consisting of n coincident planes or sheets, such that the passage from one sheet to another is made at the branch-points, and that the n sheets form together a multiply-connected surface, which can be dissected by cross-cuts into a singly-connected surface. The n-valued function w becomes thus a one-valued function. Aided by researches of Jacob Lüroth (1844–1910) of Freiburg and of R. F. A. Clebsch, W. K. Clifford brought Riemann's surface for algebraic functions to a canonical form, in which only the last two of the n leaves are multiply-connected, and then transformed the surface into the

[1] *Bull. Am. Math. Soc.*, Vol. 23, 1916, p. 8.

[2] O. Henrici "Theory of Functions," *Nature*, Vol. 43, 1891, p. 322.

surface of a solid with p holes. This surface with p holes had been considered before Clifford by A. Tonelli, and was probably used by Riemann himself.[1] A. Hurwitz of Zürich discussed the question, how far a Riemann's surface is determinate by the assignment of its number of sheets, its branch-points and branch-lines.

Riemann's theory ascertains the criteria which will determine an analytical function by aid of its discontinuities and boundary conditions, and thus defines a function independently of a mathematical expression. In order to show that two different expressions are identical, it is not necessary to transform one into the other, but it is sufficient to prove the agreement to a far less extent, merely in certain critical points.

Riemann's theory, as based on Dirichlet's principle (Thomson's theorem), is not free from objections which have been raised by L. Kronecker, K. Weierstrass, and others. In consequence of this, attempts have been made to graft Riemann's speculations on the more strongly rooted methods of K. Weierstrass. The latter developed a theory of functions by starting, not with the theory of potential, but with analytical expressions and operations. Both applied their theories to Abelian functions, but there Riemann's work is more general.[2]

Following a suggestion found in Riemann's Habilitationsschrift, H. Hankel prepared a tract, *Unendlich oft oscillirende und unstetige Funktionen*, Tübingen, 1870, giving functions which admit of an integral, but where the existence of a differential coefficient remains doubtful. He supposed continuous curves generated by the motion of a point to and fro with infinitely numerous and infinitely small oscillations, thus presenting "a condensation of singularities" at every point, but possessing no definite direction nor differential coefficient. These novel ideas were severely criticised, but were finally cleared up by K. Weierstrass' well-known rigorous example of a continuous curve totally bereft of derivatives. *Hermann Hankel* (1839–1873) satisfied at Leipzig the gymnasium requirements in ancient languages by reading the ancient mathematicians in the original. He studied at Leipzig under A. F. Möbius, at Göttingen under G. F. B. Riemann, at Berlin under K. Weierstrass and L. Kronecker. He became professor at Erlangen and Tübingen. The interest of his lectures was enhanced by his emphasis upon the history of his subject. In 1867 appeared his *Theorie der Complexen Zahlensysteme.* His brilliant *Geschichte der Mathematik in Alterthum und Mittelalter* came out in 1874 as a posthumous publication.

Karl Weierstrass (1815–1897) was born in Ostenfelde, a village in Westphalia. He attended a gymnasium at Paderborn where he became interested in the geometric researches of J. Steiner. He entered the

[1] *Math. Annalen*, Vol. 45, p. 142.

[2] O. Henrici, *Nature*, Vol. 43, 1891, p. 323.

University of Bonn as a student in law but all by himself he studied also mathematics, particularly P. S. Laplace. Wilhelm Diesterweg and J. Plücker, who lectured in Bonn, did not influence him. Seeing in a student note-book a transcript of Christof Gudermann's lectures on elliptic transcendents, Weierstrass went in 1839 to Münster, where he was during one semester the only student to attend Gudermann's lectures on this topic and on analytical spherics. *Christof Gudermann* (1798–1851) whose researches on hyperbolic functions led to a function $tan^{-1} (sinh\ x)$, called the "Gudermannian," was a favorite teacher of Weierstrass. Then he became a gymnasium teacher at Münster, then at Deutsch-Krone in western Prussia where he taught science, also gymnastics and writing, and finally at Braunsberg where he entered upon the study of Abelian functions. It is told that he missed one morning an eight-o'clock class. The director of the gymnasium went to his room to ascertain the cause, and found him working zealously at a research which he had begun the evening before and continued through the night, being unconscious that morning had come. He asked the director to excuse his lack of punctuality to his class, for he hoped soon to surprise the world by an important discovery. While at Braunsberg he received an honorary doctorate from Königsberg for scientific papers he had published. In 1855 E. E. Kummer went from Breslau to Berlin; he expressed it as his opinion that the paper on Abelian functions was not sufficient guarantee that Weierstrass was the proper man to train young mathematicians at Breslau. So Ferdinand Joachimsthal (1818–1861) was appointed there, but Kummer secured for Weierstrass in 1856 a position at the Gewerbeakademie in Berlin and at the same time an Extraordinariat at the University. The former he held until 1864 when he received an Ordinariat at the University as successor to the aged Martin Ohm. In that year E. E. Kummer and Weierstrass organized an official mathematical seminar, P. G. L. Dirichlet having held before this a private seminar. It is noteworthy that Weierstrass did not begin his university career as a professor until his forty-ninth year, a time when many scientists cease their creative work. K. Weierstrass, E. E. Kummer, and L. Kronecker added lustre to the University of Berlin which previously had been made famous by the researches of P. G. L. Dirichlet, J. Steiner, and C. G. J. Jacobi. Especially through Weierstrass unprecedented stress came to be put upon rigor of demonstration. The movement toward arithmetization of mathematics received through Kronecker and Weierstrass its greatest emphasis. The number-concept, especially that of the positive integer, was to become the sole foundation, and the space-concept was to be rejected as a primary concept.

As early as 1849 Weierstrass began to investigate and write on Abelian integrals. In 1863 and 1866 he lectured on the theory of Abelian functions and Abelian transcendents. No authorized publication of these lectures was made in his lifetime, but they became known in part through researches based upon them that were written by some of his pupils, E. Netto, F. Schottky, Georg Valentin, F. Kötter, Georg Hettner (1854–1914), and Johannes Knoblauch (1855–1915). Hettner and Knoblauch prepared Weierstrass' lectures on the theory of Abelian transcendents for the fourth volume of his collected works. In 1915 appeared the fifth volume, on elliptic functions, edited by Knoblauch. Weierstrass had selected Hettner to edit the works of C. W. Borchardt (1888), also the last two volumes of Jacobi's works. Knoblauch lectured at the University of Berlin since 1889, his chief field of activity being differential geometry. Another prominent pupil of Weierstrass was *Otto Stolz* (1842–1905) of the University of Innsbruck. The difficulty which was experienced for many years in ascertaining what were the methods and results of Weierstrass, was set forth by *Adolf Mayer* (1839–1908) of Leipzig who at one time had put at his disposal the manuscript notes of a lecture for only twenty-four hours. Mayer worked on differential equations, the calculus of variations and mechanics.

In 1861 Weierstrass made the extraordinary discovery of a function which is continuous over an interval and does not possess a derivative at any point on this interval. The function was published by P. du Bois Reymond in *Crelle's Journal*, Vol. 79, 1874, p. 29. In 1835 N. I. Lobachevski had shown in a memoir the necessity of distinguishing between continuity and differentiability.[1] Nevertheless, the mathematical world received a great shock when Weierstrass brought forth that discovery, "and H. Hankel and G. Cantor by means of their principle of condensation of singularities could construct analytical expressions for functions having in any interval, however small, an infinity of points of oscillation, an infinity of points in which the differential coefficient is altogether indeterminate, or an infinity of points of discontinuity" (J. Pierpont). J. G. Darboux gave new examples of continuous functions having no derivatives. Formerly it had been generally assumed that every function had a derivative. A. M. Ampère was the first who attempted to prove analytically (1806) the existence of a derivative, but the demonstration is not valid. In treating of discontinuous functions, J. G. Darboux established rigorously the necessary and sufficient condition that a continuous or discontinuous function be susceptible of integration. He gave fresh evidence of the care that must be exercised in the use of series by giving an example of a series always convergent and continuous, such that the series formed by the integrals of the terms is always convergent, and yet does not represent the integral of the first series.[2]

Central in Weierstrass' view-point is the concept of the "analytic function." The name, "general theory of analytic functions," says

[1] G. B. Halsted's transl. of A. Vasiliev's *Address* on Lobachevski, p. 23.

[2] *Notice sur les Travaux Scientifiques de M. Gaston Darboux*, Paris, 1884.

A. Hurwitz,[1] applies to two theories, that of A. L. Cauchy and G. F. B. Riemann, and that of K. Weierstrass. The two emanate from different definitions of a function. J. Lagrange, in his *Théorie des fonctions analytiques* had tried to prove the incorrect theorem that every continuous (stetige) function can be expanded in a power series. K. Weierstrass called every function "analytic" when it can be expanded into a power series, which is the centre of Weierstrass' theory of analytic functions. All properties of the function are contained *in nuce* in the power series, with its coefficients $c_1, c_2, \ldots, c_n, \ldots$ The behavior of a power series on the circle of convergence C had received consideration long before this time. N. H. Abel had demonstrated that the power series having a determinate value in a point on the circle of convergence C tends uniformly toward that value when the variable approaches that point along a path which does not touch the circle. If two power series involve a complex variable, whose circles of convergence overlap, so that the two series have the same value for every point common to the too circular areas, then Weierstrass calls each power series a direct continuation of the other. Using several such series, K. Weierstrass introduces the idea of a monogenic system of power series and then gives a more general definition of analytic function as a function which can be defined by a monogenic system of power series. In 1872 the Frenchman Ch. Méray gave independently a similar definition. In case of a uniform (eindeutige) function, the points in a complex plane are either within the circle of convergence of the power series in the system or else they are without. The totality of the former points constitutes the "field of continuity" (Stetigkeitsbereich) of the function. This field constitutes an aggregate of "inner" points that is dense; if this continuum is given, then there exist always single-valued analytic functions possessing this field of continuity, as was first proved by G. M. Mittag-Leffler, later by C. Runge and P. Stäckel. The points on the boundary of this field, called "singular points," constitute by themselves a set of points, by the properties of which K. Weierstrass classifies the function (1876). This classification was studied also by C. Guichard (1883) and by G. M. Mittag-Leffler, making use of theorems on point sets, as developed in 1879–1885 by G. Cantor and by I. O. Bendixson and E. Phragmén, both of Stockholm. Thus, transfinite numbers began to play a part in the theory of functions. Single-valued analytic functions resolve themselves into two classes, the one class in which the singular points form an enumerable (abzählbares) aggregate, the other class in which they do not.

Abel had proposed the problem, if one supposes the power series convergent for all positive values less than r, find the limit to which the function tends when x approaches r. The first substantial advance to a solution of Abel's problem was made in 1880

[1] A. Hurwitz in *Verh. des 1. Intern. Congr., Zürich, 1897*, Leipzig, 1898, pp. 91–112.

by G. Frobenius and in 1882 by O. Hölder, but neither of them developed conditions that are both necessary and sufficient for the establishment of the convergence of their expressions. Finally in 1892 J. Hadamard obtained expressions which include those of G. Frobenius and O. Hölder and determined the conditions under which they converge on the circle of convergence. The problem presented itself now thus: To set up analytic expressions of the complex variable x that are linear in the constants c_n and also represent the function given by the power series, or rather a branch of this function in a field D, in such a manner that they converge uniformly in the interior of D and diverge in the exterior. The first important step toward the resolution of this matter was taken in 1895 by E. Borel who proved that the expression

$$\lim_{\omega=\infty} \sum_{\nu=0}^{\infty} (c_0+c_1x+\ .\ .\ +c_\nu x^\nu)e^{-\omega}.\frac{\omega^{\nu+1}}{(\nu+1)!}$$

converges not only in all regular points (points réguliers) of the circle of convergence of the power series, but even beyond that, within a summation polygon. E. Borel held the view that his formula gave the sum of the power series even for points where it diverges. This interpretation of Borel's results was resisted by *Gösta Magnus Mittag-Leffler* (1846–1927) of Stockholm, the founder[1] of the journal *Acta mathematica*, and of a "Mathematical Institute" (in 1916) to further mathematical research in the Scandinavian countries. Mittag-Leffler conducted important researches along the above line. E. Borel's statement implies that his formula extends the boundaries of the theory of analytic functions beyond the classic region, which is denied by Mittag-Leffler. The latter published in 1898 studies on a problem more general than that of Borel. If a ray ap revolves about a through an angle 2π, the variable distance ap always exceeding a fixed value l, a surface is generated which Mittag-Leffler calls a star (Stern) with the center a. A star E is called a *convergence star* (Konvergenzstern) for a definite arithmetical expression, if the latter converges uniformly for each region within E, but diverges for every outside point. He shows that to each analytic function there corresponds a principal star, and that there is an infinite number of arithmetical expressions for a given star. Equivalent results were obtained by C. Runge. E. Borel gave in 1912 an example of an analytic function which, by an extension of the concept of a derivative so as to pass to the limit not through all the neighboring points but only through those belonging to certain dense aggregates, has a certain linear continuation beyond the domain of existence. Studies of monogenic uniform functions along the line of E. Borel and G. M. Mittag-Leffler have been made also by G. Vivanti, Marcel Riesz, Ivar Fredholm, and E. Phragmén.

[1] See G. M. Mittag-Leffler in *Atti del IV Congr. Intern, Roma, 1908*. Rome, 1909, Vol. I, p. 69. Here Mittag-Leffler gives a summary of recent results.

Interesting is the manner in which K. Weierstrass in Berlin and G. F. B. Riemann in Göttingen influenced each other. We have seen that Weierstrass defined functions of a complex variable by the power series and avoids geometrical means. Riemann begins with certain differential equations in the region of mathematical physics. In 1856 Riemann was urged by his friends to publish a *résumé* of his researches on Abelian functions, "be it ever so crude," because Weierstrass was at work on the same subject. Riemann's publication induced Weierstrass to withdraw from the press a memoir he had presented to the Berlin Academy in 1857, because, as he himself says, "Riemann published a memoir on the same problem which rested on entirely different foundations from mine, and did not immediately reveal that in its results it agreed completely with my own. The proof of this required investigations which were not quite easy, and took much time; after this difficulty had been removed a radical remodelling of my dissertation seemed necessary." In 1875 Weierstrass wrote H. A. Schwarz: "The more I ponder over the principles of the theory of functions—and I do so incessantly—the stronger grows my conviction that it must be built up on the foundation of algebraical truths, and that, therefore, to employ for the truth of simple and fundamental algebraical theorems the 'transcendental,' if I may say so, is not the correct way, however enticing *prima vista* the considerations may be by which Riemann has discovered many of the most important properties of algebraical functions." This refers mainly to the "Thomson-Dirichlet Principle," the validity of which depended on a certain minimum theorem which was shown by Weierstrass to rest upon unsound argument.

It has been objected that K. Weierstrass' definition of analytic functions is based on power series. A. L. Cauchy's definition, which was adopted by G. F. B. Riemann, is not open to this objection, but labors under the burden of requiring at the start the most difficult forms of the theory of limits. According to A. L. Cauchy a function is analytic (his "synectic") if it possesses a single-valued differential coefficient. Using Cauchy's integral theorem (Integralsatz), it follows that the synectic function admits not only of a single-valued differentiation but also of a single-valued integration. Giacinto Morera (1856–1909) of Turin showed that the synectic function might be defined by the single-valued integration. More recent researches, 1883–1895, which aim at a rigorous exposition of A. L. Cauchy's integral theorem, are due to M. Falk, E. Goursat, M. Lerch, C. Jordan, and A. Pringsheim. Cauchy's theorem may be stated: If the function $f(z)$ is synectic in a continuum in which every simply closed curve forms the boundary of an area, then the integral $\int f(z)dz$ is always zero, if it is extended over a closed curve which lies wholly within the continuum. Here the questions arise, what is a curve, a closed curve, a simply closed curve?

Analytic functions of several variables were treated by C. G. J. Jacobi in 1832, in his *Considerationes generales de transcendentibus Abelianis*, but received no attention until Weierstrass set himself the task presented to him by the study of Abelian functions, to find a solid foundation for functions of several variables that would correspond to his treatment of functions of one variable. He obtained a fundamental theorem on null-places; he also enunciated, without proof, the theorem that each single-valued (eindeutig) and in a finite region meromorphic function of several variables can be represented as the quotient of two integral functions, *i. e.* of two beständig convergent power series. This theorem was proved in 1883 by H. Poincaré for two variables and in 1895 by Pierre Cousin of Bordeaux for *n* variables. Later researches are by H. Hahn (1905), P. Boutroux (1905), G. Faber (1905), and F. Hartogs (1907).

Dirichlet's principle has repeatedly commanded attention. The question of its rigor has been put by E. Picard as follows:[1] "The conditions at the limits that one is led to assume are very different according as it is question of an equation of which the integrals are or are not analytic. A type of the first case is given by the problem generalized by P. G. L. Dirichlet; conditions of continuity there play an essential part, and, in general, the solution cannot be prolonged from the two sides of the continuum which serves as support to the data; it is no longer the same in the second case, where the disposition of this support in relation to the characteristics plays the principal rôle, and where the field of existence of the solution presents itself under wholly different conditions. . . . From antiquity has been felt the confused sentiment of a certain economy in natural phenomena; one of the first precise examples is furnished by Fermat's principle relative to the economy of time in the transmission of light. Then we came to recognize that the general equations of mechanics correspond to a problem of minimum, or more exactly of variation, and thus we obtained the principle of virtual velocities, then Hamilton's principle, and that of least action. A great number of problems appeared then as corresponding to minima of certain definite integrals. This was a very important advance, because the existence of a minimum could in many cases be regarded as evident, and consequently the demonstration of the existence of the solution was effected. This reasoning has rendered immense services; the greatest geometers, K. F. Gauss in the problem of the distribution of an attracting mass corresponding to a given potential, G. F. B. Riemann in his theory of Abelian functions, have been satisfied with it. To-day our attention has been called to the dangers of this sort of demonstration; it is possible for the minima to be simply limits and not to be actually attained by

[1] *Congress of Arts and Science*, St. Louis, 1904, Vol. I, p. 510.

veritable functions possessing the necessary properties of continuity. We are, therefore, no longer content with the probabilities offered by the reasoning long classic."

David Hilbert in 1899 spoke as follows:[1] "Dirichlet's principle owed its celebrity to the attractive simplicity of its fundamental mathematical idea, to the undeniable richness of its possible applications in pure and applied mathematics and to its inherent persuasive power. But after Weierstrass' criticism of it, Dirichlet's principle was considered as only of historical interest and discarded as a means of solving the boundary-value problem. C. Neumann deplores that this beautiful principle of Dirichlet, formerly used so much, has no doubt passed away forever. Only A. Brill and M. Noether arouse new hopes in us by giving expression to the conviction that Dirichlet's principle, being so to speak an imitation of nature, may sometime receive new life in modified form." Hilbert proceeds thereupon to rehabilitate the Principle, which involves a special problem in the calculus of variations. Dirichlet's procedure was briefly thus: On the xy plane erect at the points of the boundary curve perpendiculars the lengths of which represent the boundary values. Among the surfaces $z=f(x,y)$ which are bounded by the space-curve thus obtained, select the one for which the value of the integral $J(f)=\int\int\left\{\left(\frac{\partial f}{\partial x}\right)^2+\left(\frac{\partial f}{\partial y}\right)^2\right\}dxdy$ is a minimum. As shown by the calculus of variations, that surface is necessarily a potential surface. By reference to such a procedure G. F. B. Riemann thought he had settled the existence of the solution of boundary-value problems. But K. Weierstrass made it plain that among an infinite number of values there does not necessarily exist a minimum value; a minimum surface may therefore not exist. D. Hilbert generalizes Dirichlet's principle in this manner: "Every problem of the calculus of variations has a solution, as soon as restricting assumptions suitable to the nature of the given boundary conditions are satisfied and, if necessary, the concept of the solution receives a fitting extension." D. Hilbert shows how this may be used in finding rigorous, yet simple, existence proofs. In 1901 it was used in dissertations prepared by E. R. Hedrick and C. A. Noble.

Taking a birds' eye view of the development of the theory of functions during the nineteenth century since the time of A. L. Cauchy, James Pierpont said in 1904:[2] "Weierstrass and Riemann develop Cauchy's theory along two distinct and original paths. Weierstrass starts with an explicit analytic expression, a power series, and defines his function as the totality of its analytical continuations. No appeal is made to geometric intuition, his entire theory is strictly arithmetical. Riemann growing up under Gauss and Dirichlet, not only relies largely

[1] *Jahresb. d. d. Math. Vereinig.*, Vol. 8, 1900, p. 185.
[2] *Bull. Am. Math. Soc.*, 2. S., Vol. 11, 1904, p. 137.

on geometric intuition, but also does not hesitate to impress mathematical physics into his service. Two noteworthy features of his theory are the many-leaved surfaces named after him, and the extensive use of conformal representation. The history of functions as first developed is largely a theory of algebraic functions and their integrals. A general theory of functions is only slowly evolved. For a long time the methods of Cauchy, Riemann, and Weierstrass were cultivated along distinct lines by their respective pupils. The schools of Cauchy and Riemann were first to coalesce. The entire rigor which has recently been imparted to their methods has removed all reason for founding, as Weierstrass and his school have urged, the theory of functions on a single algorithm, viz., the power series. We may therefore say that at the close of the century there is only one theory of functions, in which the ideas of its three great creators are harmoniously united."

The study of existence theorems, particularly in the theory of algebraic functions and the calculus of variations, began with Cauchy. For implicit functions he assumed that they were expressible as power series, a restriction removed by U. Dini of Pisa. Simplifications are due to R. Lipschitz of Bonn. Existence theorems of sets of implicit functions were studied by G. A. Bliss of Chicago in the Princeton Colloquium of 1909. By means of a sheet of points Bliss deduces from an initial solution at an ordinary point a sheet of solutions somewhat analogous to K. Weierstrass' analytical continuation of a branch of a curve.

Accompanying and immediately following Riemann's time there was a development of the theory of algebraic functions, that was partly geometric in character and not purely along the line of function theory. A. Brill and M. Noether[1] in 1894 marked five directions of advance: *First*, the geometrico-algebraic direction taken by G. F. B. Riemann and G. Roch in the years 1862–1866, then by R. F. A. Clebsch 1863 to 1865, by Clebsch and P. Gordan since 1865 and since 1871 by A. Brill and M. Noether; *second*, the algebraic direction, followed by L. Kronecker and K. Weierstrass since 1860, more generally known since 1872, and in 1880 taken up by E. B. Christoffel; *third*, the invariantal direction, represented since 1877 by H. Weber, M. Noether, E. B. Christoffel, F. Klein, F. G. Frobenius, and F. Schottky; *Fourth*, the arithmetical direction of R. Dedekind and H. Weber since 1880, of L. Kronecker since 1881, of K. W. S. Hensel and others; *Fifth*, the geometrical direction taken by C. Segre and G. Castelnuovo since 1888.

Hermann Amandus Schwarz (1845–1921) of Berlin a pupil of K. Weierstrass, has given the conform representation (*Abbildung*) of various surfaces on a circle. G. F. B. Riemann had given a general theorem on the conformation of a given curve with another curve.

[1] A. Brill and M. Noether, *Jahrb. d. d. Math. Vereinigung*, Vol. 3, p. 287.

In transforming by aid of certain substitutions a polygon bounded by circular arcs into another also bounded by circular arcs, Schwarz was led to a remarkable differential equation $\psi(u', t) = \psi(u, t)$, where $\psi(u, t)$ is the expression which Cayley called the "Schwarzian derivative," and which led J. J. Sylvester to the theory of reciprocants. Schwarz's developments on minimum surfaces, his work on hypergeometric series, his inquiries on the existence of solutions to important partial differential equations under prescribed conditions, have secured a prominent place in mathematical literature.

Modular functions were at first considered merely as a by-product of elliptic functions, growing out of the study of transformations. After the epoch-making creations of E. Galois and G. F. B. Riemann, the subject of elliptic modular functions was developed into an independent theory, mainly by the efforts of H. Poincaré and F. Klein, which stands in close relation to the theory of numbers, algebra and synthetic geometry. F. Klein began to lecture on this subject in 1877; researches bearing upon this were pursued also by his then pupils W. Dyck, Joseph Gierster, and A. Hurwitz. One of the problems of modular functions is, to determine all subgroups of the linear group $x^1 = (\alpha x + \beta) / (\gamma x + \delta)$, where α, β, γ, δ are integers and $\alpha\delta - \beta\gamma \neq 0$. F. Klein's *Vorlesungen über das Ikosæder*,✡ Leipzig, 1884, is a work along this line. As an extended continuation of that are F. Klein's *Vorlesungen über die Theorie der elliptischen Modulfunctionen*, gotten out by Robert Fricke (Vol. I, 1890, Vol. II, 1892) and as a still further generalization we have the theory of the general linear automorphic functions, developed mainly by F. Klein and H. Poincaré. In 1897, under the joint authorship of Robert Fricke and Felix Klein, there appeared the first volume of the *Vorlesungen über die Theorie der Automorphen Functionen*, the second volume of which did not appear until 1912, after the theory had come under the influence of the critical tendencies due to K. Weierstrass and G. Cantor, and after E. Picard and H. Poincaré had brought out further incisive researches. It has been noted that F. Klein's own publications on these topics are in the order in which the subject itself sprang into existence. "Historically, the theory of automorphic functions developed from that of the regular solids and modular functions. At least this is the path which F. Klein followed under the influence of the well-known researches of Schwarz and of the early publications of H. Poincaré. If H. Poincaré brings in also other considerations, namely the arithmetic methods of Ch. Hermite . . . and the function-theoretical problems of Fuchs with regard to single valued inversion of the solutions of linear differential equations of the second order (eindeutige Umkehr der Lösungen . . .), these topics in turn go back to the very regions of thought from which have grown the theories of the regular solids and the elliptic modular functions." H. Poincaré published on this subject in *Math. Annalen*, Vol. 19, "Sur les fonctions uniformes qui se reproduisent par des substitutions linéaires," in the *Acta mathematica*, Vol. 1, a "Mémoire sur les fonctions fuchsiennes," and a procession of other papers extending over many years. Recently active along this same line were P. Koebe and L. E. J. Brouwer.

The question what automorphic forms can be expressed analytically by the H. Poincaré series has been investigated by Poincaré himself and also by E. Ritter and R. Fricke (1901).

After the creation of the theory of automorphic functions of a single variable, mainly by F. Klein and H. Poincaré, similar generalizations were sought for functions of several complex variables. The pioneer in this field was E. Picard; other workers are T. Levi-Civita, G. A. Bliss, W. D. MacMillan, and W. F. Osgood who lectured thereon at the Madison (Wisconsin) Colloquium in 1913. **Charles Emile Picard** (1856—1941) whose extensive researches on analysis have been mentioned repeatedly and whose *Traité d'Analyse* is well known, was born in Paris. He studied at the École Normale where he was inspired by J. G. Darboux. In 1881 he married a daughter of Hermite. Picard taught for a short time at Toulouse. Since 1881 he has been professor in Paris at the École Normale and the Sorbonne.

Uniformization

The uniformization of an algebraic or analytic curve, that is, the determination of such auxiliary variables which taken as independent variables render the co-ordinates of the points of the curve single-valued (eindeutig) analytic functions, is organically connected with the theory of automorphic functions. It was F. Klein and H. Poincaré who soon after 1880 developed the theory of automorphic functions and introduced systematically the idea of the uniformization of algebraic curves which G. F. B. Riemann had visualized upon the surfaces named after him. More recent researches on uniformization connect chiefly with the work of H. Poincaré and are due to D. Hilbert (1900), W. F. Osgood, T. Brodén, and A. M. Johanson. In 1907 followed important generalizations by H. Poincaré and by P. Koebe of Leipzig.[1] Dirichlet's Principle, having been established upon a sound foundation by D. Hilbert in 1901, was used as a starting point, for the derivation of new proofs of the general principle of uniformization, by P. Koebe of Leipzig and R. Courant of Göttingen.

Important works on the theory of functions are the *Cours de Ch. Hermite*, *J. Tannery's Théorie des Fonctions d'une variable seule*, *A Treatise on the Theory of Functions* by *James Harkness* and *Frank Morley*, and *Theory of Functions of a Complex Variable* by *A. R. Forsyth*. A broad and comprehensive treatise is the *Lehrbuch der Funktionentheorie* by W. F. Osgood of Harvard University, the first edition of which appeared in 1907 and the second enlarged edition in 1912.

[1] P. Koebe, *Atti del IV Congr., Roma, 1908*, Rome, 1909, Vol. II, p. 25.

Theory of Numbers

"Mathematics, the queen of the sciences, and arithmetic, the queen of mathematics." Such was the dictum of K. F. Gauss, who was destined to revolutionize the theory of numbers. When asked who was the greatest mathematician in Germany, P. S. Laplace answered, Pfaff. When the questioner said he should have thought Gauss was, Laplace replied, "Pfaff is by far the greatest mathematician in Germany; but Gauss is the greatest in all Europe." [1] Gauss is one of the three greatest masters of analysis,—J. Lagrange, P. S. Laplace, K. F. Gauss. Of these three contemporaries he was the youngest. While the first two belong to the period in mathematical history preceding the one now under consideration, Gauss is the one whose writings may truly be said to mark the beginning of our own epoch. In him that abundant fertility of invention, displayed by mathematicians of the preceding period, is combined with rigor in demonstration which is too often wanting in their writings, and which the ancient Greeks might have envied. Unlike P. S. Laplace, Gauss strove in his writings after perfection of form. He rivals J. Lagrange in elegance, and surpasses this great Frenchman in rigor. Wonderful was his richness of ideas; one thought followed another so quickly that he had hardly time to write down even the most meagre outline. At the age of twenty Gauss had overturned old theories and old methods in all branches of higher mathematics; but little pains did he take to publish his results, and thereby to establish his priority. He was the first to observe rigor in the treatment of infinite series, the first to fully recognize and emphasize the importance, and to make systematic use of determinants and of imaginaries, the first to arrive at the method of least squares, the first to observe the double periodicity of elliptic functions. He invented the heliotrope and, together with W. Weber, the bifilar magnetometer and the declination instrument. He reconstructed the whole of magnetic science.

Karl Friedrich Gauss [2] (1777–1855), the son of a bricklayer, was born at Brunswick. He used to say, jokingly, that he could reckon before he could talk. The marvellous aptitude for calculation of the young boy attracted the attention of Johann Martin Bartels (1769–1836), afterwards professor of mathematics at Dorpat, who brought him under the notice of Charles William, Duke of Brunswick. The duke undertook to educate the boy, and sent him to the Collegium Carolinum. His progress in languages there was quite equal to that in mathematics. In 1795 he went to Göttingen, as yet undecided whether to pursue philology or mathematics. Abraham Gotthelf Kästner (1719–1800), then professor of mathematics there, and now chiefly remembered for his *Geschichte der Mathematik* (1796), was not

[1] R. Tucker, "Carl Friedrich Gauss," *Nature*, Vol. 15, 1877, p. 534.

[2] W. Sartorius Waltershausen, *Gauss, zum Gedächtniss*, Leipzig, 1856.

a teacher who could inspire Gauss, though Kästner's German contemporaries ranked him high and admired his mathematical and poetical ability. Gauss declared that Kästner was the first mathematician among the poets and the first poet among the mathematicians. When not quite nineteen years old Gauss began jotting down in a copy-book very brief Latin memoranda of his mathematical discoveries. This diary was published in 1901.[1] Of the 146 entries, the first is dated March 30, 1796, and refers to his discovery of a method of inscribing in a circle a regular polygon of seventeen sides. This discovery encouraged him to pursue mathematics. He worked quite independently of his teachers, and while a student at Göttingen made several of his greatest discoveries. Higher arithmetic was his favorite study. Among his small circle of intimate friends was Wolfgang Bolyai. After completing his course he returned to Brunswick. In 1798 and 1799 he repaired to the university at Helmstadt to consult the library, and there made the acquaintance of J. F. Pfaff, a mathematician of much power. In 1807 the Emperor of Russia offered Gauss a chair in the Academy at St. Petersburg, but by the advice of the astonomer Olbers, who desired to secure him as director of a proposed new observatory at Göttingen, he declined the offer, and accepted the place at Göttingen. Gauss had a marked objection to a mathematical chair, and preferred the post of astronomer, that he might give all his time to science. He spent his life in Göttingen in the midst of continuous work. In 1828 he went to Berlin to attend a meeting of scientists, but after this he never again left Göttingen, except in 1854, when a railroad was opened between Göttingen and Hanover. He had a strong will, and his character showed a curious mixture of self-conscious dignity and child-like simplicity. He was little communicative, and at times morose. Of Gauss' collected works, or *Werke*, an eleventh volume was planned in 1916, to be biographical and bibliographical in character.

A new epoch in the theory of numbers dates from the publication of his *Disquisitiones Arithmeticæ*,✲ Leipzig, 1801. The beginning of this work dates back as far as 1795. Some of its results had been previously given by J. Lagrange and L. Euler, but were reached independently by Gauss, who had gone deeply into the subject before he became acquainted with the writings of his great predecessors. The *Disquisitiones Arithmeticæ* was already in print when A. M. Legendre's *Théorie des Nombres* appeared. The great law of quadratic reciprocity, given in the fourth section of Gauss' work, a law which involves the whole theory of quadratic residues, was discovered by him by induction before he was eighteen, and was proved by him one year later. Afterwards he learned that L. Euler had imperfectly enunciated that theorem, and that A. M. Legendre had attempted to prove it,

[1] *Gauss' wissenschaftliche Tagebuch*, 1796–1814. Mit Anmerkungen herausgegeben von Felix Klein, Berlin, 1901.

but met with apparently insuperable difficulties. In the fifth section Gauss gave a second proof of this "gem" of higher arithmetic. In 1808 followed a third and fourth demonstration; in 1817, a fifth and sixth. No wonder that he felt a personal attachment to this theorem. Proofs [1] were given also by C. G. J. Jacobi, F. Eisenstein, J. Liouville, Victor Amédée Lebesgue (1791–1875) of Bordeaux, Angelo Genocchi (1817–1889) of the University of Turin, E. E. Kummer, M. A. Stern, Christian Zeller (1822–1899) of Markgröningen, L. Kronecker, Victor Jacovlevich Bouniakovski (1804–1889) of Petrograd, Ernst Schering (1833–1897) of Göttingen, Julius Peter Christian Petersen (1839–1910) of Copenhagen, E. Busche, Th. Pepin, Fabian Franklin, J. C. Fields, and others. Quadratic reciprocity "stands out not only for the influence it has exerted in many branches, but also for the number of new methods to which it has given birth" (P. A. MacMahon). The solution of the problem of the representation of numbers by binary quadratic forms is one of the great achievements of Gauss. He created a new algorithm by introducing the theory of congruences. The fourth section of the *Disquisitiones Arithmeticæ*, treating of congruences of the second degree, and the fifth section, treating of quadratic forms, were, until the time of C. G. J. Jacobi, passed over with universal neglect, but they have since been the starting-point of a long series of important researches. The seventh or last section, developing the theory of the division of the circle, was received from the start with deserved enthusiasm, and has since been repeatedly elaborated for students. A standard work on *Kreistheilung* was published in 1872 by Paul Bachmann, then of Breslau.

The equation for the division of the circle and the construction of a regular polygon of n sides, n being prime, can be solved by square root extractions alone, always and only when $n-1$ is a power of 2. Hence such regular polygons can be constructed by ruler and compasses when the prime number n is 3, 5, 17, 257, 65,537, . . but cannot be constructed when n is 7, 11, 13, . . The results may be stated also thus: The Greeks knew how to inscribe regular polygons whose sides numbered 2^m, 2^m. 3, 2^m. 5 and 2^m. 15. Gauss added in 1801 that the construction is possible when the number of sides n is prime and of the form $2^{2^\mu}+1$. L. E. Dickson computed that the number of such inscriptible polygons for $n \leqq 100$ is 24, for $n \leqq 300$ is 37, for $n \leqq 1000$ is 52, for $n \leqq 100{,}000$ is 206.

Three classical constructions of the regular inscribed polygon of seventeen sides have been given: one by J. Serret in his *Algebra*, II, § 547, another by von Staudt in *Crelle*, Vol. 24, and a third by L. Gérard in *Math. Annalen*, Vol. 48 (1897), using compasses only. The analytic solution, as outlined by Gauss, was actually carried out for the regular polygon of 257 sides by F. J. Richelot of Königsberg in four articles in *Crelle*, Vol. 9. For the polygon of 65,537 sides this

[1] O. Baumgart, *Ueber das Quadratische Reciprocitätsgesetz*, Leipzig, 1885.

was accomplished after ten years of labor by Oswald Hermes (1826–1909) of Steglitz; his manuscript is deposited in the mathematical seminar at Göttingen.[1] Gauss had planned an eighth section of his *Disquisitiones Arithmeticae*, which was omitted to lessen the expense of publication. His papers on the theory of numbers were not all included in his great treatise. Some of them were published for the first time after his death in his collected works. He wrote two memoirs on the theory of biquadratic residues (1825 and 1831), the second of which contains a theorem of biquadratic reciprocity.

K. F. Gauss was led to astronomy by the discovery of the planet Ceres at Palermo in 1801. His determination of the elements of its orbit with sufficient accuracy to enable H. W. M. Olbers to rediscover it, made the name of Gauss generally known. In 1809 he published the *Theoria motus corporum cœlestium*, which contains a discussion of the problems arising in the determination of the movements of planets and comets from observations made on them under any circumstances. In it are found four formulæ in spherical trigonometry, now usually called "Gauss' Analogies," but which were published somewhat earlier by Karl Brandon Mollweide (1774–1825) of Leipzig, and earlier still by Jean Baptiste Joseph Delambre (1749–1822).[2] Many years of hard work were spent in the astronomical and magnetic observatory. He founded the German Magnetic Union, with the object of securing continuous observations at fixed times. He took part in geodetic observations, and in 1843 and 1846 wrote two memoirs, *Ueber Gegenstände der höheren Geodesie*. He wrote on the attraction of homogeneous ellipsoids, 1813. In a memoir on capillary attraction, 1833, he solves a problem in the calculus of variations involving the variation of a certain double integral, the limits of integration being also variable; it is the earliest example of the solution of such a problem. He discussed the problem of rays of light passing through a system of lenses.

Among Gauss' pupils were Heinrich Christian Schumacher, Christian Gerling, Friedrich Nicolai, August Ferdinand Möbius, Georg Wilhelm Struve, Johann Frantz Encke.

Gauss' researches on the theory of numbers were the starting-point for a school of writers, among the earliest of whom was C. G. J. Jacobi. The latter contributed to *Crelle's Journal* an article on cubic residues, giving theorems without proofs. After the publication of Gauss' paper on biquadratic residues, giving the law of biquadratic reciprocity, and his treatment of complex numbers, C. G. J. Jacobi found a similar law for cubic residues. By the theory of elliptical functions, he was led to beautiful theorems on the representation of

[1] A. Mitzscherling, *Das Problem der Kreisteilung*, Leipzig u. Berlin, 1913, pp. 14, 23.

[2] I. Todhunter, "Note on the History of Certain Formulæ in Spherical Trigonometry," *Philosophical Magazine*, Feb., 1873.

numbers by 2, 4, 6, and 8 squares. Next come the researches of P. G. L. Dirichlet, the expounder of Gauss, and a contributor of rich results of his own.

Peter Gustav Lejeune Dirichlet [1] (1805–1859) was born in Düren, attended the gymnasium in Bonn, and then the Jesuit gymnasium in Cologne. In 1822 he was attracted to Paris by the names of P. S. Laplace, A. M. Legendre, J. Fourier, D. S. Poisson, and A. L. Cauchy. The facilities for a mathematical education there were far better than in Germany, where K. F. Gauss was the only great figure. He read in Paris Gauss' *Disquisitiones Arithmeticæ*, a work which he never ceased to admire and study. Much in it was simplified by Dirichlet, and thereby placed within easier reach of mathematicians. His first memoir on the impossibility of certain indeterminate equations of the fifth degree was presented to the French Academy in 1825. He showed that P. Fermat's equation, $x^n+y^n=z^n$, cannot exist when $n=5$. Some parts of the analysis are, however, A. M. Legendre's. Dirichlet's acquaintance with J. Fourier led him to investigate Fourier's series. He became docent in Breslau in 1827. In 1828 he accepted a position in Berlin, and finally succeeded K. F. Gauss at Göttingen in 1855. The general principles on which depends the average number of classes of binary quadratic forms of positive and negative determinant (a subject first investigated by Gauss) were given by Dirichlet in a memoir, *Ueber die Bestimmung der mittleren Werthe in der Zahlentheorie*, 1849. More recently F. Mertens of Graz, since 1894 of Vienna, determined the asymptotic values of several numerical functions. Dirichlet gave some attention to prime numbers. K. F. Gauss and A. M. Legendre had given expressions denoting approximately the asymptotic value of the number of primes inferior to a given limit, but it remained for G. F. B. Riemann in his memoir, *Ueber die Anzahl der Primzahlen unter einer gegebenen Grösse*, 1859, to give an investigation of the asymptotic frequency of primes which is rigorous. Approaching the problem from a different direction, *P. L. Chebichev*, formerly professor in the University of St. Petersburg, established, in a celebrated memoir, *Sur les Nombres Premiers*, 1850, the existence of limits within which the sum of the logarithms of the primes P, inferior to a given number x, must be comprised.[2] He proved that, if $n>3$, there is always at least one prime between n and $2n-2$ (inclusive). This theorem is sometimes called "Bertrand's postulate," since J. L. F. Bertrand had previously assumed it for the purpose of proving a theorem in the theory of substitution groups. This paper depends on very elementary considerations, and, in that respect, contrasts strongly with Riemann's, which involves abstruse theorems of the integral calculus. H. Poincaré's papers, J. J. Syl-

[1] E. E. Kummer, *Gedächtnissrede auf Gustav Peter Lejeune-Dirichlet*, Berlin, 1860.

[2] H. J. Stephen Smith "On the Present State and Prospects of some Branches of Pure Mathematics," *Proceed. London Math. Soc.*, Vol. 8, 1876, p. 17.

vester's contraction of Chebichev's limits, with reference to the distribution of primes, and researches of J. Hadamard (awarded the *Grand prix* of 1892), are among the later researches in this line.

G. F. B. Riemann had advanced six properties relating to

$$\zeta(s)=\sum_{n=1}^{\infty}\frac{1}{n^s}, \text{ where } s=\sigma+ti,$$

none of which he was able to prove.[1] In 1893 J. Hadamard proved three of these, thereby establishing the existence of null-places in Riemann's zeta-function; H. von Mangoldt of Danzig proved in 1895 a fourth and in 1905 a fifth of Riemann's six properties. The remaining one, that the roots of $\zeta(s)$ in the strip $0 \leqq \sigma \leqq 1$, have all the real part $\frac{1}{2}$, remains unproved, though progress in the study of this case has been made by F. Mertens and R. v. Sterneck. If x is a positive number, and if $\pi(x)$ denotes the number of primes less than x, then what Landau calls the "prime-number theorem" (Primzahlsatz) states that the ratio of $\pi(x)$ to $x/\log x$ approaches 1 as x increases without end. A. M. Legendre, K. F. Gauss, and P. G. L. Dirichlet had guessed this theorem. As early as 1737 L. Euler[2] had given an analogous theorem, that $\Sigma 1/p$ approaches log (log p), where the summation extends over all primes not greater than p. The prime-number theorem was proved in 1896 by J. Hadamard and Charles Jean de la Vallée Poussin of Louvain, in 1901 by Nils Fabian Helge von Koch of Stockholm, in 1903 by E. Landau, now of Göttingen, in 1915 by G. H. Hardy and J. E. Littlewood of Cambridge. Hardy discovered an infinity of zeroes of the zeta-function with the real part $\frac{1}{2}$; E. Landau simplified Hardy's proof.

G. F. B. Riemann's zeta-function $\zeta(s)$ was first studied on account of its fundamental importance in the theory of prime numbers, but it has become important also in the theory of analytic functions in general. In 1909 E. Landau published his *Handbuch der Lehre von der Verteilung der Primzahlen.*✲ In 1912 he pronounced the following four questions to be apparently incapable of answer in the present state of the science of numbers: (1) Does n^2+1 for integral values of n represent an infinite number of primes? (2) C. Goldbach's theorem: Can prime values of p and p' be found to satisfy $m=p+p'$ for each even m larger than 2? (3) Has $2=p-p^1$ an infinite number of solutions in primes? (4) Is there between n^2 and $(n+1)^2$ at least one prime for every positive integral n?

The enumeration of prime numbers has been undertaken at different times by various mathematicians. Factor tables, giving the least factor of every integer not divisible by 2, 3, or 5, did not extend above 408,000 previous to the year 1811, when Ladislaus Chernac published his *Cribrum arithmeticum* at Deventer in Netherlands, which gives

[1] For details, consult E. Landau in *Proceed. 5th Intern. Congress, Cambridge*, 1912, Vol. 1, 1913, p. 97.

[2] G. Eneström in *Bibliotheca mathematica*, 3. S., Vol. 13, p. 81.

factors for numbers up to 1,020,000. J. Ch. Burckhardt (1773–1815) published factor tables in Paris, in 1817 for the numbers 1 to 1,020,000, in 1814 for the numbers 1020000 to 2028000, in 1816 for the numbers 2,028,000 to 3,036,000. James Glaisher (1809–1903) published factor tables at London, in 1879 for the numbers 3,000,000 to 4,000,000, in 1880 for numbers 4,000,000 to 5,000,000, in 1883 for the numbers 5,000,000 to 6,000,000. Zacharias Dase (1824–1861) published factor tables at Hamburg, in 1862 for the numbers 6,000,001 to 7,002,000, in 1863 for the numbers 7,002,001 to 8,010,000, in 1865 for the numbers 8,010,001 to 9,000,000. In 1909 the Carnegie Institution of Washington published factor tables for the first ten millions, prepared by D. N. Lehmer of the University of California. Lehmer gives the errors discovered in the earlier publications. Historical details about factor tables are given by Glaisher in his *Factor Table. Fourth Million*, 1879.

Miscellaneous contributions to the theory of numbers were made by *A. L. Cauchy*. He showed, for instance, how to find all the infinite solutions of a homogeneous indeterminate equation of the second degree in three variables when one solution is given. He established the theorem that if two congruences, which have the same modulus, admit of a common solution, the modulus is a divisor of their resultant. **Joseph Liouville** (1809–1882), professor at the Collège de France, investigated mainly questions on the theory of quadratic forms of two, and of a greater number of variables. A research along a different line proved to be an entering wedge into a subject which since has become of vital importance. In 1844 he proved (*Liouville's Journal*, Vol. 5) that neither e nor e^2 can be a root of a quadratic equation with rational coefficients. By the properties of convergents of a continued fraction representing a root of an algebraical equation with rational coefficients he established later the existence of numbers—the so-called transcendental numbers—which cannot be roots of any such equation. He proved this also by another method. A still different approach is due to G. Cantor. Profound researches were instituted by **Ferdinand Gotthold Eisenstein** (1823–1852), of Berlin. Ternary quadratic forms had been studied somewhat by K. F. Gauss, but the extension from two to three indeterminates was the work of Eisenstein who, in his memoir, *Neue Theoreme der höheren Arithmetic*, defined the ordinal and generic characters of ternary quadratic forms of uneven determinant; and, in case of definite forms, assigned the weight of any order or genus. But he did not publish demonstrations of his results. In inspecting the theory of binary cubic forms, he was led to the discovery of the first covariant ever considered in analysis. He showed that the series of theorems, relating to the presentation of numbers by sums of squares, ceases when the number of squares surpasses eight. Many of the proofs omitted by Eisenstein were supplied by Henry Smith, who was one of the few Englishmen who devoted themselves to the study of higher arithmetic.

Henry John Stephen Smith [1] (1826–1883) was born in London, and educated at Rugby and at Balliol College, Oxford. Before 1847 he travelled much in Europe for his health, and at one time attended lectures of D. F. J. Arago in Paris, but after that year he was never absent from Oxford for a single term. In 1849 he carried off at Oxford the highest honors, both in the classics and in mathematics, thus ranking as a "double first." There is a story that he decided between classics and mathematics as the field for his life-work, by tossing up a penny. He never married and had no household cares to destroy the needed serenity for scientific work, "excepting that he was careless in money matters, and trusted more to speculation in mining shares than to economic management of his income." [2] In 1861 he was elected Savilian professor of geometry. His first paper on the theory of numbers appeared in 1855. The results of ten years' study of everything published on the theory of numbers are contained in his Reports✡ which appeared in the British Association volumes from 1859 to 1865. These reports are a model of clear and precise exposition and perfection of form. They contain much original matter, but the chief results of his own discoveries were printed in the *Philosophical Transactions* for 1861 and 1867. They treat of linear indeterminate equations and congruences, and of the orders and genera of ternary quadratic forms. He established the principles on which the extension to the general case of n indeterminates of quadratic forms depends. He contributed also two memoirs to the *Proceedings of the Royal Society* of 1864 and 1868, in the second of which he remarks that the theorems of C. G. J. Jacobi, F. Eisenstein, and J. Liouville, relating to the representation of numbers by 4, 6, 8 squares, and other simple quadratic forms are deducible by a uniform method from the principles indicated in his paper. Theorems relating to the case of 5 squares were given by F. Eisenstein, but Smith completed the enunciation of them, and added the corresponding theorems for 7 squares. The solution of the cases of 2, 4, 6 squares may be obtained by elliptic functions, but when the number of squares is odd, it involves processes peculiar to the theory of numbers. This class of theorems is limited to 8 squares, and Smith completed the group. In ignorance of Smith's investigations, the French Academy offered a prize for the demonstration and completion of F. Eisenstein's theorems for 5 squares. This Smith had accomplished fifteen years earlier. He sent in a dissertation in 1882, and next year, a month after his death, the prize was awarded to him, another prize being also awarded to H. Minkowsky of Bonn. The theory of numbers led Smith to the study of elliptic functions. He wrote also on modern geometry. His successor at Oxford was J. J. Sylvester. Taking an anti-utilitarian view of

[1] J. W. L. Glaisher in *Monthly Notices R. Astr. Soc.*, Vol. 44, 1884.

[2] A. Macfarlane, *Ten British Mathematicians*, 1916, p. 98.

mathematics, Smith once proposed a toast, "Pure mathematics; may it never be of any use to any one."

Ernst Eduard Kummer (1810–1893), professor in the University of Berlin, is closely identified with the theory of numbers. P. G. L. Dirichlet's work on complex numbers of the form $a+ib$, introduced by K. F. Gauss, was extended by him, by F. Eisenstein, and R. Dedekind. Instead of the equation $x^4-1=0$, the roots of which yield Gauss' units, F. Eisenstein used the equation $x^3-1=0$ and complex numbers $a+b\rho$ (ρ being a cube root of unity), the theory of which resembles that of Gauss' numbers. E. E. Kummer passed to the general case $x^n-1=0$ and got complex numbers of the form $\alpha=a_1A_1+a_2A_2+a_3A_3+\ldots$, were a_i are whole real numbers, and A_i roots of the above equation. Euclid's theory of the greatest common divisor is not applicable to such complex numbers, and their prime factors cannot be defined in the same way as prime factors of common integers are defined. In the effort to overcome this difficulty, E. E. Kummer was led to introduce the conception of "ideal numbers." These ideal numbers have been applied by G. Zolotarev of St. Petersburg to the solution of a problem of the integral calculus, left unfinished by Abel.[1] J. W. R. Dedekind of Braunschweig has given in the second edition of Dirichlet's *Vorlesungen über Zahlentheorie* a new theory of complex numbers, in which he to some extent deviates from the course of E. E. Kummer, and avoids the use of ideal numbers. Dedekind has taken the roots of any irreducible equation with integral coefficients as the units for his complex numbers. F. Klein in 1893 introduced simplicity by a geometric treatment of ideal numbers.

Fermat's "Last Theorem," Waring's Theorem

E. E. Kummer's ideal numbers owe their origin to his efforts to prove the impossibility of solving in integers Fermat's equation $x^n+y^n=z^n$ for $n>2$. We premise that some progress in proving this impossibility has been made by more elementary means. For integers x, y, z not divisible by an odd prime n, the theorem has been proved by the Parisian mathematician and philosopher *Sophie Germain* (1776–1831) for $n<100$, by Legendre for $n<200$, by E. T. Maillet for $n<223$, by Dmitry Mirimanoff for $n<257$, by L. E. Dickson for $n<7000$.[2] The method used here is due to Sophie Germain and requires the determination of an odd prime p for which $x^n+y^n+z^n\equiv 0 \pmod{p}$ has no solutions, each not divisible by p, and n is not the residue modulo p of the nth power of any integer. E. E. Kummer's results rest on an advanced theory of algebraic numbers which he

[1] H. J. S. Smith, "On the Present State and Prospects of Some Branches of Pure Mathematics," *Proceed. London Math. Soc.*, Vol. 8, 1876, p. 15.

[2] See L. E. Dickson in *Annals of Mathematics*, 2. S., Vol. 18, 1917, pp. 161–187. See also L. E. Dickson in *Atti del IV. Congr. Roma*, 1908, Rome, 1909, Vol. II, p. 172.

helped to create. Once at an early period he thought that he had a complete proof. He laid it before P. G. L. Dirichlet who pointed out that, although he had proved that any number $f(\alpha)$, where α is a complex n^{th} root of unity and n is prime, was the product of indecomposable factors, he had assumed that such a factorization was unique, whereas this was not true in general.[1] After years of study, E. E. Kummer concluded that this non-uniqueness of factorization was due to $f(\alpha)$ being too small a domain of numbers to permit the presence in it of the true prime numbers. He was led to the creation of his ideal numbers, the machinery of which, says L. E. Dickson,[2] is "so delicate that an expert must handle it with the greatest care, and (is) nowadays chiefly of historical interest in view of the simpler and more general theory of R. Dedekind." By means of his ideal numbers he produced a proof of Fermat's last theorem, which is not general but excludes certain particular values of n, which values are rare among the smaller values of n; there are no values of n below 100, for which E. E. Kummer's proof does not serve. In 1857 the French Academy of Sciences awarded E. E. Kummer a prize of 3000 francs for his researches on complex integers.

The first marked advance since Kummer was made by A. Wieferich of Münster, in *Crelle's Journal*, Vol. 136, 1909, who demonstrated that if p is prime and 2^p-2 is not divisible by p^2, the equation $x^p+y^p=z^p$ cannot be solved in terms of positive integers which are not multiples of p. Waldemar Meissner of Charlottenburg found that 2^p-2 is divisible by p^2 when $p=1093$ and for no other prime p less than 2000. Recent advances toward a more general proof of Fermat's last theorem have been made by D. Mirimanoff of Geneva, G. Frobenius of Berlin, E. Hecke of Göttingen, F. Bernstein of Göttingen, Ph. Furtwängler of Bonn, S. Bohnicek and H. S. Vandiver of Philadelphia. Recent efforts along this line have been stimulated in part by a bequest of 100,000 marks made in 1908 to the Königliche Gesellschaft der Wissenschaften in Göttingen, by the mathematician F. P. Wolfskehl of Darmstadt, as a prize for a complete proof of Fermat's last theorem. Since then hundreds of erroneous proofs have been published. Post-mortems over proofs which fall still-born from the press are being held in the "Sprechsaal" of the Archiv der Mathematik und Physik.

At the beginning of the present century progress was made in proving another celebrated theorem, known as "Waring's theorem." In 1909 A. Wieferich of Münster proved the part which says that every positive integer is equal to the sum of not more than 9 positive cubes. He established also, that every positive integer is equal to the sum of not more than 37 (according to Waring, it is not more than 19) positive fourth powers, while D. Hilbert proved in 1909 that, for

[1] *Festschrift z. Feier des 100. Geburtstages Eduard Kummers*, Leipzig, 1910, p. 22.
[2] *Bull. Am. Math. Soc.*, Vol. 17, 1911, p. 371.

every integer $n>2$ (Waring had declared for every integer $n>4$), each positive integer is expressible as the sum of positive nth powers, the number of which lies within a limit dependent only upon the value of n. Actual determinations of such upper limits have been made by A. Hurwitz, E. T. Maillet, A. Fleck, and A. J. Kempner. Kempner proved in 1912 that there is an infinity of numbers which are not the sum of less than $4 \cdot 2^n$ positive 2^nth powers, $n \geqq 2$.

Other Recent Researches. Number Fields

Attracted by E. E. Kummer's investigations, his pupil, **Leopold Kronecker** (1823–1891) made researches which he applied to algebraic equations. On the other hand, efforts have been made to utilize in the theory of numbers the results of the modern higher algebra. Following up researches of Ch. Hermite, **Paul Bachmann** of Münster, now of Weimar, investigated the arithmetical formula which gives the automorphics of a ternary quadratic form.[1] Bachmann is the author of well-known texts on *Zahlentheorie*, in several volumes, which appeared in 1892, 1894, 1872, 1898, and 1905, respectively. The problem of the equivalence of two positive or definite ternary quadratic forms was solved by L. Seeber; and that of the arithmetical automorphics of such forms, by F. G. Eisenstein. The more difficult problem of the equivalence for indefinite ternary forms has been investigated by Eduard Selling of Würzburg. On quadratic forms of four or more indeterminates little has yet been done. Ch. Hermite showed that the number of non-equivalent classes of quadratic forms having integral coefficients and a given discriminant is finite, while Zolotarev and Alexander Korkine (1837–1908), both of St. Petersburg, investigated the minima of positive quadratic forms. In connection with binary quadratic forms, H. J. S. Smith established the theorem that if the joint invariant of two properly primitive forms vanishes, the determinant of either of them is represented primitively by the duplicate of the other.

The interchange of theorems between arithmetic and algebra is displayed in the recent researches of J. W. L. Glaisher (1848–1928) of Trinity College and J. J. Sylvester. Sylvester gave a Constructive Theory of Partitions, which received additions from his pupils, F. Franklin, now of New York city, and George Stetson Ely (?–1918), for many years examiner in the U. S. Patent Office.

By the introduction of "ideal numbers" E. E. Kummer took a step toward a theory of fields of numbers. The consideration of super fields (Oberkörper) from which the properties of a given field of numbers may be easily derived is due mainly to R. Dedekind and to L. Kronecker. Thereby there was opened up for the theory of numbers a new and wide territory which is in close connection with

[1] H. J. S. Smith in *Proceed. London Math. Soc.*, Vol. 8, 1876, p. 13.

algebra and the theory of functions. The importance of this subject in the theory of equations is at once evident if we call to mind E. Galois' fields of rationality. The interrelation between number theory and function theory is illustrated in Riemann's researches in which the frequency of primes was made to depend upon the zero-places of a certain analytical function, and in the transcendence of e and π which is an arithmetic property of the exponential function. In 1883–1890 L. Kronecker published important results on elliptic functions which contain arithmetical theorems of great elegance. The Dedekind method of extending Kummer's results to algebraic numbers in general is based on the notion of an *ideal*. A common characteristic of Dedekind and Kronecker's procedure is the introduction of compound moduli. G. M. Mathews says [1] that, in practice it is convenient to combine the methods of L. Kronecker and R. Dedekind. Of central importance are the Galoisian or normal fields, which have been studied extensively by D. Hilbert. L. Kronecker established the theorem that all Abelian fields are cyclotomic, which was proved also by H. Weber and D. Hilbert. An important report, prepared by D. Hilbert and entitled *Theorie der algebraischen Zahlkörper*, was published in 1894.[2] D. Hilbert first develops the theory of general number-fields, then that of special fields, viz., the Galois field, the quadratic field, the circle field (Kreiskörper), the Kummer field. A report on later investigations was published by R. Fueter in 1911.[3] Chief among the workers in this subject which have not yet been mentioned are F. Bernstein, Ph. Furtwängler, H. Minkowski, Ch. Hermite, and A. Hurwitz. Accounts of the theory are given in H. Weber's *Lehrbuch der Algebra*, Vol. 2 (1899), J. Sommer's *Vorlesungen über Zahlentheorie* (1907), and Hermann Minkowski's *Diophantische Approximationen*, Leipzig (1907). H. Minkowski gives in geometric and arithmetic language both old and new results. His use of lattices serves as a geometric setting for algebraic theory and for the proof of some new results.

A new and powerful method of attacking questions on the theory of algebraic numbers was advanced by Kurt Hensel of Königsberg in his *Theorie der algebraischen Zahlen*, 1908, and in his *Zahlentheorie*, 1913. His method is analogous to that of power series in the theory of analytic functions. He employs expansions of numbers into power series in an arbitrary prime number p. This theory of p-adic numbers is generalized by him in his book of 1913 into the theory of g-adic numbers, where g is any integer.[4]

The resolution of a given large number into factors is a difficult problem which has been taken up by Paul Seelhof, François Edouard

[1] Art. "Number" in the *Encyclop. Britannica*, 11th ed., p. 857.
[2] *Jahresbericht d. d. Math. Vereinigung*, Vol. 4, pp. 177–546.
[3] *Loc. cit.*, Vol. 20, pp. 1–47.
[4] *Bull. Am. Math. Soc.*, Vol. 20, 1914, p. 259.

Anatole Lucas (1842–1891) of Paris, Fortuné Landry (1799–?), A. J. C. Cunningham, F. W. P. Lawrence and D. N. Lehmer.

Transcendental Numbers. The Infinite

Building on the results previously reached by J. Liouville, Ch. Hermite proved in 1873 in the *Comptes Rendus*, Vol. 77, that e is transcendental, while F. Lindemann in 1882 (*Ber. Akad. Berlin*) proved that π is transcendental. Ch. Hermite reached his result by showing that $ae^m+be^n+ce^r+\ldots=0$ cannot subsist, where $m, n, r, \ldots$ $a, b, c, \ldots$ are whole numbers; F. Lindemann proved that this equation cannot subsist when $m, n, r, \ldots a, b, c \ldots$ are algebraic numbers, that in particular, $e^{ix}+1=0$ cannot subsist if x is algebraic. Consequently π cannot be an algebraic number. But, starting with two points, (0, 0) and (1, 0), a third point (a, 0) can be constructed by the aid of ruler and compasses only when a is a certain special type of algebraic number that is obtainable by successive square root extractions. Hence the point (π, 0) cannot be constructed, and the "quadrature of the circle" is impossible. The proofs of Ch. Hermite and F. Lindemann involved complex integrations and were complicated. Simplified proofs were given by K. Weierstrass in 1885, Th. J. Stieltjes in 1890, D. Hilbert, A. Hurwitz, and P. Gordan in 1896 (*Math. Annalen*, Vol. 43), F. Mertens in 1896, Th. Vahlen in 1900, H. Weber, F. Enriques, and E. W. Hobson in 1911. G. B. Halsted says of the circle, "John Bolyai squared it in non-Euclidean geometry and Lindemann proved no man could square it in Euclidean geometry."

That there are many other transcendental numbers beside e and π is evident from the researches of J. Liouville, E. Maillet, G. Faber and Aubrey J. Kempner, who give new forms of infinite series which define transcendental numbers. Of interest are the theorems established in 1913 by G. N. Bauer and H. L. Slobin of Minneapolis, that the trigonometric functions and the hyperbolic functions represent transcendental numbers whenever the argument is an algebraic number other than zero, and vice versa, the arguments are transcendental numbers whenever the functions are algebraic numbers.[1]

The notions of the actually infinite have undergone radical change during the nineteenth century. As late as 1831 K. F. Gauss expressed himself thus: "I protest against the use of infinite magnitude as something completed, which in mathematics is never permissible. Infinity is merely a *façon de parler*, the real meaning being a limit which certain ratios approach indefinitely near, while others are permitted to increase without restriction."[2] Gauss' contemporary, A. L. Cauchy, likewise rejected the actually infinite, being influenced by

[1] *Rendiconti d. Circolo Math. di Palermo*, Vol. 38, 1914, p. 353.

[2] C. F. Gauss, Brief on Schumacher, *Werke*, Bd. 8, 216; quoted from Moritz, *Memorabilia mathematica*, 1914, p. 337.

the eighteenth century philosopher of Turin, Father Gerdil.[1] In 1886 Georg Cantor occupied a diametrically opposite position, when he said: "In spite of the essential difference between the conceptions of the *potential* and the *actual* infinite, the former signifying a *variable* finite magnitude increasing beyond all finite limits, while the latter is a *fixed*, *constant* quantity lying beyond all finite magnitudes, it happens only too often that the one is mistaken for the other. . . . Owing to a justifiable aversion to such *illegitimate* actual infinities and the influence of the modern epicuric-materialistic tendency, a certain *horror infiniti* has grown up in extended scientific circles, which finds its classic expression and support in the letter of Gauss, yet it seems to me that the consequent uncritical rejection of the legitimate actual infinite is no lesser violation of the nature of things, which must be taken as they are." [2]

In 1904 Charles Emile Picard of Paris expressed himself thus:[3] "Since the concept of number has been sifted, in it have been found unfathomable depths; thus, it is a question still pending to know, between the two forms, the cardinal number and the ordinal number, under which the idea of number presents itself, which of the two is anterior to the other, that is to say, whether the idea of number properly so called is anterior to that of order, or if it is the inverse. It seems that the geometer-logician neglects too much in these questions psychology and the lessons uncivilized races give us; it would seem to result from these studies that the priority is with the cardinal number."

Applied Mathematics. Celestial Mechanics

Notwithstanding the beautiful developments of celestial mechanics reached by P. S. Laplace at the close of the eighteenth century, there was made a discovery on the first day of the nineteenth century which presented a problem seemingly beyond the power of that analysis. We refer to the discovery of Ceres by Giuseppe Piazzi in Italy, which became known in Germany just after the philosopher G. W. F. Hegel had published a dissertation proving *a priori* that such a discovery could not be made. From the positions of the planet observed by Piazzi its orbit could not be satisfactorily calculated by the old methods, and it remained for the genius of K. F. Gauss to devise a method of calculating elliptic orbits which was free from the assumption of a small eccentricity and inclination. Gauss' method was developed further in his *Theoria Motus*. The new planet was re-discovered with aid of Gauss' data by H. W. M. Olbers, an astronomer who promoted science not only by his own astronomical studies, but also by discern-

[1] See F. Cajori, "History of Zeno's Arguments on Motion," *Am. Math. Monthly*, Vol. 22, 1915, p. 114.

[2] G. Cantor, Zum Problem des actualen Unendlichen, *Natur und Offenbarung*, Bd. 32, 1886, p. 226; quoted from Moritz, *Memorabilia mathematica*, 1914, p. 337.

[3] *Congress of Arts and Science*, St. Louis, 1904, Vol. I, p. 498.

ing and directing towards astronomical pursuits the genius of F. W. Bessel.

Friedrich Wilhelm Bessel[1] (1784–1846) was a native of Minden in Westphalia. Fondness for figures, and a distaste for Latin grammar led him to the choice of a mercantile career. In his fifteenth year he became an apprenticed clerk in Bremen, and for nearly seven years he devoted his days to mastering the details of his business, and part of his nights to study. Hoping some day to become a supercargo on trading expeditions, he became interested in observations at sea. With a sextant constructed by him and an ordinary clock he determined the latitude of Bremen. His success in this inspired him for astronomical study. One work after another was mastered by him, unaided, during the hours snatched from sleep. From old observations he calculated the orbit of Halley's comet. Bessel introduced himself to H. W. M. Olbers, and submitted to him the calculation, which Olbers immediately sent for publication. Encouraged by Olbers, Bessel turned his back to the prospect of affluence, chose poverty and the stars, and became assistant in J. H. Schröter's observatory at Lilienthal. Four years later he was chosen to superintend the construction of the new observatory at Königsberg.[2] In the absence of an adequate mathematical teaching force, Bessel was obliged to lecture on mathematics to prepare students for astronomy. He was relieved of this work in 1825 by the arrival of C. G. J. Jacobi. We shall not recount the labors by which Bessel earned the title of founder of modern practical astronomy and geodesy. As an observer he towered far above K. F. Gauss, but as a mathematician he reverently bowed before the genius of his great contemporary. Of Bessel's papers, the one of greatest mathematical interest is an "*Untersuchung des Theils der planetarischen Störungen, welcher aus der Bewegung der Sonne entsteht*" (1824), in which he introduces a class of transcendental functions, $J_n(x)$, much used in applied mathematics, and known as "Bessel's functions." He gave their principal properties, and constructed tables for their evaluation. It has been observed that Bessel's functions appear much earlier in mathematical literature.[3] Such functions of the zero order occur in papers of Daniel Bernoulli (1732) and L. Euler on vibration of heavy strings suspended from one end. All of Bessel's functions of the first kind and of integral orders occur in a paper by L. Euler (1764) on the vibration of a stretched elastic membrane. In 1878 Lord Rayleigh proved that Bessel's functions are merely particular cases of Laplace's functions. J. W. L. Glaisher illustrates by Bessel's functions his assertion that mathematical

[1] *Bessel als Bremer Handlungslehrling*, Bremen, 1890.

[2] J. Frantz, *Festrede aus Veranlassung von Bessel's hundertjährigem Geburtstag*, Königsberg, 1884.

[3] Maxime Bôcher, "A bit of Mathematical History," *Bull. of the N. Y. Math. Soc.*, Vol. II, 1893, p. 107.

branches growing out of physical inquiries as a rule "lack the easy flow or homogeneity of form which is characteristic of a mathematical theory properly so called." These functions have been studied by Carl Theodor Anger (1803–1858) of Danzig, Oskar Schlömilch (1823–1901) of Dresden who was the founder in 1856 of the *Zeitschrift für Mathematik und Physik*, R. Lipschitz of Bonn, Carl Neumann of Leipzig, Eugen Lommel (1837–1899) of Munich, Isaac Todhunter of St. John's College, Cambridge.

Prominent among the successors of P. S. Laplace are the following: *Siméon Denis Poisson* (1781–1840), who wrote in 1808 a classic *Mémoire sur les inégalités séculaires des moyens mouvements des planètes. Giovanni Antonio Amaedo Plana* (1781–1864) of Turin, a nephew of J. Lagrange, who published in 1811 a *Memoria sulla teoria dell' attrazione degli sferoidi ellitici*, and contributed to the theory of the moon. **Peter Andreas Hansen** (1795–1874) of Gotha, at one time a clockmaker in Tondern, then H. C. Shumacher's assistant at Altona, and finally director of the observatory at Gotha, wrote on various astronomical subjects, but mainly on the lunar theory, which he elaborated in his work *Fundamenta nova investigationes orbitæ veræ quam Luna perlustrat* (1838), and in subsequent investigations embracing extensive lunar tables. **George Biddel Airy** (1801–1892), royal astronomer at Greenwich, published in 1826 his *Mathematical Tracts on the Lunar and Planetary Theories*. These researches were later greatly extended by him. **August Ferdinand Möbius** (1790–1868) of Leipzig wrote, in 1842, *Elemente der Mechanik des Himmels*. **Urbain Jean Joseph Leverrier** (1811–1877) of Paris wrote, the *Recherches Astronomiques*, constituting in part a new elaboration of celestial mechanics, and is famous for his theoretical discovery of Neptune. **John Couch Adams** (1819–1892) of Cambridge divided with Leverrier the honor of the mathematical discovery of Neptune, and pointed out in 1853 that Laplace's explanation of the secular acceleration of the moon's mean motion accounted for only half the observed acceleration. **Charles Eugène Delaunay** (born 1816, and drowned off Cherbourg in 1872), professor of mechanics at the Sorbonne in Paris, explained most of the remaining acceleration of the moon, unaccounted for by Laplace's theory as corrected by J. C. Adams, by tracing the effect of tidal friction, a theory previously suggested independently by Immanuel Kant, Robert Mayer, and William Ferrel of Kentucky. G. H. Darwin of Cambridge made some very remarkable investigations on tidal friction.

Sir George Howard Darwin (1845–1912), a son of the naturalist Charles Darwin, entered Trinity College, Cambridge, was Second Wrangler in 1868, Lord Moulton being Senior Wrangler. He began in 1875 to publish important papers on the application of the theory of tidal friction to the evolution of the solar system. The earth-moon system was found to form a unique example within the solar system of its particular mode of evolution. He traced back the changes in the figures of the earth and moon, until they united into one pear-shaped mass. This theory received confirmation in 1885 from a paper in *Acta math.*, Vol. 7 by H. Poincaré in which he enunciates the principle of exchange of stabilities. H. Poincaré and Darwin arrived at the same pear-shaped figure, Poincaré tracing the process of evolution forwards, Darwin proceeding backwards in time. Questions of stability of this changing pear-shaped figure occupied Darwin's later years. Researches along the same line were made by one of his pupils, James H. Jeans of Trinity College, Cambridge.

About the same time that George Darwin began his researches, **George William Hill** (1838–1914) of the Nautical Almanac Office in Washington began to study the moon. Hill was born at Nyack, New York, graduated at Rutgers College in 1859, and was an assistant in the Nautical Almanac Office from his graduation till 1892, when he resigned to pursue further the original researches which brought him distinction. In 1877 he published *Researches on Lunar Theory*, in which he discarded the usual mode of procedure in the problem of three bodies, by which the problem is an extension of the case of two bodies. Following a suggestion of Euler, Hill takes the earth finite, the sun of infinite mass at an infinite distance, the moon infinitesimal and at a finite distance. The differential equations which express the motion of the moon under the limitations adopted are fairly simple [1] and practically useful. "It is this idea of Hill's that has so profoundly changed the whole outlook of celestial mechanics. H. Poincaré took it up as the basis of his celebrated prize essay of 1887 on the problem of three bodies and afterwards expanded his work into the three volumes, *Les méthodes nouvelles de la mécanique céleste*," 1892–1899. It seems that at first G. H. Darwin paid little attention to Hill's paper; Darwin often spoke of his difficulties in assimilating the work of others. However in 1888 he recommended to E. W. Brown, now professor at Yale, the study of Hill. Nor does Darwin seem to have studied closely the "planetesimal hypothesis" of T. C. Chamberlin and F. R. Moulton of the University of Chicago. A marked contrast between G. H. Darwin and H. Poincaré lay in the fact that Darwin did not undertake investigations for their mathematical interest alone, while H. Poincaré and some of his followers in applied mathematics "have less interest in the phenomena than in the mathematical processes which are used by the student of the phenomena. They do not expect to examine or predict physical events but rather to take up the special classes of functions, differential equations or series which have been used by astronomers or physicists, to examine their properties, the validity of the arguments and the limitations which must be placed on the results" (E. W. Brown).

[1] We are using E. W. Brown's article in *Scientific Papers by Sir G. H. Darwin*, Vol. V, 1916, pp. xxxiv–lv.

Prominent in mathematical astronomy was **Simon Newcomb** (1835–1909), the son of a country school teacher. He was born at Wallace in Nova Scotia. Although he attended for a year the Lawrence Scientific School at Harvard University, he was essentially self-taught. In Cambridge he came in contact with B. Peirce, B. A. Gould, J. D. Runkle, and T. H. Safford. In 1861 he was appointed professor in the United States Navy; in 1877 he became superintendent of the American Ephemeris and Nautical Almanac Office. This position he held for twenty years. During 1884–1895 he was also professor of mathematics and astronomy at the Johns Hopkins University, and editor of the *American Journal of Mathematics*. His researches were mainly in the astronomy of position, in which line he was pre-eminent. In the comparison between theory and observation, in deducing from large masses of observations the results which he needed and which would form a basis of comparison with theory, he was a master. As a supplement to the *Nautical Almanac* for 1897 he published the Elements of the Four Inner planets, and the Fundamental Constants of Astronomy, which gathers together Newcomb's life-work.[1] For the unravelling of the motions of Jupiter and Saturn, S. Newcomb enlisted the services of G. W. Hill. All the publications of the tables of the planets, except those of Jupiter and Saturn, bear Newcomb's name. These tables supplant those of Leverrier. S. Newcomb devoted much time to the moon. He investigated the errors in Hansen's lunar tables and continued the lunar researches of C. E. Delaunay. Brief reference has already been made to G. W. Hill's lunar work and his contribution of an elegant paper on certain possible abbreviations in the computation of the long-period of the moon's motion due to the direct action of the planets, and made elaborate determination of the inequalities of the moon's motion due to the figure of the earth. He also computed certain lunar inequalities due to the action of Jupiter.

The mathematical discussion of Saturn's rings was taken up first by P. S. Laplace, who demonstrated that a homogeneous solid ring could not be in equilibrium, and in 1851 by B. Peirce, who proved their non-solidity by showing that even an irregular solid ring could not be in equilibrium about Saturn. The mechanism of these rings was investigated by James Clerk Maxwell in an essay to which the Adams prize was awarded. He concluded that they consisted of an aggregate of unconnected particles. "Thus an idea put forward as a speculation in the seventeenth century, and afterwards in the eighteenth century by J. Cassini and Thomas Wright, was mathematically demonstrated as the only possible solution."[2]

The progress in methods of computing planetary, asteroidal, and cometary orbits has proceeded along two more or less distinct lines,

[1] E. W. Brown in *Bull. Am. Math. Soc.*, Vol. 16, 1910, p. 353.

[2] W. W. Bryant, *A History of Astronomy*, London, 1907, p. 233.

the one marked out by P. S. Laplace, the other by K. F. Gauss.[1] Laplace's method possessed theoretical advantages, but lacked practical applicability for the reason that in the second approximation the results of the first approximation could be used only in part and the computation had to be gone over largely anew. To avoid this labor in finding asteroidal and cometary orbits, **Heinrich W. M. Olbers** (1758–1840) and K. F. Gauss devised more expeditious processes for carrying out the second approximation. The Gaussian procedure was refined and simplified by **Johann Franz Encke** (1791–1865), **Francesco Carlini** (1783–1862), F. W. Bessel, P. A. Hansen, and especially by **Theodor von Oppolzer** (1841–1886) of Vienna whose method has been used by practical astronomers down to the present day. Most original among the new elaborations of Gauss' method is that of *J. Willard Gibbs* of Yale, which employs vector analysis and, though rather complicated, yields remarkable accuracy even in the first approximation. Gibbs' procedure was modified in 1905 by J. Frischauf of Graz. P. S. Laplace's method has attracted mathematicians by its elegance. It received the attention of A. L. Cauchy, Antoine Yvon Villarceau (1813–1883) of the Paris Observatory, Rodolphe Radau of Paris, H. Bruns of Leipzig, and H. Poincaré. Paul Harzer of Kiel and especially Armin Otto Leuschner of the University of California made striking advances in rendering Laplace's method available for rapid computation. Leuschner adopts from the start geocentric co-ordinates and considers the effects of the perturbating body in the very first approximation; it is equally applicable to planetary and to cometary orbits.[2]

Problem of Three Bodies

The problem of three bodies has been treated in various ways since the time of J. Lagrange, and some decided advance towards a more complete solution has been made. Lagrange's particular solution based on the constancy of the relative distances of the three bodies, one from the other (called by L. O. Hesse the reduced problem of three bodies) has recently been modified by Carl L. Charlier of the observatory at Lund, in which the mutual distances are replaced by the distances from the centre of gravity.[3] This new form possesses no marked advantage. "Theoretical interest in the Lagrangian solutions has been increased," says E. O. Lovett, "by K. F. Sundman's theorem that the more nearly all three bodies in the general problem tend to collide simultaneously, the more nearly do they tend to assume one or the other of Lagrange's configurations; . . . practical

[1] We are using an article by A. Venturi in *Rivista di Astronomia*, June, 1913.

[2] For a fuller historical account, see A. O. Leuschner in *Science*, N. S., Vol. 45, 1917, pp. 571–584.

[3] We are drawing from E. O. Lovett's "The Problem of Three Bodies" in *Science*. N. S., Vol. 29, 1909, pp. 81–91.

interest in them has been revived by the discovery of three small planets, 1906 T. G., 1906 V. Y., 1907 X. M., near the equilateral triangular points of the Sun-Jupiter-Asteroid system. . . . R. Lehmann-Filhés, R. Hoppe, and Otto Dziobek, all three of Berlin, have generalized the exact solutions to cases of more than three bodies placed on a line or at the vertices of a regular polygon or polyhedron. . . . Among the most interesting extensions of Lagrange's theorem are those due to T. Banachievitz of Kasan and F. R. Moulton." In 1912 H. Poincaré indicated that on the basis of a ring representation (but in Keplerian variables), that if a certain geometric theorem (later established by G. D. Birkhoff of Harvard University) were true, the existence of an infinite number of periodic solutions would follow in the restricted problem of three bodies. These results were amplified by G. D. Birkhoff.[1] The so-called isosceles-triangle solutions of the problem of three bodies (periodic solutions in which two of the masses are finite and equal, while the third body moves in a straight line and remains equidistant from the equal bodies) received the attention of Giulio Pavanini of Treviso in Italy in 1907, W. D. MacMillan of Chicago in 1911, and D. Buchanan of Ontario in 1914. G. W. Hill, in his researches on lunar theory, added in 1877 to the Lagrangian periodic solution, which for 105 years had been the only such solution known, another periodic solution which could serve as a starting point for a study of the moon's orbit. Says E. O. Lovett: "With these memoirs he broke ground for the erection of the new science of dynamical astronomy whose mathematical foundations were laid broad and deep by Poincaré," in a research which in 1889 won a prize offered by King Oscar II, and which he developed more fully later. The original memoir of Poincaré, says Moulton, "contained an error which was discovered by E. Phragmén, of Stockholm, but it affected only the discussion of the existence of the asymptotic solutions; and in correcting this part H. Poincaré . . . confessed fully his obligations to Phragmén. . . . There is not the slightest doubt that in spite of it . . . the prize was correctly bestowed." The researches of G. W. Hill and H. Poincaré have been continued mainly by E. W. Brown, G. H. Darwin, F. R. Moulton, Hugo Gyldén (1841–1896) of Stockholm, P. Painlevé, C. L. W. Charlier, S. E. Strömgren, and T. Levi-Civita, in which questions of stability have received much attention. The general question, whether the solar system is stable, was affirmed by eighteenth century mathematicians; it was re-opened by K. Weierstrass who, in the last years of his life, devoted considerable attention to it. Expressions for the co-ordinates of the planets converge either not at all or for only limited time. In addition to the complex mixture of known cyclical changes, there might, perhaps, be a small residue of change of such a nature that the system will ultimately be wrecked. At present no rigorous answer

[1] *Bull. Am. Math. Soc.*, Vol. 20, 1914, p. 292.

has been given, but "Poincaré showed that solutions exist in which the motion is purely periodic, and therefore that in them at least no disaster of collision or indefinite departure from the central mass will ever occur" (F. R. Moulton). A startling result was Poincaré's discovery that some of the series which have been used to calculate the positions of the bodies of the solar system are divergent. An examination of the reasons why the divergent series gave sufficiently accurate results gave rise to the theory of asymptotic series now applied to the representation of many functions. Does the ultimate divergence of the series throw doubt upon the stability of the solar system? H. Gyldén thought that he had overcome the difficulty, but H. Poincaré showed that in part it still exists. Following Poincaré's lead, E. W. Brown has formulated the sufficient conditions for stability in the n-body problem. T. Levi-Civita worked out criteria in which the stability is made to depend upon that of a certain point transformation associated with the periodic function. He proved the existence of zones of instability surrounding Jupiter's orbit. The new methods in celestial mechanics have been found useful in computing the perturbations of certain small planets. Material advances in the problem of three bodies were made by Karl F. Sundman of Helsingfors in Finland, in a memoir which received a prize of the Paris Academy in 1913. This research is along the path first blazed by P. Painlevé, continued by T. Levi-Civita and others.

In the transformation and reduction of the three-body problem, "a principal rôle has been played by the ten known integrals, namely, the six integrals of motion of the centre of gravity, three integrals of angular momentum, and the integral of energy. The question of further progress in this reduction is vitally related to the non-existence theorems of H. Bruns, H. Poincaré, and P. Painlevé. H. Bruns demonstrated that the n-body problem admits of no algebraical integral other than the ten classic ones, and H. Poincaré proved the non-existence of any other uniform analytical integral." Other researches on these non-existence theorems are due to P. Painlevé, D. A. Gravé, and K. Bohlin.

E. Picard expresses himself as follows:[1] "What admirable recent researches have best taught them [analysts] is the immense difficulty of the problem; a new way has, however, been opened by the study of particular solutions, such as the periodic solutions and the asymptotic solution which have already been utilized. It is not perhaps so much because of the needs of practice as in order not to avow itself vanquished, that analysis will never resign itself to abandon, without a decisive victory, a subject where it has met so many brilliant triumphs; and again, what more beautified field could the theories new-born or rejuvenated of the modern doctrine of functions find, to essay their forces, than this classic problem of n bodies?"

[1] *Congress of Arts and Science*, St. Louis, 1904, Vol. I, p. 512.

Among valuable text-books on mathematical astronomy of the nineteenth century rank the following works: *Manual of Spherical and Practical Astronomy* by *William Chauvenet* (1863), *Practical and Spherical Astronomy* by *Robert Main* of Cambridge, *Theoretical Astronomy* by *James C. Watson* of Ann Arbor (1868), *Traité élémentaire de Mécanique Céleste of H. Resal* of the *École Polytechnique* in Paris, *Cours d'Astronomie de l'École Polytechnique* by *Faye*, *Traité de Mécanique Céleste* by *F. F. Tisserand*, *Lehrbuch der Bahnbestimmung* by *T. Oppolzer*, *Mathematische Theorien der Planetenbewegung* by *O. Dziobek*, translated into English by M. W. Harrington and W. J. Hussey.

General Mechanics

During the nineteenth century we have come to recognize the advantages frequently arising from a geometrical treatment of mechanical problems. To L. Poinsot, M. Chasles, and A. F. Möbius we owe the most important developments made in geometrical mechanics. **Louis Poinsot** (1777–1859), a graduate of the Polytechnic School in Paris, and for many years member of the superior council of public instruction, published in 1804 his *Éléments de Statique.* This work is remarkable not only as being the earliest introduction to synthetic mechanics, but also as containing for the first time the idea of couples, which was applied by Poinsot in a publication of 1834 to the theory of rotation. A clear conception of the nature of rotary motion was conveyed by Poinsot's elegant geometrical representation by means of an ellipsoid rolling on a certain fixed plane. This construction was extended by J. J. Sylvester so as to measure the rate of rotation of the ellipsoid on the plane.

A particular class of dynamical problems has recently been treated geometrically by **Sir Robert Stawell Ball** (1840–1913) at one time astronomer royal of Ireland, later Lowndean Professor of Astronomy and Geometry at Cambridge. His method is given in a work entitled *Theory of Screws*, Dublin, 1876, and in subsequent articles. Modern geometry is here drawn upon, as was done also by W. K. Clifford in the related subject of Bi-quaternions. Arthur Buchheim (1859–1888), of Manchester showed that H. G. Grassmann's Ausdehnungslehre supplies all the necessary materials for a simple calculus of screws in elliptic space. Horace Lamb applied the theory of screws to the question of the steady motion of any solid in a fluid.

Advances in theoretical mechanics, bearing on the integration and the alteration in form of dynamical equations, were made since J. Lagrange by S. D. Poisson, Sir William Rowan Hamilton, C. G. J. Jacobi, Madame Kovalevski, and others. J. Lagrange had established the "Lagrangian form" of the equations of motion. He had given a theory of the variation of the arbitrary constants which, however, turned out to be less fruitful in results than a theory advanced by S. D.

Poisson.[1] Poisson's theory of the variation of the arbitrary constants and the method of integration thereby afforded marked the first onward step since J. Lagrange. Then came the researches of Sir William Rowan Hamilton. His discovery that the integration of the dynamic differential equations is connected with the integration of a certain partial differential equation of the first order and second degree, grew out of an attempt to deduce, by the undulatory theory, results in geometrical optics previously based on the conceptions of the emission theory. The *Philosophical Transactions* of 1833 and 1834 contain Hamilton's papers, in which appear the first applications to mechanics of the principle of varying action and the characteristic function, established by him some years previously. The object which Hamilton proposed to himself is indicated by the title of his first paper, viz., the discovery of a function by means of which all integral equations can be actually represented. The new form obtained by him for the equation of motion is a result of no less importance than that which was the professed object of the memoir. Hamilton's method of integration was freed by C. G. J. Jacobi of an unnecessary complication, and was then applied by him to the determination of a geodetic line on the general ellipsoid. With aid of elliptic co-ordinates Jacobi integrated the partial differential equation and expressed the equation of the geodetic in form of a relation between two Abelian integrals. C. G. J. Jacobi applied to differential equations of dynamics the theory of the ultimate multiplier. The differential equations of dynamics are only one of the classes of differential equations considered by Jacobi. Dynamic investigations along the lines of J. Lagrange, Hamilton, and Jacobi were made by J. Liouville, Adolphe Desboves, (1818–1888) of Amiens, Serret, J. C. F. Sturm, Michel Ostrogradski, J. Bertrand, William Fishburn Donkin (1814–1869) of Oxford, F. Brioschi, leading up to the development of the theory of a system of canonical integrals.

An important addition to the theory of the motion of a solid body about a fixed point was made by Madame **Sophie Kovalevski** (1850–1891), who discovered a new case in which the differential equations of motion can be integrated. By the use of theta-functions of two independent variables she furnished a remarkable example of how the modern theory of functions may become useful in mechanical problems. She was a native of Moscow, studied under K. Weierstrass, obtained the doctor's degree at Göttingen, and from 1884 until her death was professor of higher mathematics at the University of Stockholm. The research above mentioned received the Bordin prize of the French Academy in 1888, which was doubled on account of the exceptional merit of the paper.

There are in vogue three forms for the expression of the kinetic energy of a dynamical system: the Lagrangian, the Hamiltonian, and

[1] Arthur Cayley, "Report on the Recent Progress of Theoretical Dynamics," *Report British Ass'n* for 1857, p. 7.

a modified form of Lagrange's equations in which certain velocities are omitted. The kinetic energy is expressed in the first form as a homogeneous quadratic function of the velocities, which are the time-variations of the co-ordinates of the system; in the second form, as a homogeneous quadratic function of the momenta of the system; the third form, elaborated recently by Edward John Routh (1831–1907) of Cambridge, in connection with his theory of "ignoration of co-ordinates," and by A. B. Basset of Cambridge, is of importance in hydro-dynamical problems relating to the motion of perforated solids in a liquid, and in other branches of physics.

Practical importance has come to be attached to the principle of mechanical similitude. By it one can determine from the performance of a model the action of the machine constructed on a larger scale. The principle was first enunciated by I. Newton (*Principia*, Bk. II, Sec. VIII, Prop. 32), and was derived by Joseph Bertrand from the principle of virtual velocities. A corollary to it, applied in ship-building, is named after the British naval architect William Froude (1810–1879), but was enunciated also by the French engineer Frédéric Reech.

The present problems of dynamics differ materially from those of the last century. The explanation of the orbital and axial motions of the heavenly bodies by the law of universal gravitation was the great problem solved by A. C. Clairaut, L. Euler, D'Alembert, J. Lagrange, and P. S. Laplace. It did not involve the consideration of frictional resistances. In the present time the aid of dynamics has been invoked by the physical sciences. The problems there arising are often complicated by the presence of friction. Unlike astronomical problems of a century ago, they refer to phenomena of matter and motion that are usually concealed from direct observation. The great pioneer in such problems is Lord Kelvin. While yet an undergraduate at Cambridge, during holidays spent at the seaside, he entered upon researches of this kind by working out the theory of spinning tops, which previously had been only partially explained by John Hewitt Jellett (1817–1888) of Trinity College, Dublin, in his *Treatise on the Theory of Friction* (1872), and by Archibald Smith (1813–1872).

Among standard works on mechanics of the nineteenth century are *C. G. J. Jacobi's Vorlesungen über Dynamik*, edited by *R. F. A. Clebsch*, 1866; *G. R. Kirchhoff's Vorlesungen über mathematische Physik*, 1876; *Benjamin Peirce's Analytic Mechanics*, 1855; *J. I. Somoff's Theoretische Mechanik*, 1879; *P. G. Tait and W. J. Steele's Dynamics of a Particle*, 1856; *George Minchin's Treatise on Statics; E. J. Routh's Dynamics of a System of Rigid Bodies; J. C. F. Sturm's Cours de Mécanique de l'École Polytechnique.* George M. Minchin (1845–1914) was professor at the Indian engineering college.

In 1898 Felix Klein pointed out the separation which existed between British and Continental mathematical research, as seen, for instance, by the contents of E. J. Routh's *Dynamics*, which contains the results of twenty years of research along that line in England and, in comparison with the German school, emphasizes a concrete and practical treatment. To make these treasures more readily accessible to German students, Routh's text was translated into German by Adolf Schepp (1837–1905) of Wiesbaden in 1898. Particularly strong was Routh in the treatment of small oscillations of systems; the technique of integration of linear differential equations with constant coefficients is highly developed, except that, perhaps, the extent to which the developments are valid may need closer examination. This is done in F. Klein and A. Sommerfeld's *Theorie des Kreisels*, 1897–1910. This last work gives attention to the theory of the top, the history of which reaches back to the eighteenth century.

In 1744 Serson started on a ship (that was lost), to test the practicability of the artificial horizon furnished by the polished surface of a top. This idea has been recently revived by French navigators.[1] Serson's top induced J. A. Segner of Halle in 1755 to give precision to the theory of the spinning top, which was taken up more fully by L. Euler in 1765 and then by J. Lagrange. L. Euler considers the motion on a smooth horizontal plane. Later come the studies due to L. Poinsot, S. D. Poisson, C. G. J. Jacobi, G. R. Kirchhoff, Eduard Lottner (1826–1887) of Lippstadt, Wilhelm Hess, Clerk Maxwell, E. J. Routh and finally F. Klein and A. Sommerfeld. In 1914 G. Greenhill prepared a *Report on Gyroscopic Theory*[2] which is of more direct interest to engineers than is Klein and Sommerfeld's *Theorie des Kreisels*, developed by the aid of the theory of functions of a complex variable. Among recent practical applications of gyroscopic action are the torpedo exhibited before the Royal Society of London in 1907 by Louis Brennan, also Brennan's monorail system, and the methods of steadying ships and aircraft, devised by the American engineer Elmer A. Sperry and by Otto Schlick in Germany.

Among the deviations of a projectile from the theoretic parabolic path there are two which are of particular interest. One is a slight bending to the right, in the northern hemisphere, owing to the rotation of the earth; it was explained by S. D. Poisson (1838) and W. Ferrel (1889). The other is due to the rotation of the projectile; it was observed by I. Newton in tennis balls and applied by him to explain certain phenomena in his corpuscular theory of light; it was known to Benjamin Robins and L. Euler. In 1794 the Berlin academy offered a prize for an explanation of the phenomenon, but no satisfactory explanation appeared for over half a century. S. D. Poisson in 1839 (*Journ. école polyt.*, T. 27) studied the effect of atmospheric

[1] See A. G. Greenhill in *Verhandl. III. Intern. Congr., Heidelberg, 1904*, Leipzig, 1905, p. 100. We are summarizing this article.

[2] *Advisory Committee for Aeronautics, Reports and Memoranda*, No. 146, London, 1914.

friction against the rotating sphere, but finally admitted that friction was not sufficient to explain the deviations. The difference in the pressure of the air upon the rotating sphere also demands attention. An explanation on this basis, which was generally accepted as valid was given by *H. G. Magnus* (1802–1870) of Berlin, in Poggendorff's *Annalen*, Vol. 88, 1853. In connection with golf-balls the problem was taken up by Tait.

Peter Guthrie Tait (1831–1901) was born at Dalkeith, studied at Cambridge and came out Senior Wrangler in 1854, which was a surprise, as W. J. Steele had been generally ahead in college examinations. From 1854–1860 Tait was professor of mathematics at Belfast, where he studied quaternions; from 1860 to his death he held the chair of Natural Philosophy at Edinburgh. Tait found the problem of the flight of the golf ball capable of exact statement and approximate solution. One of his sons had become a brilliant golfer. Tait at first was scoffed at when he began to offer explanations of the secret of long driving. In 1887 (*Nature*, 36, p. 502) he shows that "rotation" played an important part, as established experimentally by H. G. Magnus (1852). Says P. G. Tait: "In topping, the upper part of the ball is made to move forward faster than does the center, consequently the front of the ball descends in virtue of the rotation, and the ball itself skews in that direction. When a ball is undercut it gets the opposite spin to the last, and, in consequence, it tends to deviate upwards instead of downwards. The upward tendency often makes the path of a ball (for a part of its course) concave upwards in spite of the effects of gravity. . . ." P. G. Tait explained the influence of the underspin in prolonging not only the range but also the time of flight. The essence of his discovery was that without spin a ball could not combat gravity greatly, but that with spin it could travel remarkable distances. He was fond of the game while H. Helmholtz (who was in Scotland in 1871) "could see no fun in the leetle hole."

P. G. Tait generalized in 1898 the Josephus problem and gave the rule for n persons, certain v of which shall be left after each mth man is picked out.

The deviations of a body falling from rest near the surface of the earth have been considered in many memoirs from the time of P. S. Laplace and K. F. Gauss to the present. All writers agree that the body will deviate to the eastward with respect to the plumb-line hung from the initial point, but there has been disagreement regarding the deviation measured along the meridian. Laplace found no meridional deviation, Gauss found a small deviation toward the equator. Recently this problem has commanded the attention of writers in the United States. *R. S. Woodward*, president of the Carnegie Institution in Washington, found in 1913 a deviation away from the equator. *F. R. Moulton* of the University of Chicago found in 1914 a formula indicating a southerly deviation. *W. H. Röver* of Washington University in St. Louis has, since 1901, treated the subject in several articles which indicate southerly deviations. He declares that "no potential function is known that fits all parts of the earth," "that the formula of Gauss, the three formulæ of Comte de Sparre [Lyon, 1905], the formula of Professor F. R. Moulton, and my first formula, are all special cases of my general formula." [1]

Fluid Motion

The equations which constitute the foundation of the theory of fluid motion were fully laid down at the time of J. Lagrange, but the solutions actually worked out were few and mainly of the irrotational type. A powerful method of attacking problems in fluid motion is that of images, introduced in 1843 by G. G. Stokes of Pembroke College, Cambridge. It received little attention until Sir William Thomson's discovery of electrical images, whereupon the theory was extended by G. G. Stokes, W. M. Hicks, and T. C. Lewis.

George Gabriel Stokes (1819–1903) was born at Skreen, County Sligo, in Ireland. In 1837, the year of Queen Victoria's accession, he commenced residence at Cambridge, where he was to find his home, almost without intermission, for sixty-six years. At Pembroke College his mathematical abilities attracted attention and in 1841 he graduated as Senior Wrangler and first Smith's prizeman. He distinguished himself along the lines of applied mathematics. In 1845 he published a memoir on "Friction of Fluids in Motion." The general motion of a medium near any point is analyzed into three constituents—a motion of pure translation, one of pure rotation and one of pure strain. Similar results were reached by H. Helmholtz twenty-three years later. In applying his results to viscous fluids, Stokes was led to general dynamical equations, previously reached from more special hypotheses by L. M. H. Navier and S. D. Poisson. Both Stokes and G. Green were followers of the French school of applied mathematicians. Stokes applies his equations to the propagation of sound, and shows that viscosity makes the intensity of sound diminish as the time increases and the velocity less than it would otherwise be—especially for high notes. He considered the two elastic constants in the equations for an elastic solid to be independent and not reducible to one as is the case in Poisson's theory. Stokes' position was supported by Lord Kelvin and seems now generally accepted. In 1847 Stokes examined anew the theory of oscillatory waves. Another paper was on the effect of internal friction of fluids on the motion of pendulums. He assumed that the viscosity of the air was proportional to the density, which was shown later by Maxwell to be erroneous. In 1849 he treated the ether as an elastic solid in the study of diffraction. He favored Fresnel's wave theory of light as opposed to

[1] See *Washington University Studies*, Vol. III, 1916, pp. 153–168.

the corpuscular theory supported by David Brewster. In a report on double refraction of 1862 he correlated the work of A. L. Cauchy, J. MacCullagh, and G. Green. Assuming that the elasticity of the ether has its origin in deformation, he inferred that J. MacCullagh's theory was contrary to the laws of mechanics, but recently J. Larmor has shown that J. MacCullagh's equations may be explained on the supposition that what is resisted is not deformation, but rotation. Stokes wrote on Fourier series and the discontinuity of arbitrary constants in semi-convergent expansions over a plane. His contributions to hydrodynamics and optics are fundamental. In 1849 William Thomson (Lord Kelvin) gave the maximum and minimum theorem peculiar to hydrodynamics, which was afterwards extended to dynamical problems in general.

A new epoch in the progress of hydrodynamics was created, in 1856, by H. Helmholtz, who worked out remarkable properties of rotational motion in a homogeneous, incompressible fluid, devoid of viscosity. He showed that the vortex filaments in such a medium may possess any number of knottings and twistings, but are either endless or the ends are in the free surface of the medium; they are indivisible. These results suggested to William Thomson (Lord Kelvin) the possibility of founding on them a new form of the atomic theory, according to which every atom is a vortex ring in a non-frictional ether, and as such must be absolutely permanent in substance and duration. The vortex-atom theory was discussed by J. J. Thomson of Cambridge (born 1856) in his classical treatise on the *Motion of Vortex Rings*, to which the Adams Prize was awarded in 1882. Papers on vortex motion have been published also by Horace Lamb, Thomas Craig, Henry A. Rowland, and Charles Chree of Kew Observatory.

The subject of jets was investigated by H. Helmholtz, G. R. Kirchhoff, J. Plateau, and Lord Rayleigh; the motion of fluids in a fluid by G. G. Stokes, W. Thomson (Lord Kelvin), H. A. Köpcke, G. Greenhill, and H. Lamb; the theory of viscous fluids by H. Navier, S. D. Poisson, B. de Saint-Venant, Stokes, Oskar Emil Meyer (1834–1909) of Breslau, A. B. Stefano, C. Maxwell, R. Lipschitz, T. Craig, H. Helmholtz, and A. B. Basset. Viscous fluids present great difficulties, because the equations of motion have not the same degree of certainty as in perfect fluids, on account of a deficient theory of friction, and of the difficulty of connecting oblique pressures on a small area with the differentials of the velocities.

Waves in liquids have been a favorite subject with English mathematicians. The early inquiries of S. D. Poisson and A. L. Cauchy were directed to the investigation of waves produced by disturbing causes acting arbitrarily on a small portion of the fluid. The velocity of the long wave was given approximately by J. Lagrange in 1786 in case of a channel of rectangular cross-section, by Green in 1839 for a channel of triangular section, and by Philip Kelland (1810–1879) of Edinburgh for a channel of any uniform section. Sir George B. Airy, in his treatise on *Tides and Waves*, discarded mere approximations, and gave the exact equation on which the theory of the long wave in a channel of uniform rectangular section depends. But he gave no general solutions. J. McCowan of University College at Dundee discussed this topic more fully, and arrived at exact and complete solutions for certain cases. The most important application of the theory of the long wave is to the explanation of tidal phenomena in rivers and estuaries.

The mathematical treatment of solitary waves was first taken up by S. Earnshaw in 1845, then by G. G. Stokes; but the first sound approximate theory was given by J. Boussinesq in 1871, who obtained an equation for their form, and a value for the velocity in agreement with experiment. Other methods of approximation were given by Lord Rayleigh and John McCowan. In connection with deep-water waves, Osborne Reynolds (1842–1912) of the University of Manchester gave in 1877 the dynamical explanation for the fact that a group of such waves advances with only half the rapidity of the individual waves.

The solution of the problem of the general motion of an ellipsoid in a fluid is due to the successive labors of George Green (1833), R. F. A. Clebsch (1856), and Carl Anton Bjerknes (1825–1903) of Christiania (1873). The free motion of a solid in a liquid has been investigated by W. Thomson (Lord Kelvin), G. R. Kirchhoff, and Horace Lamb. By these labors, the motion of a single solid in a fluid has come to be pretty well understood, but the case of two solids in a fluid is not developed so fully. The problem has been attacked by W. M. Hicks.

The determination of the period of oscillation of a rotating liquid spheroid has important bearings on the question of the origin of the moon. G. H. Darwin's investigations thereon, viewed in the light of G. F. B. Riemann's and H. Poincaré's researches, seem to disprove P. S. Laplace's hypothesis that the moon separated from the earth as a ring, because the angular velocity was too great for stability; G. H. Darwin finds no instability.

The explanation of the contracted vein has been a point of much controversy, but has been put in a much better light by the application of the principle of momentum, originated by W. Froude and Lord Rayleigh. Rayleigh considered also the reflection of waves, not at the surface of separation of two uniform media, where the transition is abrupt, but at the confines of two media between which the transition is gradual.

The first serious study of the circulation of winds on the earth's surface was instituted at the beginning of the second quarter of the last century by *William C. Redfield* (1789–1857), an American meteorologist and railway projector, *James Pollard Espy* (1786–1860) of Washington, through whose stimulus the present United States Weather Bureau was started and *Heinrich Wilhelm Dove* (1803–1879) of Berlin, followed by researches by *Sir William Reid* (1791–1858) a British major-general who developed his circular theory of hurricanes while in the West Indies, *Henry Piddington* (1797–1858) a British commander in the mercantile marine who accumulated data for determining the course of storms at sea and originated the term "cyclone," and *Elias Loomis* (1811–1889) of Yale University. But the deepest insight into the wonderful correlations that exist among the varied motions of the atmosphere was obtained by William Ferrel (1817–1891). He was born in Fulton County, Pa., and brought up on a farm. Though in unfavorable surroundings, a burning thirst for knowledge spurred the boy to the mastery of one branch after another. He attended Marshall College, Pa., and graduated in 1844 from Bethany College. While teaching school he became interested in meteorology and in the subject of tides. In 1856 he wrote an article on "the winds and currents of the ocean." The following year he became connected with the *Nautical Almanac.* A mathematical paper followed in 1858 on "the motion of fluids and solids relative to the earth's surface." The subject was extended afterwards so as to embrace the mathematical theory of cyclones, tornadoes, water-spouts, etc. In 1885 appeared his *Recent Advances in Meteorology.* In the opinion of *Julius Hann* of Vienna, Ferrel has "contributed more to the advance of the physics of the atmosphere than any other living physicist or meteorologist."

W. Ferrel taught that the air flows in great spirals toward the poles, both in the upper strata of the atmosphere and on the earth's surface beyond the 30th degree of latitude; while the return current blows at nearly right angles to the above spirals, in the middle strata as well as on the earth's surface, in a zone comprised between the parallels 30° N. and 30° S. The idea of three superposed currents blowing spirals was first advanced by James Thomson (1822–1892), brother of Lord Kelvin, but was published in very meagre abstract.

W. Ferrel's views have given a strong impulse to theoretical research in America, Austria, and Germany. Several objections raised against his argument have been abandoned, or have been answered by W. M. Davis of Harvard. The mathematical analysis of F. Waldo of Cambridge, Mass., and of others, has further confirmed the accuracy of the theory. The transport of Krakatoa dust and observations made on clouds point toward the existence of an upper east current on the equator, and Josef M. Pernter (1848–1908) of Vienna has mathematically deduced from Ferrel's theory the existence of such a current.

Another theory of the general circulation of the atmosphere was propounded by Werner Siemens (1816–1892) of Berlin, in which an attempt is made to apply thermodynamics to aërial currents. Important new points of view have been introduced by H. Helmholtz, who concluded that when two air currents blow one above the other in different directions, a system of air waves must arise in the same way as waves are formed on the sea. He and Anton Oberbeck (1846–1900) of Tübingen showed that when the waves on the sea attain lengths of from 16 to 33 feet, the air waves must attain lengths of from 10 to 20 miles, and proportional depths. Superposed strata would thus mix more thoroughly, and their energy would be partly dissipated. From hydrodynamical equations of rotation H. Helmholtz established the reason why the observed velocity from equatorial regions is much less in a latitude of, say, 20° or 30°, than it would be were the movements unchecked. Other important contributors to the general theory of the circulation of the atmosphere are Max Möller of Braunschweig and Luigi de Marchi of the University of Pavia. The source of the energy of atmospheric disturbances was sought by W. Ferrel and Th. Reye in the heat given off during condensation. Max Margules of the University of Vienna showed in 1905 that this heat energy contributes nothing to the kinetic energy of the winds and that the source of energy is found in the lowering of the centre of gravity of an air column when the colder air assumes the lower levels, whereby the potential energy is diminished and the kinetic energy increased.[1] Asymmetric cyclones have been studied especially by Luigi de Marchi of Pavia. Anticyclones have received attention from Henry H. Clayton of the Blue Hill Observatory, near Boston, from Julius Hann of Vienna, F. H. Bigelow of Washington, and Max Margules of Vienna.

Sound. Elasticity

About 1860 acoustics began to be studied with renewed zeal. The mathematical theory of pipes and vibrating strings had been elaborated in the eighteenth century by Daniel Bernoulli, D'Alembert, L. Euler, and J. Lagrange. In the first part of the present century P. S. Laplace corrected Newton's theory on the velocity of sound in gases; S. D. Poisson gave a mathematical discussion of torsional vibrations; S. D. Poisson, Sophie Germain, and Charles Wheatstone studied Chladni's figures; Thomas Young and the brothers Weber developed the wave-theory of sound. **Sir J. F. W. Herschel** (1792–1871) wrote on the mathematical theory of sound for the *Encyclopædia Metropolitana*, 1845. Epoch-making were H. Helmholtz's experimental and mathematical researches. In his hands and Rayleigh's, Fourier's series received due attention. H. Helmholtz gave the mathematical theory of beats, difference tones, and summation tones. **Lord Rayleigh** (John William Strutt) of Cambridge (1842–1919) made extensive mathematical researches in acoustics as a part of the theory of vibration in general. Particular mention may be made of his discussion of the disturbance produced by a spherical

[1] *Encyklopädie der Math. Wissenschaften*, Bd. VI, 1, 8, 1912, p. 216.

obstacle on the waves of sound, and of phenomena, such as sensitive flames, connected with the instability of jets of fluid. In 1877 and 1878 he published in two volumes a treatise on *The Theory of Sound.* Other mathematical researches on this subject have been made in England by William Fishburn Donkin (1814–1869) of Oxford and G. G. Stokes. An interesting point in the behavior of a Fourier's series was brought out in 1898 by J. W. Gibbs of Yale. A. A. Michelson and S. W. Stratton at the University of Chicago had shown experimentally by their harmonic analyses that the summation of 160 terms of the series $\Sigma(-1)1^{n+1}(sin nx)/n$ revealed certain unexpected small towers in the curve for the sum, as n increased. J. W. Gibbs showed (*Nature,* Vol. 59, p. 606) by the study of the order of variation of n and x that these phenomena were not due to imperfections in the machine, but were true mathematical phenomena. They are called the "Gibbs' phenomenon," and have received further attention from Maxime Bôcher, T. H. Gronwall, H. Weyl, and H. S. Carslaw.

The theory of elasticity [1] belongs to this century. Before 1800 no attempt had been made to form general equations for the motion or equilibrium of an elastic solid. Particular problems had been solved by special hypotheses. Thus, James Bernoulli considered elastic laminæ; Daniel Bernoulli and L. Euler investigated vibrating rods; J. Lagrange and L. Euler, the equilibrium of springs and columns. The earliest investigations of this century, by Thomas Young ("Young's modulus of elasticity") in England, J. Binet in France, and G. A. A. Plana in Italy, were chiefly occupied in extending and correcting the earlier labors. Between 1820 and 1840 the broad outline of the modern theory of elasticity was established. This was accomplished almost exclusively by French writers,—Louis-Marie-Henri Navier (1785–1836), S. D. Poisson, A. L. Cauchy, Mademoiselle Sophie Germain (1776–1831), Félix Savart (1791–1841). Says H. Burkhardt: "There are two views respecting the beginnings of the theory of elasticity of solids, of which no dimension can be neglected: According to one view the deciding impulse came from Fresnel's undulatory theory of light, according to the other, everything goes back to the technical theory of rigidity (Festigkeitstheorie), the representative of which was at that time Navier. As always in such cases, the truth lies in the middle: Cauchy to whom we owe primarily the fixing of the fundamental concepts, as strain and stress, learned from Fresnel as well as from Navier."

Siméon Denis Poisson [2] (1781–1840) was born at Pithiviers. The boy was put out to a nurse, and he used to tell that when his father (a common soldier) came to see him one day, the nurse had gone out

[1] I. Todhunter, *History of the Theory of Elasticity,* edited by Karl Pearson, Cambridge, 1886.

[2] Ch. Hermite, "Discours prononcé devant le président de la République," *Bulletin des sciences mathématiques,* XIV, January, 1890.

and left him suspended by a thin cord to a nail in the wall in order to protect him from perishing under the teeth of the carnivorous and unclean animals that roamed on the floor. Poisson used to add that his gymnastic efforts when thus suspended caused him to swing back and forth, and thus to gain an early familiarity with the pendulum, the study of which occupied him much in his maturer life. His father destined him for the medical profession, but so repugnant was this to him that he was permitted to enter the Polytechnic School at the age of seventeen. His talents excited the interest of J. Lagrange and P. S. Laplace. At eighteen he wrote a memoir on finite differences which was printed on the recommendation of A. M. Legendre. He soon became a lecturer at the school, and continued through life to hold various government scientific posts and professorships. He prepared some 400 publications, mainly on applied mathematics. His *Traité de Mécanique*, 2 vols., 1811 and 1833, was long a standard work. He wrote on the mathematical theory of heat, capillary action, probability of judgment, the mathematical theory of electricity and magnetism, physical astronomy, the attraction of ellipsoids, definite integrals, series, and the theory of elasticity. He was considered one of the leading analysts of his time. The story is told that in 1802 a young man, about to enter the army, asked Poisson to take $100 in safe-keeping. "All right," said Poisson," set it down there and let me work; I have much to do." The recruit placed the money-bag on a shelf and Poisson placed a copy of Horace over the bag, to hide it. Twenty years later the soldier returned and asked for his money, but Poisson remembered nothing and asked angrily: "You claim to have put the money in my hands?" "No," replied the soldier, "I put in on this shelf and you placed this book over it." The soldier removed the dusty copy of Horace and found the $100 where they had been placed twenty years before.

His work on elasticity is hardly excelled by that of A. L. Cauchy, and second only to that of B. de Saint-Venant. There is hardly a problem in elasticity to which he has not contributed, while many of his inquiries were new. The equilibrium and motion of a circular plate was first successfully treated by him. Instead of the definite integrals of earlier writers, he used preferably finite summations. Poisson's contour conditions for elastic plates were objected to by Gustav Kirchhoff of Berlin, who established new conditions. But Thomson (Lord Kelvin) and P. G. Tait in their *Treatise on Natural Philosophy* have explained the discrepancy between Poisson's and Kirchhoff's boundary conditions, and established a reconciliation between them.

Important contributions to the theory of elasticity were made by A. L. Cauchy. To him we owe the origin of the theory of stress, and the transition from the consideration of the force upon a molecule exerted by its neighbors to the consideration of the stress upon a small plane at a point. He anticipated G. Green and G. G. Stokes in giving the equations of isotropic elasticity with two constants. The theory of elasticity was presented by Gabrio Piola of Italy according to the principles of J. Lagrange's *Méchanique Analytique*, but the superiority of this method over that of Poisson and Cauchy is far from evident. The influence of temperature on stress was first investigated experimentally by Wilhelm Weber of Göttingen, and afterwards mathematically by J. M. C. Duhamel, who, assuming Poisson's theory of elasticity, examined the alterations of form which the formulæ undergo when we allow for changes of temperature. W. Weber was also the first to experiment on elastic after-strain. Other important experiments were made by different scientists, which disclosed a wider range of phenomena, and demanded a more comprehensive theory. Set was investigated by Franz Joseph von Gerstner (1756–1832), of Prague and Eaton Hodgkinson of University College, London, while the latter physicist in England and Louis Joseph Vicat (1786–1861) in France experimented extensively on absolute strength. L. J. Vicat boldly attacked the mathematical theories of flexure because they failed to consider shear and the time-element. As a result, a truer theory of flexure was soon propounded by B. de Saint-Venant. J. V. Poncelet advanced the theories of resilience and cohesion.

Gabriel Lamé (1795–1870) was born at Tours, and graduated at the Polytechnic School. He was called to Russia with B. P. E. Clapeyron and others to superintend the construction of bridges and roads. On his return, in 1832, he was elected professor of physics at the Polytechnic School. Subsequently he held various engineering posts and professorships in Paris. As engineer he took an active part in the construction of the first railroads in France. Lamé devoted his fine mathematical talents mainly to mathematical physics. In four works: *Leçons sur les fonctions inverses des transcendantes et les surfaces isothermes; Sur les coordonnées curvilignes et leurs diverses applications; Sur la théorie analytique de la chaleur; Sur la théorie mathématique de l'élasticité des corps solides* (1852), and in various memoirs he displays fine analytical powers; but a certain want of physical touch sometimes reduces the value of his contributions to elasticity and other physical subjects. In considering the temperature in the interior of an ellipsoid under certain conditions, he employed functions analogous to Laplace's functions, and known by the name of "Lamé's functions." A problem in elasticity called by Lamé's name, viz., to investigate the conditions for equilibrium of a spherical elastic envelope subject to a given distribution of load on the bounding spherical surfaces, and the determination of the resulting shifts is the only completely general problem on elasticity which can be said to be completely solved. He deserves much credit for his derivation and transformation of the general elastic equations, and for his application of them to double refraction. Rectangular and triangular membranes were shown by him to be connected with questions in the theory of numbers. H.

Burkhardt[1] is of the opinion that the importance of the classic period of French mathematical physics, about 1810–1835, is often undervalued, but that the direction it took finally under the leadership of Lamé was unfortunate. "By his (Lamé's) taste for algebraic elegance he was misled to prefer problems which are of interest in pure rather than applied mathematics; he went so far as to require of technical men the study of number theory, because the determination of the simple tones of a rectangular plate with commensurable sides calls for the solution of an indeterminate quadratic equation."

Continuing our outline of the history of elasticity, we observe that the field of photo-elasticity was entered upon by G. Lamé, F. E. Neumann, and Clerk Maxwell. G. G. Stokes, W. Wertheim, R. Clausius, and J. H. Jellett, threw new light upon the subject of "rari-constancy" and "multi-constancy," which has long divided elasticians into two opposing factions. The uni-constant isotropy of L. M. H. Navier and S. D. Poisson had been questioned by A. L. Cauchy, and was severely criticised by G. Green and G. G. Stokes.

Barré de Saint-Venant (1797–1886), ingénieur des ponts et chaussées, made it his life-work to render the theory of elasticity of practical value. The charge brought by practical engineers, like Vicat, against the theorists led Saint-Venant to place the theory in its true place as a guide to the practical man. Numerous errors committed by his predecessors were removed. He corrected the theory of flexure by the consideration of slide, the theory of elastic rods of double curvature by the introduction of the third moment, and the theory of torsion by the discovery of the distortion of the primitively plane section. His results on torsion abound in beautiful graphic illustrations. In case of a rod, upon the side surfaces of which no forces act, he showed that the problems of flexure and torsion can be solved, if the end-forces are distributed over the end-surfaces by a definite law. R. F. A. Clebsch, in his *Lehrbuch der Elasticität*, 1862, showed that this problem is reversible to the case of side-forces without end-forces. Clebsch[2] extended the research to very thin rods and to very thin plates. B. de Saint-Venant considered problems arising in the scientific design of built-up artillery, and his solution of them differs considerably from G. Lamé's solution, which was popularized by W. J. M. Rankine, and much used by gun-designers. In Saint-Venant's translation into French of Clebsch's *Elasticität*, he develops extensively a double-suffix notation for strain and stresses. Though often advantageous, this notation is cumbrous, and has not been generally adopted. *Karl Pearson*, Galton professor of eugenics at the University of London, in his early mathematical studies, examined the permissible limits of the application of the ordinary theory of flexure of a beam.

[1] *Jahresb. d. d. Math. Vereinigung*, Vol. 12, 1903, p. 564.

[2] *Alfred Clebsch, Versuch einer Darlegung und Würdigung seiner wissenschaftlichen Leistungen von einigen seiner Freunde*, Leipzig, 1873.

The mathematical theory of elasticity is still in an unsettled condition. Not only are scientists still divided into two schools of "rari-constancy" and "multi-constancy," but difference of opinion exists on other vital questions. Among the numerous modern writers on elasticity may be mentioned Émile Mathieu (1835–1890), professor at Besançon, Maurice Levy (1838–1910) of the Collège de France in Paris, Charles Chree, superintendent of the Kew Observatory, A. B. Basset, Lord Kelvin of Glasgow, J. Boussinesq of Paris, and others. Lord Kelvin applied the laws of elasticity of solids to the investigation of the earth's elasticity, which is an important element in the theory of ocean-tides. If the earth is a solid, then its elasticity co-operates with gravity in opposing deformation due to the attraction of the sun and moon. P. S. Laplace had shown how the earth would behave if it resisted deformation only by gravity. G. Lamé had investigated how a solid sphere would change if its elasticity only came into play. Lord Kelvin combined the two results, and compared them with the actual deformation. Kelvin, and afterwards G. H. Darwin, computed that the resistance of the earth to tidal deformation is nearly as great as though it were of steel. This conclusion was confirmed more recently by Simon Newcomb, from the study of the observed periodic changes in latitude and by others. For an ideally rigid earth the period would be 360 days, but if as rigid as steel, it would be 441, the observed period being 430 days.

Among the older text-books on elasticity may be mentioned the works of G. Lamé, R. F. A. Clebsch, A. Winckler, A. Beer, E. L. Mathieu, W. J. Ibbetson, and F. Neumann, edited by O. E. Meyer.

In recent years the modern analytical developments, particularly along the line of integral equations, have been brought to bear on theories of elasticity and potential. The solution of the static problem of the theory of elasticity of a homogeneous isotropic body under certain given surface conditions has been taken up particularly by E. I. Fredholm of Stockholm, G. Lauricella of the University of Catania, R. Marcolongo of Naples and Hermann Weyl of Zurich, and by a somewhat different mode of procedure, by A. Korn of Berlin and T. Boggio of Turin.[1]

Closely connected with researches on attraction and elasticity is the development of spherical harmonics. After the initial paper of A. M. Legendre on zonal harmonics applied by him to the study of the attraction of solids of revolution, and after the remarkable memoir of 1782 by P. S. Laplace who used spherical harmonics in finding the potential of a solid nearly spherical, the first advance was made by Olinde Rodrigues (1794–1851), a French economist and reformer, who in 1816 gave a formula for P^n which later was derived independently by J. Ivory and C. G. J. Jacobi. The name "Kugelfunktion" is due to K. F. Gauss. Important contributions were made in Ger-

[1] See *Rendiconti del Circolo Math. di Palermo*, Vol. 39, 1915, p. 1.

many by C. G. J. Jacobi, L. Dirichlet, Franz Ernst Neumann (1798–1895) who was professor of physics and mineralogy in Königsberg, his son Carl Neumann (1832–1925), Elwin Bruno Christoffel (1829–1900) of the University of Strassburg, R. Dedekind, Gustav Bauer (1820–1906) of Munich, Gustav Mehler (1835–1895) of Elbing in West Prussia, and Karl Baer (1851–) of Kiel. Especially active was Eduard Heine (1821–1881) of the University of Halle, the author of the *Handbuch der Kugelfunktionen*, 1861, 2. Ed. 1878–1881. The chief representative in the cultivation of this subject, in Switzerland, was L. Schläfli of the University of Bern; in Belgium, was Eugène Catalan of the University of Liege; in Italy, was E. Beltrami; in the United States, was W. E. Byerly of Harvard University. In France there were S. D. Poisson, G. Lamé, T. J. Stieltjes, J. G. Darboux, Ch. Hermite, Paul Mathieu, Hermann Laurent (1841–1908), Professor at the Polytechnic School in Paris whose researches gave rise to contests of priority with German writers. In Great Britain spherical harmonics received the attention of Thomson, (Lord Kelvin) and P. G. Tait in their *Natural Philosophy* of 1867, and of Sir William D. Niven of Manchester, Norman Ferrers (1829–1903) of Cambridge, E. W. Hobson of Cambridge, A. E. H. Love of Oxford, and others.

Light, Electricity, Heat, Potential

G. F. B. Riemann's opinion that a *science* of physics only exists since the invention of differential equations finds corroboration even in this brief and fragmentary outline of the progress of mathematical physics. The undulatory theory of light, first advanced by C. Huygens, owes much to the power of mathematics: by mathematical analysis its assumptions were worked out to their last consequences. **Thomas Young**[1] (1773–1829) was the first to explain the principle of interference, both of light and sound, and the first to bring forward the idea of transverse vibrations in light waves. T. Young's explanations, not being verified by him by extensive numerical calculations, attracted little notice, and it was not until **Augustin Fresnel** (1788–1827) applied mathematical analysis to a much greater extent than Young had done, that the undulatory theory began to carry conviction. Some of Fresnel's mathematical assumptions were not satisfactory; hence P. S. Laplace, S. D. Poisson, and others belonging to the strictly mathematical school, at first disdained to consider the theory. By their opposition Fresnel was spurred to greater exertion. D. F. J. Arago was the first great convert made by Fresnel. When polarization and double refraction were explained by T. Young and A. Fresnel, then P. S. Laplace was at last won over. S. D. Poisson drew from Fresnel's formulæ the seemingly paradoxical deduction

[1] Arthur Schuster, "The Influence of Mathematics on the Progress of Physics," *Nature*, Vol. 25, 1882, p. 398.

that a small circular disc, illuminated by a luminous point, must cast a shadow with a bright spot in the centre. But this was found to be in accordance with fact. The theory was taken up by another great mathematician, W. R. Hamilton, who from his formulæ predicted conical refraction, verified experimentally by Humphrey Lloyd. These predictions do not prove, however, that Fresnel's formulæ are correct, for these prophecies might have been made by other forms of the wave-theory. The theory was placed on a sounder dynamical basis by the writings of A. L. Cauchy, J. B. Biot, G. Green, C. Neumann, G. R. Kirchhoff, J. MacCullagh, G. G. Stokes, B. de Saint-Venant, Émile Sarrau (1837–1904) of the Polytechnic School in Paris, Ludwig Lorenz (1829–1891) of Copenhagen, and Sir William Thomson (Lord Kelvin). In the wave-theory, as taught by G. Green and others, the luminiferous ether was an incompressible elastic solid, for the reason that fluids could not propagate transverse vibrations. But, according to G. Green, such an elastic solid would transmit a longitudinal disturbance with infinite velocity. G. G. Stokes remarked, however, that the ether might act like a fluid in case of finite disturbances, and like an elastic solid in case of the infinitesimal disturbances in light propagation. A. Fresnel postulated the density of ether to be different in different media, but the elasticity the same, while C. Neumann and J. MacCullagh assumed the density uniform and the elasticity different in all substances. On the latter assumption the direction of vibration lies in the plane of polarization, and not perpendicular to it, as in the theory of A. Fresnel.

While the above writers endeavored to explain all optical properties of a medium on the supposition that they arise entirely from difference in rigidity or density of the ether in the medium, there is another school advancing theories in which the mutual action between the molecules of the body and the ether is considered the main cause of refraction and dispersion.[1] The chief workers in this field were J. Boussinesq, W. Sellmeyer, H. Helmholtz, E. Lommel, E. Ketteler, W. Voigt, and Sir William Thomson (Lord Kelvin) in his lectures delivered at the Johns Hopkins University in 1884. Neither this nor the first-named school succeeded in explaining all the phenomena. A third school was founded by C. Maxwell. He proposed the electromagnetic theory, which has received extensive development recently. It will be mentioned again later. According to Maxwell's theory, the direction of vibration does not lie exclusively in the plane of polarization, nor in a plane perpendicular to it, but something occurs in both planes—a magnetic vibration in one, and an electric in the other. G. F. Fitzgerald and F. T. Trouton in Dublin verified this conclusion of C. Maxwell by experiments on electro-magnetic waves.

Of recent mathematical and experimental contributions to optics,

[1] R. T. Glazebrook, "Report on Optical Theories," *Report British Ass'n* for 1885, p. 213.

mention must be made of *Henry Augustus Rowland* (1848–1901), who was professor of physics at the Johns Hopkins University and his theory of concave gratings, and of A. A. Michelson's work on interference, and his application of interference methods to astronomical measurements.

A function of fundamental importance in the mathematical theories of electricity and magnetism is the "potential." It was first used by J. Lagrange in the determination of gravitational attractions in 1773. Soon after, P. S. Laplace gave the celebrated differential equation,

$$\frac{\partial^2 V}{\partial x^2}+\frac{\partial^2 V}{\partial y^2}+\frac{\partial^2 V}{\partial z^2}=0,$$

which was extended by S. D. Poisson by writing $-4\pi k$ in place of zero in the right-hand member of the equation, so that it applies not only to a point external to the attracting mass, but to any point whatever. The first to apply the potential function to other than gravitation problems was **George Green** (1793–1841). He introduced it into the mathematical theory of electricity and magnetism. Green was a self-educated man who started out as a baker, and at his death was fellow of Caius College, Cambridge. In 1828 he published by private subscription at Nottingham a paper entitled *Essay on the application of mathematical analysis to the theory of electricity and magnetism.* About 100 copies were printed. It escaped the notice even of English mathematicians until 1846, when William Thomson (Lord Kelvin) had it reprinted in *Crelle's Journal*, vols. xliv. and xlv. It contained what is now known as "Green's theorem" for the treatment of potential. Meanwhile all of Green's general theorems had been rediscovered by William Thomson (Lord Kelvin), M. Chasles, J. C. F. Sturm, and K. F. Gauss. The term *potential function* is due to G. Green. W. R. Hamilton used the word *force-function*, while K. F. Gauss, who about 1840 secured the general adoption of the function, called it simply *potential*. G. Green wrote papers on the equilibrium of fluids, the attraction of ellipsoids, on the reflection and refraction of sound and light. His researches bore on questions previously considered by S. D. Poisson. K. F. Gauss proved what C. Neumann has called "Gauss' theorem of mean value" and then considered the question of maxima and minima of the potential.[1]

Large contributions to electricity and magnetism have been made by **William Thomson** later **Sir William Thomson** and **Lord Kelvin** (1824–1907). He was born at Belfast, Ireland, but was of Scotch descent. He and his brother James studied in Glasgow. From there he entered Cambridge, and was graduated as Second Wrangler in 1845. William Thomson, J. J. Sylvester, C. Maxwell, W. K. Clifford, and J. J. Thomson are a group of great men who were Second Wranglers at Cambridge. At the age of twenty-two W. Thomson was elected professor

[1] For details see Max Bacharach, *Geschichte der Potentialtheorie*, Göttingen, 1883.

of natural philosophy in the University of Glasgow, a position which he held till his death. For his brilliant mathematical and physical achievements he was knighted, and in 1892 was made Lord Kelvin. He was greatly influenced by the mathematical physics of J. Fourier and other French mathematicians. It was Fourier's mathematics on the flow of heat through solids which led him to the mastery of the diffusion of an electric current through a wire and of the difficulties encountered in signalling through the Atlantic telegraph. In 1845 W. Thomson visited Paris. P. S. Laplace, A. M. Legendre, J. Fourier, Sadi Carnot, S. D. Poisson, and A. Fresnel were no longer living. W. Thomson met J. Liouville to whom he gave the now famous memoir of G. Green of the year 1828. He met M. Chasles, J. B. Biot, H. V. Regnault, J. C. F. Sturm, A. L. Cauchy, and J. B. L. Foucault. A. L Cauchy tried to convert him to Roman Catholicism. One evening. J. C. F. Sturm called upon him in high excitement. "Vous avez le mémoire de Green," he exclaimed. The *Essay* was produced; Sturm eagerly scanned its contents. "Ah! voilà mon affaire," he cried, jumping from his seat as he caught sight of the formula in which G. Green had anticipated his theorem of the equivalent distribution. Kelvin's researches on the theory of potential are epoch-making. What is called "Dirichlet's principle" was discovered by him in 1848, somewhat earlier than by P. G. L. Dirichlet. Jointly with P. G. Tait he prepared the celebrated *Treatise of Natural Philosophy*, 1867. As a mathematician he belonged most decidedly to the intuitional school. Purists in mathematics often carped at Kelvin's "instinctive" mathematics. "Do not imagine," he once said, "that mathematics is hard and crabbed, and repulsive to common sense. It is merely the etherealization of common sense." Yet even in mathematics he had his dislikes. When in 1845 he met W. R. Hamilton at a British Association meeting, who then read his first paper on Quaternions, one might have thought that W. Thomson would welcome the new analysis: but it was not so. He did not use it. On the merits of quaternions he had a thirty-eight years' war with P. G. Tait.[1] We owe to W. Thomson new synthetical methods of great elegance, viz., the theory of electric images and the method of electric inversion founded thereon. By them he determined the distribution of electricity on a bowl, a problem previously considered insolvable. The distribution of static electricity on conductors had been studied before this mainly by S. D. Poisson and G. A. A. Plana. In 1845 F. E. Neumann of Königsberg developed from the experimental laws of Lenz the mathematical theory of magneto-electric induction. In 1855 W. Thomson predicted by mathematical analysis that the discharge of a Leyden jar through a linear conductor would in certain cases consist of a series of decaying oscillations. This was first established experimentally by Joseph Henry of Washington. William Thomson worked out the

[1] S. P. Thompson, *Life of William Thomson*, London, 1910, pp. 452, 1136–1139.

electro-static induction in submarine cables. The subject of the screening effect against induction, due to sheets of different metals, was worked out mathematically by Horace Lamb and also by Charles Niven. W. Weber's chief researches were on electro-dynamics. H. Helmholtz in 1851 gave the mathematical theory of the course of induced currents in various cases. **Gustav Robert Kirchhoff** [1] (1824–1887), who was professor at Breslau, Heidelberg and since 1875 at Berlin, investigated the distribution of a current over a flat conductor, and also the strength of current in each branch of a network of linear conductors.

The entire subject of electro-magnetism was revolutionized by **James Clerk Maxwell** (1831–1879). He was born near Edinburgh, entered the University of Edinburgh, and became a pupil of Kelland and Forbes. In 1850 he went to Trinity College, Cambridge, and came out Second Wrangler, E. Routh being Senior Wrangler. Maxwell then became lecturer at Cambridge, in 1856 professor at Aberdeen, and in 1860 professor at King's College, London. In 1865 he retired to private life until 1871, when he became professor of physics at Cambridge. Maxwell not only translated into mathematical language the experimental results of Michael Faraday, but established the electro-magnetic theory of light, since verified experimentally by H. R. Hertz. His first researches thereon were published in 1864. In 1871 appeared his great *Treatise on Electricity and Magnetism.* He constructed the electro-magnetic theory from general equations, which are established upon purely dynamical principles, and which determine the state of the electric field. It is a mathematical discussion of the stresses and strains in a dielectric medium subjected to electro-magnetic forces. The electro-magnetic theory has received developments from Lord Rayleigh, J. J. Thomson, H. A. Rowland, R. T. Glazebrook, H. Helmholtz, L. Boltzmann, O. Heaviside, J. H. Poynting, and others. **Hermann von Helmholtz** (1821–1894) was born in Potsdam, studied medicine, was assistant at the charity hospital in Berlin, then a military surgeon, a teacher of anatomy, a professor of physiology at Königsberg, at Bonn and at Heidelberg. In 1871 he went to Berlin as successor to Magnus in the chair of physics. In 1887 he became director of the new Physikalisch-Technische Reichsanstalt. As a young man of twenty-six he published the now famous pamphlet *Ueber die Erhaltung der Kraft.* His work on *Tonempfindung* was written in Heidelberg. After he went to Berlin he was engaged chiefly on inquiries in electricity and hydrodynamics. Helmholtz aimed to determine in what direction experiments should be made to decide between the theories of W. Weber, F. E. Neumann, G. F. B. Riemann, and R. Clausius, who had attempted to explain electro-dynamic phenomena by the assumption of forces acting at a distance between two portions of the hypothetical electrical fluid,—

[1] W. Voigt, *Zum Gedächtniss von G. Kirchhoff*, Göttingen, 1888.

the intensity being dependent not only on the distance, but also on the velocity and acceleration,—and the theory of M. Faraday and C. Maxwell, which discarded action at a distance and assumed stresses and strains in the dielectric. His experiments favored the British theory. He wrote on abnormal dispersion, and created analogies between electro-dynamics and hydrodynamics. Lord Rayleigh compared electro-magnetic problems with their mechanical analogues, gave a dynamical theory of diffraction, and applied Laplace's coefficients to the theory of radiation. H. Rowland made some emendations on G. G. Stokes' paper on diffraction and considered the propagation of an arbitrary electro-magnetic disturbance and spherical waves of light. Electro-magnetic induction has been investigated mathematically by Oliver Heaviside, and he showed that in a cable it is an actual benefit. O. Heaviside and J. H. Poynting have reached remarkable mathematical results in their interpretation and development of Maxwell's theory. Most of Heaviside's papers have been published since 1882; they cover a wide field.✡

One part of the theory of capillary attraction, left defective by P. S. Laplace, namely, the action of a solid upon a liquid, and the mutual action between two liquids, was made dynamically perfect by K. F. Gauss. He stated the rule for angles of contact between liquids and solids. A similar rule for liquids was established by Franz Ernst Neumann. Chief among more recent workers on the mathematical theory of capillarity are Lord Rayleigh and E. Mathieu.

The great principle of the conservation of energy was established by **Robert Mayer** (1814–1878), a physician in Heilbronn, and again independently by Ludwig A. Colding of Copenhagen, J. P. Joule, and H. Helmholtz. **James Prescott Joule** (1818–1889) determined experimentally the mechanical equivalent of heat. H. Helmholtz in 1847 applied the conceptions of the transformation and conservation of energy to the various branches of physics, and thereby linked together many well-known phenomena. These labors led to the abandonment of the corpuscular theory of heat. The mathematical treatment of thermic problems was demanded by practical considerations. Thermodynamics grew out of the attempt to determine mathematically how much work can be gotten out of a steam engine. **Sadi Nicolas Léonhard Carnot** (1796–1832) of Paris, an adherent of the corpuscular theory, gave the first impulse to this. The principle known by his name was published in 1824. Though the importance of his work was emphasized by *B. P. E. Clapeyron*, it did not meet with general recognition until it was brought forward by William Thomson (Lord Kelvin). The latter pointed out the necessity of modifying Carnot's reasoning so as to bring it into accord with the new theory of heat. William Thomson showed in 1848 that Carnot's principle led to the conception of an absolute scale of temperature. In 1849 he published "an account of Carnot's theory of the motive power of heat, with numerical results deduced from Regnault's experiments." In February, 1850, **Rudolph Clausius** (1822–1888), then in Zürich (afterwards professor in Bonn), communicated to the Berlin Academy a paper on the same subject which contains the Protean second law of thermodynamics. In the same month **William John M. Rankine** (1820–1872), professor of engineering and mechanics at Glasgow, read before the Royal Society of Edinburgh a paper in which he declares the nature of heat to consist in the rotational motion of molecules, and arrives at some of the results reached previously by R. Clausius. He does not mention the second law of thermodynamics, but in a subsequent paper he declares that it could be derived from equations contained in his first paper. His proof of the second law is not free from objections. In March, 1851, appeared a paper of William Thomson (Lord Kelvin) which contained a perfectly rigorous proof of the second law. He obtained it before he had seen the researches of R. Clausius. The statement of this law, as given by Clausius, has been much criticised, particularly by W. J. M. Rankine, Theodor Wand, P. G. Tait, and Tolver Preston. Repeated efforts to deduce it from general mechanical principles have remained fruitless. The science of theormodynamics was developed with great success by W. Thomson, Clausius, and Rankine. As early as 1852 W. Thomson discovered the law of the dissipation of energy, deduced at a later period also by R. Clausius. The latter designated the non-transformable energy by the name *entropy*, and then stated that the entropy of the universe tends toward a maximum. For entropy Rankine used the term *thermodynamic function*. Thermodynamic investigations have been carried on also by Gustav Adolph Hirn (1815–1890) of Colmar, and H. Helmholtz (monocyclic and polycyclic systems). Valuable graphic methods for the study of thermodynamic relations were devised by J. W. Gibbs of Yale College.

Josiah Willard Gibbs (1839–1903) was born in New Haven, Conn., and spent the first five years after graduation mainly in mathematical studies at Yale. He passed the winter of 1866–1867 in Paris, of 1867–1868 in Berlin, of 1868–1869 in Heidelberg, studying physics and mathematics. In 1871 he was elected professor of mathematical physics at Yale. "His direct geometrical or graphical bent is shown by the attraction which vectorial modes of notation in physical analysis exerted over him, as they had done in a more moderate degree over C. Maxwell." Greatly influenced by Sadi Carnot, by William Thomson (Lord Kelvin) and especially by R. Clausius, Gibbs began in 1873 to prepare papers on the graphical expression of thermodynamic relations, in which energy and entropy appeared as variables. He discusses the entropy-temperature and entropy-volume diagrams, and the volume-energy-entropy surface (described in C. Maxwell's *Theory of Heat*). Gibbs formulated the energy-entropy criterion of equilibrium and stability, and expressed it in a form applicable to complicated problems of dissociation. That chemistry has tended to take a mathematical turn, says E. Picard, is evident from "the celebrated memoir of J. W. Gibbs on the equilibrium of chemical systems, so analytic in character, and where is needed some effort on the part of the chemists to recognize, under their algebraic mantle, laws of high importance."

In 1902 appeared J. W. Gibbs' *Elementary Principles in Statistical Mechanics*, developed with special reference to the rational foundation of thermodynamics. The modern kinetic theory of gases was mainly the work of R. Clausius, C. Maxwell, and Boltzmann. "In reading Clausius we seem to be reading mechanics; in reading Maxwell, and in much of L. Boltzmann's most valuable work, we seem rather to be reading in the theory of probabilities." C. Maxwell, and L. Boltzmann are the creators of "statistical dynamics." While they treated of molecules of matter directly, J. W. Gibbs considers "the statistics of a definite vast aggregation of ideal similar mechanical systems of types completely defined beforehand, and then compares the precise results reached in this ideal discussion with the principles of thermodynamics, already ascertained in the semi-empirical manner."[1] Important works on thermodynamics were prepared by R. Clausius in 1875, by R. Rühlmann in 1875, and by H. Poincaré in 1892.

In the study of the law of dissipation of energy and the principle of least action, mathematics and metaphysics met on common ground. The doctrine of least action was first propounded by P. L. M. Maupertuis in 1744. Two years later he proclaimed it to be a universal law of nature, and the first scientific proof of the existence of God. It was weakly supported by him, violently attacked by König of Leipzig, and keenly defended by L. Euler. J. Lagrange's conception of the principle of least action became the mother of analytic mechanics, but his statement of it was inaccurate, as has been remarked by Josef Bertrand in the third edition of the *Mécanique Analytique*. The form of the principle of least action, as it now exists, was given by W. R. Hamilton, and was extended to electro-dynamics by F. E. Neumann, R. Clausius, C. Maxwell, and H. Helmholtz. To subordinate the principle to all reversible processes, H. Helmholtz introduced into it the conception of the "kinetic potential." In this form the principle has universal validity.

An offshoot of the mechanical theory of heat is the modern kinetic theory of gases, developed mathematically by R. Clausius, C. Maxwell, Ludwig Boltzmann of Vienna, and others. The first suggestions of a kinetic theory of matter go back as far as the time of the Greeks. The earliest work to be mentioned here is that of Daniel Bernoulli, 1738. He attributed to gas-molecules great velocity, explained the pressure of a gas by molecular bombardment, and deduced Boyle's law as a consequence of his assumptions. Over a century later his ideas were taken up by J. P. Joule (in 1846), A. K. Krönig (in 1856), and R.

[1] *Proceed. of the Royal Soc. of London*, Vol. 75, 1905, p. 293.

Clausius (in 1857). J. P. Joule dropped his speculations on this subject when he began his experimental work on heat. A. K. Krönig explained by the kinetic theory the fact determined experimentally by Joule that the internal energy of a gas is not altered by expansion when no external work is done. R. Clausius took an important step in supposing that molecules may have rotary motion, and that atoms in a molecule may move relatively to each other. He assumed that the force acting between molecules is a function of their distances, that temperature depends solely upon the kinetic energy of molecular motions, and that the number of molecules which at any moment are so near to each other that they perceptibly influence each other is comparatively so small that it may be neglected. He calculated the average velocities of molecules, and explained evaporation. Objections to his theory, raised by C. H. D. Buy's-Ballot and by Emil Jochmann, were satisfactorily answered by R. Clausius and C. Maxwell, except in one case where an additional hypothesis had to be made. C. Maxwell proposed to himself the problem to determine the average number of molecules, the velocities of which lie between given limits. His expression therefor constitutes the important law of distribution of velocities named after him. By this law the distribution of molecules according to their velocities is determined by the same formula (given in the theory of probability) as the distribution of empirical observations according to the magnitude of their errors. The average molecular velocity as deduced by C. Maxwell differs from that of R. Clausius by a constant factor. C. Maxwell's first deduction of this average from his law of distribution was not rigorous. A sound derivation was given by O. E. Meyer in 1866. C. Maxwell predicted that so long as Boyle's law is true, the coefficient of viscosity and the coefficient of thermal conductivity remain independent of the pressure. His deduction that the coefficient of viscosity should be proportional to the square root of the absolute temperature appeared to be at variance with results obtained from pendulum experiments. This induced him to alter the very foundation of his kinetic theory of gases by assuming between the molecules a repelling force varying inversely as the fifth power of their distances. The founders of the kinetic theory had assumed the molecules of a gas to be hard elastic spheres; but Maxwell, in his second presentation of the theory in 1866, went on the assumption that the molecules behave like centres of forces. He demonstrated anew the law of distribution of velocities; but the proof had a flaw in argument, pointed out by L. Boltzmann, and recognized by C. Maxwell, who adopted a somewhat different form of the distributive function in a paper of 1879, intended to explain mathematically the effects observed in Crookes' radiometer. L. Boltzmann gave a rigorous general proof of Maxwell's law of the distribution of velocities.

None of the fundamental assumptions in the kinetic theory of gases leads by the laws of probability to results in very close agreement with observation. L. Boltzmann tried to establish kinetic theories of gases by assuming the forces between molecules to act according to different laws from those previously assumed. R. Clausius, C. Maxwell, and their predecessors took the mutual action of molecules in collision as repulsive, but L. Boltzmann assumed that they may be attractive. Experiments of J. P. Joule and Lord Kelvin seem to support the latter assumption.

Among the later researches on the kinetic theory is Lord Kelvin's disproof of a general theorem of C. Maxwell and L. Boltzmann, asserting that the average kinetic energy of two given portions of a system must be in the ratio of the number of degrees of freedom of those portions.

In recent years the kinetic theory of gases has received less attention; it is considered inadequate since the founding of the quantum hypothesis in physics.

Relativity

Profound and startling is the "theory of relativity." On the theory that the ether was stationary it was predicted that the time required for light to travel a given distance forward and back would be different when the path of the light was parallel to the motion of the earth in its orbit from what it was when the path of the light was perpendicular. In 1887 A. A. Michelson and E. W. Morely found experimentallythat such a difference in time did not exist. More generally, the results of this and other experiments indicate that the earth's motion through space cannot be detected by observations made on the earth alone. In order to explain Michelson and Morley's negative result and at the same time save the stationary-ether theory, H. A. Lorentz constructed in 1895 a "contraction hypothesis," according to which a moving solid contracts slightly longitudinally. This same idea occurred independently to G. F. Fitzgerald. In 1904 and in his *Columbia University Lectures* Lorentz aimed to reduce the electromagnetic equations for a moving system to the form of those that hold for a system at rest. Instead of x, y, z, t he introduced new independent variables, viz., $x'=\lambda\gamma(x-vt)$, $y=\lambda y$, $z'=\lambda z$, $t'=\lambda\gamma(t-\frac{v}{c^2}x)$, where γ depends upon velocity of light c and of the moving body v, and λ is a numerical coefficient such that, $\lambda=1$ when $v=0$. His fundamental equations turned out to be invariant under this now called "Lorentz transformation." In 1906 H. Poincaré made use of this transformation for the treatment of the dynamics of the electron and also of universal gravitation.[1] In 1905 A. Einstein published a paper on the electrodynamics of moving bodies in *Annalen der Physik*, Vol. 17, aiming at perfect

[1] L. Silberstein, *The Theory of Relativity*, London, 1914, p. 87.

reciprocity or equivalence of a pair of moving systems, and investigating the whole problem from the bottom, carefully considering the matter of "simultaneous" events in two distant places; he has succeeded in giving plausible support to and a striking interpretation of Lorentz's transformations. Einstein opened the way to the modern "theory of relativity." He developed it somewhat more fully in 1907. A fundamental point of view in his theory was that mass and energy are proportional.☼ For the purpose of taking account of gravitational phenomena, Einstein generalized his theory by assuming that mass and weight are also proportional, so that, for example, a ray of light is attracted by matter.☼ The mathematical part of Einstein's theory, as developed by M. Grossmann in 1913, employs quadratic differential forms and the absolute calculus of Gregorio Ricci of Padua. Another remarkable speculation was brought out in 1908 by Hermann Minkowski who read a lecture on *Raum und Zeit*, in which he maintained that the new views of space and time, developed from experimental considerations, are such that "space by itself and time by itself sink into the shadow and only a kind of union of the two retains self-dependence." No one notices a place, except at some particular time, nor time except at a particular place. A system of values x, y, z, t he calls a "world point" (Weltpunkt); the life-path of a material point in four-dimensional space is a "world line." The idea of time as a fourth dimension had been conceived much earlier by J. Lagrange in his *Théorie des fonctions analytiques* and by D'Alembert in his article "Dimension" in Diderot's *Encyclopédie*,[1] 1754. H. Minkowski considers the group belonging to the differential equation for the propagation of waves of light. *Hermann Minkowski* (1864–1909), was born at Alexoten in Russia, studied at Königsberg and Berlin, held associate professorships at Bonn and Königsberg and was promoted to a full professorship at Königsberg in 1895. In 1896 he went to the polytechnic school at Zurich and in 1903 to Göttingen. The importance which H. Minkowski, starting with the principle of relativity in the form given it by Einstein, has given to the Lorentzian transformations by the introduction of a four dimensional manifoldness or space-time-world, has been made intuitively evident by a number of writers, particularly F. Klein (1910), L. Heffter (1912), A. Brill (1912), and H. E. Timerding (1912). F. Klein said: "What the modern physicists call 'theory of relativity' is the theory of invariants of the fourth dimensional space-time-region x, y, z, t (Minkowski's world) in relation to a definite group of collineations, namely the 'Lorentz-group.'"[2] A novel presentation aiming at great precision was given in 1914 by Alfred A. Robb who on the idea of "conical order" and 21 postulates builds up a system in which the theory of space becomes absorbed in the theory of time. A philosophical discussion of relativity, mechan-

[1] R. C. Archibald in *Bull. Am. Math. Soc.*, Vol. 20, 1914, p. 410.
[2] Klein in *Jahresb. d. d. Math. Verein.*, Vol. 19, 1910, p. 287.

ics and geometrical axioms is given by Federigo Enriques of Bologna in his *Problems of Science* (1906), which has been translated into English by Katharine Royce in 1914. F. Enriques argues that certain optical and electro-optical phenomena seem to lead to a direct contradiction of the principles of classic mechanics, especially of Newton's principle of action and reaction. "Physics," says Enriques, "instead of affording a more precise verification of the classic mechanics, leads rather to a correction of the principles of the latter science, taken a priori as rigid."

The Croatian mathematician, Vladimir Varicâk found that the Lobachevskian geometry presented itself as the best adapted for the mathematical treatment of the physics of relativity. He enters upon optical phenomena and the resolution of paradoxes due to Ehrenfest and Bonn. Starting from this point of view, E. Borel in 1913 was able to deduce new consequences of the theory of relativity. One advantage of Varicâk's presentation is that it safeguards the parallelism between the old enunciations of physics and the new. L. Rougier of Lyon[1] asks the question, is then the Lobachevskian geometry physically true and the Euclidean wrong? No. One may keep, says he, the ordinary geometry for the discussion of the physics of relativity, as is done by H. A. Lorentz and A. Einstein, or one may add a fourth imaginary dimension to our three dimensions in the manner of H. Minkowski, or one may use the non-Euclidean geometries of mechanics and electromagnetics developed by E. B. Wilson and G. N. Lewis,[2] then of Boston. Each of these interpretations enjoys some particular advantages of its own.

Nomography

The use of simple graphic tables for computation is encountered in antiquity and the middle ages. The graphic solution of spherical triangles was in vogue in the time of Hipparchus,[3] and in the seventeenth century by W. Oughtred,[4] for instance. Edmund Wingate's *Construction and Use of the Line of Proportion*, London, 1628, described a double scale upon which numbers are indicated by spaces on one side of a straight line and the corresponding logarithms by spaces on the other side of the line.[5] Recently this idea has been carried out by A. Tichy in his *Graphische Logarithmentafeln*, Vienna, 1897. The Longitude Tables and Horary Tables of Margetts, London, 1791, were graphical. More systematic use of this idea was made by Pouchet in his *Arithmétique linéaire*, Rouen, 1795. In 1842 appeared the *Anamorphose logarithmique* of the Parisian engineer Léon Lalanne

[1] *L'Enseignement Mathématique*, Vol. XVI, 1914, p. 17.

[2] *Proceed. Am. Acad. of Arts and Science*, Vol. 48, 1912.

[3] A. von Braunmühl, *Geschichte der Trigonometrie*, Leipzig, Vol. I, 1900, pp. 3, 10, 85, 191.

[4] F. Cajori, *William Oughtred*, Chicago and London, 1916.

[5] F. Cajori, *Colorado College Publication*, General Series 47, 1910, p. 182.

(1811–1892) in which the distances of points from the origin are not necessarily proportional to the actual values of the data, but may be other functions of them, judiciously chosen. In the product $z_1z_2=z_3$, the variables z_1 and z_2 are brought in correspondence, respectively, with the straight lines $x=\log z_1$, $y=\log z_2$, so that $x+y=\log z_3$, which represents the straight lines perpendicular to the bisectors of the angle between the co-ordinate axes. Advances along this line were made by J. Massau of the University of Ghent, in 1884, and E. A. Lallemand in 1886. The Scotch Captain Patrick Weir in 1889 gave an azimuth diagram which was an anticipation of a spherical triangle nomogram. But the real creator of nomography is *Maurice d'Ocagne* of the École Polytechnique in Paris, whose first researches appeared in 1891; his *Traité de nomographie* came out in 1899. The principle of anamorphosis, by successive generalizations, "has led to the consideration of equations representable not only by two systems of straight lines parallel to the axes of co-ordinates and one other unrestricted system of straight lines, but by three systems of straight lines under no such restrictions." D'Ocagne also studied equations representable by means of systems of circles. He has introduced the method of collinear points by which "it has been possible to represent nomographically equations of more than three variables, of which the previous methods gave no convenient representation." [1]

Mathematical Tables

The increased accuracy now attainable in astronomical and geodetic measurements and the desire to secure more complete elimination of errors from logarithmic tables, has led to recomputations of logarithms. Edward Sang of Edinburgh published in 1871 a 7-place table of common logarithms of numbers to 200,000. These were mainly derived from his unpublished 28-place table of logarithms of primes to 10,037 and composite numbers to 20,000, and his 15-place table from 100,000 to 370,000.[2] In 1889 the Geographical Institute of Florence issued a photographic reproduction of G. F. Vega's *Thesaurus* of 1794 (10 figures). Vega had computed A. Vlacq's tables anew, but his last figure was unreliable. In 1891 the French Government issued 8-place tables which were derived from the unpublished *Tables du Cadastre* (14-places, 12 correct) which had been computed near the close of the eighteenth century under the supervision of G. Riche de Prony. These tables give logarithms of numbers to 120,000, and of sines and tangents for every 10 centesimal seconds, the quadrant being divided centesimally.[3] Prony consulted A. M. Legendre and other

[1] D'Ocagne in *Napier Tercentenary Memorial Volume*, London, 1915, pp. 279–283. See also D'Ocagne, *Le calcul simplifié*, Paris, 1905, pp. 145–153.

[2] E. M. Horsburgh, *Napier Tercentenary Celebration Handbook*, 1914, pp. 38–43.

[3] This and similar information is drawn from J. W. L. Glaisher in *Napier Tercentenary Memorial Volume*, London, 1915, pp. 71–73.

mathematicians on the choice of methods and formulas, and entrusted the computation of primary results to professional calculators, while the task of filling the rest of the columns beyond the primary results was performed by assistants "apt merely in performing additions" by the use of the method of differences. "It is curious," says D'Ocagne "to note that the majority of these assistants had been recruited from among the hair-dressers whom the abandonment of the powdered wig in men's fashion had deprived of a livelihood."

In 1891 *M. J. de Mendizâbel-Tamborrel* published at Paris tables of logarithms of numbers to 125,000 (8 places) and of sines and tangents (7 or 8 places) for every millionth of the circumference, which were almost wholly derived from original 10-place calculations. *W. W. Duffield* published in the Report of the U. S. Coast and Geodetic Survey, 1895–1896, a 10-place table of logarithms of numbers to 100,000, in 1910, 8-place tables of numbers to 200,000 and trigonometric tables to every sexagesimal second were published by *J. Bauschinger* and *J. Peters* of Strassburg. A special machine was constructed for the computation of these tables. In 1911 *H. Andoyer* of Paris published a 14-place table of logarithms of sines and tangents to every 10 sexagesimal seconds. "This table was derived from a complete recalculation, made entirely by M. Andoyer himself, without any assistance, personal or mechanical."

In recent years a demand has arisen for tables giving the natural values of sines and cosines. In 1911 *J. Peters* published in Berlin such a table,✡ extending from 0° to 90°, and carried to 21 decimals, for every 10 sexagesimal seconds (and for every second of the first six degrees). Extensive tables of natural values, first computed by Rhaeticus and published in 1613, were abandoned after the invention of logarithms, but are now returning in use again, since they are better fitted for the growing practice of calculating directly by means of machines and without resort to logarithms.

The decimal division of angles has been agitated again in recent years. In 1900 R. Mehmke made a report to the German Mathematiker Vereinigung.[1] Why are degrees preferred to radians in practical trigonometry? Because, on account of the periodicity of the trigonometric functions, we frequently would have to add and subtract π or 2π which are irrational numbers and therefore objectionable. The sexagesimal subdivision of the degree which resulted in great harmony among the Babylonians who used the sexagesimal notation of numbers and fractions, and the sexagesimal divisions of the day, hour and minute, is less desirable now that we have the decimal notation of numbers. There has been some difference of opinion among advocates of the decimal system in angular measurement, what unit should be chosen for the decimal subdivision. In 1864 *Yvon*

[1] See *Jahresb. d. d. Math. Vereinigung*, Leipzig, Vol. 8, Part 1, 1900, p. 139.

Villarceau, at a meeting of the Bureau of Longitudes in Paris, suggested the decimal subdivision of the entire circumference, while in 1896 *Bouquet de la Grye* preferred the semi-circumference. *R. Mehmke* argues that whatever the unit may be that is subdivided, the four arithmetical operations with angles would be materially simplified, interpolation in the use of trigonometric tables would be easier, the computation of the lengths of arcs would be shorter. If the right angle is the unit that is subdivided, then the reduction of large angles to corresponding acute angles can be effected merely by the subtraction of the integers 1, 2, 3, . . The determination of supplementary or complementary angles is less laborious. A more convenient arrangement of trigonometric tables was claimed by G. J. Hoüel and greater comfort in taking observations was promised by J. Delambre. Nevertheless, no decimal division of angles is at the present time threatened with adoption, not even in France.

A very specialized kind of logarithms, the so-called "Gaussian logarithms," which give log $(a+b)$ and log $(a-b)$, when log a and log b are known, were first suggested by the Italian physicist *Guiseppe Zecchini Leonelli* (1776–1847) in his *Thêorie des logarithmes*, Bordeaux 1803; the first table was published by K. F. Gauss in 1812 in *Zach's Monatliche Korrespondenz.* It is a 5-place table. More recent tables are the 6-place tables of Carl Bremiker (1804–1877) of the geodetic institute of Berlin, Siegmund Gundelfinger (1846–1910) of Darmstadt, and George William Jones (1837–1911) of Cornell University, and the 7-place table of T. Wittstein.

Proceeding to hyperbolic and exponential functions, we mention the 7-place tables of $\log_{10} \sinh x$ and $\log_{10} \cosh x$ prepared by Christoph Gudermann of Münster in 1832, the 5-place tables by Wilhelm Ligowski (1821–1893) of Kiel in 1890, the 5-place tables by G. F. Becker and C. E. Van Orstrand in their *Smithsonian Mathematical Tables*, 1909. Tables for $\sinh x$ and $\cosh x$ were published by Ligowski (1890), Burrau (1907), Dale, Becker, and Van Orstrand. In the *Cambridge Philosophical Transactions*, Vol. 13, 1883, there are tables for $\log e^x$ and e^x by J. W. L. Glaisher, for e^{-x} by F. W. Newman. G. F. Becker and C. E. Van Orstrand also give tables for these functions.

An isolated matter of interest is the origin of the term "radian," used with trigonometric functions. It first appeared in print on June 5, 1873, in examination questions set by James Thomson at Queen's College, Belfast. James Thomson was a brother of Lord Kelvin. He used the term as early as 1871, while in 1869 Thomas Muir, then of St. Andrew's University, hesitated between "rad," "radial" and "radian." In 1874 T. Muir adopted "radian" after a consultation with James Thomson.[1]

[1] *Nature*, Vol. 83, pp. 156, 217, 459, 460.

Calculating Machines, Planimeters, Integraphs

The earliest calculating machine, invented by Blaise Pascal in 1641, was designed only to effect addition. Three models of Pascal's machine are kept in the Conservatoire des arts et métiers in Paris. It was G. W. Leibniz who conceived the idea of adapting to a machine of this sort a mechanism capable of repeating several times, rapidly, the addition of one and the same number, so as to effect multiplication mechanically. Of the two Leibniz machines said to have been constructed, one (completed 1694?) is preserved in the library of Hanover. This idea was re-invented and worked out for actual use in practice in 1820 by Ch. X. Thomas de Colmar in his *Arithmomètre*. More limited practical use was given to the machine of Ph. M. Hahn, first constructed in Stuttgart in 1774.

A machine effecting multiplication, not by repeated additions, but directly by the multiplication table, was first exhibited at the universal exposition in Paris in 1887. This decidedly original design was the invention of a young Frenchman, Léon Bollée, who took also a prominent part in the development of the automobile. In his computing machine there are calculating plates furnished with tongues of appropriate length which constitute a kind of multiplication table, acting directly on the recording apparatus of the machine. A somewhat simpler elaboration of the same idea is due to O. Steiger (1892) in a machine called the *millionnaire*.[1] In 1892 a Russian engineer, W. T. Odhner, invented and constructed a widely used machine, called the *Brunsviga Calculator*, which is of the "pin wheel and cam disc" type, the first idea of which goes back to Polenus (1709) and to Leibniz. Of American origin are the Burroughs Adding and Listing machine, and the Comptometer invented about 1887 by Dorr E. Felt of Chicago.

The first idea of automatic engines calculating by the aid of functional differences of various orders goes back to J. H. Müller (1786), but no steps towards definite plans and actual construction were taken before the time of Babbage. *Charles Babbage* (1792–1871) invented a machine, called a "difference-engine," about 1812. Its construction was begun in 1822 and was continued for 20 years. The British Government contributed £17,000 and Babbage himself £6000. Through some misunderstanding with the Government, work on the engine, though nearly finished, was stopped. Inspired by Ch. Babbage's design, Georg und Eduard Scheutz (father and son) of Stockholm made a difference engine which was acquired by the Dudley Observatory in Albany.

In 1833 Ch. Babbage began the design of his "analytical engine";

[1] D'Ocagne in the *Napier Tercentenary Memorial Volume*, London, 1915, pp. 283–285. For details, see D'Ocagne, *Le calcul simplifié*, Paris, 1905, pp. 24–92; *Encyklopädie d. Math. Wiss.*, Bd. I, Leipzig, 1898–1904, pp. 952–982; E. H. Horsburgh, *Napier Tercentenary Celebration Handbook*, Edinburgh, 1914, "Calculating Machines" by J. W. Whipple, pp. 69–135.

a small portion of it was put together before his death. This engine was intended to evaluate any algebraic formula, for any given values of the variables. In 1906 H. P. Babbage, a son of Charles Babbage, completed part of the engine, and a table of 25 multiples of π to 29 figures was published as a specimen of its work.[1]

Planimeters have been designed independently and in many different ways. It is probable that J. M. Hermann designed one in 1814. Planimeters were devised in 1824 by Gonella in Florence, about 1827 by Johannes Oppikoffer (1783–1859) [2] of Bern and constructed by Ernst in Paris, about 1849 by Wetli of Vienna and improved by the astronomer Peter Andreas Hansen of Gotha, about 1851 by Edward Sang of Edinburgh and improved by Clerk Maxwell, J. Thomson and Lord Kelvin. All of these were rotation planimeters. Most noted of polar planimeters are that of *Jakob Amsler* (1823–1912) and those constructed by Coradi of Zurich. J. Amsler was at one time privatdocent at the University of Zurich, later manufacturer of instruments for precise measurements. He invented his polar planimeter in 1854; his account of it was published in 1856.

Another interesting class of instruments, called "integraphs" has been invented by Abdank Abakanovicz (1852–1900) in 1878 and by C. Vernon Boys [3] in 1882. These instruments draw an "integral curve" when a pointer is passed round the periphery of a figure whose area is required. More recently numerous integraphs have been invented through the researches of E. Pascal of the University of Naples. Thus in 1911 he designed a polar integraph for the quadrature of differential equations.

[1] *Napier Tercentenary Celebration Handbook*, 1914, p. 127.

[2] Morin, *Les Appareils d'Intégration*, 1913. See E. M. Horsburgh, *op. cit.*, p. 190.

[3] Boys in *Phil. Mag.*, 1882; Abdank Abakanowicz, *Les Intégraphes*, Paris, 1886. See also H. S. Hele Shaw, "Graphic Methods in Mechanical Science" in *Report of British Ass'n* for 1892, pp. 373–531; E. M. Horsburgh, *Handbook*, pp. 194–206.

EDITOR'S NOTES

EDITOR'S NOTES TO THE FOURTH EDITION

Page 4, line 1. This chapter is in large measure based on Otto Neugebauer, *The Exact Sciences in Antiquity,* 2nd Ed., New York, Dover Publications, 1969, Chapters I, II, III. The reader is referred to this work for further information on Babylonian mathematics and astronomy and for detailed references to the sources. It is, besides, highly readable, having some of the fascination of a detective story, as Professor Neugebauer shows us how the investigator works in unraveling the meaning of original source material.

Page 101, line 22. The given dates are, of course, the dates of the Caliphate.

Page 111, line 8. According to Professor Jacques Dutka, this represents a misinterpretation of a remark of Günther.

Page 141, line 42. That is, excluding the Greeks, and others, of the pre-Christian era.

Page 156, line 2. Equations having real roots. [*Author.*]

Page 185, line 10. The term *hypergeometric series* was coined by Wallis. It was Pfaff who applied it to the important series that has carried the name ever since. The series itself had been introduced into Analysis by Gauss, who gave it no particular name.

Page 198, line 9. Newton's hesitation in stating clearly what constituted the basis for the Calculus, his vacillation among three possible bases—infinitesimals, limits, and basic physical intuition—is understandable in light of subsequent events. The period of Newton and his immediate successors has often been called the period of indecision—even the period of confusion. But by the end of the nineteenth century, in consequence of the work of Cauchy, Cantor, Dedekind, Weierstrass, and others, the matter seemed to have been resolved: The true basis of the Calculus is the theory of limits. This answer to an important and vexing problem, achieved after close to three centuries, seemed to many to be a definitive answer. Then, in the 1960's, a new development: "In the fall of 1960 it occurred to [Abraham Robinson] that the concepts and methods of contemporary Mathematical Logic are capable of providing a suitable framework for the development of the Differential and Integral Calculus by means of infinitely small and infinitely large numbers." Thus came into existence Non-standard Analysis, in which the Calculus was based in a rigorous manner upon infinitesimals. In consequence, some three hundred years after Newton the Calculus can be justified on either of *two* bases: the theory of limits, and infinitesimals. The logical machinery that made this second justification possible did not come into existence until well into the twentieth century; it was not available even to the mathematicians of the late nineteenth century.

For a brief and very clear exposition of Non-standard Analysis, see J. Barkley Rosser, *Logic for Mathematicians,* 2nd Ed., New York, 1979, Appendix D. See also Abraham Robinson, *Non-standard Analysis,* 2nd Ed., Amsterdam, 1974 (1st Ed., 1966), especially pages 1-2 and 260-282 ('Concerning the History of the Calculus').

Page 219, line 36. Berkeley's objections were well taken and could not be dealt with until 1966, with the creation of Non-standard Analysis. Non-standard Analysis has built a new structure for the Calculus within the framework of which infinitesimals are accommodated and in which Berkeley's objections no longer apply. (See the Editor's Note to page 198.)

Page 226, line 41. This is not to say that nothing along these lines had not been done before him. Results in finite differences had been obtained by Briggs, James Gregory, and Isaac Newton.

Page 230, line 7. Until 1924, the discovery of the normal curve of errors was attributed to Gauss. But (*Isis,* Vol. 8, pp. 671 ff.; *Biometrika,* Vol. 16, p. 402) Karl Pearson pointed out that De Moivre had made the discovery in a paper published in 1733, 76 years before Gauss. This first appeared in a supplement to the *Miscellanea Analytica* and later, in English, on pages 243-254 of *The Doctrine of Chances* (3rd Ed., 1756).

Page 233, line 31. Publication, begun in 1911, is not yet complete. Seventy substantial volumes (most between 400 and 600 pages) have already (1985) been published. Eighteen more volumes are planned.

Page 234, footnote 1. A third edition was published in 1924. It has been published in English translation under the title *Elementary Mathematics from an Advanced Standpoint,* Vol. I.

Page 243, line 37. I.e., for the privilege of playing the game under the conditions stated.

Page 262, lines 11 and 14. Bowditch published the work at his own expense (roughly $350,000 in 1985 dollars), refusing all offers of subsidy. The English translation comes to over 4,000 printed pages, half of each page, on average, consisting of text and half of running commentary. The commentary fills in the missing details and clarifies obscure points. Whatever errors Bowditch discovered, he communicated to Laplace, who incorporated the corrections in later editions but without offering thanks or making any acknowledgement.

Page 265, line 37. But the idea goes back further; see page 148, line 2.

Page 266, line 40. Gauss did, indeed, do some research on this subject, but he did not publish it.

Page 270, line 24. A translation into English has been published under the title *Analytic Theory of Heat.*

Page 270, line 34. For example (and quite astonishingly), functions representable by *entirely different* analytic functions in different subintervals. According to R. Courant (*D. and I. Calculus,* Vol. I. pages 439 and 455), functions representable by Fourier series "have a very high degree of arbitrariness."

Page 279, line 41. By 1979, according to *Mathematical Reviews,* the number of such periodicals *then in print* had increased to 1,544, 365 of them devoted exclusively to mathematics. Specialization has increased to the point where over 50 periodicals are now devoted to one or another particular *branch* of mathematics. And the trend is continuing.

Page 302, line 3. *How to Draw a Straight Line* has been reprinted in *Squaring the Circle, and Other Monographs,* by E. W. Hobson, et al.

Page 305, footnote 1. A translation into English, with the original Latin on facing pages, was published at Chicago in 1920. A reprint is in preparation.

Page 306, line 20. *Ueber die Hypothesen, welche der Geometrie zu Grunde leigen.* This seminal work was translated by W. K. Clifford in 1873 and is included in Clifford's *Mathematical Papers.* A German-language edition, with an exegesis by Hermann Weyl has been published (3rd ed., Zürich, 1923; included in *Das Kontinuum, and andere Monographien,* New York, 1960 and 1973).

Page 320, line 28. A revised and enlarged French-language edition has been published in three volumes: *Traité des Courbes Spéciales Remarquables Planes et Gauches*, Coimbra, 1908-1915. A third edition (with corrections indicated by the author) has been published in three volumes: New York, 1971.

Page 325, line 31. A second edition was published in 1972, edited by Mrs. R. H. Tanner (the authors' daughter) and I. Grattan-Guinness, on the basis of a copy annotated by the authors.

Page 327, line 8. It is not clear that Hilbert had any *direct* knowledge of the work of the Italian school.

Page 327, line 10. An eleventh edition was published in 1972. The tenth edition has been published in English translation, Chicago, 1971.

(The first edition was published in English translation in 1902.)

Page 333, line 23. Second edition, London, 1899 and 1901. A third (corrected) edition has been published: New York, 1969.

Page 310, line 4. Also by G. Lamé and J. D. Gergonne. [*Author.*]

Page 315, line 11. A second, revised and enlarged edition of the first two volumes was published in 1914 and 1915, together with a list of errata for the third and fourth volumes, which were reprinted unchanged. The fourth volume contains the index to all four volumes, so that the revision of the earlier volumes left the unchanged index partly in error. All of this was taken care of in a third edition, published at New York in 1972.

Page 316, line 35. That is, in two volumes. Reprinted, New York, 1971, in one volume.

Page 337, line 15. Because of the ill-success of this difficult work, Grassmann published a rewritten and, he hoped, a more accessible version in 1862. When this, too, met with little success, he went back to the original work by reissuing it in 1878 with only the correction of errata and some interpolated material. Each interpolation was printed in smaller type and narrower 'measure' and each was labelled 1877; the first four words of the title of this 1878 work was *Die Ausdehnungslehre von 1844.*

Page 340, footnote 1. This was extended to five volumes, the fifth under the title *Contributions to the History of Determinants, 1900-1920.*

Page 342, line 3. In the early 1920's. The word *modern* acquires a somewhat different meaning from time to time.

Page 347, line 28. A translation into English has been published, *Lectures on the Icosahedron.*

Page 356, line 4. His baptismal name was *Christian Felix,* but he did not use the former of the two names and always called himself and was known as, simply, *Felix Klein.*

Page 367, line 25. Srinivassa Ramanujan (1877-1920), a self-taught Indian mathematician, who became a protégé and then a colleague of G. H. Hardy, stated without proof many truly remarkable identities. The proof of these identities has proved a challenge to mathematicians, and the proofs of some have still (1985) not yet been found. Hardy called him "the most romantic figure in the recent history of mathematics."

Page 369, line 18. The word *emphasized* applies properly only to D'Alembert (see page 243). For Newton, see page 198; for Jurin and Robins, see pages 219-220; and for Maclaurin, see page 228.

Page 373, line 17. See the note to page 185, line 10.

Page 377, lines 27 and 29. This was true up to the early 1930's. The progress since then has been enormous.

Page 378, line 38 and **379, line 6.** Also called 'geometrical probability.'

Page 391, line 16. A second edition and also a large supplementary volume, both edited by Isaac Todhunter, were published after Boole's death. A fifth edition, containing both volumes, was published at New York in 1959.

Page 397, line 11. Both of these celebrated works have been translated into English and published under the title *Essays on the Theory of Numbers.*

Page 409, line 15. The complete work is in three volumes: *Principia Mathematica,* Cambridge, 1910-1913; 2nd ed., 1925-1927.

Page 410, line 16. This assessment is emphatically no longer valid. With the later formalization of logical systems, logic and metalogic (the study of such systems) has answered some of the most profound questions about mathematics. See also the Editor's Note to page 198, line 9.

Page 412, footnote. See also Oystein Ore, *Niels Henrik Abel: Mathematician Extraordinary,* Minneapolis, 1957 and New York, 1974.

Page 414, line 21. The fact that Jacobi and Sylvester (see page 343) are explicitly mentioned as being of Jewish parentage should not be taken to mean that there are no others. There are, in fact, many score among those named in this *History,* among them G. Eisenstein, L. Kronecker, L. Cremona, M. Pasch, M. Noether, G. Cantor, A. Pringsheim, A. M. Schönflies, A. Hurwitz, V. Volterra, H. Minkowski, J. Hadamard, T. Levi-Civita, H. Lebesgue, E. Landau, G. Ascoli, R. Courant, N. Wiener.

Page 421, line 23. See the Editor's Note to page 306.

Page 432, line 21. See the Editor's Note to page 347.

Page 435, line 33. A second edition, edited by E. C. J. Schering, was published at Leipzig in 1870. A translation into French has been published. A translation into German containing in addition all of Gauss's writings on number theory has been published: *Arithmetische Untersuchungen,* Berlin, 1899, reprinted, New York, 1981. A translation into English has been published under the Latin title, *Disquisitiones Arithmeticae,* New Haven, 1966.

Page 439, line 30. A second edition, containing added papers by Landau and an appendix by Paul T. Bateman on developments up to 1953, has been published, New York 1953 and 1974.

Page 441, line 16. These reports are included in his collected papers and have also been published separately: *Report on the Theory of Numbers,* New York, 1965.

Page 475, line 17. Heaviside's papers have been published in five volumes: *Electrical Papers,* in two volumes, London, 1892 and *Electromagnetic Theory,* in three volumes, London, 1893-1912. These have been reprinted several times.

Page 480, lines 8 and 11. More, exactly, energy has mass and mass represents energy; a body when hot has greater mass than when it is cold, each erg of energy increasing its mass by the minuscule amount of 1/900,000,000,000,000,000,000 grams; similarly, when a gram of mass is annihilated (as in a thermonuclear explosion), it is converted into 900,000,000,000,000,000,000 ergs of energy. The bending of light in a gravitational field is due to the fact that light carries energy and energy has mass.

Page 483, line 25. This table is reproduced in Jean Peter's *Eight-Place Tables of Trigonometric Functions,* New York, 1963.

INDEX

A star ⋆ before a person's name indicates that his collected papers have been published.

INDEX

CHELSEA

SCIENTIFIC

BOOKS

THEORIE DES OPERATIONS LINEAIRES

By S. BANACH

—1933-64. xii + 250 pp. 5⅜x8. 8284-0110-1.

DIFFERENTIAL EQUATIONS

By H. BATEMAN

CHAPTER HEADINGS: I. Differential Equations and their Solutions. II. Integrating Factors. III. Transformations. IV. Geometrical Applications. V. Diff. Eqs. with Particular Solutions of a Specified Type. VI. Partial Diff. Eqs. VII. Total Diff. Eqs. VIII. Partial Diff. Eqs. of the Second Order. IX. Integration in Series. X. The Solution of Linear Diff. Eqs. by Means of Definite Integrals. XI. The Mechanical Integration of Diff. Eqs.

—1917-67. xi + 306 pp. 5⅜x8. 8284-0190-X.

MEASURE AND INTEGRATION

By S. K. BERBERIAN

A highly flexible graduate level text. Part I is designed for a one-semester introductory course; the book as a whole is suitable for a full-year course. Numerous exercises.

Partial Contents: PART ONE: I. Measures. II. Measurable Functions. III. Sequences of Measurable Functions. IV. Integrable Functions. V. Convergence Theorems. VI. Product Measures. VII. Finite Signed Measures. PART TWO: VIII. Integration over Locally compact Spaces (. . . The Riesz-Markoff Representation Theorem, . . .). IX. Integration over Locally Compact Groups (Topological Groups, . . . , Haar Integral, Convolution, The Group Algebra, . . .). BIBLIOGRAPHY. INDEX.

—1965-70. xx + 312 pp. 6x9. 8284-0241-8.

L'APPROXIMATION

By S. BERNSTEIN and CH. de LA VALLÉE POUSSIN

TWO VOLUMES IN ONE:

Leçons sur les Propriétés Extrémales et la Meilleure Approximation des Fonctions Analytiques d'une Variable Réelle, *by Bernstein.*

Leçons sur l'approximation des Fonctions d'une Variable Réelle, *by Vallée Poussin.*

—1925/19-65. 363 pp. 6x9. 8284-0198-5. 2 v. in 1.

LECTURES ON THE CALCULUS OF VARIATIONS

By O. BOLZA

A standard text by a major contributor to the theory. Suitable for individual study by anyone with a good background in the Calculus and the elements of Real Variables.

—2nd (c.) ed. 1961-71. 280 pp. 5x8. 8284-0145-4. Cl.
8284-0152-7. Pa.

VORLESUNGEN UEBER VARIATIONSRECHNUNG

By O. BOLZA

A standard text and reference work, by one of the major contributors to the theory.

—1909-63. ix + 715 pp. 5⅜x8. 8284-0160-8.

THEORIE DER KONVEXER KOERPER

By T. BONNESEN and W. FENCHEL

"Remarkable monograph."

—*J. D. Tamarkin, Bulletin of the A. M. S.*

—1954-71. 171 pp. 5½x8½. 8284-0054-7.

THE CALCULUS OF FINITE DIFFERENCES

By G. BOOLE

A standard work on the subject of finite differences and difference equations by one of the seminal minds in the field of finite mathematics.

Numerous exercises with answers.

—5th ed. 1970. 341 pp. 5⅜x8. 8284-1121-2. Cloth
—4th ed. 1958. 336 pp. 5⅜x8. 8284-0148-9. Paper

A TREATISE ON DIFFERENTIAL EQUATIONS

By G. BOOLE

Including the Supplementary Volume.

—5th ed. 1959. xxiv + 735 pp. 5⅜x8. 8284-0128-4.

MATHEMATICAL PAPERS

By W. K. CLIFFORD

One of the world's major mathematicians, Clifford's papers cover only a 15-year span, for he died at age 34. [Included in this volume is Clifford's English translation of an important paper of Riemann.]

—1882-67. 70 + 658 pp. 5⅜x8. 8284-0210-8.

ESSAI SUR L'APPLICATION DE L'ANALYSE AUX PROBABILITÉS

By M. J. CONDORCET

A photographic reproduction of a very rare and historically important work in the Theory of Probability. An original copy brings many hundreds of dollars in the rare book market.

—1785. Repr. 1971. 191 + 304 pp. 6x9.

A TREATISE ON THE CIRCLE AND THE SPHERE

By J. L. COOLIDGE

—2nd (c.) ed. 1916-71. 602 pp. 5⅜x8. 8284-0236-1.

MODERN PURE SOLID GEOMETRY

By N. A. COURT

—2nd ed. 1964. xiv + 353 pp. 5½x8¼. 8284-0147-0.

SPINNING TOPS AND GYROSCOPIC MOTION

By H. CRABTREE

Partial Contents: Introductory Chapter. CHAP. I. Rotation about a Fixed Axis. II. Representation of Angular Velocity. Precession. III. Discussion of the Phenomena Described in the Introductory Chapter. IV. Oscillations. V. Practical Applications. VI-VII. Motion of Top. VIII. Moving Axes. IX. Stability of Rotation. Periods of Oscillation. APPENDICES: I. Precession. II. Swerving of "sliced" golf ball. III. Drifting of Projectiles. IV. The Rising of a Top. V. The Gyro-compass. ANSWERS TO EXAMPLES.

—2nd ed. 1914-67. 203 pp. 6x9. 8284-0204-3.

FAMOUS PROBLEMS, and other monographs
By KLEIN, SHEPPARD, MacMAHON, and MORDELL

FOUR VOLUMES IN ONE.

FAMOUS PROBLEMS OF ELEMENTARY GEOMETRY, by *Klein.* A fascinating little book. A simple, easily understandable, account of the famous problems of Geometry—The Duplication of the Cube, Trisection of the Angle, Squaring of the Circle—and the proofs that these cannot be solved by ruler and compass—presentable, say, before an undergraduate math club (no calculus required). Also, the modern problems about transcendental numbers, the existence of such numbers, and proofs of the transcendence of *e*.

FROM DETERMINANT TO TENSOR, by *Sheppard.* A novel and charming introduction. Written with the utmost simplicity. PT I. Origin of Determinants. II. Properties of Determinants. III. Solution of Simultaneous Equations. IV. Properties. V. Tensor Notation. PT II. VI. Sets. VII. Cogredience, etc. VIII. Examples from Statistics. IX. Tensors in Theory of Relativity.

INTRODUCTION TO COMBINATORY ANALYSIS, by *MacMahon.* A concise introduction to this field. Written as introduction to the author's two-volume work.

THREE LECTURES ON FERMAT'S LAST THEOREM, by *Mordell.* This famous problem is so easy that a high-school student might not unreasonably hope to solve it; it is so difficult that tens of thousands of amateur and professional mathematicians, Euler and Gauss among them, have failed to find a complete solution. Mordell's very small book begins with an impartial investigation of whether Fermat himself had a solution (as he said he did) and explains what has been accomplished. This is one of the masterpieces of mathematical exposition.

—2nd ed. 1962. 350 pp. 5⅜x8. Four vols. in one.
8284-0108-X.

VORLESUNGEN UEBER NICHT-EUKLIDISCHE GEOMETRIE
By F. KLEIN

—1928-59. xii + 326 pp. 5⅜x8. 8284-0129-2.

LEHRBUCH DER THETAFUNKTIONEN

By A. KRAZER

"Dr. Krazer has succeeded in the difficult task of giving a clear deductive account of the complicated formal theory of multiple theta-functions within the compass of a moderate sized text-book . . . Distinguished by clearness of style and general elegance of form."—*Mathematical Gazette.*

—1903-71. xxiv + 509 pp. 5⅜x8. 8284-0244-2.

GROUP THEORY

By A. KUROSH

Translated from the second Russian edition and with added notes by PROFESSOR K. A. HIRSCH.

Partial Contents: PART ONE: The Elements of Group Theory. Chap. I. Definition. II. Subgroups (Systems, Cyclic Groups, Ascending Sequences of Groups). III. Normal Subgroups. IV. Endomorphisms and Automorphisms. Groups with Operators. V. Series of Subgroups. Direct Products. Defining Relations, etc. PART TWO: Abelian Groups. VI. Foundations of the Theory of Abelian Groups (Finite Abelian Groups, Rings of Endomorphisms, Abelian Groups with Operators). VII. Primary and Mixed Abelian Groups. VIII. Torsion-Free Abelian Groups. Editor's Notes. Bibliography.

Vol. II. PART THREE: Group-Theoretical Constructions. IX. Free Products and Free Groups (Free Products with Amalgamated Subgroup, Fully Invariant Subgroups). X. Finitely Generated Groups. XI. Direct Products. Lattices (Modular, Complete Modular, etc.). XII. Extensions of Groups (of Abelian Groups, of Non-commutative Groups, Cohomology Groups). PART FOUR: Solvable and Nilpotent Groups. XIII. Finiteness Conditions, Sylow Subgroups, etc. XIV. Solvable Groups (Solvable and Generalized Solvable Groups, Local Theorems). XV. Nilpotent Groups (Generalized, Complete, Locally Nilpotent Torsion-Free, etc.). Editor's Notes. Bibliography.

—Vol. I. 2nd ed. 1959. 271 pp. 6x9. 8284-0107-1.
—Vol. II. 2nd ed. 1960. 308 pp. 6x9. 8284-0109-8.
—Vol. III. Approx. 200 pp. 6x9. **In prep.**

THE DOCTRINE OF CHANCES

By A. DE MOIVRE

In the year 1716 Abraham de Moivre published his *Doctrine of Chances*, in which the subject of Mathematical Probability took several long strides forward. A few years later came his *Treatise of Annuities*. When the third (and final) edition of the *Doctrine* was published in 1756 it appeared in one volume together with a revised edition of the work on Annuities. It is this latter two-volumes-in-one that is here presented in an exact photographic reprint, with an afterword by Prof. H. M. Walker.

—3rd ed. 1756-1967. xii + 368 pp. 6x9. 8284-0200-0.

COLLECTED MATHEMATICAL PAPERS

By L. E. DICKSON

—5 vols. Approx. 3,400 pp. 6½x9¼.

HISTORY OF THE THEORY OF NUMBERS

By L. E. DICKSON

"**A monumental work** . . . Dickson always has in mind the needs of the investigator . . . The author has [often] expressed in a nut-shell the main results of a long and involved paper *in a much clearer way than the writer of the article did himself*. The ability to reduce complicated mathematical arguments to simple and elementary terms is highly developed in Dickson."—*Bulletin of A. M. S.*

—Vol. I (Divisibility and Primality) xii + 486 pp. Vol. II (Diophantine Analysis) xxv + 803 pp. Vol. III (Quadratic and Higher Forms) v + 313 pp. 5⅜x8. 1971. 8284-0086-5. Three vol. set.

STUDIES IN THE THEORY OF NUMBERS

By L. E. DICKSON

A systematic exposition, starting from first principles, of the arithmetic of quadratic forms, chiefly (but not entirely) ternary forms, including numerous original investigations and correct proofs of a number of classical results that have been stated or proved erroneously in the literature.

—1930-62 viii+230 pp. 5⅜x8.

THEORY OF PROBABILITY

By B. V. GNEDENKO

This textbook, by a leading Russian probabilist, is suitable for senior undergraduate and first-year graduate courses. It covers, in highly readable form, a wide range of topics and, by carefully selected exercises and examples, keeps the reader throughout in close touch with problems in science and engineering.

"extremely well written . . . suitable for individual study . . . Gnedenko's book is a milestone in the writing on probability theory."—*Science.*

Partial Contents: I. The Concept of Probability (Various approaches to the definition. Space of Elementary Events. Classical Definition. Geometrical Probability. Relative Frequency. Axiomatic construction . . .). II. Sequences of Independent Trials. III Markov Chains IV. Random Variables and Distribution Functions (Continuous and discrete distributions. Multidimensional d. functions. Functions of random variables. Stieltjes integral). V. Numerical Characteristics of Random Variables (Mathematical expectation. Variance... Moments). VI. Law of Large Numbers (Mass phenomena. Tchebychev's form of law. Strong law of large numbers...). VII. Characteristic Functions (Properties. Inversion formula and uniqueness theorem. Helly's theorems. Limit theorems. Char. functs. for multidimensional random variables...). VIII. Classical Limit Theorem (Liapunov's theorem. Local limit theorem). IX. Theory of Infinitely Divisible Distribution Laws. X. Theory of Stochastic Processes (Generalized Markov equation. Continuous S. processes. Purely discontinuous S. processes. Kolmogorov-Feller equations. Homogeneous S. processes with independent increments. Stationary S. process. Stochastic integral. Spectral theorem of S. processes. Birkhoff-Khinchine ergodic theorem). XI. Elements of Queueing Theory (General characterization of the problems. Birth-and-death processes. Single-server queueing systems. Flows. Elements of the theory of stand-by systems). XII. Elements of Statistics (Problems. Variational series. Glivenko's Theorem and Kolmogorov's criterion. Two-sample problem. Critical region . . . Confidence limits). TABLES. BIBLIOGRAPHY. ANSWERS TO THE EXERCISES.

—4th ed. 1968. 527 pp. 6x9. 8284-0132-2.

GRUPPEN VON LINEAREN TRANSFORMATIONEN

By B. L. VAN DER WAERDEN

—1935-48. 94 pp. 5½x8½. 8284-0045-8.

SYMBOLIC LOGIC

By J. VENN

Venn's style is to take his readers very much into his confidence: as he builds the theory, he carefully points out the alternative paths he might have taken, the alternative definitions he might have used, shows what the implications of these alternatives are, and justifies his choice on the broadest possible grounds.

—2nd ed. 1894-1971. 38+540 pp. 5⅜x8. 8284-0251-5.

THE LOGIC OF CHANCE

By J. VENN

One of the classics of the theory of probability. Venn's book remains unsurpassed for clarity, readability, and sheer charm of exposition. No mathematics is required.

CONTENTS: PART ONE: Physical Foundations of the Science of Probability. CHAP. I. The Series of Probability. II. Formation of the Series, III. Origin, or Causation, of the Series. IV. How to Discover and Prove the Series. V. The Concept of Randomness. PART TWO: Logical Superstructure on the Above Physical Foundations. VI. Gradations of Belief. VII. The Rules of Inference in Probability. VIII. The Rule of Succession. IX. Induction. X. Causation and Design. XI. Material and Formal Logic . . . XIV. Fallacies. PART THREE: Applications. XV. Insurance and Gambling. XVI. Application to Testimony. XVII. Credibility of Extraordinary Stories. XVIII. The Nature and Use of an Average as a Means of Approximation to the Truth.

—1962. 4th ed. (Repr. of 3rd ed.) xxix + 508 pp. 5⅜x8.

ANALYTIC THEORY OF CONTINUED FRACTIONS

By H. S. WALL

Partial Contents: CHAP. I. The c. f. as a Product of Linear Fractional Transformations. II. Convergence Theorems. IV. Positive-Definite c. f. VI. Stieltjes Type c. f. VIII. Value-Region Problem. IX. J-Fraction Expansions. X. Theory of Equations. XII. Matrix Theory of c. f. XIII. C. f. and Definite Integrals. XIV. Moment Problem. XVI. Hausdorff Summability. XIX. Stieltjes Summability. XX. The Padé Table.

—1948-67. xiv + 433 pp. 5⅜x8. 8284-0207-8.

LOGIC, COMPUTERS, AND SETS

By H. WANG

Partial Contents: GENERAL (EXPOSITORY) SKETCHES (*Chaps. I-VI*) : I. The Axiomatic Method (§ 2. The problem of adequacy, . . . , §6. Gödel's Theorems, . . .). II. Eighty Years of Foundational Studies. III. The Axiomatization of Arithmetic (§2. Grassmann's calculus, . . . , § 6. Dedekind and Frege). V. Computation (§ 1. Concept of computability, . . . , § 6. Control of errors in calculating machines). CALCULATING MACHINES (*Chaps. VI-X*) : VI. A Variant to Turing's Theory. VII. Universal Turing Machines. VIII. The Logic of Automata. IX. Toward Mechanical Mathematics. FORMAL NUMBER THEORY (*Chaps. XI-XV*) : XII. Many-Sorted Predicate Calculi. XIII. Arithmetization of Metamathematics (§ 1. Gödel numbering, . . . , § 4. Arithmetic translations of axiom systems). XIV. Ackermann's Consistency Proof. . . . IMPREDICATIVE SET THEORY (*Chaps. XVI-XX*) : XVI. Different Axiom Systems . . . XVIII. Truth Definitions and Consistency Proofs . . . PREDICATIVE SET THEORY (*Chaps. XXI-XXV*) : XXII. Undecidable Sentences. XXIII. Formalization of Mathematics. . . .

Originally published under the title: *A Survey of Mathematical Logic.*

—1962-71. x + 651 pp. 6x9. 8284-0245-0.